Volcanoes

Volcanoes

A Planetary Perspective

Peter Francis

Reader in the Department of Earth Sciences
at the Open University
and Visiting Professor at the University of Hawaii

Clarendon Press

OXFORD

UNIVERSITY PRESS

Great Clarendon Street, Oxford OX2, 6DP

Oxford University Press is a department of the University of Oxford.
and furthers the University's aim of excellence in research, scholarship,
and education by publishing worldwide in

Oxford New York

Athens Auckland Bangkok Bogotá Buenos Aires Calcutta
Cape Town Chennai Dar es Salaam Delhi Florence Hong Kong Istanbul
Karachi Kuala Lumpur Madrid Melbourne Mexico City Mumbai
Nairobi Paris São Paulo Singapore Taipei Tokyo Toronto Warsaw
with associated companies in Berlin Ibadan

Oxford is a registered trade mark of Oxford University Press

Published in the United States
by Oxford University Press Inc., New York

A catalogue record for this book is available from the British Library

Library of Congress Cataloging Publication Data
Francis, Peter.
Volcanoes: a planetary perspective/Peter Francis.
p.   cm.
1. Volcanoes. I. Title.
QE522.F73   1993   551.2'1—dc20   92–29756
ISBN 0 19 854033 7

Printed and bound in Hong Kong

For Mary

who proved that Disraeli was wrong

.

Preface

Sir William Hamilton, the first volcanologist, concluded a letter describing an eruption of Vesuvius in 1767 with an apology: 'I fear I have tired you; but the subject of volcanoes is so favourite a one with me, that it has led me on I know not how'

Like Sir William, I have studied volcanoes because I have found them endlessly engrossing. I hope that this book will lead readers on to share this absorption. If I tire them, I will have failed. But natural processes are not simple, and cannot always be conveniently encapsulated in a few pithy words. It is all too easy to be led on too far.

There have been many books on volcanoes since Hamilton described the eruptions of Vesuvius. These have been of three kinds: popular treatments; scholarly monographs; and books intended to appeal to everyone, from the man in the street to the university student and his professors. Like many other impoverished academic authors, I have tried to write for this general readership. I hope that this book will be especially useful to readers without specialist knowledge of volcanoes, and perhaps little background in geology, but who have a lively interest in natural phenomena. But trying to appeal to everyone always runs the risk of satisfying no one: stifling general readers with too much detail, while appearing frivolous to committed scholars. Why take such an obvious risk? Why not just produce a straightforward textbook? Simply because the kind of book that I take most pleasure in writing is the sort that I would enjoy reading; a book to be *read*, rather than consulted. If this book fills its intended purpose, it will end up bedraggled in bathrooms or dog-eared in hotel bedrooms, rather than primly pristine on a shelf.

Then why not a popular account? There were two reasons for this. First, I felt that I had already had one enjoyable bite at that cherry. Second, the niche is already well filled by two delightful books by Robert and Barbara Decker (*Volcanoes* and *Mountains of fire*). These books are so entertaining, informative, and up to date that it would be pointless to duplicate them.

Inevitably, this book presents a personal view of volcanism. It is not a book on planetary volcanism, neither is it a comprehensive treatment of terrestrial volcanology. Faced with a huge range of absorbing volcanological subjects, I decided that it was best to write about topics I know best, or which most appeal to me. Many of these are aspects of volcanology that I learned at first hand, sitting on a lava flow and discussing them with volcanologists who did the definitive work. Much of the book was written in Honolulu, Hawaii, with the verdant cone of Koko crater looming steeply in the background. Thus, a fair amount of Hawaii has seeped through into the text. Similarly, much of my research over the past twenty years has been carried out among the austerely beautiful high volcanoes of the Central Andes. Readers may therefore find that these, too, have coloured the text.

In order to keep the text readable, I have avoided the style of formal referencing favoured by scientific texts. Long strings of references break up the flow of narrative text, and often seem to be used to create an illusion of serious scholarship, rather than providing evidence to support the case being argued. In an earlier book on volcanoes, however, I was berated for not supplying references to help students to follow up on particular topics. Here, therefore, I have cited some key works, either by mentioning them directly, or by flagging them in the text and providing a list of references at the end of each chapter. These are not intended to be exhaustive bibliographies, but should provide points of entry into the literature. In several places in the text, I have followed closely or paraphrased the primary work. I hope that the original authors will understand my intentions, and not feel that their contributions have been slighted by sparse referencing. Rather, I hope that they will accept such use of their material as a measure of its value.

A preface provides an opportunity to acknowledge the contributions of those who have led to the completion of a book. In my case, I can identify three major groups. First, and most important, there are my volcanological friends and colleagues in many countries who have over the years enriched my knowledge of the subject, guided me around many volcanoes, and answered patiently many questions. Secondly, I should acknowledge the indulgence of the three institutions who have made it possible for me to pursue my work in volcanology: the Open University in England; the Lunar and Planetary Institute in Houston, Texas, and the Planetary Geosciences Division of the University of Hawaii. I am particularly grateful to that magnificent, munificent, but much maligned agency, NASA, whose financial support enabled me to spend a wonderful decade studying the volcanoes of the central Andes.

Finally, there are all those who were instrumental in preparing the book. For inspiration, helpful discussions or for commenting on parts of the text I thank Rodey Batiza, Catherine Hickson, John Mahoney, Alex Malahoff, Clive Oppenheimer, Dave Pieri, David Pyle, Steven Self, Josie Titcomb, Geoff Wadge, and George Walker. Colin Wilson scrutinized the pyroclastic chapters. Marylyn Moore in the Geology and Geophysics Library at the University of Hawaii tracked down many vaguely remembered references. For their generous help in supplying photographs I thank Rodey Batiza, Steve Carey, Shan de Silva, Mike Garcia, Ian Gass, Lori Glaze, Ron Greeley, Nancy Green, R. V. Fisher, Catherine Hickson, Carol Ann Hodges, Peter Hooper, Wally Johnson, Terry Keith, Jorg Keller, Jack Lockwood, Alex McEwen, Peter Mouginis-Mark, Duncan Munro, Clive Oppenheimer, David Pyle, Bill Rose, Mark Robinson, John Roobol, Dave Rothery, Lee Siebert, John Spencer, USGS Volcano, and Tony Waltham.

Milton Keynes   P.W.F.
November 1992

Contents

1

The basics: isotopes and green cheese

A volcano is a mountain with a Jekyll and Hyde personality. For most of its life, it is inactive. One's image is that of a graceful cone, capped with snow, commanding the cherry-blossom-draped landscape of a Japanese travel poster. This serene mood may continue for millennia, until the darker side of the volcano's character is abruptly manifested in a violent eruption. Convoluted clouds climb many kilometres into the atmosphere, bringing tenebrous darkness at noon, and raining hot ashes on towns and villages beneath. Tongues of molten rock ooze down the slopes of the volcano, engulfing the flimsy structures standing in their way.

In primitive societies, volcanoes were often identified with deities or evil spirits, and regarded with superstitious fear. Volcanoes are still regarded with religious awe in many third world countries, while in more sophisticated societies they command respect for their prodigious destructive potential. This respect is often confused, since volcanoes and earthquakes are often linked indiscriminately in the media. Eruptions are widely publicized, and rightly so, because their raw physical power makes them compelling natural spectacles. But these spectacles may also be lethal. Several of the greatest natural disasters in history were caused by volcanic eruptions; some directly through explosions; others indirectly through the effects of giant waves (*tsunamis*) as at Krakatau in 1883; and yet others more indirectly through starvation, as in Iceland in 1783. Some eruptions influenced the course of human civilization. Many others have had global environmental consequences. These dramatic aspects of volcanoes will, of course, be discussed in this book—they are, after all, what gives volcanology its spice. But many more subtle aspects will also be explored, to enquire why volcanoes exist at all, why they occur where they do, and how they fit into the history of the Earth and other planets.

William Smith began to draw the first geological maps of England in 1815, when the Industrial Revolution was creating a demand for a network of canals. It was in the sticky blue clays and yellow marly limestones of the canal cuttings that Smith perceived the orderly succession of strata and fossils that formed the starting point for geology as a science. In the hundred and fifty years that followed, geologists became intimately acquainted with the rocks of their planet. By the middle of the twentieth century, their science had become rather stagnant. Geologists thought they understood how the Earth functioned. Although much remained to be done in delineating the history of the Earth, the work had become rather mundane, comparable with the dust-dry scholarship of archaeologists cataloguing pottery shards from a routine excavation. All of this changed in the 1960s. A double revolution took place in geology, which propelled it further forward than in all the decades since Smith began to contemplate what lay beneath the tranquil pastures of England.

First came a new understanding of the way our planet works: the so-called *plate tectonic revolution*. Plate tectonics made intelligible the interrelationships of a vast range of geological phenomena, by resolving the enduring paradox of continental drift: the complementary coastlines of Africa and America and other geological dovetails showed that the continents must have split apart, but no plausible mechanism could explain how it happened. Controversy smouldered for years, but Fred Vine and Drummond Matthews at Cambridge University finally set the fire of revolution ablaze in the mid-1960s, when they showed how magnetic patterns of the ocean floor could be interpreted.

Hard on the heels of this revolution came the second: men landed on the Moon in 1969, giving geologists a new world to explore. Long before study of the rocks collected during the Apollo programme had been completed, unmanned missions began to return a flood of data from Mercury, Venus, Mars, Jupiter, Saturn, Uranus, and Neptune. By the late 1980s, the flood had diminished to a trickle, but will swell to a torrent once more in the 1990s as new space craft arrive at Venus, Mars, and Jupiter. Geologists—and volcanologists—can no longer sit back cosily contemplating a single planet. They must now think about three other planets, and an entire squadron of satellites.

A single conclusion can be distilled from the torrent of new planetary data: there are only two fundamental geological processes—impact cratering and volcanism. The surfaces of *all* the solid bodies in the solar system have been shaped by these two processes, modified sometimes by erosion. This truth happens to be least evident on our own planet, and is one reason why traditional geologists had difficulties in coming to terms with the revolutionary changes in Earth sciences. Evidence for impact cratering on the Earth is not conspicuous, but it is there, none the less. Its importance in the history of the Earth is slowly dawning on even the most conservative geologists, who have had to accept that the Earth's peaceful progression through space and time has been shattered more than once by the impact of an asteroid.

Although volcanism is much more obvious than cratering on Earth, its importance has been masked because most of the Earth is covered by water. Drain away the oceans, and vast basalt plains left low and dry would demonstrate that the Earth is not so different from Mars or Venus. It is striking that we have begun to explore the Earth's submarine volcanoes through sonar techniques at just the same time that imaging radars on space craft have revealed Venus' cloud-shrouded volcanoes.

— 1.1 Origins of volcanism

First, a definition. Earlier books have not had to grapple with the issue of what a volcano is—everyone has an idea of what a terrestrial volcano looks like. Planetary developments dictate, however, that we must have a term broad enough to cover *all* types of volcanism. Volcanism may be defined, accurately but ponderously as: *the manifestation at the surface of a planet or satellite of internal thermal processes through the emission at the surface of solid, liquid, or gaseous products.* This definition enables us to include phenomena as varied as the familiar eruptions of silicate magmas on Earth, lavas of sulphur and plumes of sulphur dioxide on Io, 'lava' flows of water on the surfaces of icy satellites, and fountains of nitrogen from Triton. A 'volcano' therefore is simply a site at which material reaches the surface of the planet from the interior. Such sites can assume remarkably diverse shapes and forms, as later sections will reveal.

Two prerequisites must exist before there can be any volcanism: a source of heat, and something to melt. Although perhaps blindingly obvious at first sight, these two concepts require exploration, the first into the realms of physics, the second of chemistry.

1.1.1 Sources of heat

In contemplating any volcanic eruption, even a quiet effusion of lava, one cannot escape the observation that huge amounts of thermal energy are involved. This fact is rawly tangible in the fierce radiance that prevents one from approaching too close to a lava flow. There is a tale that an old fabric-covered aircraft flying over a lava lake in Hawaii narrowly escaped disaster when radiant heat from the lava began to bubble the paintwork and scorch the wing.

Where does all this thermal energy come from? And can it be quantified? To heat one kilogram of rock through one degree Celsius requires about 1200 joules (J) of energy, roughly a quarter of that required to do the same for water. The *latent heat of fusion* for basalt (the amount of heat required to melt one kilogram of rock at its melting point) is about 400 times greater. Thus, to raise one kilogram of basaltic rock from room temperature (say 20°C) to its melting point (say 1200°C) and then melt it completely requires about 1.9×10^6 J. A single barrel of oil has an energy content of 5.7×10^9 J. *Total* world energy production in 1980 was about 1.4×10^{20} J[1]. Thus, *all* the world's energy supplies for 1980 would be sufficient to melt only 7.3×10^{13} kg of rock. At average densities for basalts, this corresponds with about 27 cubic kilometres of rock, enough to build only a miniscule volcano, less than 1000 m high.

All the world's volcanoes on dry land erupt, on average, two cubic kilometres of magma per year, or about 5×10^{12} kg. (Individual major eruptions may erupt much larger volumes, but they are rare.) Although the rate of volcanic activity on Earth may therefore seem modest, large amounts of heat are none the less required to drive the phenomena that we observe, and much more those that we do *not* observe. Far more material is added to the oceanic crust each year than is erupted on dry land, perhaps as much as 20 cubic kilometres per year. Furthermore, the rate of volcanic productivity on Earth is probably far less today than it was early in the history of the solar system. This is also true of the other planets and satelites. So what were the energy sources that made volcanism occur in the past, and what drives it today? There are many possible sources of heat, some of which are overwhelmingly more important than others at the present day.

1.1.2 Long-lived radioactivity

Geophysical studies show that heat is leaking away from the interior of the Earth at an average rate of about 82 milliwatts per square metre— this is the *mean heat flow*. (It would require an area of about one square kilometre to provide enough energy to light a 100 watt light bulb.) In 1864 the distinguished British physicist Lord Kelvin (who gave his name to the Kelvin temperature scale) attempted to calculate the age of the Earth from its present temperature, and the surface heat flow, assuming that it had started completely molten, and cooled continuously.[1] He came up with an age of 100 million years, an underestimate of about 4500 million years! Kelvin had not taken into account a fact of which nineteenth century science was unaware: decay of radioactive isotopes within the Earth generates heat continuously. We have all reluctantly become familiar with the inventory of man-made 'hot' isotopes used in nuclear weapons. Four other naturally occurring radioactive isotopes of uranium, thorium, and potassium generate virtually all the radiogenic heat in rocks today: ^{235}U, ^{238}U, ^{232}Th, and ^{40}K. Each of these isotopes has a different heat production, and a different half-life.

Table 1.1 shows that kilogram for kilogram, ^{235}U is much the most important heat producer. But there are two additional factors which affect the relative importance of isotopes which keep planets hot: how long they last, and how abundant they are. Although ^{235}U is an intense heat producer, its half-life is much shorter than those of the other isotopes. Furthermore, uranium and thorium are rare elements in nature, whereas potassium is between a hundred and a thousand times more abundant in ordinary rocks. Thus, its radioactive isotope ^{40}K is relatively abundant (Table 1.2). Taken together, these factors add up to the relationships expressed in Fig. 1.1: there has been a sharp change in the relative importance of the uranium and thorium isotopes through geological time.

1.1.3 Short-lived radioactivity

Scientists have appreciated the importance of long-lived isotopes in heating planetary bodies since the pioneering work of Arthur Holmes in 1913.[3] Not until 1976 was the role of short-lived radioactive isotopes discovered, when anomalously large amounts of the isotope ^{26}Mg were found in a calcium- and aluminium-rich chondrule (a small spherical grain) in a meteorite

Table 1.1 Heat production and half-lives of the major radioactive isotopes

Isotope	Heat production (mW kg^{-1}*)	Half-life (yr)	Isotope/ element ratio†	Present terrestrial abundance (kg $\times 10^{17}$)
^{238}U	9.6×10^{-2}	4.5×10^9	0.9	1.82
^{235}U	56×10^{-2}	7.1×10^8	0.007	0.013
^{232}Th	2.6×10^{-2}	1.4×10^{10}	1.0	6.49
^{40}K	2.8×10^{-2}	1.3×10^9	1.2×10^{-4}	84

*Milliwatts per kilogram
†This indicates the fraction of the element constituted by the isotope in question, for example ^{238}U forms 99 per cent of all uranium at the present day

Table 1.2 Concentrations of heat-producing isotopes in some common rocks in parts per million, and the resulting heat productions of the rocks

Rock type	^{238}U	^{235}U	^{232}Th	K (total)	^{40}K	Heat production mW kg$^{-1} \times 10^{-8}$
Continental rocks (granodiorites)	3.9	0.03	18	35 000	3.5	96
Ocean basalts	0.79	0.006	3	9 600	0.96	18
Mantle nodules	0.01	7×10^{-5}	0.06	12	1.2×10^{-3}	0.26
Carbonaceous chondrite meteorites	0.01	7×10^{-5}	0.38	700	0.10	0.50

which had fallen at Pueblo de Allende, Mexico in 1969[4] (Fig. 1.2). Nuclear physics shows that the *only* source of 'excess' ^{26}Mg is radioactive decay of a parent isotope, ^{26}Al, which is a half-life of only 720 000 years. Seven hundred and twenty thousand years may seem long in human terms, but in only ten half-lives (7.2 million years), the isotope had decayed to vanishing point, and it is therefore extinct in the solar system today. Aluminium is an abundant element, so its radioactive isotope must have been a powerful heat source in the early solar system. Thus, a young planetary body with isotopic ratios similar to those in the Allende meteorite would only have needed to be a few kilometres in diameter for it to have contained enough ^{26}Al to have melted completely.

The difficulty with ^{26}Al as a potential planetary heat source, however, arises from its extremely short half-life. Isotopes like ^{26}Al can be manufactured only in stars, not planets. Even so, violently explosive stellar eruptions (supernovae) are required. Having created a 'hot' isotope, the problem then is to get it quickly enough from the supernova debris into a newly forming planetary system, *before* its radioactivity has burned out. Ten million to a hundred million years are thought to be needed to accrete the solid bodies of planets from the dust of a primitive solar nebula; rather a long time compared with the

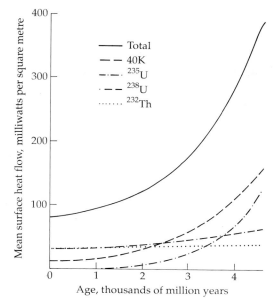

Fig. 1.1 Variations in heat production through geological time of the four important naturally occurring radioactive isotopes, expressed in mean surface heat flow. ^{40}K and ^{235}U have both declined sharply in importance.

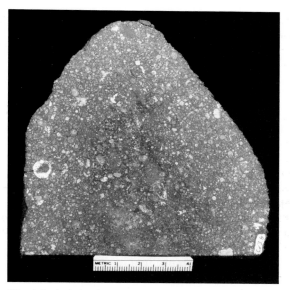

Fig. 1.2 A slab of the Allende meteorite, showing abundant small spherical chondrules. Some of the white chondrules (e.g. that near the left edge) are rich in calcium and aluminium, and contain excess quantities of the isotope ^{26}Mg, relics of short-lived radioactive isotopes which contributed to early heating of the Earth. Photo: courtesy of Ed Scott, Planetary Geosciences, University of Hawaii.

half-life of ^{26}Al. The isotope could, however, have been important in melting small, rapidly accreted bodies, and also in the *early* history of larger bodies. Since large planetary bodies cool slowly, the effects of such early heating could persist for long periods. Apart from ^{26}Al, there are other short-lived hot isotopes such as ^{129}I and ^{244}Pu which could have contributed heat to young planets, but these are so scarce in solar system materials that their effects are negligible.

1.1.4 Accretional heating

Two decades of space exploration have shown that asteroid bombardment has been crucially important in the histories of all the solid bodies in the solar system. Apart from shaping the surfaces of planets and statellites, the *kinetic energy* liberated during impact events was also critical in their thermal histories. Naturally, this effect was most intense during the earliest days of the solar system, when the various planetary bodies were accreting into individual entities from a whirling cloud of smaller particles. As the gravitational attractions of the growing planets increased, bombardment became self-perpetuating, until the solar system had been effectively swept clean. Most remaining stray chunks of material were locked up by orbital mechanics in the asteroid belt; a few of these (the Apollo and Amor asteroids) still occupy Earth-crossing orbits. Occasional major impacts can still be expected. (It really is not worth worrying about these—they are quite rare, and nothing whatever can be done to stop them.)

Although the kinetic energy of a moving body is straightforward ($1/2\ mv^2$), the thermal effects of impacts and accretion are difficult to quantify, because it is hard to determine what fraction of the kinetic energy of an impacting body remains as heat in the target body, and what fractions are lost in the ejection of debris back into space, and by radiation. A single iron–nickel object 30 metres in diameter impacting the Earth at a velocity of 15 kilometres per second would impart about 1.7×10^{16} J of energy, equivalent to the explosion of about 4 million tonnes of TNT.[5]

This is roughly comparable with the energy released during the formation of the Earth's best-known impact site, Meteor Crater in Arizona (Fig. 1.3).

For bodies larger than about 1000 kilometres in radius, the energy acquired from the innumerable impacts on them was sufficient to cause a rise in temperature of about 1000°C. W. M. Kaula calculated that the primitive, growing Earth would get hot enough to melt almost completely, but the much smaller Moon would not.[6] As with short-lived radioactivity, this thermal input takes place only in the earliest stages of planet evolution, but its effects on a planet's subsequent career may be profound. Furthermore, as an accreting body grows in size, the material at depth becomes more highly compressed by the weight of material above. Just as the compression of air in a cycle pump causes heating, so too does the process of gravitational self-compression. Thermodynamic considerations enable this to be modelled fairly well for planetary interiors. From an initial temperature of 500°C, self-compression would lead to an additional temperature rise of 77, 738, 965, 119, and 25 degrees for Mercury, Venus, Earth, Mars, and the Moon, respectively.

1.1.5 Core formation

When initially homogeneous planets sort themselves out into their familiar onion-skin layered structures of core, mantle, and crust, energy is liberated, because the segregation of dense material to the core involves the release of gravitational *potential energy*. Heat generated by accretion, short-lived radioactivity, and self-compression causes a blast furnace process to operate, metallic iron separating out downwards to form the core, while lighter silicate materials migrate upwards to form the mantle. In the Earth, this process took place surprisingly rapidly, within a few million years. Although core formation may not be an immediately obvious source of heat, the temperature increases that it causes can be significant. The figures shown in Table 1.3 have been suggested for the Earth, Mars, Mercury, and the Moon.

1.1.6 Dissipation of tidal energy

The gentle ebb and flood of the ocean tides are a far cry from the violence of a volcanic eruption. Tidal energy, however, is an important source of heat, especially in satellites too small to contain

Fig. 1.3 Meteor Crater, Arizona, was formed by a hyper-velocity impact about 50 000 years ago. Shattered and melted rocks fill the bottom of the crater which is 1.2 kilometres in diameter, and two hundred metres deep. Small fragments found scattered around the crater show that the impacting body was an iron meteorite, now called Canyon Diablo. It was about 40 metres in diameter, with a mass of about 10 million kilograms.

Table 1.3 Energy release and temperature rise due to core formation in originally homogeneous planets[7]

Planet	Core radius (km)	Energy release (J)	Temperature rise (°C)
Earth	3485	1.5×10^{31}	2300
Mars	1400–2100	$1.8–2.3 \times 10^{29}$	300–330
Mercury	1840	2×10^{29}	680
Moon	<400	1×10^{27}	12

much radioactive material. Since the days of Newton, it has been known that the Sun and Moon raise large tides in the Earth's oceans. Similar, but much smaller, tidal distortions take place in the interiors of both the Earth and Moon. The highest tide in the Earth's mantle has a height of only about 11 cm, so it is not easy to detect. Over the course of geological history, however, the effect of tides on the Earth has been to slow down its rotation rate, and to cause the Moon to recede from the Earth. All the energy used in 'braking' the Earth's rotation about its axis is manifested as heat.

Present estimates suggest that the energy liberated in this way in the Earth amounts to about 3×10^{19} J per year. This is about the same amount as that liberated by the isotope ^{232}Th, so it is not negligible. There is no evidence for tidal

Fig. 1.4 One of the epoch-making images of active volcanism on Io, obtained by the Voyager 1 space craft in March 1979. Two volcanic plumes are visible: one is seen faintly in profile against the blackness of space, the other is the bright spot near the centre. This plume rose so high above the dark side of the satellite (260 km) that its crown caught the full light of the Sun.

heating on Venus and Mars, both of which lack large satellites. Mercury, which is close to the Sun's vast mass and is locked into a complexly resonant orbit, must experience some tidal heating, but not enough to result in any volcanic manifestations today.

Voyager space craft images of Io, Jupiter's innermost large satellite, revealed by far the most magnificent display of tidal heating in the solar system. Orbiting only 421 000 km away from the massive planet, Io is locked into a complex orbital resonance with both Jupiter and the next satellite outwards, Europa. Io revolves around Jupiter in 1.77 days, keeping the same face turned towards it at all times, while Europa makes the same revolution in exactly twice the time, 3.55 days. So much work is done on Io in maintaining this resonance and in continually 'kneading' its interior, that enough heat is liberated to maintain the satellite in a state of constant, violent volcanic activity (Fig. 1.4). Intellectually, the most exciting aspect of this exotic form of volcanic activity was that it had been *predicted* in a brilliant theoretical analysis *before* space craft first observed the eruptions[8].

Although not nearly so spectacular as the active plumes on Io, there is evidence for similar tidal heating in the histories of many icy satellites of the outer planets. Of course, the amounts of energy required to produce melting in an icy body are less than those required in a silicate body, but they are significant, none the less.

In 1989, the far-travelled Voyager 2 space craft returned images of the most bizarre volcanic activity of all: plumes of gas exhaled from Neptune's satellite Triton, which is so frigid that *nitrogen* freezes on its surface! This remarkable phenomenon may not be due to tidal heating, but may instead be the result of a curious 'greenhouse' effect (Section 18.8.2).

— 1.2 Heat production through geological time —

This is the way the world ends
Not with a bang but a whimper.
(T. S. Eliot, *The Hollow Men*)

Long-lived radioactive isotopes are the most important continuing sources of heat for volcanic processes but they, like all the others, cannot last for ever. Fig. 1.5. demonstrates eloquently the decline in radioactive heat production through time, expressed in surface heat flow, and points to the inexorable thermal death of all planetary bodies. It is easy to see why volcanism should have been an important process in the early solar system, when there was so much heat about. It may be more pertinent to enquire how it is that volcanic activity continues anywhere today, not least on the Earth.

There are three reasons for this. First, rocks are poor conductors of heat. Thus, it takes geologically long periods for primordial heat to seep out from the interior of a large planet. Second, a large body has a smaller surface area in proportion to its radius than a smaller one. Radiative heat losses through the surface are thus proportionally smaller, so a large, hot body will cool more slowly than a small one. Third, once a volume of material (of whatever composition) has been melted, it contains a great deal of 'stored' thermal energy in the form of latent heat of fusion. On crystallizing to the solid state, latent heat is liberated once more, buffering the temperature of the system. (Latent heat of fusion will be exasperatingly familiar to anyone who has tried to thaw a container of frozen soup from the freezer in a hurry—huge quantities of heat have to be poured into it.) Thus, even without inputs from radioactive isotopes, crystallization of a large body of melt such as the iron in the Earth's core is bound to be a long, slow process.

These observations raise an important question about the thermal state of the Earth. Is it in equilibrium, so that heat loss is balanced by radioactive heat generation, or is it cooling? As we saw earlier, surface heat flow averages 82 milliwatts per square metre. But is this heat *flow* balanced by the rate of heat *production*? This

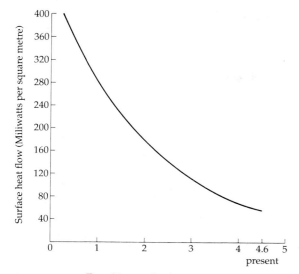

Fig. 1.5 Decline of the Earth's surface heat flow through geological time; a consequence of decreasing radioactive heat production, and indicative of the Earth's eventual thermal extinction.

isotopes in rocks at the Earth's surface, and in the meteorites that fireball in unexpectedly from space. Abundances in the bulk Earth are much more difficult to derive. In many respects, the Earth's bulk composition is close to that of some meteorites (discussed later), and these give a guide to the radioactive isotope content. Best estimates of the bulk Earth's potassium content are between 100 and 200 parts per million. Now the ratio of potassium to uranium in *all* terrestrial rocks is around 10 000:1, so this indicates that the bulk Earth's uranium content is about 0.01–0.02 parts per million. It is possible to calculate what the heat flow at the surface of the Earth should be if these potassium and uranium figures are correct. This figure turns out to be about half of the *observed* heat flow, implying that the other half must be due to heat from sources other than radioactive isotopes; probably primordial heat, left over from the Earth's accretion and core formation. A clear implication is that the Earth is cooling down.[9] This is nothing we need to worry about on a day-to-day basis; the rate of *secular cooling*, as it is formally known, is something like 100 °C per billion years. No need to think about cranking up the living-room thermostat just yet . . .

requires knowledge of the Earth's total content of radioactive isotopes. Unfortunately, all that one can measure directly is the abundance of these

1.3 Compositions of the planets . . . the material that melts

There is an abundance of geophysical evidence for the Earth's threefold structure of core, mantle, and crust. When we begin to probe the *compositions* of these units, however, the situation becomes less cut and dried. While there are many data on the compositions of upper crustal rocks, the composition of the lower contintental crust is highly controversial, and so too is the composition of the core. Geophysical data, which are all we have to go on, show that the density of the core is too low to be a simple mixture of iron and nickel: there must be a significant proportion of a light component. Elements as geochemically diverse as carbon, silicon, sulphur, oxygen, and potassium have all been seriously proposed as the light component,

which represents no less than four per cent of the total mass of the Earth. Given, then, that our knowledge of the Earth's composition is so deficient, how can we say anything serious about the compositions of planets so distant that they appear only as small discs, even in powerful telescopes? How can we be *sure* they are not made of green cheese?

Fortunately, the laws of physics rule out bizarre possibilities such as hollow planets made of roquefort. There are also plenty of observational data on the relative abundances of different elements in the Universe. Classical Newtonian mechanics enable us to establish straight forwardly the sizes, masses, and densities of the planets. Mercury, Mars, Venus, and the

Moon have densities close to the Earth's (see Table 1.4), and therefore must be made of basically similar stuff.

1.3.1 The 'chondritic Earth' model

Meteorites have long been used as pointers to the composition of the planets. Until Neil Armstrong picked up the first rock from the surface of the Mare Tranquillitatis in 1969, they were the *only* materials available from extraterrestrial sources. All but a tiny number of meteorites are extremely old, with radiometric ages close to 4.6 billion years, the age of the solar system. They represent samples which have remained substantially unchanged since the formation of the solar system, and in a sense represent bits left over from the initial assembly of the planets. Most importantly, while many meteorites appear to have been derived from small planets (asteroids) which had evolved metallic cores and stony mantles, there is one important group whose

members never formed part of larger bodies and which have 'primitive' compositions, close to that of the nebula of dust and gas from which the solar system condensed. These are the C1 chondrites. (A *chondrule* is a small mineral spherule, thought to originate in the process of formation of chondritic meteorites; see Fig. 1.2.) Early ideas for the bulk composition of the Earth simply assumed that it was the same as that of these primitive C1 chondritic meteorites, a view enshrined as the *chondritic Earth model*.

It was a logical enough starting point, but does it stand up to close scrutiny? Much the largest fraction of the Earth's mass—about 66 per cent—is represented by its mantle of silicate rocks. Thus, if we could determine the composition of the mantle, we could go a long way towards testing the chondritic Earth model. Fortunately, although we are denied direct access to the mantle, small chunks of it are available in the form of *xenoliths* (Greek, 'foreign

Table 1.4 Some basic solar system statistics

Name	Distance from Sun (km × 10⁶)	Rotation period (days)	Diameter (km)	Mass (kg)	Density (kg³m⁻³)	Surface gravity (m s⁻²)
Mercury	57.8	58.7	4878	3.3×10^{23}	5.42	3.5
Venus	108.1	243R*	12 104	4.87×10^{24}	5.25	8.5
Earth	149.5	1	12 756	5.98×10^{24}	5.52	9.8
Moon		27.32	3476	7.35×10^{22}	3.34	1.6
Mars	227.8	1.02	6787	6.42×10^{23}	3.94	3.7
Asteroids						
Ceres	413.8	0.38	1020	?	?	
Vesta	353.1	0.22	549	2.5×10^{20}	2.90	
Pallas	414.5	0.33	538	?	?	
Jupiter	777.8	0.41	143 800	1.9×10^{27}	1.3	25.9
Galilean satellites						
Io		1.77	3640	8.89×10^{22}	3.5	
Europa		3.55	3130	4.79×10^{22}	3.03	
Ganymede		7.15	5280	1.48×10^{23}	1.93	
Callisto		16.68	4840	1.08×10^{23}	1.79	
Saturn	1426.0	0.43	120 660	5.69×10^{26}	0.69	11.2
Uranus	2867.7	0.7R*	52 290	8.6×10^{25}	1.19	10.8
Neptune	4571.1	0.7	49 500	1.03×10^{26}	1.6	9.8
Pluto	5896.2	6.4?	3100	9×10^{21}	1.2	?

R* = retrograde rotation

stones') carried up with some basalt magmas and in the rare explosive events that produce minera-logically complex rocks called *kimberlites*. There is convincing evidence that these exotic xenoliths are indeed derived from deep within the mantle. Not least is the presence in some kimberlites of the costly, high-pressure form of carbon known as *diamond*. (Kimberlites are named after the town of Kimberley in South Africa where the Premier diamond mine is located.) Xenoliths suggest that the upper mantle is made mostly of the silicate minerals olivine and pyroxene which together form a dense, greenish rock called *peridotite*. (*Peridot* is gem-quality olivine.)

But how representative of the mantle can xenoliths be? The very fact that they have reached the Earth's surface through volcanic vents means that they may come from anomalous parts of the mantle. And furthermore, although petrological studies suggest that such xenoliths were formed at depths of around 100 km, this is definitely the *uppermost* part of the mantle. This part of the mantle is described as *depleted*, since geochemists have determined that

the elements required to make the Earth's crust have been extracted from it. Depleted mantle is deficient in the so-called *incompatible* elements such as potassium and rubidium. (They are termed incompatible because their ions are too large to fit into the atomic structures of common minerals.) To work out the composition of truly *primitive* or pristine mantle the amounts of the various elements extracted to make the crust have to be added back to the xenolith composi-tions. Unfortunately, many of these estimates are more like guesses, and involve a degree of circular argument. For most elements such as calcium and aluminium, the exercise shows good agreement with the chondritic Earth model, but some elements show definite discrepancies. Potassium is an example; it appears to be significantly depleted in the Earth with respect to chondritic meteorites (Table 1.5). In the Earth, the potassium to uranium ratio is around 10 000 to 1; whereas in chondrites it is 64 000:1. Now because it is difficult to fractionate potassium from uranium via conventional igneous process (the two elements are both highly incompatible,

Table 1.5 Composition of the Orgueil C1 chondrite meteorite, spinel lherzolite and 'undepleted' mantle[10]

Component*	Orgueil meteorite	Orgueil water-free	Spinel lherzolite	Undepleted mantle
SiO_2	21.7	26.8	45.1	45.1
TiO_2	0.1	0.1	0.1	0.2
Al_2O_3	1.6	1.9	2.2	3.3
Cr_2O_3	0.4	0.4	0.5	0.4
MnO	0.2	0.2	0.1	0.14
FeO	22.9	28.3	8.3	8.0
NiO	1.2	1.4	0.3	
MgO	15.2	18.8	41.5	38.1
CaO	1.2	1.5	2.0	3.1
Na_2O	0.7	0.8	0.2	0.4
K_2O	0.07	0.08	0.1	0.03
P_2O_5	0.3	0.4	0.1	
H_2O	19.2	0	0.3	
Organics	9.7	11.9		
S	5.7	7.0		
Totals	100.17	100.0	100.9	98.77

*For historical reasons, compositions of rocks are reported in terms of element oxides, rather than elements.

so they behave similarly), these differences can only have arise during accretion of the Earth, when the more volatile potassium was lost to space.

Various models for the compositions of the mantles of the other terrestrial planets have been advanced. They differ widely, especially in their estimates of heat-producing elements. The debate has been at times a passionate one, but one whose flames have been fuelled more by enthusiasm than data. At present, the best one can say is that the mantles of the terrestrial planets are of approximately chondritic composition, but some are more chondritic than others. Mars, for example, may have a more nearly chondritic potassium to uranium ratio than the Earth. Further problems arise with the numerous satellites of the solar system. Some of these, like our Moon, are rocky, silicate bodies, but most of the satellites of the giant planets are made of water ice, and a range of other more exotic volatile compounds such as ammonia and methane.

Scholarly controversies about the compositions of ancient meteorites may seem impossibly arcane, far removed from the fire and spectacle of volcanism, and about as useful as theological debates about the number of angels that could dance on the head of a pin. They are critical, however, to understanding the history of volcanism on the Earth and other planets, because they provide bounds on the compositions of the source materials from which all volcanic rocks are derived—they tell us about the raw material that is melted, to be spewed up as lavas through volcanoes. They also give pointers to the thermal, and therefore volcanic, histories of the planets, discussed in Chapter 18.

— 1.4 Volcanic rocks

Readers with some geological knowledge will be familiar with petrological terminology, and could profitably skip this section, but for non-geologists, the complexity of geological nomenclature can be dismaying. Many hundreds of different rock types have been described and named. Take courage, however. For chemists, the apparently boundless diversity of plant life disappears when they recall that *all* plants are composed of carbon, hydrogen, oxygen, and a few less important elements—a daisy is chemically almost indistinguishable from an oak tree. Much the same is true of volcanic rocks—they are merely variations on a common theme. For volcanological purposes, less than a dozen rock names are all that are essential. *Lava* is the most familiar, generic kind of volcanic rock: it barely needs definition, except to say that it should be confined to material that flows in a molten form. *Pyroclastic* rocks (from Greek, meaning 'fire-broken') may be slightly less familiar; the term is used for material ejected from volcanoes as *solid* fragments, loosely termed ash or pumice. Pyroclastic rocks which fall to ground from eruption clouds are often collectively called *tephra* (Greek, meaning 'ashes'). Both lavas and pyroclastic rocks which have a fragmented, cindery texture are called *scoria* (Greek, 'refuse'). Volcanic rocks are self-evidently those erupted by volcanoes at the surface; *plutonic* rocks are those that do not make it to the surface. Many volcanoes are probably underlain at depth by *plutons;* large bodies of now crystalline rocks that once formed parts of the volcano's plumbing system.

Rocks are made up of minerals. A mineral may be defined as '*a naturally formed inorganic substance with a definite chemical composition and a definite atomic structure*'. Many thousands of natural minerals have been discovered since the nineteenth century, but fortunately, all but a few volcanic rocks are made of various combinations of only seven different mineral families:

Olivines
Pyroxenes
Amphiboles
Micas
Feldspars (and feldspathoids)
Quartz
Oxides.

The first six are all silicate minerals; oxides typically form a tiny percentage of volcanic rocks and are almost exclusively oxides of iron and titanium. Olivines, pyroxenes, amphiboles, micas, and feldspars form true families, not individual minerals; each family is defined by a particular atomic structure, but the chemical composition of minerals within each family may vary widely, within certain limits. Olivines, for example, range from a magnesium end-member (Mg_2SiO_4) to an iron end-member (Fe_2SiO_4). These limits are set by the atomic structures, which represent different ways of assembling together the basic building block in geology, the SiO_4 tetrahedron (Fig. 1.6). In pyroxenes, for example, the tetrahedra are joined in single chains; in amphiboles in double chains, and in micas in sheets.

Volcanic rocks are composed of mixtures of only a few mineral families, and there is a steady change in the proportions of the different families present as one goes from one end of the rock spectrum to another. Examined under the microscope, a typical lava consists of a variable proportion of large, easily visible mineral crystals—*phenocrysts*—set in a ground mass of tiny, interlocking needle-like crystals felted together in the *groundmass*. One way of visualizing phenocrysts is to think of the pips in raspberry jam (Fig. 1.7). Groundmass crystals form after eruption, when the lava chills quickly on the surface.

Sometimes, chilling takes place without any crystallization at all, and natural glass results. Much the best known is the rhyolitic glass *obsidian*, but black (*tachlytes*) and brown (*sideromelane*) glasses of basaltic composition are also common. The ease with which a glass forms depends on the degree to which the silicate molecules in the melt are *polymerized* or linked together (see section 5.1). Rhyolitic melts are more strongly polymerized than basaltic ones. This accounts for the fact that rhyolitic lava flows are often glassy from top to bottom, whereas basaltic glass is usually found as thin, chilled skin surrounding better-crystallized material.

1.4.1 Nomenclature of volcanic rocks

Classification systems are often vexatious. Rock classification can be particularly irksome, but a simple solution is to consider rocks as varying chemically in two ways: in silica content and in alkali content. For historical reasons, rocks with a lot of SiO_2 are often termed *acidic*. *Silicic* is a more useful term. They consist mostly of quartz and feldspars, with a little mica or amphibole. *Basic* rocks (another historic term) lie at the other end of the spectrum. They contain much less SiO_2, so there is no free quartz (all the silica is present as silicates), but there is a lot of feldspar and pyroxene, with variable amounts of olivine and oxide minerals. Most of these minerals are dark in colour, so basic rocks with no quartz tend to be dark coloured, while acid rocks are much lighter in colour, and usually only have isolated crystals of dark minerals. Basic rocks are often termed *mafic*, in recognition of the high magnesium and iron contents that their mineralogy yields.

Basalt, the universal volcanic rock type, exemplifies basic compositions. *Rhyolite* is the most silicic volcanic rock. Intermediate between rhyolite and basalt are *andesites* and intermediate between andesites and rhyolites are *dacites*. So far, so good. A complication to this simple system of nomenclature is that natural volcanic rocks show large variations in their content of alkali elements, usefully expressed by the total of sodium and potassium ($Na_2O + K_2O$). Thus, while 'normal' rocks, such as those from a typical Caribbean volcano, plot along one trend on

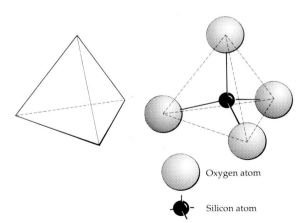

Oxygen atom

Silicon atom

Fig. 1.6 The silicon tetrahedron, basic building block of all the silicate minerals making up volcanic rocks.

Fig. 1.7 Thin section of a typical basaltic lava, showing ground mass minerals and phenocrysts (large black and white bodies). Here, the phenocrysts are mostly plagioclase, distinguished in polarized light by 'pyjama striped' twinning.

Fig. 1.8, others such as those from the Red Sea, plot along a trend parallel to this, but displaced to the more alkalic side. Such rocks are conveniently termed *alkali* basalts, andesites, or rhyolites, but many other names are embedded in the literature. *Phonolites* and *trachytes*, for example, are similar to andesites and rhyolites, but have more alkalis. Fig. 1.9 summarizes the nomenclature of common volcanic rocks; all of those with which we need to concern ourselves.[11] Within

the field of 'basalts' on the diagram, two varieties are especially important. *Tholeiites* are relatively silica rich and contain calcium-rich minerals such as plagioclase and pyroxene, while *alkali basalts* have more sodium and potassium, and typically contain olivine.

Even with the added complication of alkali content, this nomenclature may seem unreasonably simple. So it is. While it works for most common rocks (certainly the ones discussed in

Fig. 1.8 Volcanic rocks range in silica content from basalts through to andesites, dacites, and rhyolites, and form suites with different alkali contents ($Na_2O + K_2O$), characteristic of different tectonic settings. Rocks from the Caribbean volcano, Mount Misery, have relatively low alkalis; those from the Red Sea span a similar range of silica contents, but have more alkalis. After Cox *et al.* 1979.

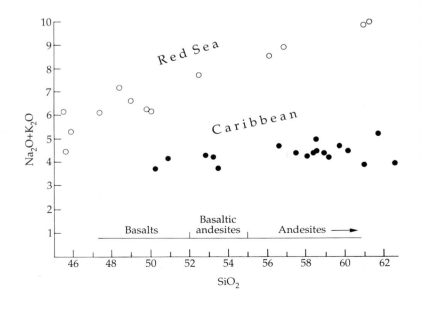

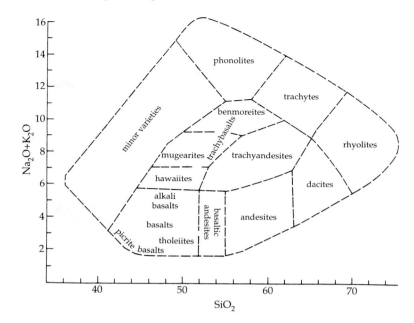

Fig. 1.9 A convenient system for classifying volcanic rocks, devised by Cox *et al.* The field that is labelled 'minor varieties' includes a number of geochemically interesting volcanic rock types which do not occur in large volumes and are not discussed in this book.

this book) there are many exotic volcanic rocks which almost defy classification. Most bizarre are the lavas of Oldoinyo Lengai, Tanzania (East Africa) which consist largely of sodium carbonate, washing soda (Fig. 1.10). Some geologists were a little sceptical about the existence of these lavas until the French volcanologist Maurice Krafft delighted a 1987 international volcanological conference in Japan with the first movie film

of a carbonate lava erupting. Intrusive carbonate rocks are more common, and are associated with suites of alkalic volcanic rocks whose mineralogy is so complex that their nomenclature is tangled enough to make even a hardened geologist weak at the knees: ijolites, uncompahgrites, urtites . . .

Weak-kneed geologists and other readers need not fear. Only the simplest petrological terms are used in this book.

Fig. 1.10 From a distance, Oldoinyo Lengai volcano looks unremarkable. Its erupted products, however, are exceptional: they consist of alkaline *carbonates* such as nyerereite $(Na_{0.82}K_{0.18})_2Ca(CO_3)_2$, similar to washing soda. Photo: courtesy of W. I. Rose.

— Notes —

1. US Department of Energy (1983). *Annual Review of Energy*.

2. Kelvin, Lord (1864). On the secular cooling of the Earth. *Trans. Roy. Soc. Edin.* **23**, 157–70.

3. Holmes, A. (1931). Radioactivity and the Earth's thermal history. *Geol. Mag.* **62**, 102–12.

4. Lee, T., Papanastassiou, D. A., and Wasserburg, G. J. (1976). Demonstration of ^{26}Mg excess in Allende and evidence for ^{26}Al. *Geophys. Res. Lett.* **3**, 109–12.

5. Greeley, R. (1985). *Planetary landscapes*. Allen and Unwin, London, 265 pp.

6. Kaula, W. M. (1979). Thermal evolution of the Earth and Moon growing by planetismal impacts. *J. Geophys. Res.* **84**, 999–1008.

7. Solomon, S. C. *et al.* (1981). Thermal histories of the terrestrial planets. In *Basaltic volcanism on the terrestrial planets*, Basaltic Volcanism Study Project. Pergamon Press, New York, 1286 pp.

8. Peale, S. J., Casson, P., and Reynolds, R. T. (1979). Melting of Io by tidal dissipation. *Science* **203**, 892–4.

9. Schubert, G., Stevenson, D., and Cassen, P. (1980). Whole planet cooling and the radiogenic heat source contents of the Earth and Moon. *J. Geophys. Res.* **85**, 2531–38.

10. Ringwood, A. E. (1977). *Composition and origin of the Earth*. Res. School Earth Sci. Australian National Univ. Canberra, Publ. No. 1299, 65 pp.

11. Cox, K. G., Bell, J. D., and Pankhurst, R. J. (1979). *The interpretation of igneous rocks*. George Allen and Unwin, London.

2

Keeping planets cool: volcanoes, hot-spots, and plate tectonics

A cup of coffee whitened with the appalling 'non-dairy creamer' beloved by Americans is not simply a static volume of liquid. Tiny particles of powdered 'creme' swirl around beneath the surface in startlingly rapid currents. Think about this when confronted with your next Styrofoam cup of coffee. Think also about this fact: almost all of the world's active volcanoes are located either in the oceans, or very close to them. These two apparently unrelated facts tell us a lot about the way that the Earth works as a planet.

First, however, a semantic point. While the term 'active volcano' is universally used, it is often not clear what is meant, because the term is applied indiscriminately to volcanoes that are actually erupting in a business-like way; those

that are steaming gently; and those which only *might* erupt at some unspecified future date. At one time, the term was formally applied to volcanoes with documented records of eruptions, but because so many volcanoes are located in remote regions where there is no recorded history, this is not realistic. It is best to regard as *potentially active* any volcano that preserves geological evidence of eruptions within the last 10 000 years. In any one year, about 50 individual volcanoes erupt. Approximately 500 have erupted within historic times, and a further 2500 have been active within the last 10 000 years. Of course, a few volcanoes may have *repose times* between eruptions of more than 10 000 years, but 10 000 years is a convenient cut-off.

2.1 A package tour of the Earth's volcanoes

A good atlas will show that volcanoes are found in three distinct environments: first, along the margins of continents and on strings of islands straggling away from them; second, in the middle of the oceans, and third, in odd places in the middle of the continents. There are rather few of the last group. There are none at all in Australia or South America, but there are some important examples in Africa.

For many decades, geologists had to accept this erratic distribution at face value. Hints of an underlying pattern came in the nineteenth

century when oceanographers began to survey the ocean floors. They found that the volcanic islands dotted along the length of the Atlantic oceans were outlying pinnacles of a submarine mountain range extending for thousands of kilometres, from the Arctic to the Antarctic, tracing an outline that curiously mirrors the coastlines of Africa and South America. For obvious reasons, this ridge became known as the *Mid-Atlantic Ridge*. Other surveys showed similar ridges beneath the other oceans, and rock samples dredged from them showed that they were all made of basaltic lavas. Later studies showed that ocean ridges are not simple, continuous features, but are interrupted and offset by great curving transform faults and fractures (Fig. 2.1). It became clear that the ocean ridges form collectively by far the longest system of mountain ranges on Earth. Increasing knowledge of their topography, however, did not resolve the key question of how they formed. It was not until the 1960s that Vine and Matthews's interpretation of the patterns of the magnetism of the rocks of the ocean floor confirmed what some scientists had long suspected: new oceanic crust is generated at ocean ridges, while 'continental drift' is the consequence of the widening of oceans in this way.

This realization was the core of the 'Plate Tectonic Revolution' which interrupted the scholarly slumbers of distinguished geologists around the world. The events which led up to the revolution have been retold many times, so most readers will be familiar with them. Here, therefore, we merely make a brief world tour of the key environments where volcanoes occur, to set some of the phenomena described later in a global context (Fig. 2.2). For an excellent technical review, the synthesis by Peter Wylie is recommended.[1]

2.1.1 Continental margin and island arc volcanoes

Much the most notorious of this group is the so-called 'Ring of Fire', which loops erratically around the Pacific Ocean (Fig. 2.3). It begins amidst the monochromatic beauty of the Graham Land Peninsula, the long finger of land stretching northwards from the Antarctic continent. An eruption of Deception Island there in

1969 wrecked British and Chilean research bases. Hapless penguins still scald their feet in boiling hot springs. From there, the Ring swings eastwards in an arc of active volcanoes forming the desolate South Sandwich islands, heading first north and then west, describing a great loop through South Georgia (non-volcanic) before making for the extreme south of the South American continent. Here, in the tangle of islands and fjords that is Patagonia, is Mt. Burney, the southernmost volcano on the Andean continental margin: an obscure volcano, rarely visited by geologists. Some 600 km further north is Mt. Hudson, which erupted on 12 August 1991. A great plume of ash rose to an altitude of 16 km, to be borne away by brisk north-westerly winds. Ash fall brought misery to ranching areas of the Patagonian pampas, taking the lives of a million sheep, 40 per cent of the total in the area. Ash fell as far distant as the Falkland Islands, one thousand kilometres to the east. By 20 August, the plume had reached Australia, 15 000 kilometres further east. Passengers and air-crew on a flight from Melbourne to Sidney noticed a strange, hazy cloud with a brownish orange tinge, and detected the acrid smell of sulphur dioxide.

From Mt. Hudson northwards, volcanoes occur along almost the entire 7000 km length of the Andes as far as the Caribbean. There are thousands of volcanoes along the cordillera, but only about one hundred are potentially active. North Chile, Peru, and Ecuador boast the highest volcanoes in the world: dozens exceed 6000 m in height (Fig. 2.4). Chimborazo and Cotopaxi in Ecuador were thought to be the world's highest mountains until the Himalayas were surveyed in the nineteenth century. They inspired one of mountaineering's classic works,

Fig. 2.1 A 25° section of the Mid-Atlantic Ridge, showing the ridge axis offset by a multitude of fractures. Each fracture zone is curved, centred about a pole of rotation, accommodating differential movements of rigid lithospheric plates.

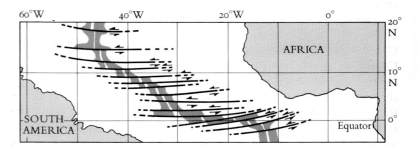

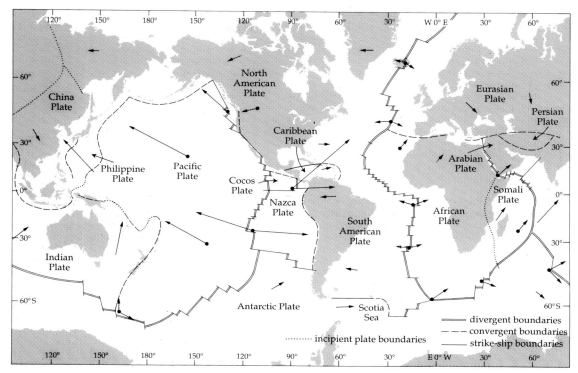

Fig. 2.2 Boundaries of the main lithospheric plates and their movement directions, related to a 'hot-spot' frame of reference. Plates include both oceanic and continental lithosphere. Length of arrows indicates relative velocities of plates in that direction. (Divergent boundaries are synonymous with mid-ocean ridges.) After an original by Seija Uyeda, University of Tokyo.

Edward Whymper's *Travels amongst the Great Andes of the Equator* (1891), and their romance 'stole the soul' of the poet W. J. Turner. In Colombia and Venezuela, the Andean mountain chain and volcanic Ring become geologically intricately interwoven. In this previously obscure volcanic region, 25 000 people in the town of Armero, Colombia, were killed in 1985 by mud-flows spawned from the ice-capped volcano Nevado del Ruiz. It was the world's most lethal eruption in fifty years.

After a short gap in the jungles of the Isthmus of Panama, the Ring reappears in full vigour in Costa Rica, extending northwards through the Central American republics of Nicaragua and Guatemala, which are as explosive politically as the many volcanoes they contain. One reason that the Panama Canal was built in Panama rather than Nicaragua—a better choice in some ways—was that contemporary Nicaraguan post-

age stamps displayed a volcano. Adroit supporters of the Panamanian cause used this to bolster their argument that Nicaragua was far too unstable a place to build the canal. Further north, central Mexico contains major volcanoes, some burdened with tortuous Aztec names such as Popocatepetl and Iztacchautl, but the country is chiefly remembered by volcanologists for another previously obscure volcano, El Chichón, which took many hundreds of lives during its brief eruption in 1982.

Continuing northwards, the United States has fewer active volcanoes. None the less, the State of Washington was the site of the most closely observed eruption in history: the May 1980 eruption of Mt. St Helens, which startled volcanologists by its violence (Fig. 2.5). We shall return to Mt. St Helens many times in later chapters. Further north still, Canada is not usually thought of as a volcanic region, but there

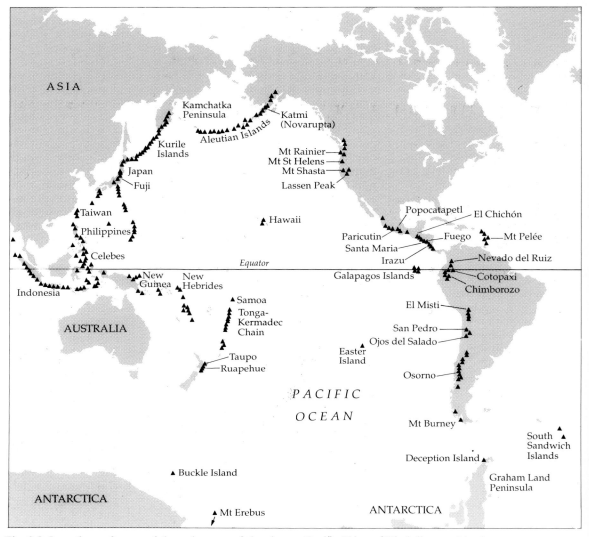

Fig. 2.3 Locations of some of the volcanoes of the circum-Pacific 'Ring of Fire' discussed in the text.

were many eruptions in prehistoric times. An eruption in northernmost British Colombia in about AD 700 may have caused wholesale disruption of the Athapaskan Indian community. In Alaska, the Ring of Fire picks up again strongly. Novarupta, on the continental margin near Mt. Katmai, was in 1912 the site of the largest eruption this century. Fortunately it is so remote that there were few casualties. Westwards from Katmai, volcanoes swing out to form the long Aleutian and Kurile island arcs. These curve northwards towards the Kamchatka Peninsula

of Russia, another highly active volcanic province, and the Ring then sweeps southwards to Japan, a country whose landscape is so dominated by volcanoes that they form an integral part of the national culture. One of the Japanese works of art best loved in the west is *The Great Wave at Kanagana* by the 19th century artist Katushika Hokusai. This picture is dominated by a splendidly crested tsunami with Mt Fuji diminutive in the background, but it is actually only one of a series of no less than 46 views devoted to Mt Fuji. In the later, slightly less

Fig. 2.4 Nevado Ojos del Salado, the world's highest active volcano, 6885 m high, located in an hyper-arid part of the Atacama desert on the frontier between Chile and Argentina. Much of the light-toned material in the photograph is not snow, but an air-fall pumice deposit. Base level of the volcano is 4500 m.

Fig. 2.5 A huge convecting eruption column developed after Mount St Helen's initial violently explosive blast on 18 May 1980. Ash rose more than 20 km, and fell out over much of the central USA. This view is from the south. US Geological Survey photograph.

famous views, the waves have died down and Mt Fuji is depicted in all its symmetrical elegance.

From Japan, volcanoes continue through Taiwan to the Philippines, where Taal and Mayon volcanoes have long and lethal histories, and Pinatubo gave vent to the third-largest eruption of the twentieth century on 15 June 1991. From there, the volcanic zone continues on to Sulawesi and south-eastwards through New Guinea, the New Hebrides, and a scatter of small Melanesian volcanic islands before making an abrupt dog's leg to the Solomon islands and the Tonga–Kermadec Archipelago, which continues southwards again into New Zealand.

North Island, New Zealand, was the locus of the most violent eruption known to volcanologists, the Taupo paroxysm of AD 186. There are no volcanoes on the South Island, and the Ring of Fire that we have been following for 40 000 km around the Pacific fizzles out there. To find more southerly volcanoes, we must look to Antarctica, over 1500 km away. Buckle Island lies just off the Antarctic coast, while Mt. Erebus fumes persistently above Scott's original base camp at McMurdo Sound, from which he set out his heroic journey to the pole. Strictly speaking, Mt. Erebus is not part of the Ring of Fire—it is more like the exotic volcanoes of Africa—but it does bring us back conveniently to Antarctica, where we started.

To sum up, throughout all its lethal length, the Ring of Fire is characterized by tall, conical volcanoes that occur on continental margins or island arcs, which erupt rather infrequently, but which have caused many of the greatest volcanic catastrophes in history. If we include the other volcanoes of this class which lie on continental margins or island arcs away from the circum-Pacific region, notably those in the Mediterranean, we can conclude that these volcanoes have caused almost *all* the most destructive eruptions in history.

2.1.2 *Volcanoes in the middle of the oceans— the Atlantic*

Our second volcanic package tour begins nearly at the opposite pole, at Jan Mayen, an ice-gripped island in the Arctic Ocean, half-way between Greenland and northernmost Norway (Fig. 2.6). Dominating this bleak scrap of land is Beerenberg volcano, which last erupted in 1984. South-west of Jan Mayen is Iceland, the largest volcanic island in the world, and famous for its volcanoes and geysers. Icelanders christened their finest gusher of boiling water *Geysir*. This name was applied subsequently to those depressing pieces of gas-fired plumbing which used to be familiar parts of British bathrooms, as well as to other natural jets of ebullient water around the world.

Iceland has twenty-two active volcanoes. Those on the Westmann Islands, off the southern coast, are best known. A new island, Surtsey, was spectacularly born in full view of the media during an eruption in 1963. In February 1973, another eruption burst out on the nearby island of Heimaey. For a time it seemed that Heimaey, Iceland's largest fishing port, would be destroyed. Fortunately, the eruption petered out in July 1973, after only a part of the town had been engulfed (Fig. 2.7).

South of Iceland, the Atlantic Ocean stretches out devoid of even the smallest of islands for thousands of kilometres. On 29 December 1884, Captain Perry of the British steamship *Bulgarian* reported sighting a submarine eruption in the middle of the ocean, half-way between Ireland and Newfoundland (49°N, 34°30'W). On hearing of this report, the British oceanographer Sir John Murray remarked that he hoped that it signified the emergence of a new island, because the Royal Navy needed a coaling station in that area. Unfortunately for imperial strategic designs, this eruption, if it *was* an eruption, came to nothing and no new island succeeded in raising its head above water, to be added to Queen Victoria's already extensive collection.

At the latitude of the Straits of Gibraltar, and 1300 kilometres from the Portuguese coast, are the Azores, a cluster of volcanic islands where eruptions have taken place as recently as 1957. South of the Azores, but hugging the African coast, are the Canary Islands, beloved of topless tourists, and also the scene of recent eruptions of a more volcanic kind. El Teide, whose soaring 3700-m-high cone dominates the island of Tenerife, was erupting as Columbus passed the island on his epic voyage in 1492. South and west of the

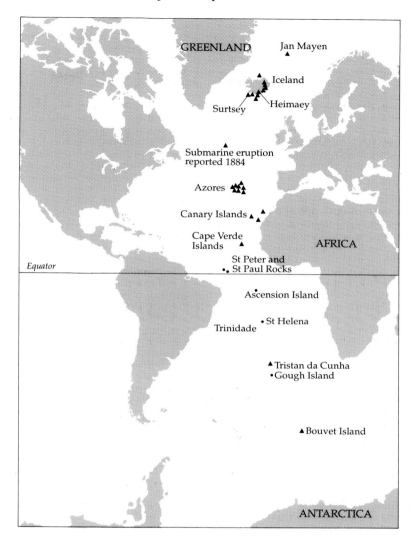

Fig. 2.6 Volcanoes and volcanic islands of the Atlantic Ocean. The Mid-Atlantic Ridge does not break surface except in Iceland; the widely-spaced active volcanic islands generally lie slightly off the ridge axis.

Canaries, the Atlantic stretches for thousands of kilometres, broken only by the Cape Verde Islands, again volcanic, and the St Peter and St Paul rocks. South of this tiny pair are a few lonely islands separated from one another by hundreds of kilometres of open sea: first, Ascension Island; then St Helena, site of Napoleon's second and terminal exile. Both are young volcanic complexes. Further south still is Tristan da Cunha, notorious for its eruption in 1962 which precipitated the evacuation of the island's tiny, ethnically unique population (Fig. 2.8). Tristan has two near neighbours, both of them uninhabited (Nightingale and Inaccessible Islands) and a

third, Gough Island, a couple of hundred kilometres further south. All three are the wave-battered remains of volcanoes whose activity was extinguished about 2–3 million years ago.

Further south still, the Roaring Forties begin, the latitudes at which the expanses of the Southern Ocean extend emptily right around the world and where unrestrained westerly gales whip the sea surface into waves of terrifying scale. Bouvet Island, the last link in our tenuous volcanic chain lies even further south; south of latitude 50°. Almost never visited, little is known about the desolate, ice-bound 780-m-high-volcano although a minor steam emission was noted in

Fig. 2.7 Basaltic lava flows advancing through the Icelandic port of Heimaey in April 1973. Although the town was extensively damaged, lavas did not advance much further than their position here. Fortuitously, the lavas improved the harbour, making it more sheltered. Much of the town is now supplied with geothermal heat from the volcano. (Photo: S. Thorarinsson.)

Fig. 2.8 Lava emerging from a small vent on the southern flanks of the Tristan da Cunha volcano in 1962. Parts of the active vent are visible at the bottom. Some of the islanders' fields, boats, and houses can be seen on the left. The islanders were evacuated, because the eruption threatened their settlement, but returned later. (Photo: courtesy of I. G. Gass.)

the 1980s by a passing solo yachtsman of extraordinary fortitude.

2.1.3 Continental rift valley volcanoes

A glance at a map of Africa shows that its topography is different from the other continents: it has no long mountain ranges such as the Alps, Andes, or Rockies. Africa's highest places are almost all young volcanoes. They occur in two distinct environments: along the East African rift valley system, and, most remarkable of all, as isolated massifs deep in the heart of the Sahara. On a map, the most obvious features of the Rift system are the long, narrow lakes that lie within its bounding faults (Fig. 2.9). One distinct

chain of lakes starts at Lake Malawi, a few hundred kilometres from the Indian Ocean, then winds up through Lakes Tanganyika, Kivu, Edward, and Albert (source of the Nile). These lakes lie within the *Western Rift*. Rising above Lake Kivu are some of the most exotic mountains in the world, the Virunga Mountains, first sighted by John Hanning Speke in 1861. They lie almost on the equator in the heart of Africa, but rise so high (4500 m) that they form ecological oases, which support exotic flora and fauna, including the only remaining population of mountain gorillas. Two volcanoes, Nyamuragira (3056 m) and Nyiragongo (3470 m) have been among the most active in the world. Nyiragongo

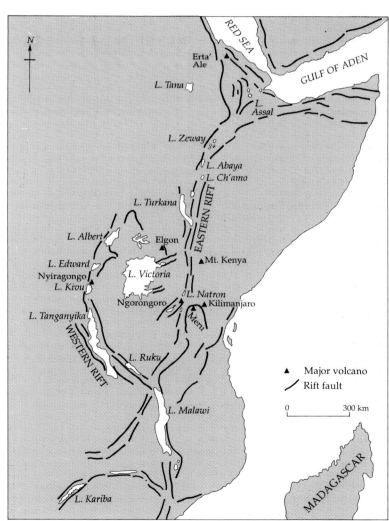

Fig. 2.9 The main lakes, rift valleys, and volcanoes of the East African Rift system. Rifts are shown schematically; in detail, the faulting is complex. Not all volcanoes are shown.

contained a lava lake in its crater when it was discovered by the German explorer von Götzen in 1894, and the lake remained as a persistent feature through the present century. Catastrophe struck in 1977, when the lava lake abruptly drained through a series of fissures. During the course of less than an hour on the morning of 10 January, about 22 million cubic metres of fluid basalt lava flooded down the flanks of Nyiragongo, flowing at about 60 kilometres per hour, and taking the lives of 70 people.

On the eastern side of Lake Victoria, a second set of lakes marks the floor of the *Eastern Rift*, starting in the south at Lake Manyara and extending northward via Lakes Natron and Naivasha to Lake Turkana. A chain of smaller lakes (Ch'amo, Abaya, Zeway) leads northeastwards from Lake Turkana through Ethiopia to the Red Sea. Mounts Kilimanjaro and Kenya, both volcanoes, rise above the eastern walls of the rift valley to form the highest points in Africa, 5895 m and 5199 m high. Some of the smaller volcanoes are even better known: the Ngorongoro Crater, for example, made famous by a plethora of wildlife movies. The Eastern Rift is also the site of the strange volcano Oldoinyo Lengai, which supplies caustic Lake Natron with alkali.

In Ethiopia, the Eastern Rift opens out into the harsh world of the Danakil Depression, a scorching lowland, too hot for permanent lakes but containing a few salty *playas*. These include the lowest point in Africa, Lake Assal in Djibouti, 150 m *below* sea-level. In the heart of the Danakil Depression is Erta' Ale volcano, active, with two permanent lava lakes. North of Erta' Ale, the Danakil Depression merges into the Red Sea, remarkable for its rectilinear shape. At its northern end, the Red Sea terminates in a prominent fork; one branch forming the Gulf of Suez, the other the valley of the River Jordan. It does not take a great conceptual leap to perceive that the Red Sea and the African rift valleys are intimately linked.

Africa is not the only continent to possess rift valley volcanoes, although they are best developed there. In Europe, the Eifel region of Germany is the site of young volcanism on the Rhine Rift. A major eruption took place only 11 000 years ago to form the Laachersee, a 1-km-diameter crater, now containing a picturesque lake. In Asia volcanism was associated with the rift now occupied by Lake Baikal, at 1200 m deep the world's deepest lake, and reservoir of one-fifth of the world's fresh water. Seismic activity in both regions continues to the present day. In North America, the Rio Grande river, best known from Western movies, flows in a magnificent rift valley for hundreds of kilometres of its journey towards the Gulf of Mexico. Many young volcanoes are dotted along its course, but there have been no eruptions in modern times. Both the Baikal and the Rio Grande rifts occupy the crests of broad topographic swells: gently uplifted domes that developed before and during rifting.

2.1.4 Volcanoes in the middle of nowhere

North Africa's Sahara Desert is not the endless 'sand sea' of popular fiction. Several rugged mountain massifs reach up thousands of metres into cooler air, providing welcome relief from the glaring lowland heat. Three massifs are noteworthy: Jebel Marra in Sudan, Tibesti in Chad, and Ahoggar in Algeria. These massifs are isolated from one another, and have little in common except that all three are major volcanic complexes. Half a dozen young volcanoes rise above 3000 metres in the Tibesti Complex. Astronauts report that one of them, Pic Tousside volcano (3265 m), is the easiest of all the Earth's landmarks to identify from orbit, because its dark massif rises up in such stark contrast against the ochreous wastes of the desert. Paradoxically, it is scarcely known to the world at large. Jebel Marra in the Sudan is less outstanding, but still reaches 3042 m above sea-level. Its 5-km-diameter caldera was formed by a colossal eruption about 3500 years ago, an event that may have showered ash on Pharaonic Egypt (Fig. 2.10).

A world away, in more senses than one, are the Sandwich (more commonly, but less properly called the *Hawaiian*) Islands, in the centre of the Pacific Ocean. A scatter of small scraps of land (Midway Island, Kure Atoll) marks the western limit of the archipelago, which extends almost 2000 m eastwards along a chain of atolls (Laysan Island, French Frigate Shoals), until the first

Fig. 2.10 Oblique aerial photograph of the Deriba Caldera, Darfur Province, western Sudan. Highest point on the rim (right) is Jebel Marra, which rises 3042 m above sea-level. Rarely visited by geologists, this remote caldera probably formed about 3000 years. Age of the younger cone containing small lake is not known.

substantial island, Kauai, is reached. Eastwards from Kauai, the islands get bigger and higher via Oahu, Molokai and Maui, until Hawaii—the Big Island as *haole* residents call it—is reached. Hawaii is the site of two active volcanoes, one of which (Mauna Loa) rises 4170 m above the ocean. Kilauea is much lower (1200 m), but is the most continuously active volcano on Earth; lava streaming into the ocean at a steady rate of about five cubic metres per second for years on end. A third major volcano, Mauna Kea (4206 m high) is inactive, and is the site of the world's largest cluster of astronomical observatories. The gleaming white geometry of their high-tech domes makes an odd contrast with the reddish curves of the scoria cones surrounding the summit. A short distance off the south coast of Hawaii, a submarine volcano, Loihi, is also active, but has another 1000 m to go before it breaks surface.

— 2.2 Plate tectonics encapsulated

According to plate tectonic theory, the Earth's crust is made up of seven large, rigid plates and many smaller ones, which shift around over the surface of the Earth. Three different types of plate margins separate the plates: *constructive* plate margins, where oceanic crust is formed; *destructive* margins, where it is consumed again; and *conservative* or *transform* margins, where plates slide passively past one another (Fig. 2.2). The fragmentation of the crust into plates which jostle against one another on the surface is an expression of something much deeper seated: the Earth's internal heat. A planet as large as the Earth contains tremendous amounts of heat, both primordial heat left over from its formation, and the heat generated by decay of radioactive isotopes. There are three different ways in which this internal heat can escape, and the various planets which we will explore express different combinations of these three.

Straightforward *conduction* through solid rock is the simplest, commonest, and least volcanically interesting mechanism of heat loss—there is nothing to see at the surface. It is the only mechanism that small bodies such as the Moon display. A second mechanism may operate on planets big enough to contain enough heat for large amounts of mantle peridotite to be melted. Huge volumes of basalt lavas may then be pumped to the surface of the planet through a few massive volcanoes. *Hot-spot volcanism* of this sort has been important on several bodies in the solar system, and it plays a non-negligible role on Earth. However, the Earth conveys most of its heat to the surface via a third mechanism, *plate recyling*, or plate tectonics. Basalt magmas generated at ocean ridges form oceanic crust, which then cools as heat is dumped to sea-water. Millions of years later, the oceanic crust is reabsorbed into the mantle in a continuous cycle. In this sense, basalt magmas and the crust that they generate act like water circulating in an automobile radiator: they form a working fluid which transfers heat from the hot centre to the outside world. Convection currents swirling in a coffee cup do the same thing in microcosm.

2.2.1 The mantle

There is one essential difference between convection in coffee and in the mantle: convection in the mantle takes place in the *solid* state. Although temperatures in the mantle are higher than the melting point of mantle materials at the surface, pressures are so great that the material remains solid. If the idea of convection taking place in the solid state seems paradoxical, recall that the process is infinitely slow in human terms. Pitch

provides an analogy to mantle materials. If bludgeoned with a hammer, a solid block of pitch will shatter into glossy black fragments. But if the same hammer is left on top of the same block for a long period in a warm room, it will sink imperceptibly into the pitch, which slowly oozes around and swallows up the hammer head.

Mantle materials convect by creep deformation (a progressive distortion of crystal structures) rather than fluid flow, and move only a few centimetres per year. Most of what we know about the mantle comes from seismic evidence, which demonstrates that it is indeed solid. The velocities of seismic shock waves propagated by earthquakes through the mantle are controlled by the physical properties of the material that they travel through—the denser and more rigid the material, the higher the seismic wave velocities.

In 1909, the Croatian seismologist Andrija Mohorovičic identified the boundary between the Earth's crust and mantle, now called the *Mohorovičic discontinuity*, or Moho, which is defined by an abrupt increase in density and seismic velocities. In recent years, seismology has progressed rapidly, advancing our knowledge of the mantle enormously. Sophisticated computing techniques such as seismic tomography enable seismologists to construct three-dimensional images of the mantle and its convection cells, much in the way that modern radiologists use CAT scanners to study the interior of a patient's skull. Whereas it was thought early on that convection affected only the *upper* mantle, we now know that the whole mantle is convecting. What remains is to determine how the upper and lower mantle interact.

2.2.2 Lithosphere

In plate tectonic terms, a 'plate' does not consist of a fragment of crust alone; and neither need it be made exclusively of oceanic or continental crustal material. A plate consists of crustal *and* mantle material as far down as the *asthenosphere*, the hot, mobile interior of the Earth. Between the asthenosphere and the surface is the *lithosphere*, the rigid outermost shell of the Earth. 'Continental drift' focuses attention on the continents, but it is actually lithospheric plates that shift around in the movements that are expressed in the wanderings of continents.

Formally, lithosphere is defined in terms of thermal gradients. At the Earth's surface, rocks are cool and rigid, and they therefore deform elastically under loads—when stressed, they first give a bit, and then break abruptly in *brittle failure*. Earthquakes are one result of rocks failing under stress. More importantly, heat can be transported through rigid rocks only by *conduction*. Thus, temperature increases steadily with depth, and thermal gradients are linear. Different geological provinces are characterized by different heat flows, and therefore by different thermal gradients.

Temperatures cannot increase infinitely without something happening. Depending on the thermal gradient, a depth is ultimately reached where temperatures are so high that the rocks approach their melting point (*c*. 1350°C), become ductile, and have lower seismic wave velocities. This is where the asthenosphere starts. It is easily deformable, and therefore *convection* is the dominant mode of heat flow in the asthenosphere, and because convection tends to stir things up, the temperature in the asthenosphere increases much less with increasing depth. A useful way of defining the base of the lithosphere, therefore, is the point of inflexion on a temperature/depth plot, where the linear conductive gradient gives way to the convective gradient (Fig. 2.11). Above the point of inflexion is the lithosphere, below it, the asthenosphere. Beneath the continents, the lithosphere is between 100 and 150 km thick, but beneath the oceans it varies with age, young lithosphere being as little as 10 km thick, and old lithosphere about 120 km thick.

It is the contrast in long-term mechanical properties between the rigid lithosphere and the ductile asthenosphere that permits lithospheric plates to move around on the surface of the Earth. Seismic data also reveal a number of other important boundaries within the mantle.[2] Much the most important zone for plate tectonics is the *Low Velocity Zone* (LVZ), roughly between the depths of 50 and 150 kilometres. As its name implies, the LVZ is a zone of reduced seismic wave velocities, indicating that the mantle materials there are much less rigid than those above and below, and probably contain some partially molten mantle material. Because of this, the term LVZ is often used synonymously (but erroneously) with the term asthenosphere. Below the LVZ there is an abrupt change in seismic wave velocities at about 400 km, probably corresponding with an important mineralogical change: the ubiquitous olivine of the upper mantle changing to denser spinel. Below 670 km, the spinel structure gives way to perovskite (Fig. 2.12).

Fig. 2.11 'Plate tectonics' concerns lithospheric plates, rather than drifting continents. Within the rigid lithosphere (upper part of the diagram), heat is transferred to the surface by conduction, and temperature increases progressively with depth. At a depth of about 120 km, the temperature reaches about 1350°C, rocks are no longer rigid, heat is transferred by convection, so temperature increases much less rapidly with depth. It is convenient to define the boundary between lithosphere and asthenosphere at the intersection between the two thermal gradients. (After D. McKenzie and M. J. Bickle.)

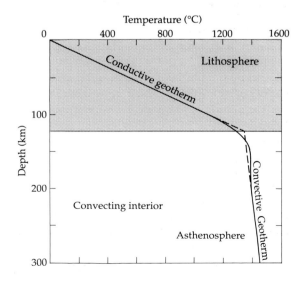

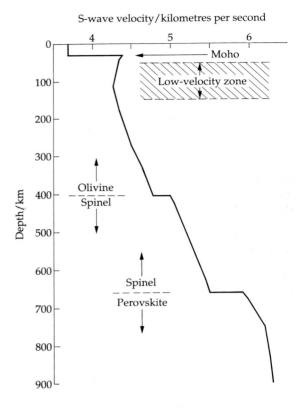

Fig. 2.12 Variations of seismic wave velocities with depth define the main layers within the mantle. Moho is at 30 km depth below continents, where denser, olivine-dominated mantle begins. The S-wave (shear wave) velocities are at a minimum between roughly 50 and 150 km defining the Low Velocity Zone, where partial melting may be present. At 400 km, olivine gives way to spinel, and at 670 km spinel to perovskite. (After S. P. Grand and D. V. Helmberger.)

— 2.3 Ocean ridge volcanism

Ocean ridges are the essence of plate tectonics. It used to be thought that they sat directly above upwelling mantle convection cells, which were supposed to be the immediate causes of sea-floor spreading. Work done in the 1980s shows that it is more likely that ocean ridges result from *passive* spreading—the oceanic lithosphere is thinned by tectonic forces until it splits, stretched by the pull of an older, denser lithosphere sinking into the mantle at a subduction zone (Fig. 2.13). Bruce Heezen, a distinguished marine geologist, aptly likened this passive *sea-floor spreading* process to a 'wound that never heals'. It is a continuous process, operating at rates of a few centimetres per year. Thinning of the oceanic lithosphere due to extension allows mantle isotherms to rise, permitting mantle material that was stable at temperatures and pressures equivalent to more than 100 km to rise to depths of 50 km or less. Enough thermal energy is con-

tained in the rising material to cause widespread melting at the shallower level (Fig. 2.14). Interpretation of ocean ridges as passive phenomena helps to explain a number of previously enigmatic features, such as the absence of consistent gravity anomalies over them, and the huge fractures that displace ridge axes, sometimes for hundreds of kilometres. It was always difficult to envisage how mantle convection cells could be consistent with such vast offsets.

2.3.1 Ophiolites—fossils of the oceanic crust

Ophiolites are chunks of ancient oceanic crust which have been pushed up on to continental crust. They provide unique insights into processes taking place below sea-level at spreading ridges. Oceanic crust is consistently thin: about 6–7 km. At the bottom of an ophiolite sequence, coarse-grained plutonic igneous rocks (peridotites) are found, which were once located in the

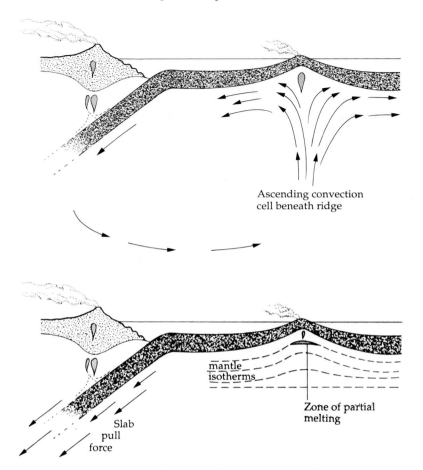

Fig. 2.13 Cartoon views illustrating the shift in thinking about the forces driving plate motions. At first (top), ocean spreading ridges were perceived as direct expressions of upwelling mantle convection cells which drove sea-floor spreading via 'ridge push' forces. Today the same relationships are interpreted differently: the 'pull' of the cold, dense, descending slab drives spreading (below). Thinning of the oceanic lithosphere, in response to extension, causes mantle isotherms to rise, so that melting and volcanism result.

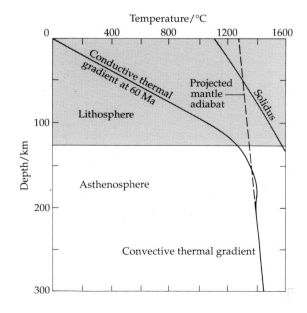

Fig. 2.14 This diagram is the key to ocean ridge volcanism. It is similar to Fig. 2.11, but has two extra lines. The dashed line 'projected mantle adiabat' shows the temperature which a mass of deep mantle material would have if the pressure on it were decreased. At depth of about 50 km, this intersects the peridotite solidus, which expresses the pressure/temperature conditions at which solid peridotite begins to melt. At depths shallower than 50 km, melting takes place when mantle materials are depressurized, following the adiabat.

mantle, beneath the Moho. Overlying these mantle-derived rocks are rocks which formed the oceanic crust proper, and which were derived by partial melting of mantle peridotites. Ideally, these form a three-layered sequence.

Lowermost of the three layers are plutonic rocks of gabbroic composition, sometimes called the *cumulate sequence*. These represent the 'magma chambers' which supplied overlying volcanoes through the second unit, called *sheeted dyke complexes*. Overlying the sheeted dyke

complex is the truly volcanic material: piles of *pillow lavas*, which were erupted from the dykes and accumulated to form a layer several hundred metres thick (Figs. 2.15–2.16).

Dykes are crucial components of ocean ridge volcanism. Addition of material to the oceanic crust in the form of dykes is the mechanism by which sea-floor spreading takes place. Sheeted dyke complexes consist of nothing but dykes: thin, parallel sheets of rock intruded successively into one another; so abundantly that it is often difficult to match the edges of a single dyke together—half a dozen other dykes may have been intruded subsequently between them (Fig. 2.17). All this dyke intrusion at the ridge axis is the consequence of extension: new material being intruded to fill what would otherwise be a gaping void, the wound that never heals. An individual dyke may be only a metre thick, but a thousand such dykes represents extension of one kilometre.

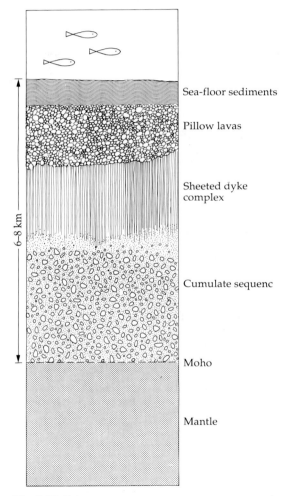

Sea-floor sediments

Pillow lavas

Sheeted dyke complex

Cumulate sequenc

Moho

Mantle

6–8 km

Fig. 2.15 Schematic section through typical oceanic crust. Total thickness of *crust* is less than 8 km, but oceanic *lithosphere* more than 60 million years old may be as much as 120 km thick, since it includes both crust and mantle components.

Fig. 2.16 A world-famous outcrop of pillow lavas of Cretaceous age forming part of an ophiolite sequence, Wadi Jizi, Oman. (Photo: courtesy of D. A. Rothery.).

Fig. 2.17 Hundreds of vertical dykes are exposed in the hills forming the background of this view of the sheeted dyke complex of the Oman Ophiolite. Dykes form 50 per cent of the volume of rock here; the remainder is gabbro. (Photo: courtesy of D. A. Rothery.)

2.3.2 MORB magma chambers

Mid-ocean ridge basalt (*MORB*) is by far the commonest volcanic rock on Earth. Modern marine acoustic tomographic techniques have provided some insights into the anatomy of the magma chambers underlying an active ridge, the East Pacific Rise, where MORBs are being formed at the present day.[3] These studies suggest that the magma chambers, consisting of more than 50 per cent melt, are quite small, but are surrounded by a wider reservoir of hot rock, at a temperature of more than 1000°C, containing a few per cent of melt. This halo of hot, partly molten rock is more than 6 km across, and supports isostatically a topographic ridge 200–400 m high along the axis of the East Pacific Rise. A thin pool of nearly 100 per cent melt approximately 4 km wide but only a few hundred metres thick caps the top of the broader reservoir of hot rock. Thus, the axial magma chamber which supplies the myriads of dykes may resemble a mushroom in cross-section, with a narrow stalk of partial melt feeding a thin lens of pure melt four kilometres wide (Fig. 2.18).

2.3.3 Spreading at the surface—Iceland

To see ocean ridge volcanism at work, we need look no further than Iceland, sitting astride the Mid-Atlantic Ridge. Because it is above sea-level, Iceland is not typical of submarine ridge axes, but the differences are of degree, rather than funda-

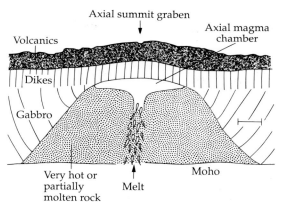

Fig. 2.18 Structure of a ridge such as the East Pacific Rise according to K. C. Macdonald, derived by seismic tomographical methods. A four-kilometre-wide mushroom-shaped cupola of mostly molten rock (more than 50 per cent melt) is surrounded by a much larger halo of intensely hot, but only partially molten material (less than 5 per cent melted). Ridge sketched here is 200–400 m high, and 8 km wide, and is supported isostatically by the hot, low density rocks. (After Macdonald, K. C. (1989). Anatomy of the magma reservoir. Reprinted with permission from *Nature* **339**, p.178.)

mental. Volcanic activity on Iceland has been so prolific that it has been able to keep itself above sea-level. Much of the bleak, beautiful island is covered with basalt lavas so fresh that they convince even the geologically innocent that Iceland is a young landmass, newly created from the spreading zone that runs through it from south-west to north. Here, the land is broken up by innumerable long, straight-sided fissures and rents which run for kilometres (Fig. 2.19). These are the surface expression of dykes. It is easy to relate mentally this much fissured land surface to the sheeted dyke complexes of ophiolite terrains. Every few years, a new dyke breaks the surface at volcanoes such as Krafla. Lava is sprayed out along a straight, narrow fissure many hundreds of metres long, forming a veritable 'wall of fire' before activity focuses on a single centre, and the fireworks fade away from the rest of the fissure. Rapid spreading rates have been measured in Iceland; the island is widening each year overall by a centimetre or two, but in the centre of the active zone, the rate is much faster.

Although Iceland's volcanoes are very active, none is as high as Mt. Fuji or Mauna Loa. One reason for this is precisely because Iceland is spreading so fast: a volcano initiated over the active zone in the centre of the island will soon (geologically) be pushed sideways, until it is away from the active zone, cut off from the supply of magma, and thus condemned to extinction.

2.3.4 Oceanic lithosphere away from the ridge axis

Oceanic lithosphere does not consist only of the few kilometres thickness of dykes and pillow lavas that form the crust—it incorporates a good deal of mantle material as well. Recall that the base of the oceanic lithosphere is defined as the depth at which rocks soften as their temperature approaches $c.\,1350°C$, and the asthenospheric mantle begins to deform viscously. Near the ridge axis, the oceanic plate is thin, perhaps only 10 km thick—the 1350°C isotherm is near the surface. As the plate moves further away from the ridge axis, it loses heat to the waters of the ocean, mostly through the circulation of sea-water fluids through cracks and fissures within the uppermost kilometre or two.

Hydrothermal cooling has profound effects: when first formed, the rocks of the oceanic lithosphere are hot and buoyant, but they get denser and less buoyant as they cool. As it cools, the lithospheric plate thickens and sags into the mantle to maintain isostatic equilibrium. Old, cold lithosphere reaches a maximum thickness of about 125 km. As a result of cooling, the depth of the oceans increases smoothly and progressively away from ridge axes: the depth at a given distance varying with the square root of the age of the plate at that point (Fig. 2.20). *Heat flow* through the plate, a measure of its cooling, also decreases with the square root of the age of the plate.

Fig. 2.19 Long, linear fractures and gaping fissures are prominent in the actively spreading part of Iceland. These are near Thingvellir, site of Iceland's ancient parliament. (Photo: courtesy of P. Mouginis-Mark.)

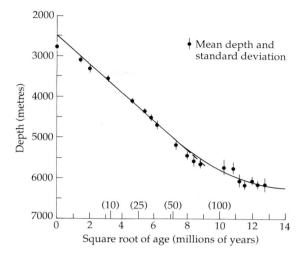

Fig. 2.20 Ocean depth increases predictably with age and distance from the ridge crest. Here, depth of the north Pacific Ocean is plotted against the square root of the age of the oceanic crust (actual ages in brackets). Ocean floor older than about 60 million years does not increase in depth so rapidly, so curve flattens off to a maximum depth of about 6 km. (After B. Parsons and D. McKenzie.)

— 2.4 Destructive plate margin volcanism —

Although the Earth is a big place, its surface area is not nearly big enough to accommodate all the oceanic lithosphere that has been generated over the last 4600 million years. Oceanic lithosphere *must* be consumed as fast as it is created. If it were not, the Earth would be constantly expanding—an argument that actually finds the occasional passionate advocate. In the early days of Earth history, before there were any continents, plate consumption was straightforward: as oceanic lithospheric moved away from the ridge axis that gave birth to it, it grew so thick and dense that it could no longer 'float' on the asthenosphere beneath. It therefore sank downwards, diving into the mantle, the descending slab pulling the oceanic lithosphere along behind it. (Recall that the *slab pull* force is probably the most important driving force behind plate tectonics.) One can see the same processes at work in miniature on the restlessly shifting surface of a basaltic lava lake.

2.4.1 Island arcs

Several things happen when a slab of lithosphere descends into the mantle. Most obvious are the mechanical interactions between the descending slab and the mantle it grinds against, manifested in earthquakes. Many of the world's most dangerous areas of seismicity are located above subduction zones, where earthquake foci define clearly the zones of contact, named *Benioff* zones, after Hugo Benioff, the American geophysicist who first mapped them. The descending slab is cool and dense, and so cools the warm mantle into which it is sinking—mantle isotherms are pushed downwards (Fig. 2.21). Correspondingly, the cool slab itself warms up. Ultimately of course, when it has descended deep enough into the mantle—several hundred kilometres—it will reach the same temperature and become part of the mantle again. At depths of more than 600 km the earthquakes of the Benioff zone fade away as the mechanical contrasts between slab and mantle disappear, but compositional differences extend much deeper.

At shallow depths, between 50 and 150 km, several changes which have volcanic implications take place to the downgoing lithospheric slab. First, recall what the slab is made of. The lowermost part of it, below the Moho, is made of mantle material—peridotites, while the crustal part is made of the cumulate gabbro/sheeted dyke/pillow lava trinity *plus* a quantity of sedimentary material that has accumulated during its ride along the ocean floor conveyor belt. This consists most of fine-grained sticky clays and oozes deposited in the abyssal ocean depths.

Early hypotheses for volcanism at island arcs

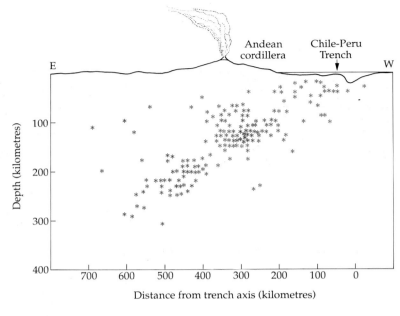

Fig. 2.21 (above) Distribution of earthquake foci beneath the central part of the South American plate margin, revealing a well-defined Benioff zone dipping eastwards at about 30°. Most earthquakes are shallow; few occur at depths greater than 400 km. (below) Schematic diagram showing deflection of mantle isotherms caused by descent of the cold oceanic plate into the mantle at a long-established subduction zone. (After A. Sugimura and S. Uyeda.)

invoked straightforward melting of the basaltic crust which was being subducted along the arc. To manufacture basalt from basaltic crust would require almost complete melting, but this is difficult to reconcile with geochemical data. It is now thought that as the downgoing slab is heated and stewed by its descent into the mantle, much

of the water it contains is driven off, rising into the mantle wedge (Fig. 2.22). Introduction of water into the overlying wedge of mantle promotes melting, ultimately leading to the eruption of lavas similar to those erupted at ocean ridges. They are still basalts, but differ slightly in their chemistry, so they are called *island arc basalts*, or

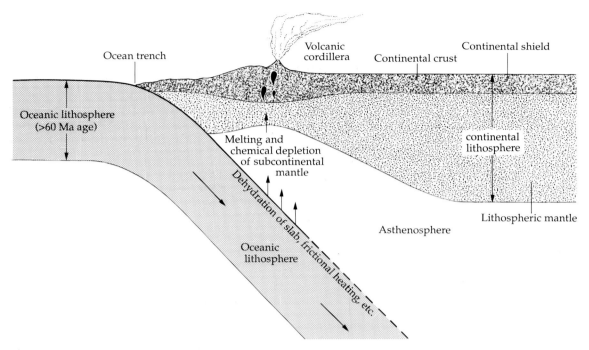

Fig. 2.22 Cartoon summarizing subduction zone processes at a continental margin. As the descending slab sinks into the mantle, it is dehydrated. Upwards migration of fluids promotes melting in the overlying mantle, ultimately leading to volcanism at the surface. Usually, volcanoes are sited about 150 km above the top of the descending slab.

IABs, because island arcs are where they are found—in particular, the curving archipelagos of volcanic islands that dominate maps of the western Pacific.

Island arcs rarely consist of basalt alone. Although the geochemical differences between IABs and MORBs may seem slight (Chapter 3), they are cumulative. Thus, as an island arc matures, and more melt is extracted from the descending slab and overlying mantle wedge, the bulk composition of the island arc moves away from basalt. If the island arc gets old and thick enough, the early formed rocks at its base may get partially melted themselves, giving rise to magmas with more evolved compositions. *Partial melting* is a vitally important process in petrology. Every time a melting process takes place, elements which are incompatible with the source rock composition are mobilized and escape. Typically, these are elements with large ionic radius—potassium is an outstanding example. Other elements, such as iron and magne-

sium which form minerals stable at higher temperature and reluctant to melt—such as olivine and pyroxene—tend to remain in the solid rock. This has the effect of concentrating compounds such as SiO_2 in the melt formed. If rocks were wholly melted, then the melt would obviously have the same composition as the original rock. In practice, the smaller the extent of partial melting, the more fractionation takes place between compatible and incompatible elements. Thus, a kind of distillation process takes place, giving rise to magmas richer in silica, potassium and other incompatible elements.

A second process, *fractional crystallization*, leads to similar results. When crystallization of large igneous intrusions commences, minerals such as olivine which are stable at higher temperatures crystallize first, forming *cumulate* rocks. By removing iron and magnesium from the magma, crystallization of olivine and pyroxene has the effect of enriching the remaining liquid in silica and the large ion incompatible elements. Although it rarely happens in a single stage, this process of *fractional crystallization*, carried to extremes, can produce rocks of *rhyolitic* (granitic) composition. In combination, these igneous processes lead ultimately to the formation of the continental crust, the real estate on which we live. This does not happen in a single stage: throughout the history of the Earth, island arcs have been accreting together, first forming mini-continents, and later full-scale continents.

From a volcanic point of view, the significant characteristic of a maturing island arc is that relatively undramatic basaltic volcanism gives way to andesitic and rhyolitic volcanism, with explosive consequences for the style of eruptive behaviour. Much of this book will be devoted to exploring these vigorous types of eruption. One consequence is visibly obvious: whereas mid-ocean volcanoes such as those on Iceland are topographically modest, many island arc andesitic volcanoes are high, sweeping cones; recognizably 'volcanoes' to everyone from kindergarten upwards.

2.4.2 Continental margin processes

Subduction of oceanic lithosphere takes place beneath continental margins as well as island arcs. Overall, the process is the same, and starts with new basaltic magmas being formed from the descending ocean plate and from the intervening wedge of mantle. Recall that these magmas start off enriched in incompatible elements. The difference now is that there is a large thickness of continental crust between the magma source and the surface, and it is difficult for magmas to arrive at the surface without being extensively modified on their way. Much of the modification takes place through a complex process called *assimilation–fractional crystallization*, the details of which need not detain us.

As continental crust gets thicker, there is less chance of true basalts being erupted at the surface. In the central Andes, where the continental crust is more than 60 km thick, basalts are effectively absent. Basaltic andesites, andesites, and dacites are widespread. It will surprise no one to learn that Andean volcanoes are characterized by andesites. Such andesitic volcanoes are comparable in most ways with their counterparts in island arcs. The chief difference between island arc and continental margin volcanoes lies in the varying proportions of highly evolved rocks of dacitic or rhyolitic compositions.

At a continental margin such as the central Andes, subduction of oceanic lithosphere and generation of basaltic magmas has been continuous for many millions of years. This has led to the underplating of large amounts of basaltic material at the base of the crust, and to the introduction of large amounts of heat into the lower crust. As heating of the lower crust continues, the overlying continental crustal material begins to melt. Some may be assimilated by basaltic magmas, so that the resulting andesites retain geochemical fingerprints of both mantle and crustal source regions. Large amounts of melting produce huge volumes of silicic magmas which balloon buoyantly upwards to the surface in great intrusive bodies known as *batholiths*. If these magma bodies come close enough to the surface that the strength of the overlying crustal rocks can no longer contain them, volcanic eruptions of great magnitude ensue, ejecting thousands of cubic kilometres of magmatic material in the space of a few hours. To replace the huge volumes lost on eruption, the crust at the surface founders downwards to form huge subsidence volcanoes, or *calderas*, up to 100 kilometres across (Chapter 14).

—— 2.5 Within-plate volcanoes, 'hot-spots' and mantle plumes -

Although it is a considerable over-simplification, it is convenient to think of ocean ridges and subduction zones as complementary parts of a convective cycle. How, then, do within-plate volcanoes fit into this neat picture; those enig-matic volcanoes sticking up in the middle of plates, such as Hawaii in the mid-Pacific, and Tibesti, in the heart of Africa?

Taking the oceanic examples first, these volcanoes are now recognized as 'hot-spot volcanoes';

surface expressions of thermal 'plumes' in the mantle, which may be initiated as deep down as the core–mantle boundary. One instructive example is the Cape Verde volcano group, located mid-way between the coast of Africa and the Mid-Atlantic Ridge. Around the Cape Verde Islands, the sea-floor is elevated into a 1500 km wide swell centred around the islands, showing that it is buoyed up by a plume of warm, low-density material (Fig. 2.23). Furthermore, the heat flow measured on the ocean floor is anomalously high, showing that the mantle beneath must be 100–150°C hotter than normal.[4] Iceland is a more complicated example, where the 'hot-spot' actually underlies the Mid-Atlantic Ridge. Abnormally warm mantle beneath Iceland manifests itself again in a topographic swelling: the ocean floor for 1000 km around Iceland is about a kilometre shallower than normal, buoyed up by the mushroom-shaped head of a plume of unusually hot mantle.

Where a mantle plume has persisted for geologically long periods, the overlying lithospheric plate may move over it. In the Hawaiian island group, the currently active hot-spot volcano (Kilauea) is merely the newest-forged link in an 80 million year old chain (the Hawaii–Emperor chain) extending thousands of kilometres westwards. The thermal plume has acted as if it were a sort of blow torch in the mantle, burning a trace through the overlying Pacific plate as it shifted slowly above. Even the marked dog leg at 33°N, 172°W can be explained in terms of a shift in the movements of the Pacific plate as a whole (Fig. 2.24, Fig. 2.25, and Fig. 2.26). Many other hot-spot tracks are known, such as the Cook–Austral island chain in the South Pacific. J. Tuzo Wilson, who made many seminal contributions to plate tectonic theory, first recognized the age progressions in chains of hot-spot volcanoes. In his picturesque words, 'the islands are in fact arranged like plumes of smoke . . . carried downwind from their sources'.[5]

2.5.1 Continental hot-spots

Mid-oceanic plate volcanoes are almost exclusively basaltic—with only mantle sources to draw from, there is little alternative. And, of course, their lavas appear indistinguishable from ordinary oceanic basalts. Mantle hot-spots also operate beneath continental lithosphere, however, producing a wider range of rocks. This is the case with the remote volcanic massifs in the central Sahara. Unlike the other continents, Africa has been stationary relative to the mantle for tens of millions of years, so hot-spots in the underlying mantle have been able to construct long-lived volcanic edifices. Africa's present

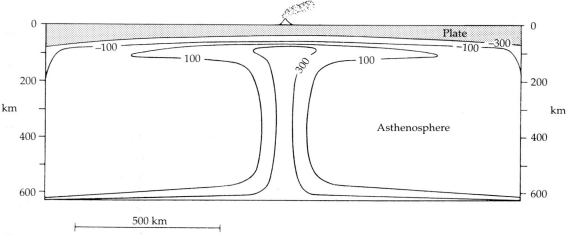

Fig. 2.23 White and McKenzie's model of the thermal plume beneath the Cape Verde Island. Isotherms show temperature anomalies in degrees Celsius relative to the mean asthenosphere temperature. The narrow plume spreads out into a broad mushroom head beneath the overlying oceanic plate. From White, R. and McKenzie, D. (1989). Magmatism at rift zones: the generation of volcanic continental margins and flood basalts. *J. Geophys. Res.* **94**, 7685–29.

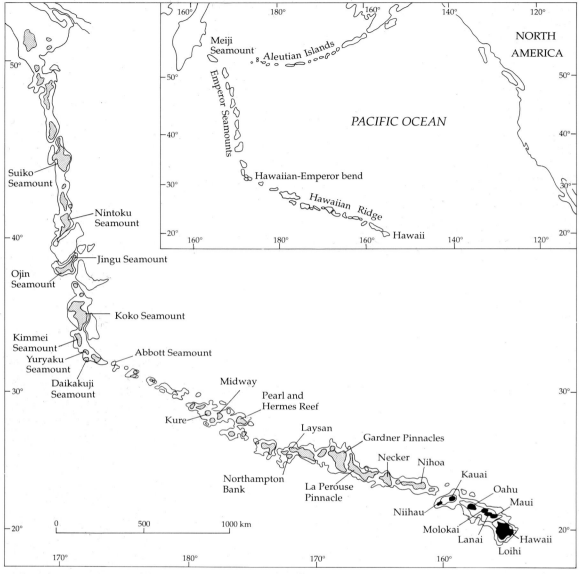

Fig. 2.24 The Hawaii–Emperor seamount and island chain is a beautiful example of a 'hot-spot' track across a lithospheric plate. Contours are shown for 1 and 2 km depths in the area of the chain only. Note the prominent 'dog-leg at 33°N, 172°W. Only the features east of Midway Island emerge above the sea-level, the remainder are seamounts. After Clague, D. A., and Dalrymple, G. B. (1987). The Hawaiian–Emperor volcanic chain, Part 1, Geologic Evolution In *Volcanism in Hawaii*, US Geol. Surv. Prof. Pap. No. 1350, pp. 5–54, Washington, DC, 1987.

topography represents an image of mantle convection where it impinges on the under surface of the lithosphere—the volcanic high points sit above uprising plumes.[6] Much basalt has been erupted from such edifices, but because the volcanism has been taking place within continental crust, partial melting and assimilation of continental rocks at different levels has taken place, leading to eruption of rocks with a range of silica and alkali contents. Thus, the mid-Saharan

Fig. 2.25 Evolution of the Hawaii–Emperor chain. Radiometric ages of samples from the islands and sea-mounts in Fig. 2.21 are plotted against their distance from the currently active volcano, Kilauea. There is an elegant correlation between age and distance, even across the dog leg, consistent with plate movement of about 8.2 cm per year over the hot-spot. (The solid line represents a movement of *exactly* 8.6 cm per year.) Clague, D. A. and Dalrymple, G. B. (1987). The Hawaiian–Emperor volcanic chain, Part 1, Geologic evolution. In *Volcanism in Hawaii*, US Geol. Surv. Prof. Pap. No. 1350, pp5–54, Washington DC.

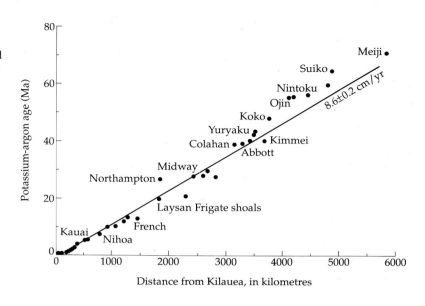

Fig. 2.26 Mauna Loa from Kilauea. It may seem difficult to credit, but the gently rising ridge in the background is actually the 4000-m-high active volcano Mauna Loa, located at the youngest end of the Hawaii–Emperor chain.

volcanoes such as Emi Koussi and Jebel Marra are sharply different from those of basaltic, oceanic hot-spots such as Hawaii.

Where a continental plate moves over a hot-spot, a track will be 'burned' through the continent, like the brand scorched into the Pacific plate to form the Hawaiian chain. An outstanding example is the Snake River Plain province of North America. Here, a hot-spot which first burned into the west coast 15 million years ago, has blow-torched an 80-km-wide swath across 450 kilometres of continental crust, erupting huge quantities of basalt on the way.[7] At present, the hot-spot is located beneath

Yellowstone National Park, Wyoming, but at its present rate of progress of about 3.5 cm per year, it will have burned its way to the Canadian frontier in 20 million years. Yellowstone, home of the Old Faithful Geyser, is the site of a huge silicic caldera, which erupted catastrophically six hundred thousand years ago, showering ash over much of North America. Minor eruptions have continued since then, and others will certainly happen in the future. Today, Yellowstone is the site of by far the largest heat flow recorded in North America and is arguably the most important volcanic province in North America, although there have been no eruptions in historic times. In petrological detail, the province is notable for its *bimodal* character—huge quantities of both basalt and rhyolite (involving melting of continental crust) have been erupted (Fig. 14.11).

2.5.2 Continental flood basalts

Continental flood basalts (CFBs) dominate the landscape in places like the Deccan (north-west India), the Drakensberg (South Africa) and the Columbia River Plateau (north-west USA). They are so widespread that they are hard to ignore. Although not so spectacular there, they are also important in the Scottish Hebrides, where basalt petrology was born. In 1903, the eminent British geologist Sir Archibald Geikie wrote:

There have been periods in the Earth's history when the crust was rent into innumerable fissures over areas of thousands of square miles in extent, and when the molten rock instead of issuing, as it does in most modern volcanoes, in narrow streams from a central elevated cone, welled out from these vents or from numerous small vents along their course and flooded enormous tracts of country without forming any mountain or conspicuous volcanic cone.[8]

When deeply dissected by erosion, thick piles of basalt lavas give rise to a characteristic 'stepped' topography, the soft, scoriaceous upper parts of each flow eroding more rapidly than the more solid lower parts. The hard lavas therefore tend to form cliff-like 'risers' while the soft parts form flat 'treads'. When repeated throughout the thickness of the lava pile, a complete 'stair case' results. This easily recognizable feature of basalt

plateaux led to their being called collectively *traps*, after an old Swedish word for a staircase. Geologists still use the term, commonly talking about the 'Deccan Traps' and 'Siberian' Traps, for example.

Some of the statistics are impressive: India's 65-million-year-old Deccan Traps cover half a million square kilometres today, and may have covered a million and a half when first erupted. They have an average thickness of at least one kilometre, and locally two. Most of their huge volume may have been erupted in less than half a million years, although this is controversial (Fig. 2.27). In the Columbia River province 170 000 cubic kilometres of basalt flooded out between 17 and 15 million years ago, in individual flows 20–30 m thick (Fig. 2.28). The largest of these are of astounding dimensions, having volumes of several hundred cubic kilometres and extending hundreds of kilometres. Some 99 per cent of the volume of the entire province may have been erupted in less than 2 million years.[9] (For those who would like to explore the complexities of the Columbia River basalts further, a recent book is devoted to them.)[10]

One of world's largest and most enigmatic flood basalt provinces is the 245-million-year-old Siberian province, which covers today more than 300 000 square kilometres of arctic wilderness east of the Yenisey River. Unfortunately, it is relatively little known, at least to Western geologists, few of whom have penetrated the region's poorly charted expanses.

There may even be examples of submarine flood basalt provinces: recent research suggests that enigmatic submarine plateaux such as the Ontong–Java Plateau in the western Pacific may be analogous to CFB provinces. Submarine plateaux are the least-well understood features of the ocean floor; they are extensive elevated regions 2–3 km higher than the surrounding seafloor, characterized by crust 20–40 km thick and containing huge thicknesses of lava which appear to have been erupted very rapidly, within 5–10 million years. In Alaska and British Columbia, sequences of basalt reaching up to 6 kilometres thickness are exposed, forming a vast terrain that has been christened Wrangellia.

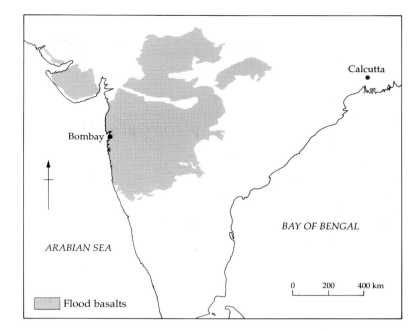

Fig. 2.27 A significant fraction of India's land surface is covered by the outcrop of the 65-million-year-old Deccan traps, which are more than a kilometre thick in places. Redrawn from: Mahoney, J. (1988). Deccan Traps. In *Continental flood basalts* (ed. J. D. Macdougall), pp. 151–94, Kluwer, Dordrecht.

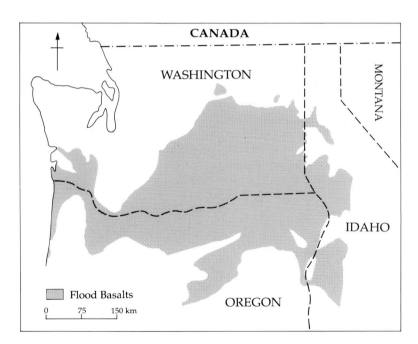

Fig. 2.28 Outcrop of the Columbia River flood basalt province in Washington, Oregon and Idaho. Note the difference in scale between this figure and Fig. 2.27. Redrawn from Hooper, P. R. (1988). The Columbia River Basalt. In *Continental flood basalts* (ed. J. D. Macdougall), pp. 1–33, Kluwer, Dordrecht.

These flood basalts, which are of Triassic age (about 240 million years old) are thought to have formed originally as a submarine plateau, which was accreted on to the North American continent about 100 million years ago.[11]

More about mantle plumes

How are we to understand the sudden, extraordinary outpourings of flood basalts that form continental flood basalts and submarine plateaux? This is a challenging problem for petrologists and geophysicists.[12] Readers old enough to recall the vulgar 'lava lamps' that were fashionable in the 1960s will be well placed to grasp one hypothesis. In these lamps, two immiscible liquids of different colours and carefully selected densities (such as aniline and water) are contained in a transparent cylinder. Gentle heat is applied to the base of the cylinder, warming the denser liquid at the bottom, and reducing its density until it becomes buoyant. An instability first forms on the interface between the two liquids, then a glutinous spherical blob forms and begins to rise, drawing out a thread-like stalk behind it. Ultimately, it detaches itself from the interface, its tail breaking up into a series of smaller beads, and floats to the top of the lamp. After cooling, the blob sinks downwards again, completing the convective cycle. Some volcanologists tolerated these lamps in their homes long after their unspeakable garishness had made them kitsch to more refined tastes . . .

A similar process may operate in the mantle on a scale hundreds of times larger, according to Richards et al.[13] They envisage the entire mantle as a vast lava lamp: a thermal instability develops (for reasons not well understood) at the core–mantle boundary, triggering a plume hundreds of kilometres in diameter which ascends right through to the surface of the Earth. When it arrives at the base of the lithosphere, the deep mantle material making up the plume will be 250–300 degrees hotter than the surrounding upper mantle, so 10–20 per cent melting would rapidly take place. It is this melting that supplies the basalt lavas that gush out abruptly on to the surface of the Earth to form CFB provinces such as the Deccan and the Columbia River Province.

There is more, though. Just as the rising globules in lava lamps draw out stalks behind them, so it is argued, do mantle plumes. After fuelling the copious Deccan lavas, the 'tail' of the plume responsible was not completely extinguished, but kept on operating as the Indian subcontinent drifted northwards. It is still active today, stoking up one of the world's most active volcanoes: Piton de la Fournaise on Réunion Island in the Indian Ocean (Fig. 2.29). An important implication of this model is that the *rates* of eruption from the plume head and tail should be quite different. Richards et al. suggest an eruption rate of the order of 1.5 cubic kilometres per year for the Deccan, but only 0.04 cubic kilometres per year for the Réunion hot spot track.

2.5.3 Continental rifts

It is a basic tenet of plate tectonics that new oceans form when continents are rifted apart. On every flight of the Space Shuttle, astronauts photograph compelling evidence for this when they orbit over the umistakably rectilinear coastlines of the Red Sea. Oceanographic studies show that there is a spreading ridge along the axis of the Sea. In the area of the Red Sea, there are in fact three rifts which meet at a triple point where the Red Sea makes an abrupt dog's leg bend: the Red Sea itself, the Gulf of Aden, and the portion of the African rift system that runs through the Danakil Depression into the East Rift of Kenya. These three form an evolutionary progression. In the Danakil, rupture of continental lithosphere is just commencing; in the Red Sea, extension began 20–23 million years ago, but subsequently slowed. Over the last 5 million years, renewed spreading has been taking place at about 0.8 centimetres per year, but the Red Sea is not yet an ocean. In the Gulf of Aden, by contrast, the ruptured continental blocks of Africa and Saudi Arabia have been moving rapidly apart for the last ten million years. The active spreading ridge in the Gulf is an extension of the major spreading ridge in the Indian Ocean, the Carlsberg Ridge. If astronauts had been around in the Cretaceous, one hundred million years ago, the Atlantic Ocean would have looked exactly as the Red Sea and Gulf of Aden do from the Space Shuttle today.[14]

There is a chicken and egg debate about the initiation of rifting. For rifting to take place, there has to be extension: the continental crust and lithosphere have to be stretched and thinned. Beneath a rift, buoyant mantle material rises towards the surface, and the buoyancy of this warm material can cause considerable uplift of the surface, perhaps as much as 2 km. This explains why rifts often occupy broad topographical rises, like the Rio Grande and Baikal Rifts.

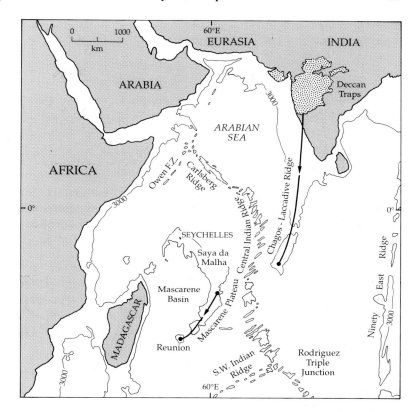

Fig. 2.29 White and McKenzie's postulated track of the Réunion hot-spot track from its inception beneath the Deccan to its present site at Piton de la Fournaise, offset by the Central Indian Ridge. From White, R. and McKenzie, D. (1989). Magmatism at rift zones: the generation of volcanic continental margins and flood basalts. *J. Geophys. Res.* **94**, 7685–729.

But the problem is this: does extension come first, *initiating* volcanism, or is it a thermal anomaly in the mantle which promotes both volcanism and extension? There are arguments on both sides, but it is clear that *some* rifts are sites of major hot-spots and of extraordinarily rapid effusions of lava; others are not. Voluminous volcanism in *active* rifts is associated with considerable extension, and with mantle that is unusually warm—100° to 150°C warmer than average. In such cases, the mantle hot-spot may drive the extension. Non-volcanic or *passive* rifts are associated with slight to moderate extension, and with fairly normal underlying mantle—some parts of the African rift system are like this. They may result from extensional forces acting on the lithosphere over large distances.

2.5.4 *The relationships between rifting and continental flood basalts*

A characteristic feature of the volcanism associated with rifting is that it takes place quickly, often in less than a million years. Eruptions of flood basalts are often intimately related to rifting. According to Robert White and Dan McKenzie of Cambridge University there is only one explanation of such rapid volcanism. They argue that when rifting takes place, the lithosphere is thinned and buoyant mantle material rises from depth into regions of lower pressure.[15] At shallower levels and lower pressures, the melting point of mantle materials is lower, and the hot material rising to the surface cannot lose heat sufficiently by conduction to remain solid, so *decompression melting* begins at depths of about 50 km.[16] Taken to the extreme, rapid decompression melting can act like a volcanic fire-hydrant pumping out colossal volumes of lava. According to White and McKenzie, the difference between rift volcanism and CFBs is merely one of scale, and is the result of anomalously hot mantle beneath CFB source regions. They suggested that mantle 'hot-spots' about 100°C hotter than 'normal' mantle temperatures of *c.* 1340°C are required to start the basalt fire-hydrant gushing.

A critical weakness in White and McKenzie's argument is that lithospheric thinning and extension *must* precede CFB eruption. But this is clearly not always the case. Important CFB provinces such as Siberia and the Columbia River Plateau are clearly *not* the sites of even incipient rifting—they are just accumulations of basalt lavas. Furthermore, where rifting is associated with CFBs, rifting often *follows* rather than precedes the volcanism. There is also a marked difference in the nature of the volcanism: CFBs are hugely monotonous outpourings of tholeiitic basalts, whereas the lavas associated with rifting are less voluminous and more diverse chemically, including alkali basalts and more intermediate, trachytic compositions.[17]

Although these are critical points, they are details. It is more important to consider the larger picture. Some highly instructive matching suites of CFBs are found on opposite sides of the South Atlantic in Brazil and South Africa: the Parana and Etendeka provinces respectively. They show that the first events leading to the separation of South America and Africa were rifting and voluminous basalt eruption. As lithospheric extension and thinning continued, the ocean opened, and about 125 million years ago the two continents began to become unzipped from one another, the split progressing implacably northwards until after about five million years it was all over between them. Since then, the divorced couple have drifted thousands of kilometres apart, but evidence of their fiery parting is plainly written in the huge stacks of basalt lava on the coasts of Africa and Brazil (Fig. 2.30). Similar relationships can be seen in the Brito-Arctic province. Sixty million years ago the opening of the North Atlantic fragmented this province into the small chunks now preserved in Greenland, Scotland, the Faroe Islands, and Iceland.

Note, however, that CFBs are not found in matched sets all along the Atlantic margins. We can conclude that the CFB provinces were the sites of powerful thermal anomalies—mantle plumes—but that while in some places they have

Fig. 2.30 Eruption of voluminous flood basalts to form the Paranà (Brazil) and Etendeka (Namibia) provinces preceded the opening of the Atlantic about 125 million-years ago. According to White and McKenzie, a 1000 -km-diameter thermal plume head (shaded) underlay the site of initial opening. Significantly, flood volcanism did not take place along the entire length of the Atlantic—there is none, for example, where South America and Africa interlock at the top of the diagram. From White, R. and McKenzie, D. (19989). Magmatism at rift zones: the generations of volcanic continental margins and flood basalts. *J. Geophys. Res.* **94**, 7688–729.

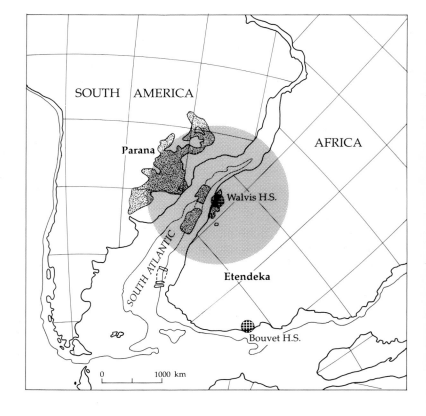

been implicated in rifting and ocean formation, they are not essential to it.

2.5.5 Problem areas

The natural world is rarely straightforward. This chapter may have given the impression that any volcano in the world can be dovetailed neatly into its own plate tectonic setting. This is not the case. There are many anomalies around the world. Outstanding among these are many small—and some not so small—volcanic centres of the Basin and Range province of western North America (eastern California, Nevada, and Arizona). The tectonics of this area merit an essay in themselves, but for now it will be

adequate to think of the region as being one where the crustal lithosphere has been thinned by almost a half, as though rifting had been distributed over an enormously wide zone. Thinning of the crustal part of the lithosphere brings mantle isotherms to shallow levels, so it is relatively easy for basaltic magmas to rise into the crust and erupt there. There are also major young volcanoes on the Tibet Plateau, the roof of the world, far from any modern plate margin. These little-studied volcanoes present such wonderfully stimulating problems that discussion of them had better be deferred, perhaps thereby tantalizing the reader into seeking fresh insights into volcanism and tectonics.

Notes

1. Wylie, P. J. (1988). Magma genesis, plate tectonics and the chemical differentiation of the Earth. *Rev. Geophys.* **26**, 370–404.
2. Olson, P., Silver, P. G., and Carlson, R. W. (1990). The large scale structure of convection in the Earth's mantle. *Nature* **344**, 209–15.
3. Macdonald, K. C. (1989). Anatomy of the magma reservoir. *Nature* **339**, p. 178.
4. Courtney, R. C. and White, R. S. (1986). Anomalous heat flow and geoid across the Cape Verde Rise: evidence for dynamic support from a thermal plume in the mantle. *Geophys. J. Roy. Astron. Soc.* **87**, 815–67.
5. Wilson, J. T. (1963). Continental drift. *Sci. Am.* **208**, 86–100.
6. Burke, K. and Wells, G. L. (1989). Trans-African drainage system of the Sahara—was it the Nile? *Geology* **17**, 743–7.
7. Greeley, R. (1982). The Snake River Plain, Idaho: representative of a new category of volcanism. *J. Geophys. Res.* **87**, 2705–12.
8. Geikie, A. (1903). *Textbook of geology*. Macmillan, London, 1147 pp.
9. Swanson, D. A., Wright, T. H., and Helz, R. T. (1975). Linear vent systems and estimated rates of magma production and eruption for the Yakima basalt on the Columbia Plateau. *Am. J. Sci.* **275**, 877–905.
10. Reidel, S. P. and Hooper, P. R. (eds) (1989). *Volcanism and tectonism in the Columbia River Basalt Province*, Geol. Soc. Am. Spec. Publ. No. 239, 386 pp. Geological Society of America, Boulder, Colorado.
11. Richards, M. A., Jones, D. L., Duncan, R. A., and Depaolo, D. J. (1991). A mantle plume initiation model for the Wrangellia flood basalt and other oceanic plateaux. *Science* **254**, 263–7.
12. Macdougall, J. D. (ed.) (1988). *Continental flood basalts*. Kluwer Academic Publishers, Dordrecht, Netherlands, 341 pp.
13. Richards, M. A., Duncan, R. A., and Courtillot, V. E. (1989). Flood basalts and hot-spot tracks: plume heads and tails. *Science* **246**, 103–7.
14. Bonatti, E. (1987). The rifting of continents, *Sci. Am.* **256**, 3, 97–103.
15. White, R. and McKenzie, D. P. (1989). Magmatism at rift zones: the generation of volcanic continetal margins and flood basalts. *J. Geophys. Res.* **94**, 7685–729.
16. White, R. S. and McKenzie, D. P. (1989). Volcanism at rifts. *Sci. Am.* **261**, 7, 44–59.
17. Hooper, P. R. (1990). The timing of crustal extension and the eruption of continental flood basalts. *Nature* **345**, 246–9.

3

Basically basalt

Basalt is a dark igneous rock characterized by small grain sizes (less than about 1 mm) and containing roughly equal proportions of plagioclase feldspar and calcium rich pyroxene, with less than 20 per cent of other minerals (typically olivine, calcium poor pyroxene, and iron titanium oxides). The name is derived indirectly from the Greek basanos, *meaning touchstone (which is an entirely different rock: a dark flinty stone used to test gold and silver alloys). Basalt is by far the commonest igneous rock on the surface of the solid Earth, and occurs abundantly on other terrestrial planets.*

These are the first words of a 1286-page treatise entitled *Basaltic volcanism on the terrestrial planets*; a compendium of almost everything known about basalts up until 1981.[1] Mere existence of such a lengthy work is a sure guide to the importance of basalts. As a definition, the paragraph that introduces this rambling, 101-author tome is as informative as one could devise for this vast group of rocks. Nearly 70 per cent of the Earth's surface and 17 per cent of the Moon's surface are covered by basalt. On the other planets, the proportions are less well known, but most of Venus and large areas of Mars are certainly covered by basalt. A few asteroids are covered by basalt and there are even basalt meteorites, derived from asteroids.

In this chapter, we review the principal types of basalts on the Earth, Moon, and elsewhere. Much has been written about basalts over the decades, including a great deal on the minutiae of basalt petrology. There is a vast literature dealing with the nomenclature of basalt types, studded with references to things such as *mugearites* and

benmoreites: obscure varieties of basalt cropping out on dismal moorlands in the Scottish Hebridean province (Fig. 1.9). While these rocks were important in their day, when petrologists such as Norman Bowen[2] were first working out how igneous rocks crystallize, they are not particularly helpful in understanding basalts in a planetary context.

According to one school of thought, *all* volcanism is fundamentally basaltic. Because basalts are so crucial to volcanology, it is appropriate to introduce the main varieties and their origins. In the early days of geology, the 'bowels of the Earth' figured prominently in debates about where magmas came from. Today, we can go further, and discuss more precisely where in the Earth's anatomy the source regions of different basalt types are located. We can also dissect the interiors of other planets in similar terms. Much of this chapter is geochemically based; some readers may prefer to move on to more directly volcanological material in Chapter 4.

— 3.1 The most primitive basalts —

It is not difficult to see why basalts are ubiquitous in the solar system. All the terrestrial planets have silicate mantles with compositions similar to chondritic meteorites, and basalt is the result of partial melting of chondritic silicates. *Primitive* basalts are those which formed most directly by partial melting of chondritic material, the basic raw material of the solar system. A logical question to ask, therefore, is: what are the most primitive basalts on Earth?

Mid-ocean ridge basalts (MORBs) are manufactured in such huge volumes at ocean ridges that it was supposed for a long while that they are formed by simple, single-stage melting of mantle materials, and that they are therefore 'primitive'. This reasonable supposition is now known to be wrong: MORBs form by multi-stage processes and are not primitive at all. At present, the most primitive known basalts are some scarce lavas, richer in magnesium and the alkalis than MORBs, which contain nodules of mantle materials. Such lavas are widely, if thinly, distributed around the world, in ocean islands, rift zones and continental interiors. Some of the nodules are *spinel peridotites*, stable at high

pressure, which shows that the magma ascended directly from depths greater than 30 km, without getting involved in the shallow-level processes involved in the generation of MORBs (Table 3.1).

3.1.1 *Komatiites*

Another way of thinking about the most 'primitive' basalt types is to consider those which formed at the highest temperatures. All the evidence presented in the previous chapter shows that the Earth and other planets had much more thermal energy in their early days than they do now. A natural consequence of this would be higher mantle temperatures and higher degrees of partial melting, leading to production of lavas with compositions closer to bulk mantle compositions than those of today. Such lavas do indeed exist in the terrestrial geological record. Rather odd lavas called *komatiites* (named after a locality near the Komati River in South Africa) turn up in a number of places around the world, in Canada and Australia as well as South Africa. A common feature of komatiites is that they contain large blade-like crystals of olivine many

Table 3.1 Compositions of undepleted mantle and some important basalt types

	Undepleted mantle	Primitive basalt (1)	MORB (2)	Komatiite (3)	Lunar mare (4)	Shergotty meteorite (Mars) (5)
SiO_2	45.1	44.2	50.5	45.2	43.6	50.1
TiO_2	0.2	3.7	1.6	0.2	2.6	0.9
Al_2O_3	3.3	12.1	15.3	3.7	7.9	6.7
FeO	8.0	10.9	10.5	11.0	21.7	18.7
MnO	0.1	0.2	—	0.2	0.3	0.5
MgO	38.1	13.1	7.5	32.2	14.9	9.4
CaO	3.1	10.1	11.5	5.3	8.3	10.0
Na_2O	0.4	3.6	2.6	0.4	0.2	1.3
K_2O	0.03	1.3	0.2	0.2	0.1	0.2
P_2O_5		0.8	0.1	0.02	—	0.7
Cr_2O_3	0.4	0.1			0.9	0.2

Sources: (1) Primitive basanite, Ross Island, Antarctica; BV, p. 413; (2) Average basaltic glass from Atlantic, Pacific, and Indian Ocean spreading centres, BV, p. 139; (3) Olivine spinifex-textured sample, Barberton, BV p. 16; (4) Apollo 12 olivine basalt, no. 6 on p. 239 of BV; (5) Shergotty, BV, p. 221. BV = Basaltic Volcanism Study Project, 1981.

centimetres long, forming a sheaf-like *spinifex* texture (spinifex is a spikey Australian grass). Their high olivine content means that komatiites are rich in magnesium, sometimes containing over 30 per cent, indicating that they were as much as 200°C hotter than typical basalts when erupted. As Table 3.1 shows, komatiite compositions are closer to that of the mantle than basalts. Some may have been formed by as much as 50 per cent partial melting. Komatiites are found on Earth *only* in Archaean rocks (those more than 2500 million years old), and therefore it might seem logical to conclude that higher mantle temperatures prevailed then.

However, the situation is actually more complicated. Komatiite lavas themselves are not truly primitive: that is they were not formed by extensive melting of mantle materials in a single stage. Instead, most seem to have formed by remelting of *previously* melted mantle, which was therefore already depleted of many basalt-forming elements. Furthermore, komatiites are not the only lava types in the Archaean. Far from it. Basalt lavas with compositions indistinguishable from modern lavas are more abundant than komatiites, and they occur within the same rock sequences as komatiites. Thus, the high temperatures and high degrees of partial melting involved in the formation of komatiites were not all-pervasive in the Archaean mantle. Heat production in the Archaean was undoubtedly greater than it is now, but it seems that rapid convection was able to keep temperatures down, and only occasionally did temperatures rise to yield thermal gradients sufficient to produce komatiites.

You might be forgiven, gentle reader, for supposing that komatiites are as obscure and dismal a group of rocks as benmoreites. In some ways, they are indeed dismal, because they are usually heavily altered and therefore uninspiring to contemplate. But komatiites are not only valuable for what they tell us about Archaean volcanism on Earth, but also for what they tell us about other planets, especially Mars. Because they are erupted at high temperatures, their viscosities are much lower than ordinary basalts. This gives rise to some profound consequences for planetary volcanism, as we shall see.

— 3.2 Some important terrestrial basalt types —

3.2.1 More about MORB

One basalt lava looks boringly like another. It is hard to distinguish between a lunar and a terrestrial one merely from their appearance. Under the microscope, some mineralogical differences between basalt types emerge, but many basalts are so fine grained that optical studies are difficult. Geochemistry is therefore used to discriminate between different basalt types. For our purposes, only two are essential: *tholeiites* and *alkali basalts*. Tholeiitic basalts (including most MORBs) contain a calcium-poor pyroxene, while alkali basalts lack these pyroxenes and usually have olivine. They can be distinguished on the plot of sodium plus potassium against silica[3] (Fig. 1.9).

MORBs must have been derived from the mantle: they are so voluminous that there is simply no alternative, and the plumbing that brings MORB magmas to the surface is evident.

Large degrees of partial melting of the mantle source material at a depth of about 70 km probably form a magnesium-rich *primary* magma from which MORBs are subsequently derived, via further complex steps which need not detain us here.

MORBs are relatively depleted in incompatible elements such as rubidium and thorium, whereas these same elements are abundant in the continental crust. Continental crust is manufactured from the mantle at subduction zones, and therefore it follows that the lack of elements such as rubidium and thorium in the upper mantle is due to their having been extracted from it when the crust was made. This conclusion begs two other questions: when did the depletion take place, and did it affect the whole of the mantle, or just the MORB source region?

Isotopic studies provide some of the answers. Measurements of the strontium isotope ratios in ancient basalts show that the removal of rubidium from the mantle (and formation of the continents) had begun long before 3000 million years ago. It is much more difficult, however, to determine how the rate of depletion (and of continental crust formation) has varied through geological time.

3.2.2 Ocean island basalts

Ocean island basalts (OIBs) help to illuminate the question of whether crust formation depleted the whole mantle, or just a part of it. As their name implies, ocean-island volcanoes are situated in ocean areas, far from any continent, and away from ocean ridges. They consist mostly of tholeiites, but show a wide range of compositions—many islands are made mainly or completely of alkali basalts. On a global scale, OIBs are volumetrically insignificant. As Fig. 3.1 shows, OIBs differ in their trace-element chemistry from the MORBs of the ocean floor surrounding them.[4]

These differences cannot be easily explained away, for instance by postulating that OIBs and MORBs are derived from the same mantle sources merely by different physical means, such as different degrees of partial melting. Isotopic studies shows that they must have formed from *physically separate* mantle source regions; the OIB source being enriched in incompatible elements such as rubidium relative to the MORB source. Furthermore, geochronological evidence shows that the two source regions must have been distinct for periods of the order of 2000 million years. A subtle but important conclusion follows from this: the mantle, which forms such a large part of the Earth, cannot be completely homogeneous.

At present, there is no universally accepted explanation of these relationships. Many geochemists believe that the mantle contains two layers. According to this hypothesis, the upper mantle is kept convectively stirred and has been depleted through time of its incompatible, crust-forming elements: it provides the MORB source region. Below the spinel–perovskite transition at a depth of about 650–700 km is a lower, more primitive (undepleted) mantle, which forms a different convective regime because it is more viscous. It is presumed to be relatively enriched in crust-forming elements, more nearly chondritic in composition, and provides the OIB source region. Other geochemists believe that convection affects the whole mantle, and that the source of OIBs may be in thermal plumes rising from as deep as the core–mantle boundary. In both models, the source regions for the voluminous MORBs is shallow, depleted mantle (Fig. 3.2).

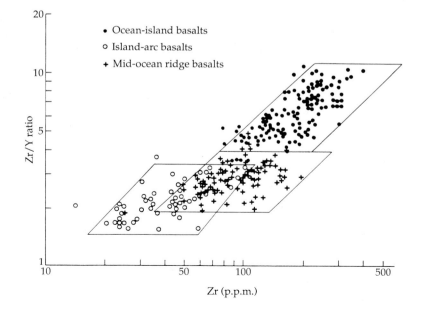

Fig. 3.1 Although all basalts may look the same, they can be fingerprinted by their trace-element chemistry. Basalts from mid-ocean ridges, ocean islands, and island arcs are distinguished here by their zirconium and yttrium contents. From Pearce, J. A. and Norry, M. J. (1979). Petrogenetic implications of Ti, Zr, Y, and Nb variations in volcanic rocks. *Contr. Mineral. Petrol.* **69**, 33–47.

• Ocean-island basalts
o Island-arc basalts
+ Mid-ocean ridge basalts

Zr/Y ratio

Zr (p.p.m.)

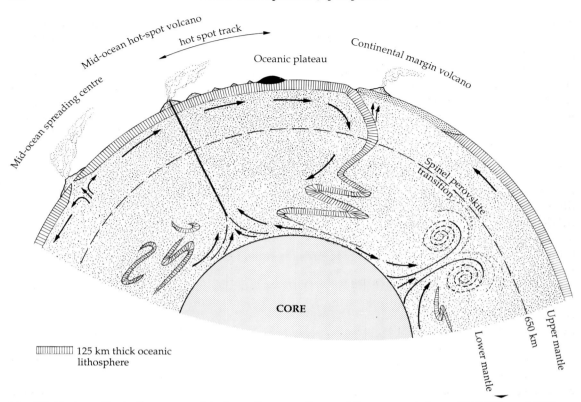

Mid-ocean hot-spot volcano
hot spot track
Continental margin volcano
Oceanic plateau
Mid-ocean spreading centre
Spinel perovskite transition
Upper mantle
650 km
Lower mantle
CORE

IIIIIIIIII 125 km thick oceanic lithosphere

Fig. 3.2 Cartoon illustrating one model of mantle convection, and the source regions of MORBs and OIBs. At far left, MORBs and oceanic lithosphere are formed at a spreading centre from a shallow, depleted mantle source; to the right, a long-established thermal plume rooted at the core mantle boundary feeds an ocean hot-spot volcano supplying OIBs; to the right of this are older, extinct volcanoes in the hot-spot track, terminating in an oceanic plateau. This marks the site where the plume head first reached the surface. Oceanic lithosphere is subducted beneath a continental margin at centre. Below 650 km, subducted oceanic lithosphere becomes progressively more plastic and is stretched, thinned, and deformed in the lower mantle, but still retains some geochemical identity. Small fragments of older subducted plates are also present. At far right, a new thermal plume is rising from the core-mantle boundary, but has not yet risen through the 650 km transition. From Davies, G. F. (1990). Mantle plumes, mantle stirring and hot-spot chemistry. *Earth Planet. Sci. Lett.* **99**, 94–109.

3.2.3 Island arc basalts

Island arc basalts (IABs) are erupted from the numerous island arc volcanoes around the world. No two arcs are the same, but most contain at least 20 per cent basalt. Fig. 3.1 and Table 3.2 show that IABs have different trace-element compositions from MORBs and OIBs. In other respects, IABs have higher and more variable silica contents than MORBs (48–53 per cent); and different strontium isotope ratios.[5] As described in Section 2.4.1, dehydration of the subducted slab and introduction of fluids into the

Table 3.2 Some important differences in major and trace elements between MORBs and typical island arc basalt[6]

	MORB	Island arc basalt
K_2O(%)	0.2	0.9
TiO_2(%)	1.4	1.0
Rb (ppm)	2	23
Ba	20	260
Zr	90	71
Hf	2.4	2.2

overlying mantle wedge play a key part in the manufacture of island arc rocks, but the petrological details need not detain us. We need merely note that geochemists interpret the differences between MORBs and IABs as suggesting that the source regions for the IABs were depleted in several elements relative to MORB, and hence that some basalt *had already been extracted* from the IAB sources.

3.2.4 Continental rift basalts

Basalts which floor the new oceans produced by rifting carried to its conclusion are typical MORB tholeiites. In the early stages of continental rifting, however, some more exotic rock types are produced. Volcanism associated with rifts tends to be geographically symmetrical about the rift axis, with older, more alkaline rocks on the flanks, and younger tholeiitic basalts at the centre. Alkaline rocks are the result of fractionation processes taking place at relatively deep levels, while tholeiites are formed at shallower levels. In the simplest case, this geographical variation may be matched by the evolution of the rift through time, the earliest rocks being formed by the smallest degrees of partial melting of mantle materials. In the East African rift, the oldest rocks are diamond-bearing *kimberlites*, 41 million years old, derived from a depth in excess of 150 km; 31 million years ago the Tororo

carbonatite was erupted; 22 million years ago flood basalts were erupted. Later, some 65 000 cubic kilometres of *phonolitic* and *trachytic* rocks were erupted, which may have originated from depths less than 40 km.[7] Alkali-rich basalts and trachytes, produced by small degrees of partial melting of mantle rocks plus a contribution from the continental crust are fairly straightforward, but other rift-related rocks are more bizarre. Oldoinyo Lengai's lavas of sodium carbonate are the most peculiar of all (see Section 1.4).

3.2.5 Continental flood basalts

Broadly speaking, flood basalts are tholeiitic, similar to their ocean floor counterparts. However, their trace-element abundances vary widely, and their strontium isotope ratios indicate an input from the continental crust through which the rising magmas penetrated. In general, it appears that the mantle source regions of continental flood basalts are less depleted in incompatible, crust-forming elements than the MORB source, and are broadly similar to those of the OIBs. Like OIBs, continental flood basalts are probably derived ultimately from mantle source regions in thermal plumes rising from the core–mantle boundary. Their chemistry thus only incidentally reflects the rocks through which they have passed. (Figs 3.3–3.4).

Fig. 3.3 Plenty of basalt here. In the foreground is the mighty Columbia River, and behind are some of the huge sequence of basalt lavas forming the eponymous flood basalt province. Stepped topography is typical of basalt 'traps'. Individual flows range in thickness from a few metres to tens of metres.

(a)

(b)

— 3.3 Lunar basalts

In one sense, lunar basalts are the most familiar of all: one only has to go out on a moonlit night to see the dark circular splotches on the Moon, which are huge impact basins filled with basalt lavas (Fig. 3.5). Lunar *mare basalts* closely match Giekie's description of terrestrial flood basalts (Section 2.5.2): they extend over huge areas, flooding the underlying topography to great depths, and yet seem to have welled up to the surface through inconspicuous fissures. Because they form such expansive plains, free of obstructions and only lightly peppered with craters, the mare basins formed natural landing sites for the

Apollo missions. Thus, many of the samples collected on the Moon were mare basalt lavas. Worth many times their weight in gold, these samples have been exhaustively examined—at one time their mineralogy, petrology, and geochemistry were more minutely known than those of *any* terrestrial rock.

All the sampled lunar basalts were erupted between 3.95 and 3.15 thousand million years ago. This distinguishes them immediately from terrestrial basalts, most of which are less than 100 million years old. Apart from their extreme age, the lunar basalts held few surprises for the first

Fig. 3.5 Plenty of basalt here. A photograph of the full Moon (14 days old) taken with a 100-inch telescope at the Hale Observatory. All of the familiar dark blotches making up the Man (or Rabbit) in the Moon are circular impact basins hundreds of kilometers in diameter, filled with basalt lavas. Pale-toned, heavily cratered areas are the lunar highlands, composed largely of anorthosites.

Fig. 3.4 (a) Plenty of basalt here. Basalt lavas total nearly 1000 metres in thickness in this deeply dissected escarpment, bounding the Drakensberg Plateau in southern Africa. These 193-million-year-old lavas form part of the Karoo flood basalt province, which may originally have covered as much as three million square kilometres. Photo: courtesy of Keith Cox, Oxford University.
(b) Plenty of basalt here: the Deccan Traps in the Western Ghats. Although the details of timing are controversial, most of these basalts are thought to have erupted 65 million years ago, at the time of the Cretaceous–Tertiary boundary. Topographically, the area has much in common with Fig. 3.4(a). Photo: courtesy A. V. Murali, Lunar and Planetary Institute.

investigators, so similar were they to other basalts (Fig. 3.6). One discovery was of a new pyroxene mineral *pyroxferroite* ($Fe_{0.85}CaO_{0.15}$) SiO_3; abundant in some samples, and difficult to miss because of its bright yellow colour. An inconspicuous opaque mineral *armalcolite* (named from the three Apollo 11 astronauts, *Arm*strong, *Al*drin and *Col*lins) was also found in some rocks.

Dozens of different lunar lavas have been analysed in the post-Apollo period. One thing emerges clearly: lunar basalts are chemically varied. It is difficult to pin down a 'typical' lunar mare basalt. Some have amazingly high titanium contents (13 per cent TiO_2), others have low titanium; some have high potassium contents, while others have low potassium and so on. Some of these gross differences are significant enough to show up in remotely sensed images: when infrared and ultraviolet telescopic images are combined, some parts of the maria appear 'blue', others 'orange', and others 'red'. The blue signature is apparently associated with high-titanium basalts; the red with low-titanium basalts.

Fig. 3.6 Abundant vesicles in this spongy Apollo 15 basalt lava confirm that sufficient volatiles (probably carbon monoxide) were present to produce scoriaceous lava, closely similar to terrestrial basalt. Note the rather uniform size of the bubbles. Clast is about 10 cm across. NASA S 71 45243, courtesy of Graham Ryder, LPI.

There are some subtle but important geochemical differences between lunar mare basalts and our familiar MORBs. Lunar basalts are primitive compared with MORB: they have more iron, are less oxidized, and have less alkalis (Na_2O and K_2O; Table 3.1). Most importantly, *they contain no water at all*. This accounts for the extreme freshness of lunar basalt samples—in the absence of water, no weathering whatever can take place. These factors indicate that the lunar basalts were derived from a source region with a low oxidizing potential, and one which was markedly depleted in volatile components. They were formed by relatively high degrees of partial melting of the lunar mantle, and are actually closer in composition to komatiites than MORB, although they are richer in iron and poorer in magnesium than terrestrial komatiites.

How were such voluminous and varied lunar mare basalts derived from the lunar mantle, which was presumably initially homogeneous and of near-chondritic composition? Current thinking suggests that soon after the proto-Earth formed from the solar nebula, it was struck by a huge impacting body, about the size of Mars. A large mass of material was blasted off, perhaps vaporized. This ejected debris settled into orbit round the Earth, and rapidly accreted to form the Moon (Fig. 3.7). Initially, it was probably entirely molten, but about 4.4 thousand million years ago, a *global magma ocean* with a depth of some hundreds of kilometres had formed over a denser solid core. A chilled skin crusted over this ocean, which subsequently began to crystallize. Dense olivines segregated downward, followed by olivine and pyroxene, then by pyroxenes, to form a layered stack of cumulates (Fig. 3.8). Lighter plagioclase crystals accumulated at the top to form a thick scum, which eventually became the anorthositic lunar highlands (the pale-toned terrain surrounding the maria). The layered stack of cumulates was sharply differentiated in trace elements, with most of the incompatible elements concentrating into a residual melt which gave rise to form a component rejoicing in the name KREEP (potassium, K; rare-earth elements, REE; and phosphorus, P).

Later, when radiogenic heat began to warm the upper lunar mantle, partial melting of the

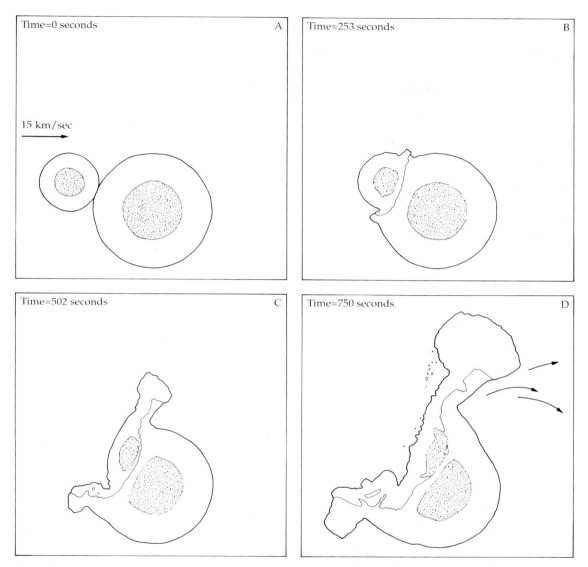

Fig. 3.7 The first twelve minutes in the history of the Moon. In this computer simulation, a Mars-sized impactor strikes the proto-Earth early in the history of the Solar system (A and B). Both the impactor and the proto-Earth must already have differentiated into crust and mantle. A jet of high-velocity, high-temperature material is flung off by the massive impact (C). Material at the head of the jet may be travelling as fast as 20 kilometres per second, and is lost in space. Material in the neck of the jet is moving slowly enough (about 10 km per second) to be 'captured' by the battered Earth, and goes into orbit around it (D). Temperature of this material was probably of the order of 6000 C; enough to vaporize it. This scenario accounts for the Moon's lack of volatiles—they were lost on vaporization. From Kipp, M. E. and Melosh, H. J. (1986). Short note: a preliminary modelling study of colliding planets. In *Origin of the Moon. Papers presented at the Conference on the Origin of the Moon, Kona, Hawaii, 1984*, pp. 643–8, LPI, Houston.

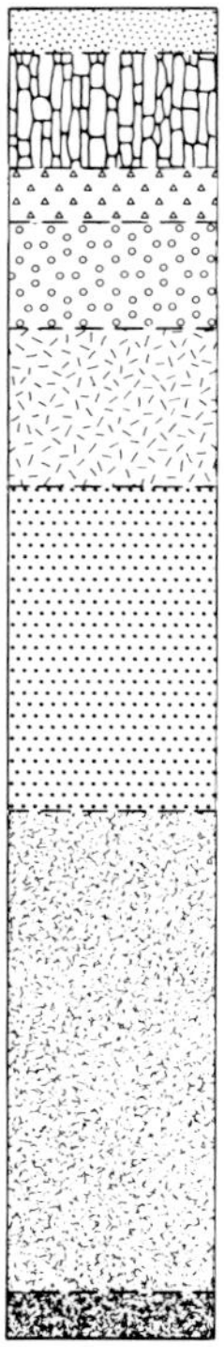

Dominant
mineralogy

Plagioclase — 'Primitive' chilled crust / Anorthosite ≡highlands

KREEP source region

Clinopyroxene-
plagioclase — High-alumina basalt source

Clinopyroxene-
orthopyroxene-
olivine-ilmenite — High-titanium basalt source

Olivine-
orthopyroxene
clinopyroxene — Low-titanium basalt source

'Green glass' source

Olivine-
orthopyroxene

Olivine — 'Whole Moon composition'

Fig. 3.8 A 'global magma ocean' several hundreds of kilometres deep probably existed on the Moon shortly after its formation. As it cooled and crystallized. layers dominated by different mineralogies formed as olivine sank downwards, and lighter plagioclase floated upwards (left side of column). Subsequent massive impact events partially homogenized this early layering. Later. partial melting events affecting zones of the lunar mantle dominated by different mineralogies gave rise to the range of different basalt types found in samples collected at the surface (right side). 'Green glass' refers to a primitive basalt composition. found only in beads of green glass collected at the Apollo 15 landing site (cf. Fig. 18.8). From Fig. 1.2.9.14 in Basaltic Volcanism Study Project (1986). *Basaltic volcanism on the terrestrial planets.* Pergamon Press. New York. 1286 pp.

cumulate layers took place, producing the range of basalt types found. There may have been some chronological pattern to this, the basalts derived from shallower levels (some of the high-titanium basalts) being erupted first; those from deeper levels with more magnesium rich (primitive) compositions being erupted last, as radiogenic heat production waned and melting was confined to progressively deeper levels in the crust.

An elegant observation which underpins this outline of lunar evolution is that the mare basalts all show a conspicuous depletion in europium, relative to other rare-earth elements. Europium is an element which can slot comfortably into the plagioclase feldspar crystal lattice, and the lunar highlands crust (mostly anorthositic plagioclase) is *enriched* in europium. Evidently, when the feldspar that went to form the highlands separated from the initial melt constituting the global magma ocean, it left the residue *depleted* in europium. Consequently, the basalts ultimately derived from the residue were themselves depleted in europium. Thus, whatever else happened, the lunar highlands and the mare basalts are firmly linked.

— 3.4 Basaltic meteorites

Apart from lunar basalts, there is only one other kind of extraterrestrial basalt that we can study directly. These are the basalts found in meteorites, which are the oldest melts of pristine solar system materials available. Basaltic meteorites are rare—fewer than 100 samples are known— and they include a subclass which is extremely tantalizing in its implications.

Most meteorites contain chondrules (Section 1.3.1) but *basaltic achondrites*, as their name suggests, lack chondrules. They possess instead an 'ordinary' igneous texture, indicating that they formed as melts from an earlier source, then cooled quickly to form the fine-grained texture typical of lavas. *Eucrites* and *howardites* are plagioclase–pyroxene achondrites, which have thoroughly basaltic appearances and are the commonest varieties. It may be supposed that meteoritic basalts could easily be confused with terrestrial basalts. How does one know that a particular small lump of basalt is actually a meteorite, and not just a lump of ordinary lava? The best way, of course, is to see the object fall from the sky. A memorable instance was on the afternoon of 2 August 1946, when 70 kg of an enstatite achondrite splashed into a swimming pool at Peña Blanca, Texas! Sadly, such sightings are rare. (Perhaps, on reflection, this is a good thing—swimming could become a bit of a risky business.) Eyewitness evidence apart, the best evidence for an extraterrestrial origin of a basaltic meteorite lies in its oxygen isotope ratios. Each of the bodies in the solar system appears to have its own isotopic signature.

Much work has been done on the source of basaltic achondrites. (The *Eucrite Parent Body* is spoken of as casually as if it resided in a local old peoples' home; often visited and taken out for air). The important thing about all but a tiny group of basaltic meteorites is that they have ancient crystallization ages, around 4.6 thousand million years, close to the age of the solar system itself. Whatever melting events took place to form these basalts, they clearly happened immediately after the initial accretionary events in the solar system. But where and what were the sources which melted to form the basalts?

Petrological studies show that their source regions were dominated by olivine (50–80 per cent) and lesser amounts of pyroxene; were poor in alkalis and other volatile elements; had a high iron to magnesium ratio; and contained perhaps about 10 weight per cent of free metal. These observations show that, broadly speaking, the parent bodies were chondritic, approximating quite closely to the lunar bulk composition.

A second important conclusion stems from analyses of basaltic achondrites: they must have been derived from rather small bodies. This follows because the meteoritic basaltic magmas must have evolved in equilibrium with a source containing plagioclase minerals. On Earth, pressures at comparable temperatures in the source region would be so great that minerals such as spinel or garnet would be present in place of plagioclase at depths greater than about 30 km. Thus, melting of primitive near-chondritic material to form eucrites must have taken place at low pressures. This could only have taken place in a small body, a few hundred kilometres in diameter, with a steep (if only ephemeral) thermal gradient. Although small, the body must none the less have been large enough to differentiate into the basic planetary structure of core, mantle and crust. Melting must have taken place within 100 million years of its formation.

Asteroids are the only bodies in the solar system which are the right size to be the parent bodies of basaltic meteorites. Telescopic observations have shown that there are one or two bodies in the asteroid belt with spectral signatures of pyroxene basalts like the eucrites. Most favoured candidate for the Eucrite Parent Body is the asteroid Vesta, 540 km in diameter, whose surface appears to be entirely covered by basalt.

(Iron meteorites are the best known of all extraterrestrial objects. They provide samples of the cores of asteroidal bodies, fragmented by later impacts. Many asteroids have spectra which match those of iron meteorites. Basaltic meteorites provide samples of the compositions of the crusts of asteroids such as Vesta. But curiously, while many asteroids have spectra characteristic of olivine-dominated mineralogies, corresponding with the mantles of asteroidal bodies, meteorites made mostly of olivine are rare. Given that samples of asteroidal cores and crusts arrive on Earth as meteorites, it is not clear why meteoritic samples of asteroidal *mantles* should be so scarce.)

Cosmic-ray exposure ages of basaltic meteorites show that only a few million years have elapsed since they were removed from their parent body. The only way that these objects could have reached the Earth from the asteroid belt is if they were blasted off their parent bodies by impacts of one asteroid on another, at closing velocities of several kilometres per second. No one quite knows how to explain the mechanics of this process, but the splash of the basaltic meteorite into the Texas swimming-pool demonstrates that it *does* happen.

3.4.1 The case of the dog that died. Are SNC meteorites samples from Mars?

To begin this investigation, it is essential to recall the death of the Egyptian dog. In this case, the killer is much more interesting than the dog, which, so far as we know, died nameless and unmourned. There are two remarkable things about the dog that died at Nakhla, Egypt in 1911. First, it was apparently killed by a falling meteorite, one of a shower of about 40 that fell at Nakhla el Baharia. Second, and even more remarkable, is that the meteorite *probably came from Mars*. This elementary deduction arises from the observations that the Nakhla meteorite has an obvious igneous texture, far different from the majority of meteorites; and that it has an apparent crystallization age of about 1.3 thousand million years, much younger than other meteorites. Where in the solar system could there have been a crystallization event 1.3 thousand

million years ago? As Sherlock Holmes would have argued, once the impossible has been eliminated, whatever remains, however improbable, is the solution. Improbable though it may sound, Mars is the *only* plausible source of the Nakhla meteorite (and a handful of others known as the SNCs, for Shergotty–Nakhla–Chassigny). These samples are exceptionally valuable, because of what they can reveal about the geological history of Mars, without the expense of actually going there.

There is plenty of evidence, based on crater-size/frequency distribution studies, that there was extensive basaltic volcanism on Mars at about the right time, and what we can deduce about the composition of the Martian mantle is consistent with the observed composition of the SNC meteorites (Fig. 3.9). While the SNC meteorites closely resemble terrestrial igneous rocks in many ways, they clearly did not originate on Earth, because (apart from having fallen out of the skies and dispatched the unlamented hound at Nakhla) their oxygen isotope signatures are different from those of the Earth and Moon.[8] Clinching evidence that the SNC meteorites were derived from Mars lies in the trace contents of noble gases such as neon

Fig. 3.9 Plenty of basalt here. This image was made on the surface of Mars by the Viking 2 lander on 7 September 1976. Remote sensing and surface sample chemistry suggest that the boulders are most likely basalt lavas; fragmented and ejected by meteorite impacts on the smooth plains of Mars's northern hemisphere. Many boulders exhibit bubbly cavities, possibly vesicles (cf. Fig. 3.6); further evidence of an origin as basalt lavas.

and argon they contain. The ratios of the abundances of different noble gases in the meteorites are closely similar to those actually measured in the Mars atmosphere by the Viking landers in 1976, and are different from those in any other planetary atmosphere.[9]

When the Martian origin of these meteorites was first proposed by McSween[10] it met with bitter resistance. Objections centred around the mechanical difficulty of ejecting rock samples from the surface of Mars and delivering them to Earth. Most of the postulated hypotheses involved impacts which excavated craters and ejected material into space, but the early cratering models suggested that all the ejecta would be melted. In the face of the convincing noble gas data, though, the sceptics were forced to retreat, and to look for cratering mechanisms that do not melt all the ejecta. Further support for a Martian origin came from the discovery on the Antarctic ice of several meteorites which *must* have been derived from the Moon, because they closely resemble the Apollo samples, and no terrestrial rocks are remotely comparable. If rock samples can be blasted off the Moon and delivered unmelted to the Earth, a similar process could happen on Mars.

All of this is a long-winded lead up to the concept that in the rather drab-looking meteorites of the SNC group, we may actually possess samples of Martian rocks, a conclusion of the first importance for cosmochemists and planetary volcanologists (Fig. 3.10).

Study of the SNC meteorites is complicated by the effects on their mineralogy caused by the enormous shock pressures they suffered on impact, but two points are unambiguous.[10] First, the three groups (shergottites, nakhlites, and Chassigny) are *different* from one another, showing that Mars is petrologically diverse. Second, of the three groups, the shergottites look most like ordinary terrestrial basalt lavas. Shergotty itself consists of 75 per cent clinopyroxene grains with 25 per cent plagioclase. Nakhlites are not simple lavas but cumulate rocks, while Chassigny appears to have been derived from a small *intrusive* body. Current thinking suggests that the shergottites and nakhlites were derived from thick lavas on the surface of Mars, which ponded deeply in places and allowed crystals to settle downwards to form cumulates. Analogous features can sometimes be seen in terrestrial lavas where olivine crystals cumulate to form *picritic* lavas.

In geochemical terms, the Martian rocks are similar to terrestrial tholeiitic lavas (Shergotty) and peridotites (the nakhlites). Because some SNCs contain trace quantities of hydrous minerals such as amphibole, it follows that the Martian mantle source region contains volatiles, in sharp

Fig. 3.10 'Elephant Moraine 79001' is the name given to this unimpressive looking lump of rock, which is actually a meteorite of shergotty type, recovered from the Antarctic ice-cap in 1979. Its age and composition suggest that it was a volcanic rock blasted off the surface of Mars by an impact. The smooth, dark surface is the skin of the melt formed during passage through Earth's atmosphere. Photo: courtesy Robby Score, Johnson Space Center, Houston.

contrast with the desiccated Moon, but in agreement with everything else we know about Mars. Lavas such as Shergotty probably resembled terrestrial basalts, containing less than 1 per cent water. Overall, the mantle of Mars as inferred from the SNCs appears to be similar to the Earth's, except that, like the Eucrite Parent Body, it is much more iron rich than the Earth. Estimates of the compositions of Mars' core permitted by the composition of its mantle suggest that it cannot be dense iron–nickel alloy like the Earth's, but is probably mostly composed of iron sulphide, and occupies a larger fraction of the planet. One estimate suggests that the radius of the core may be as much as 60 per cent of the radius of the entire planet.

Although most cosmochemists now accept that the SNCs came from Mars, these remarkable rocks are frustrating for volcanologists, because they provide many hints about the nature of Mars' volcanism without actually resolving anything decisively. So there remain plenty of reasons for volcanologists to include the Red Planet in their future field-work plans.

— Notes

1. Basaltic Volcanism Study Project (1981). *Basaltic volcanism on the terrestrial planets*. Pergamon Press, New York, 1286 pp.
2. Bowen, N. L. (1928). *The evolution of the igneous rocks*. Princeton University Press, Princeton, New Jersey.
3. Carmichael, I. S. E., Turner, F. J., and Verhoogen, J. (1974). *Igneous petrology*. McGraw-Hill, New York.
4. Pearce, J. A. and Cann, J. R. (1973). Tectonic setting of basic volcanic rocks determined using trace element analysis. *Earth. Planet. Sci. Lett.* **19**, 290–300.
5. Pearce, J. A. and Norry, M. J. (1979). Petrogenetic implications of Ti, Zr, Y and Nb variations in volcanic rocks. *Contrib. Mineral. Petrol.* **69**, 33–47.
6. Pearce, J. A. (1982). Trace element characteristics of lavas from destructive plate boundaries. In *Andesites: orogenic andesites and related rocks* (ed. R. S. Thorpe, pp. 525–48, Wiley, London.
7. Wendlandt, R. F. and Morgan, P. (1982). Lithospheric thinning associated with rifting in East Africa. *Nature* **298**, 734–6.
8. Clayton, R. N. and Mayeda, T. K. (1983). Oxygen isotopes in eucrites, shergottites, nahklites and chassignites. *Earth Planet. Sci. Lett.* **62**, p. 106.
9. Bogard, D. D. and Johnson, P. (1983). Martian gases in an Antarctic meteorite. *Science* **221**, 651–3.
10. McSween, H. Y. (1985). SNC meteorites: clues to Martian petrologic evolution. *Rev. Geophys.* **23**, 391–415.

4

Four classic eruptions

There have been innumerable eruptions during the few thousand years that civilized man has lived on Earth. Most were insignificant, but some made such an impact that they were momentous historical events. A few changed the course of civilization. One which influenced European civilization of classical antiquity was the Minoan eruption of Thera (Santorini) in the Aegean, which may have extinguished the sophisticated Minoan civilization centred on Crete in about 1620 BC. It also gave rise to the enduring legends of Atlantis, the 'lost' continent, because the eruption caused the submergence beneath the sea of much of Thera. Evidence for the legends is drawn from some vague references by the Greek philosopher Plato, in particular a passage in his *Timaeus* which reads: 'But at a later time there occurred portentous earthquakes and floods, and one grievous night and day befell them when the whole body . . . of warriors was swallowed up by the earth, and the island of Atlantis in like manner was swallowed up by the sea and vanished.'

Although this account is tenuous, and the whole Atlantis legend threadbare, there is no doubt at all that a catastrophic volcanic eruption took place on Thera about 1620 BC. Almost 20 cubic kilometres of tephra fell into the sea and on to the adjacent islands; *tsunamis* ravaged coastal towns around the eastern Mediterranean, and a spectacular caldera was formed (Fig. 4.1 and Fig. 4.2). On Thera itself, archaeological excavations have revealed a story rivalling the celebrated AD 79 eruption of Vesuvius. Akrotiri, a city with a population of 30 000, was buried—and thus

beautifully preserved—beneath many metres of ash. The town appears to have been efficiently evacuated, because it had been stripped of movable objects, and no bodies were found in the ruins. Evidence of earthquake damage to buildings suggests that the inhabitants had received warning of impending doom, giving them time to flee. It is likely, too, that the eruption was preceded by some minor explosive activity, leaving the citizens in no doubt that the volcano had awoken. Although valuable artefacts had been removed prior to the eruption, Akrotiri preserves some superb aesthetic and archaeological treasures. Wall paintings of extraordinary sensitivity and accomplishment found in the ruins are among the most important artistic discoveries of the century.

Important though it was in cultural terms, there are no written records of the eruption, so all that we know has to be deduced from the geological record.[1] Although it was of great magnitude, there are doubts about how severely the eruption affected Minoan civilization. These doubts stem from discrepancies in the dates assigned to the eruption, and to the end of the Minoan civilization. In recent years, estimates of the eruption date have been derived from sources as different as the Greenland ice-cap, tree-rings in North American bristle cone pines and radiocarbon dates on artefacts from Akrotiri itself. These give a range of dates between 1675 and 1525 BC. Archaeologists, however, have long accepted that the Minoan civilization on Crete was destroyed much later; about 1450 BC. This date, however, is also imprecisely known, and depends

Fig. 4.1 Sketch map of the islands of Thera, Therasia, and Aspronisi, relics of the great eruption of Santorini about 1620 BC. Steep cliffs which define the caldera wall are hatched; sites of the ancient Minoan town of Akrotiri and new city of Phira are shown. Eruptions in 1570, 1707, and 1925 gave birth to the new island of Nea Kameni.

Fig. 4.2 Stunning cliffs outline the caldera wall formed by the Minoan eruption of Santorini, near the modern city of Phira. Light layer capping the cliffs is pumice from the eruption.

on factors such as correlation of pottery styles with dated sequences in the Egyptian archaeological record. Some scholars argue that the eruption provides a more reliable date for the end of the Minoan civilization than pottery correlations, and thus that a link between the eruption and the end of the Minoans is inescapable. Others argue that it is unlikely that the conventional archaeological record could be so far askew, and that the great eruption had little direct effect on Minoan culture.

The debate continues and widens. Even if the eruption did not *immediately* extinguish the Minoan civilization, it may have had such far-reaching environmental effects that the civilization declined and fell. Traces of Minoan ash found in the Nile Delta region have been used to suggest that biblical accounts of one of the ten plagues at the time of the Exodus sprang from folk memories of the dense ash cloud: 'there was a thick darkness in all the land of Egypt for three days' (Exodus 10:21). While this historic link can never be proved, even today a massive eruption taking place in the eastern Mediterranean would cause such widespread havoc that its long-term political consequences are impossible to predict.

Our cultural traditions incline us, of course, to think about Western civilization. Other cultures have suffered similar, or worse, disasters as the result of volcanic events. One little-known eruption which took place in the south-western Yukon Territory of Canada 1400 years ago probably caused the wholesale migration of the Athapaskan Indians southwards and away from

their ancestral homelands around the White River. In Papua New Guinea, folk legends still speak of a 'Time of Darkness' when a great eruption brought devastation, perhaps 400 years ago. In Japan, archaeologists are beginning to unearth ruins of an extensive agricultural community which have lain beneath two metres of ash from the Haruna volcano since the sixth century AD. In Central America, a great eruption taking place in about AD 300, centred near Lake Ilopango in El Salvador, caused a huge disruption to the highland Maya indians. Seventy-five kilometres from the lake, a half-metre thickness of ash fall accumulated at the Mayan city of Chalchupa. In the wake of the eruption, a massive shift of population appears to have taken place, perhaps leading ultimately to the flowering of the classic Mayan civilization in the tropical lowlands of the Peten and Yucatán.[2] Doubtless, many other eruptions with far-reaching cultural consequences remain to be documented from archaeological and ethno-historical records around the world.

In this chapter, four eruptions more completely documented than that of Thera are described. These are classics of their kind, selected not only because of their role in history, but because each casts light on different facets of volcanology. This chapter emphasizes narrative aspects of the eruptions: volcanological processes will be scrutinized in later chapters. All are destructive plate margin volcanoes: Vesuvius in the Mediterranean; Krakatau in Indonesia; Mt. Pelée in the Caribbean; and Mt. St Helens in North America.

4.1 Vesuvius, AD 79

On 5 February AD 62 an earthquake jolted the area around what is now Naples, on the west coast of Italy. In Roman times, many thriving towns clustered around the bay of Naples (Fig. 4.3, Fig. 4.4). Two of these, Pompeii and Herculaneum, located on the lower flanks of Vesuvius, were badly damaged by the tremors, but the townspeople were not especially alarmed. They were enjoying the prosperity of the Roman

empire at its height and simply rebuilt fallen walls, repaired the Temple in Pompeii, and carried on with their normal well-ordered lives. They did not associate the earthquake with the volcano, and neither was there any particular reason why they should. Although Strabo had noted in 30 BC that the rocks of Vesuvius resembled the lavas of persistently erupting Etna, the residents did not think of Vesuvius as a

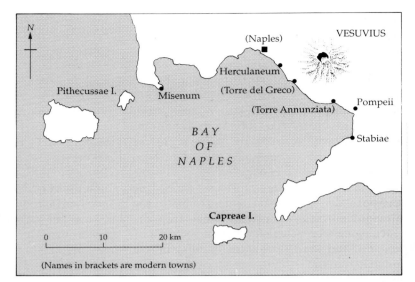

Fig. 4.3 Sites of principal towns around the Bay of Naples mentioned by Pliny the Younger in his account of the AD 79 eruption. Names of modern towns are shown in brackets.

Fig. 4.4 Vesuvius seen from present-day Castellammare (Stabiae). The profile of the volcano has changed greatly since AD 79, the prominent central cone has grown since the eruption; low peak on the right is part of the older edifice known as Monte Somma.

potentially active volcano because it had been inactive for centuries previously, and they had no first-hand experience of its malevolent character. So peaceful had the volcano been that its fertile soils were extensively cultivated, and vineyards flourished on favoured sunny slopes. They attached little broader significance to the nuisance of the earthquake, because earthquakes have never been rare events in Italy. During the next sixteen years, minor earthquakes continued to shake the area spasmodically. We now know that major eruptions are typically preceded by seismic activity, so that tremors might have provided warning of the forthcoming eruption, had anyone been aware of their significance (Fig. 4.5). Two thousand years ago this awareness did not exist, so life went on normally until the last moment. On 24 August AD 79, its preliminary stirrings ignored, Vesuvius burst into life, initiating one of the most momentous eruptions in history. It lasted two days. At the end of it, both Pompeii and Herculaneum had ceased to exist, and thousands of people had died appalling deaths.

One of those to die was a naturalist and an Admiral in the Roman navy, Caius Plinius or Pliny the Elder, a much respected man, famous in his own time. It was the death of this great man that led to our having a vivid eyewitness account of the events 1900 years ago. Pliny the Elder had been so widely esteemed that the historian Tacitus was anxious to discover the circum-

stances of his death. Tacitus therefore asked Pliny's nephew, who was seventeen at the time and survived the eruption, to tell him what had happened. Pliny the Younger's account took the form of two letters to Tacitus, in which he described his uncle's fate, and how he and his mother had fared. These letters are remarkably valuable historical documents, because Pliny was aware of the need to describe objectively what he had experienced, and to resist the temptation to exaggerate or to embroider. As he wrote himself: '*Nec defuerunt qui fictis mentionis-que terroribus vera pericula augerunt*'.* Well, of course, there always are! The following description is based closely on Pliny's letters.

4.1.1 Pliny's account

At about 1 p.m. on the 24 August, when the various Plinys were all in Misenum, a small town 30 kilometres across the Bay of Naples from Vesuvius (Fig. 4.3), a curious large cloud appeared in the sky above the volcano. At first, it looked harmless enough, and did not even seem to be coming from the mountain. It grew rapidly, however, soon dispelling all doubts. Rising initially in a vertical plume for many thousands of metres and then spreading out laterally in the atmosphere, it resembled a Mediterranean pine tree, with bare trunk and a crown of branches

*There were people too, who added to the real perils by inventing fictitious dangers.

Fig. 4.5 Vesuvius looms dimly in the background urban haze in this view of the forum of Pompeii. Excavations have revealed that many buildings had been damaged by earthquakes and repaired prior to AD 79.

higher up. Pliny the Elder was full of curiosity at first, and planned to sail across the bay in his galley for a closer look. His nephew was not so keen on the idea, and pleaded pressure of work in declining his uncle's invitation to go along as well. Before he had had time to organize his galley and set off, the elder Pliny received urgent requests for help from people living nearer the volcano, so the trip which had been planned as something of a scientific investigation rapidly became a rescue mission.

Pliny intended to try to evacuate people living on the coast immediately beneath the volcano, in the area which is now Torre del Greco, but as his galley approached the coast it was showered with hot ashes and sizeable lumps of pumice. The shore was already becoming inaccessible as piles of tephra accumulated, and so he was forced to abandon his attempt, and turned south-east-wards, running before the wind to escape from the increasingly heavy rain of pumice. He made landfall at Stabiae (near present Castellammare), where conditions were still tolerable. There he encountered a friend of his, one Pomponianus, who was making frantic attempts to escape, fretting for a favourable wind so that he could put to sea. Full of confidence, Pliny tried to reassure the overwrought Pomponianus. To demonstrate his own unconcern, he went off to freshen up in the local baths, and later sat down to a hearty meal.

As night fell on Stabiae, Vesuvius presented an awesome sight. Many fires had been started by the fall of hot pyroclasts, lighting up the heavy ash cloud above the volcano with a baleful red glare. Remarkably, Pliny slept calmly during the early part of the night, although tephra had begun to pile up outside the house, but woke later as the situation began to deteriorate. Pomponianus and his companions were badly worried, both by the ash fall, and by the frequent tremors shaking the house. After a discussion, they decided that their best hope for safety would be to leave the house, make for the shore, and attempt to get away in a ship. Tying pillows to their heads to protect themselves from the larger falling lumps of pumice, they set off. It was totally dark, blacker than the darkest night, even after dawn should have come. Dense ash clouds hung over them, cutting off all light from the Sun. Carrying torches to light their way, they all reached the shore safely, but to their dismay found that the wind was still unfavourable, blowing onshore, and the sea too rough to enable them to get away. At this time, apparently, Pliny became unwell; he lay down on a cloth spread out for him, and twice asked for water to drink. Later, when wafts of choking, sulphurous fumes from the volcano made conditions still more unpleasant, most of the people with him fled. He rose, tried to follow them, but fell dead immediately. His contemporaries thought that he had been poisoned by the volcanic gases, since when his body was found three days later, it was unmarked, looking more like that of a sleeping man than a dead one. (It now seems unlikely that the gases alone were responsible, since Stabiae is so far from the volcano, and his companions survived. Pliny, was was a corpulent man, may have died from a heart attack brought on by the stress of the eruption, aggravated by the acrid fumes.)

On the morning when Pliny the Elder died in Stabiae, conditions were also becoming alarming for Pliny the Younger and his mother on the opposite side of the Bay of Naples, at Misenum. Pliny records that he was reading Livy, but he must have found it hard to concentrate. Misenum was upwind from the volcano, and thus suffered less severely than the areas downwind, such as Stabiae. Frequent tremors shook the house they were in, and a dense pall of ash obscured the sun, so that the light was faint and uncertain. They decided for safety's sake to leave the town and set off in chariots, with a large crowd of panic-stricken fellow refugees jostling and harassing them. They stopped in open country just outside the town, clear of danger from collapsing buildings, but found that tremors were still so frequent that their chariots were constantly on the move, despite having their wheels chocked with stones. Overhead, lightning flickered frequently as static electricity accumulating in the ash cloud discharged. Once, they noticed the sea receding from nearby beaches, leaving quantities of sea creatures stranded, and then surging back in *tsunamis* generated by the earthquakes.

Ash began to sift down around them, lightly at

first, but sufficient to add to the horror of their situation. Conditions were not so bad, however, that Pliny and his mother could not have reached safety by travelling further away from the volcano, but they were reluctant to leave because of their alarm and uncertainty about what had become of the elder Pliny. They remained therefore, on the outskirts of the town, until, terrified of being crushed by the fear-stricken mob, they decided to seek refuge in open fields. They had hardly agreed on this when fresh ash clouds, denser than ever, overwhelmed them and brought total darkness. So complete was the darkness that Pliny compared it with a sealed room in which the lamp had been put out. It was much more frightening than that, however, for they were not in a sealed room, but out in the open in daytime, with the air rent with cries from the crowd—screams of naked fear and prayers for deliverance. Ash began to fall more heavily, piling up around them, so that they had periodically to shake themselves clear of it.

This waking nightmare lasted for many hours until a sickly daylight showed again, and a dim sun could be seen again through the pall of ash. Exhausted by their ordeal, survivors were able to collect themselves, find their friends and look about. Great changes had been wrought by the eruption. When the air had cleared sufficiently to see across the bay, it was apparent that where once the smooth cone of Vesuvius had once risen, only an awful stump now remained; where once there had been fields and woods and vineyards and all the comfortable clutter of the countryside, there now stretched an unbroken grey carpet of ash, mantling everything as uniformly as a dirty snowfall. Amidst it all, not a bird or insect stirred.

4.1.2 The archaeological story

Pliny's account is so illuminating that similar eruptions, blasting several cubic kilometres of pumice into the air, are still called *plinian* eruptions (Chapter 9). Pompeii, only nine kilometres downwind from the volcano was buried under almost 3 metres of tephra, so that only the upper parts of tall buildings protruded above it. With most of its inhabitants either dead or ruined by the eruption, and the surrounding countryside an uninhabitable wasteland of ash, there was no

hope of salvaging the buried town, so it was abandoned. So completely was it forsaken that with the passage of time and the weathering away of the few surface remnants, its location was forgotten, athough the facts of its existence and fate were known to scholars through the writings of historians such as Tacitus.

Centuries passed. The Roman Empire declined and fell. The Dark Ages came and went. New peoples settled round Vesuvius, and ultimately new towns sprang up. In 1595, some remains of a city came to light during excavation work for a new aqueduct at Civita—a few coins, and fragments of a marble tablet containing inscriptions referring to Pompeii. Eventually, it was realized that the remains of Pompeii lay preserved beneath the blanket of pumice. In the seventeenth and eighteenth centuries, when post-Renaissance noble families all over Europe became intensely art conscious, and anxious to fill their palaces with pieces of classical statuary, the remains of Pompeii were ravaged haphazardly. Innumerable random pits were dug in the hope of turning up items of value, such as bronze sculptures, vases and pieces of jewellery. There was no pretence of archaeological intention—the subject had not evolved by then. Pompeii was simply a large open-pit mine for works of art. Sir William Hamilton, Britain's ambassador to Naples, was once deeply involved in this pillaging, giving some of his acquisitions to his young wife, Emma, better known as Nelson's mistress. While later generations may deplore Hamilton's plundering, he redeemed himself in the eyes of volcanologists by writing the first books on volcanology, *Observations on Mount Vesuvius, Mount Etna and other volcanoes* (1772) and *Campi Phlegræi* (1776).

In the nineteenth century, more co-ordinated methods began to be employed in the Pompeii excavations. Digging began to be directed at finding out about the town, rather than merely treasure seeking. It has been said that modern archaeology was born in the ruins of Pompeii. Even today, the excavations are not wholly complete, but many hectares of the town have been uncovered, and the abandoned streets are bustling once more, as thousands of visitors stroll amongst the remains of shops, taverns, and villas. Exploring the ruins creates an eerie impression, more reminiscent of a city bombed in World War II than of one abandoned near 2000 years ago (Fig. 4.6). Originally, the town had a population of around 20 000. Although the remains of many hundreds of individuals have been found, it is not clear how many townspeople died during the eruption. Undoubtedly, many fled during the early stages of the eruption, but some of these may have met their deaths in the surrounding country.

Uneaten food laid out on tables in some houses showed that normal life continued in the town until the very last moment. Disaster came suddenly and unexpectedly. Many human fossils have been found. After death, the victims' bodies were rapidly buried by the accumulating tephra, and rain falling on the ash cemented it into a hard mass before the bodies had decayed. As a result of the wetting, the ash swelled slightly and set hard around the bodies, making perfect natural moulds of them, preserving in some cases even the imprint of clothing and facial expressions. Nineteenth-century excavators found the moulds as cavities in the ash, and obtained three-dimensional casts of the original corpses by pouring plaster of Paris into the cavities.

These fossils tell us a good deal about the horrific final hours of the lives of the people of Pompeii. Most victims were killed when they were engulfed in clouds of searing hot dust and gas. Some of the plaster casts demonstrate this with unpleasant clarity: the victims' hands are still pressed to the mouth, preserving their last futile efforts to breathe (Fig. 4.7). One of the most pathetic casts is that of a pet dog, which died still chained to a post. It survived the tephra fall for many hours, somehow remaining above the accumulating thickness of tephra until ultimately it succumbed, its dreadful last moments expressed in its twisted body, arched back, and straining neck (Fig. 4.8). Some of the human victims reduced their chances of survival by leaving their homes burdened with bags of gold and jewellery. They died still clutching them. At least one person died through coming back into the town during a lull in the eruption, to rescue his own little hoard of gold (or to loot from abandoned houses), only to be killed by fresh fiery blasts from the volcano.

Within the last five years, detailed volcanological studies have cast more light on exactly what happened at Pompeii. Pre-eminent among these is the work of the Icelandic volcanologist Haraldur Sigurdsson and his colleagues, summarized here.[3-4]

Events began with a heavy fall of tephra, which accumulated to a depth of about two metres. This deposit of light-coloured pumice is conspicuous in many parts of the ruins at Pompeii. People probably began to flee from their houses when the weight of tephra building up on their houses

Fig. 4.6 One of the streets of Pompeii. Citizens crossed the streets using the large blocks as stepping stones, without treading in the ordure left by the many horses and mules pulling carts along the rutted streets.

Fig. 4.7 One of the human victims of the eruption, preserved as a plaster cast.

Fig. 4.8 The dog.

started to cause the roofs to fall in, probably causing some fatalities. This simple tephra fall was bad enough, but it was not what killed most victims. On several occasions, steady fall of tephra was interrupted by *pyroclastic surges*. These are incandescent, ground-hugging clouds of pumice and dust that collapse downward from eruption columns, and race along the surface at speeds as much as 100 kilometres per hour (Chapters 10 and 11). Nothing could survive in the face of the surges that swept through Pompeii. Human victims perished in the hot blasts of dust and gas. Those parts of the buildings which protruded above the ash-fall

deposit were knocked flat. Pompeii ceased to exist (Fig. 4.9).

4.1.3 Herculaneum

Pompeii was not the only town to suffer from the rain of pumice and ash, although it was the largest. Tephra fall covered many hundreds of square kilometres, burying several other smaller settlements, which remain entombed to this day. Pompeii was particularly badly hit because it was close to the volcano, and directly downwind of it, so the densest part of the ash cloud was carried over it by the south-easterly wind. Neighbouring Herculaneum suffered differently. It was located just as close to the volcano, but on its south-west flank. While Pompeii was obliterated relatively slowly through the accumulation of air-fall tephra and by surges over a period of two days, Herculaneum was overwhelmed in a matter of minutes by pyroclastic surges and flows which rushed through the town, burying it more than twenty metres deep. Some buildings were demolished, but most were simply engulfed in dusty ash and pumice. When indurated and compacted, this material poses much more difficult problems for archaeologists than the unconsolidated pumice of Pompeii. This factor, coupled with the awkward presence of a new town above it, makes it unlikely that Herculaneum will ever be completely excavated (Fig. 4.10). About

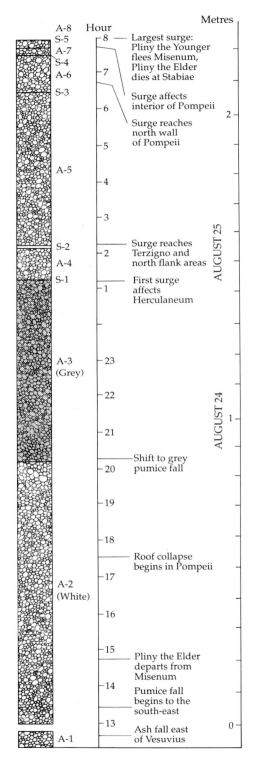

A-8
S-5
A-7
S-4
A-6
S-3

A-5

S-2
A-4
S-1

A-3
(Grey)

A-2
(White)

A-1

Hour

Metres

8 — Largest surge:
Pliny the Younger
flees Misenum,
Pliny the Elder
dies at Stabiae

7

6 — Surge affects
interior of Pompeii

— Surge reaches
north wall
of Pompeii

5

2

4

3

2 — Surge reaches
Terzigno and
north flank areas

1 — First surge
affects
Herculaneum

23

22

21

20 — Shift to grey
pumice fall

19

18

17 — Roof collapse
begins in Pompeii

16

1

15 — Pliny the Elder
departs from
Misenum

14 — Pumice fall
begins to the
south-east

13 — Ash fall east
of Vesuvius

0

AUGUST 25

AUGUST 24

Fig. 4.10 A colonnade exhumed at Herculaneum. Herculaneum was deeply buried under many metres of material, and presents formidable problems for excavators, particularly since new buildings (background) have been constructed on top of the volcanic deposits.

40 000 square metres, corresponding with about eight city blocks have been uncovered, and a larger area probed with tunnels and borings. Elaborate tunnelling was needed to extract artefacts, and therefore Herculaneum survived the depredations of eighteenth-century treasure-seekers rather better than Pompeii, and has furnished an even greater wealth of detail on the daily life of a Roman town.

For many years, the lack of human remains in Herculaneum suggested that the population had escaped unscathed. Recent excavations have revealed a woefully different story. These excavations have focused on what was the shoreline in Roman times. Buildings along the shore were

Fig. 4.9 A chronological and stratigraphical summary of the sequence of ash-fall and surge events at Pompeii during 24–25 August. Most victims were killed by the brief surge events (S1–5), not the prolonged ash fall. Details are discussed in later chapters. (From Sigurdsson, H., Carey, S., Cornell, W., and Pescatore, T. (1985). The eruption of Vesuvius in A.D. 79. *Nat. Geog. Res.* **1**, 332–87.)

supported by a number of arched chambers, open to the beach, in which boats and fishing tackle were kept. When these chambers were uncovered in the 1980s, hundreds of skeletons were found, huddled in pathetic groups, clearly those of townspeople who had fled to the beach, vainly hoping that the shore-front chambers would offer refuge from the volcano. Much more remains to be excavated, and more poignant evidence of Herculaneum's final hours will surely come to light, bringing fresh sorrow to future archaeologists and volcanologists.

4.2 Krakatau, 1883

Between the Indonesian islands of Java and Sumatra is the Sunda Strait, one of the great sea lanes of the world. Even in the nineteenth century, hundreds of ships passed through the Strait each year, many of them large vessels trading between Europe and the prosperous Dutch colonies of the East Indies. Only 24 km across at its narrowest point, the Strait averages about 200 metres deep. Thirty-two kilometres west of the narrowest point of the Strait there lay a small group of rather oddly named islands: Krakatau, Verlaten, Lang, and—oddest of all—Polish Hat. Although coastal details of these islands were known to the hydrographers of both the British and Dutch navies, little was known of their inland topography. British Admiralty charts showed that the largest, Krakatau, about nine kilometres from north to south, consisted of several volcanic cones arranged roughly in line. At the southern end of the line was a prominent cone, 800 m high, known by its Javanese name of Rakata (Fig. 4.11). At the northern end was a much lower cone, Perbuwatan, whose crater wall had been breached at some time in the past by a large lava flow.

None of the islands was inhabited, and neither were they often visited, except by woodcutters from both shores of the Strait who landed periodically to fell timber in the luxuriant forests that clothed the islands. From time to time, fishermen would anchor in the sheltered bays to take on water, or ride out storms. Naval survey parties may also have landed briefly on Krakatau, because a hot spring is marked on some maps. Apart from this, we know remarkably little about the islands prior to 1883. There are vague reports of an eruption taking place between May 1680 and November 1681, which stripped bare the vegetation and ejected vast quantities of pumice that covered the surrounding sea, but it is not even known where on Krakatau the eruption was centred. It may have been Perbuwatan, since lava flows in the crater there seemed fresh when examined in the nineteenth century. Apart from

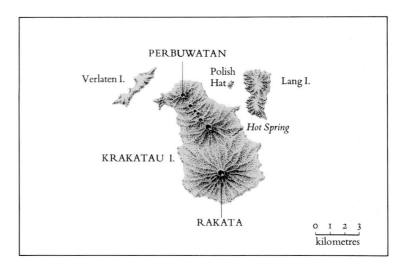

Fig. 4.11 Krakatau and adjacent islands before the eruption of 1883.

this, the islands had been tranquil. Their beauty was well known to the passengers of steamships passing through the Strait. On many a sultry tropical evening, travellers must have leaned contentedly on the warm mahogany of their ships' rails, enjoying the moonlight glittering off the water around the islands. None of them could have conceived that they were looking at the future site of the most catastrophic event of modern times.

4.2.1 Prelude

In the late 1870s, frequent minor earthquakes began to shake the area around the Strait. They may have been the first intimations of disaster. A powerful shock on 1 September 1880 demolished the top of an important lighthouse on the Java coast, and was felt as far away as northern Australia. Like Vesuvius and all destructive plate margin volcanoes, however, Krakatau is located in an area of frequent seismic activity, so local people did not attach special significance to the tremors. After a period of increasing seismic activity, Krakatau came abruptly to life on 20 May 1883, announcing its reawakening with a series of explosions audible over 150 kilometres distant. Very long wavelength atmospheric pressure waves from the explosions were energetic enough to stop clocks, rattle windows, and dislodge hanging lamps. Since these pressure waves were inaudible, their effects were often mistaken for those of earthquakes. On the following day, a sprinkling of ash fell over a wide area, and a great column of steam rose above Krakatau, leaving no doubt that an eruption was under way. Vigorous activity continued for a few days, yielding a column of steam and ash climbing eleven kilometres high above the volcano, which showered ash over points nearly 500 km away.

By 27 May things had quietened down sufficiently for a party of 86 hardy souls to charter a steamboat from Batavia (now Jakarta) to see what was going on. Their boat was the *Governeur Generaal Loudon*, later to be involved in the thick of the eruption. While the party was approaching Krakatau, the noise was deafening. In the words of one of them (obviously a sophisticated, party-going individual), the background din was so loud that a rifle shot sounded like 'the popping of a champagne cork amid the hubbub of a banquet'. From their boat, the party saw that the island was covered with fine white dust, like snow, and that trees on the northern part of the island had been stripped bare of their foliage by falling ash. Scrambling ashore and scuffing their way through ankle-deep ash, the party found that the centre of all the excitement was the Perbuwatan crater. This was in a state of semi-continuous activity, with minor explosions taking place every five or ten minutes, showering ash 200 metres into the air, and occasionally revealing the cherry-red glow of lava in the crater. A great banner of steam rose continuously 3000 m into the air. The visiting party, who were the first and last ever to get a good look at the crater of Perbuwatan, found it to be about a kilometre in diameter and fifty metres deep, with a small pit about fifty metres deep in the centre of the crater floor. It was from this pit that the steam column was escaping with a great roar (Fig. 4.12).

After this remarkable visit, the volcano continued to be active for a week or so, but the explosive activity died down somewhat, suggesting that the eruption would fade away and be forgotten. On 9 June, however, things began to warm up again, and the column of steam and ash began to rise higher and higher as more and more powerful explosions rent the air. By the end of June, observers on Sumatra reported that the higher parts of Perbuwatan had been blown away, and that a *second* eruption column was now rising from the centre of the island. During July, many areas of Java and Sumatra were rocked by explosions of exceptional violence, and by many minor earthquakes. Even this severe shaking, however, failed to alarm the local people, who had by that time been living with the eruption for many weeks. It is remarkable the extent to which familiarity with even something as exceptional as a violent eruption can breed contempt.

On 11 August, a Dutch government surveyor, Captain Ferzenaar, made another examination of the island, landing briefly on a beach to collect ash samples. He reported that all the formerly rampant vegetation on Krakatau had now been destroyed, only a few of the thicker tree trunks still protruding above the heavy mantle of

Krakatoa. Rep. Roy. Soc. Com. Plate

View of Krakatoa during the Earlier Stage of the eruption.
from a Photograph taken on Sunday the 27th of May, 1883.

Fig. 4.12 Drawing from the Royal Society's report of the eruption on Krakatau, based on a photograph taken by a member of the last party to visit the island, on 27 May 1883. Stumps of trees are shown on the slopes. No photographs exist of the catastrophic phase of the eruption.

tephra, and that there were *three* active eruption columns carrying clouds of dust and ash high into the air. One of these was the original vent at Perbuwatan; the other two seemed to be nearer the centre of the island. From the north-east of the island, Captain Ferzenaar reported that he could see no less than eleven other sites of minor activity, which were either emitting small steam columns or occasionally ejecting ash in small explosive bursts. There may well have been other active vents, but the heavy curtain of ash and fumes prevented the captain from sailing all round the island. His observations, therefore, were made only from the upwind side, but they remain valuable because they provide the last

account of the situation before the culminating events which ensued fifteen days later.

4.2.2 *Crescendo*

Wagner's exhilarating 'Ride of the Valkyries' has been used effectively as the background to films of lurid lava eruptions. But Wagner's operas are remarkable for their length and extreme tedium, as well as the magnificence of their occasional climaxes. In this respect the Krakatau eruption has some Wagnerian parallels, because after an impressive overture, incidental activity dragged on for a full three months before the climax on 26 and 27 August 1883. That climax, however, was truly worthy of *Götterdammerung*. So extensive was the havoc wrought in those two days that it was not until many months afterwards that a picture began to emerge of what had happened. A fact-finding Scientific Commission was appointed by the Dutch government in October 1883. It was led by Rogier D. M. Verbeek, a mining engineer and geologist, who visited the scene for the first time on 15 October 1883 and repeatedly thereafter. A preliminary report was published some six months later.[5] In Britain, the Royal Society also set up an investigative committee, which published a weighty tome in 1888.[6] These two reports remain the prime sources of almost all information about the eruption. The Dutch report concentrated on the local effects of the eruption, whereas the Royal Society report emphasized the more distant effects. The Society went to great lengths to amass every possible scrap of information, and even inserted a notice in the London *Times* requesting anyone who had seen or heard anything to come forward. Copies of the original Dutch and British reports are difficult to obtain, but fortunately, the Smithsonian Institution has published a superb compilation of the contemporary reports and modern interpretations.[7]

Part of the difficulty for the investigators was that there were so few survivors from the coastal towns along the Sunda Strait. Dutch officials living in Batavia and Buitenzorg were able to provide useful eyewitness reports, while the instruments at the Batavia gasworks provided a unique chronological record of the pressure waves from the major explosions. Officers on board the various vessels on passage through the Strait also provided vital observations, which were the more valuable because of the nautical practice of logging the time of observations. Three ships were right in the thick of things. A British ship, the *Charles Bal*, heading for Hong Kong, was sailing eastwards through the Strait on 26 August, passing about sixteen kilometres south of Krakatau. The *Gouveneur-General Loudon* was plying back and forth across the Strait between Anjer in Java and Telok Betong in Sumatra. She passed about forty-eight kilometres north of Krakatau on the evening of the 26th, spent the night of the 26th/27th anchored in Telok Betong and tried to sail again for Anjer in

the morning but was prevented from doing so by the violence of the eruption. Another British vessel, the *Sir Robert Sale* was at the eastern, narrower end of the Strait, 64 km from Krakatau, on the 26th and attempted to sail westwards (towards Krakatau) on the 27th, but was unable to do so. Apart from these three, reports came in to the Royal Society from more than fifty other vessels at various distances from Krakatau (Fig. 4.13).

After sifting through scores of reports from observers at sea and on dry land, Verbeek was able to piece together a perceptive account of the events of the two fateful days, Sunday 26 and Monday 27 August. All the reports agreed that there had been a gradual but marked increase in the intensity of activity on Krakatau during the three days preceding the 26th. At 1 p.m. on the 26th, explosions loud enough to be heard over 150 kilometres away were taking place at intervals of about 10 minutes. At about 2 p.m., a British ship 120 km from the scene sighted a black cloud rising to an altitude estimated to be no less than 25 km above the volcano. By 3 p.m. the explosions were so loud they were audible 240 km away; by 5 p.m. they were so stupendous that the sound was carrying all over Java. In Batavia, 160 km from Krakatau, the din was terrific, the noise being compared with 'the discharge of artillery close at hand . . . causing rattling of windows and shaking of pictures, chandeliers and other hanging objects'. Similar activity continued throughout Sunday evening and most of the night. The *Charles Bal*, which was at its closest to the volcano at this time, reported:

. . . sounds like discharges of artillery at intervals of a second of time, and a cracking noise, probably due to the impact of fragments in the atmosphere . . . the whole commotion increasing towards 5 p.m. when it became so intense that the captain feared to continue his voyage, and began to shorten sail. From 5 to 6 p.m. a rain of pumice in large pieces, quite warm, fell upon the ship.

Captain Woolridge on the *Sir Robert Sale*, rather further away, reported seeing the eruption column rising above the volcano: '. . . a most terrible appearance, the dense mass of clouds

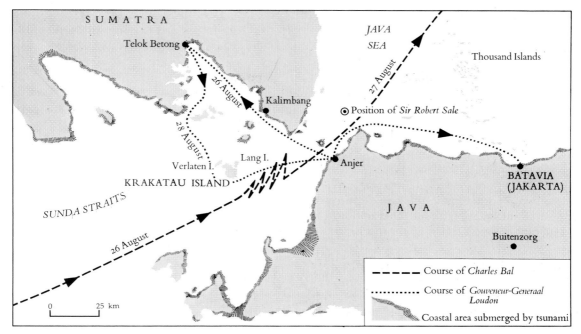

Fig. 4.13 The Sunda Straits in 1883, showing the courses of the ships *Charles Bal* and *Gouveneur Generaal Loudon* and the location of the *Sir Robert Sale*, which all survived the paroxysmal phase of the eruption at close quarters. Coastal areas devastated by tsunamis are shaded.

being covered with a murky tinge, with fierce flashes of lightning'. A little later, at 7 p.m., the whole scene was lit up from time to time by electrical discharges. At one time the cloud above Krakatau presented 'the appearance of an immense pine tree, with the stem and branches formed with volcanic lightning', a description which echoes Pliny's description of the Vesuvius eruption column 1800 years earlier.

Things became so bad for the *Charles Bal* later on in the evening that she had to spend the entire night tacking back and forth south-east of Krakatau, probably remaining within 20 km of the volcano. Captain Watson could not see well enough through the murk to steer away to safety, but ironically the glare from the volcano provided a terrifying lighthouse with which to check his bearings. A less pleasant night can scarcely be imagined, with a rain of hot ash falling on the ship, and the air laden with choking fumes of sulphurous gases. As Captain Watson was to write after the eruption: 'The night was a fearful one: the blinding fall of sand and stone, the

intense blackness above and around us, broken only by the incessant glare of various kinds of lightning and the continued explosive roars of Krakatau, made our situation a truly awful one'. He did not exaggerate. To make matters worse, peculiar pinky glows of static electricity lit up the mastheads and rigging of the ship with an unearthly light, known to mariners as St Elmo's Fire. On the *Governeur Generaal Loudon* this phenomenon was even more extensive, and the terrified native crew: 'engaged themselves busily in putting out this phosphorescent light with their hands . . . and pleaded that if this light . . . made its way below, a hole would burst in the ship; not that they feared the ship taking fire, but they thought the light was the work of evil spirits'.

St Elmo's fire is a peculiar phosphorescent glow, caused by the atmosphere around the ship becoming highly positively charged with static electricity, which is generated by the rush of steam and ash through the volcanic vent, and by interactions between the myriads of fragmentary

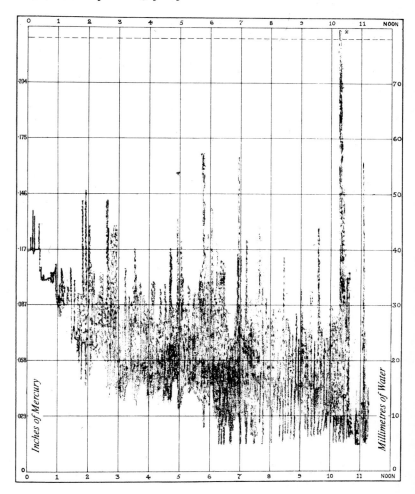

Fig. 4.14 Part of the pressure record from the Batavia gasworks, showing pressure changes caused by the air waves from explosions at Krakatau. The wave at 10.15 a.m. local time is far larger than the others, going off scale, and caused the gas-holder to jump out of its well. (The sound waves took 5 minutes to travel from Krakatau.)

particles being swept around within the eruption cloud by powerful convective forces.

4.2.3 Climax

After 4 a.m. on the morning of the 27th, the eruption appeared to die down a little, but the grandest moments were yet to come. They arrived in the form of a series of explosions on a far greater scale even than any of the preceding ones, and far greater than any other that man has recorded. According to Verbeek's analysis of the pressure gauge at the Batavia gas-works (which was sensitive to the transient atmospheric pressure changes caused by the huge explosions) the largest explosions took place at 05.30, 06.44, 10.02, and 10.52 (Krakatau time) on the morning of Sunday 27 August. Of these, the third was

much the more powerful (Fig. 4.14). Reverberations from these great explosions rumbled over a large part of the Earth's surface: at Elsey Creek in South Australia, 3224 km from Krakatau, the noise was loud enough to wake sleeping people, who described it as being similar to the sound of rock being blasted. At Diego Garcia, 3647 km distance in the Indian Ocean, the explosions were at first thought to be from a ship in distress, firing

Fig. 4.15 (*opposite*) Tide gauge record from Tandjong Priok Harbour at Batavia (Jakarta), showing the arrival times of the great sea waves. Largest wave was near noon on 27 August; its amplitude was so great that the instrument went off scale. Inset shows the Batavia gasometer pressure trace.

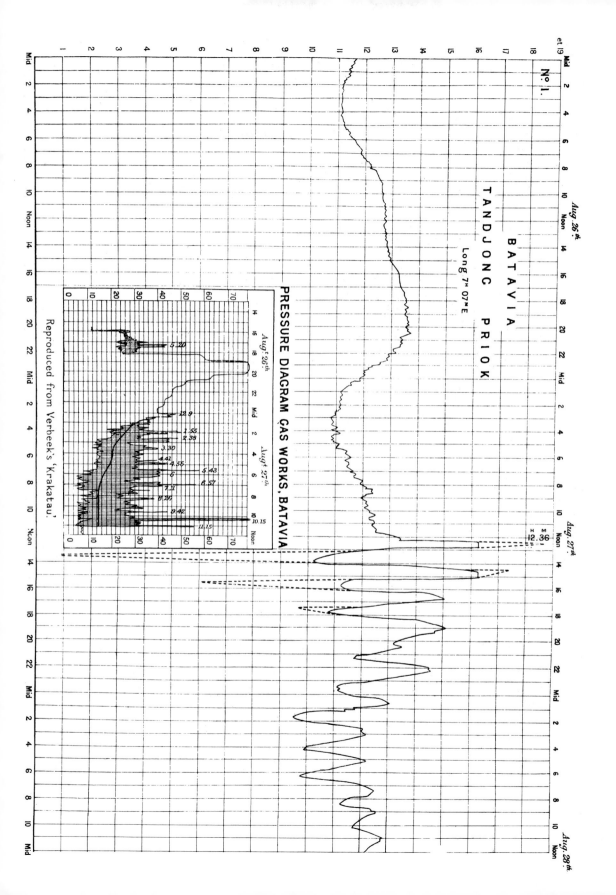

PRESSURE DIAGRAM GAS WORKS, BATAVIA

BATAVIA

TANDJONG PRIOK

Long 7ʰ 07ᵐ E

Reproduced from Verbeek's 'Krakatau'.

its guns to attract attention, so people ran to cliff-top vantage points to try to see it. Rodriguez Island, near Mauritius in the Indian Ocean was the furthest point at which the sounds were audible, 4811 km distant. Here again the rumbles were at first taken for gunfire. Although inaudible, lower-frequency atmospheric pressure waves from the explosions were detectable on sensitive barographs *all over the world*. Tokyo, 5863 km distant registered a transient pressure increase of 1.45 millibars.

Nearer Krakatau, the effects were of course much more serious. At Batavia and Buitenzorg, the blast blew in dozens of windows, and even cracked walls. At the Batavia gas-works, the blast was so great that the gas-holder leapt out of its well, causing the gas to escape. Much the most destructive effects of the eruption, however, were a series of *tsunamis* which ravaged the shores of the Sunda Strait. It was these seismic sea waves or 'tidal' waves, rather than the awesome explosions, that caused almost all of the 36 000 fatalities during the eruption. Low-lying areas all along the coast were devastated, and complete towns and villages were overwhelmed. The port of Anjer simply ceased to exist as great waves washed over it, carrying away the flimsy wooden buildings that made up the town. One of the tsunamis, which arrived in the harbour of Tandjong Priok (Batavia) at 12.36 on the 27th was far larger than any of the others, and indicated that some extraordinary event had occurred. More than a hundred years later, the precise nature of this event is still debated (Fig. 4.15).

A huge eruption column was generated by the climactic explosions of 27 August. Early estimates suggested that it rose as much as 80 km into the air. Tephra raining down from it added to the misery of the people in the area—the *Sir Robert Sale* reported lumps the size of pumpkins falling on her decks, and she was at least 40 km distant. Ash fall was reported as far away as the Cocos Islands, 1850 km distant.

At Batavia, the pall of ash took a long while to manifest itself fully. In the early morning of the 27th, the sky was clear, but by 10.15 it had become lurid and yellowish as ash spread overhead; by 10.30 the first fine ash was sifting softly down on to the streets; at 11.00 ash was falling heavily; by 11.20 the ash pall was so dense that the sun was blotted out. Total darkness fell on the city, remaining until 1 p.m. Ash ceased falling about two hours later. Since Batavia was more than 160 km from Krakatau, it escaped lightly. Nearer the volcano, unrelieved darkness continued for nearly two days in some places. Such stygian conditions naturally made it impossible to determine what was actually happening on Krakatau, but it is thought that some milder explosive activity was continuing. The *Charles Bal* and *Sir Robert Sale* were beating about in tenebrous darkness for the whole of the 27th. Ash rained down on them so steadily that the crews had to continually shovel it off the decks and shake it clear of sails and rigging. On the *Governeur Generaal Loudon*, it was reported that at one time dust and water were falling together, as mud, and that a thickness of 15 centimetres accumulated in only 10 minutes. Fortunately, this soon declined to a more tolerable rate.

At 7 p.m. on the 27th, another outbreak of minor explosive activity occurred, getting progressively more vigorous until 11 p.m. when it started to decline. These outbursts marked the end of the entire eruption, for at 2.30 p.m. on the 28th, after one hundred days of activity, the last mild explosion rumbled out over Krakatau, and silence returned.

4.2.4 Aftermath

Slowly, life returned to something like normal around the Strait. Bewildered survivors were able to bury their dead, and salvage what remained of their homes in coastal towns and

Fig. 4.16 (*opposite*) Krakatau before and after the eruption, reconstructed from contemporary hydrographical charts. Pyroclastic flows deposited a layer of pumice on the sea-floor as much as 40 m thick. These flows extended preferentially to the north and east, perhaps because the high cone of Rakata to the south acted as a barrier. Two new islands, Steers and Calmeyer, were formed where the pumice deposits reached above sea-level, and were sites of continuing secondary eruptions. Sertung and Rakata Kecil islands may have been fragments of an older caldera; a deep, new caldera was formed by the eruption.

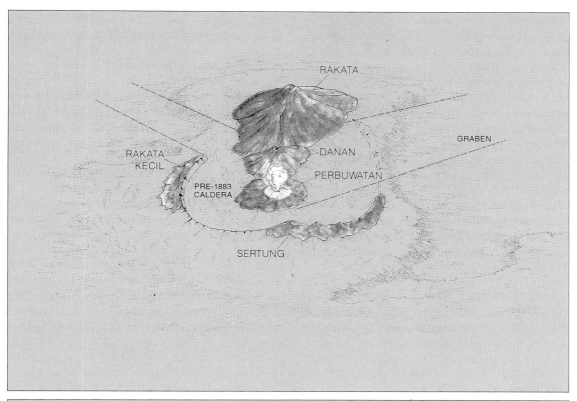

RAKATA

GRABEN

RAKATA
KECIL

DANAN

PERBUWATAN

PRE-1883
CALDERA

SERTUNG

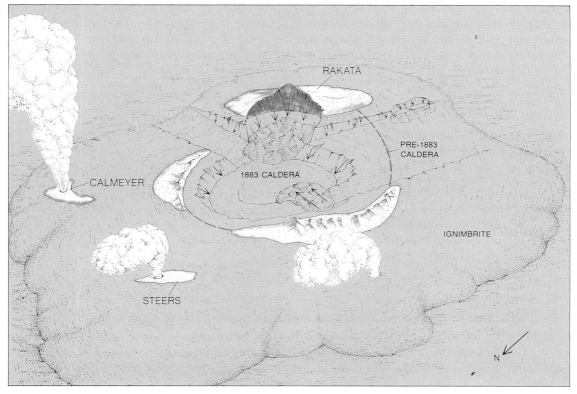

RAKATA

PRE-1883
CALDERA

CALMEYER

1883 CALDERA

IGNIMBRITE

STEERS

N

villages. Initially, great masses of floating pumice made it difficult for ships to push their way through the water. Rafts of pumice three feet thick were reported in places. Eventually, scientific parties were able to reach the islands and determine what changes had taken place. A detailed survey immediately after the eruption by the Royal Dutch Navy revealed some startling contrasts. They found that the northern part of the island of Krakatau had disappeared, with the exception of a bank of pumice and a small isolated rock about ten metres square. Not a vestige remained of the Perbuwatan cone, and the whole of the northern part of the Rakata cone had vanished, leaving a soaring semi-vertical cliff. A great crater (caldera) had been formed below sea-level—soundings showed that where land had once stood 300 m above sea-level, the water was now 300 m deep (Fig. 4.16). In all, about two-thirds of Krakatau had disappeared. Most of the island foundered into the sea when the roof of the underlying magma chamber collapsed, but the cone of Rakata was apparently left perched on the southern rim of the caldera, its northern face virtually unsupported. This side of the volcano then slumped into the sea, leaving behind a spectacularly bisected volcanic cone, which is all that remains today of the island of Krakatau (Fig. 4.17).

Verbeek estimated that about 15 cubic kilometres of matter had been ejected by the eruption, close to the modern estimate of about 20 cubic kilometres. Some of this fell out as tephra over an area of nearly four million square kilometres, but most was emplaced in the immediate vicinity of Krakatau. So while the island of Krakatau ended up considerably smaller, Verlaten and Lang Islands ended up much bigger, buried and surrounded by enormous volumes of pyroclastic deposits, which formed extensive 'sand' banks all around Krakatau. Two brand-new islands, Steers and Calmeyer, were formed by these deposits. And Polish Hat? Not a trace remained.

In the century since Verbeek published his meticulous study, it has been difficult to improve on his work. Wave action rapidly eroded away much of the newly deposited ash, and tropical vegetation quickly recolonized the dry land, so that it is now difficult for volcanologists to find accessible outcrops of the 1883 ash to study. A new volcano, Anak Krakatau, has also been built up. Some new insights into the eruption have been gained, however, from field-work and comparative studies of other great eruptions.[8-12] It now seems likely, for example, that the massive tsunami of 27 August was triggered by the collapse of half of the Rakata

Fig. 4.17 All that remained of Krakatau Island after the eruption. Two-thirds of the island foundered into the sea, bisecting the Rakata peak, leaving this spectacular cliff face which exposes the interior of the old volcanic cone. A layer of white 1883 pumice deposits covers both flanks of the cone. From Verbeek's *Album of Krakatau*.

cone into the sea, in a giant avalanche similar to that at Mt. St Helens, described later in this chapter. And while Verbeek had no knowledge of pyroclastic flows, it is now plain that most of the ejecta from Krakatau was emplaced in flows, rather than as tephra fall-out from the eruption cloud. Evidence that some of these flows travelled remarkably long distances over water is hinted at in the contemporary reports. There are accounts of people being burned by hot ash from the area around Kalimbang in southern Sumatra, 40 km north-east of Krakatau. These burns must have been due to horizontally travelling flows, rather than vertical ash fall, because in one instance the survivors described hot gases and ash *blowing upwards* through the floorboards of a house.

Krakatau, then, was a complex eruption, remarkable for its magnitude and violence. It has become almost a caricature of volcanic violence, and it deserves its reputation. As the remaining examples in this chapter show, more recent eruptions have killed many thousands of people and have been impressively destructive, but none comes close to Krakatau. We have not even touched yet on the world-wide atmospheric and climatic effects caused by the eruption. That must wait until Chapter 17.

4.3 Mt. Pelée, 1902

Krakatau's long drawn-out eruption was compared to a Wagnerian opera. In the case of the 1902 eruption of Mt. Pelée such artificial comparisons are unnecessary, because the eruption actually inspired an opera, *The Violins of St Jacques*. This seldom-performed work was based on a novel by Patrick Leigh Fermor, who wove a complicated plot of romance and intrigue against the background of a sultry Caribbean town in the midst of a hectic carnival. A volcano rumbled threateningly above the town, until it ultimately destroyed the island. Mt. Pelée is thinly disguised as the volcano Salpetrière, while the real town of St Pierre appears as St Jacques. However, the real events of Thursday 8 May 1902 were so dramatic, and the tragedy so complete, that it is unnecessary to dress up the facts in a romanticized account.

At the time of the eruption, St Pierre was the principal town on Martinique, a small island in the Caribbean, which was then a prosperous French colony. Six kilometres north of the town rose the gentle slopes of Mt. Pelée, 1400 metres high, and often wreathed in clouds (Fig. 4.18). Both island and volcano were renowned for their beauty. To quote a contemporary description, which reads like a modern travel brochure:

It has the softest of summer zephyrs blowing across its fields and hillsides; swift and tumbling waters break through forest and plain; and mountain heights rise to where they can gather the island's mists to their crowns. There are pretty thatched cottages, nestling in the shade of coco-nut, mango and breadfruit, and decked out with bright hibiscus and bougainvillaea.

St Pierre itself was no less appealing, stretching for three kilometres along the shores of the Caribbean (Fig. 4.19; Fig. 4.20). It boasted a number of imposing public buildings, including a town hall, cathedral, military hospital, and a theatre, and even had electric lighting, unusual in those days. Most of the town consisted of a picturesque maze of narrow, rambling streets, lined with old-fashioned houses with steeply pitched red-tiled roofs. In the official census of 1894, the population of this delightful town was estimated to be 19 722, most of them Martiniquans, with a few French government officials and civil servants.

Like Krakatau, Mt. Pelée had been active for weeks before disaster struck. But whereas Krakatau's outburst was the first for two centuries, Mt. Pelée had erupted within the living memory of the older residents of St Pierre; a mild affair in 1851 which did little apart from sprinkling a bit of ash about. Hints of renewed activity were first noted in 1889, when fumaroles appeared in the summit crater. In January 1902, the fumaroles increased in number and strength. On 2 March 1902, a party of ramblers made an ascent of the volcano. Looking down from the top, they saw

Fig. 4.18 Mt. Pelée and St Pierre in 1987. Today, the scene appears as alluring and tranquil as it must have done before May 1902. But history will inevitably repeat itself.

that a small crater containing a dried-up lake was emitting sulphurous vapours from several points. This crater lake was known as the Étang Sec, a name to remember, because it later became the focus of the eruption proper. At 9 p.m. on 23 April, things began to liven up. Minor explosive activity in the Étang Sec hurled ashes and bits of rock into the air, and the eruption gradually increased in vigour.

On 27 April, an investigating party climbed up from St Pierre to find out what was happening. They discovered that the Étang Sec now contained a small lake, in the middle of which a small ash cone had begun to grow. Activity steadily increased in the following days, showering St Pierre with light falls of ash and enveloping it with unpleasant wafts of sulphurous fumes. Mrs Prentiss, the wife of the American consul in St Pierre wrote in a letter home: 'The smell of sulphur is so strong that horses in the street stop

and snort, and some of them drop in their harness and die of suffocation. Many of the people are obliged to wear wet handkerchiefs to protect themselves from the strong fumes of sulphur.'

There was a sharp increase in activity on 2 May, and by 4 May the situation was really bad, with frequent loud explosions, and substantial amounts of ash accumulating in the streets. *Les Colonies*, the St Pierre local newspaper carried an article saying: 'The rain of ashes never ceases . . . the passing of carriages is no longer heard in the streets. The wheels are muffled . . . puffs of wind sweep the ashes from roofs and awnings, and blow them into rooms, the windows of which have imprudently been left open.'

On 5 May the first lethal blow fell, when water which had been ponded in the Étang Sec burst through the crater walls, and rushed down the valley of the Rivière Blanche. Intense fumarolic activity associated with a mass of hot rock

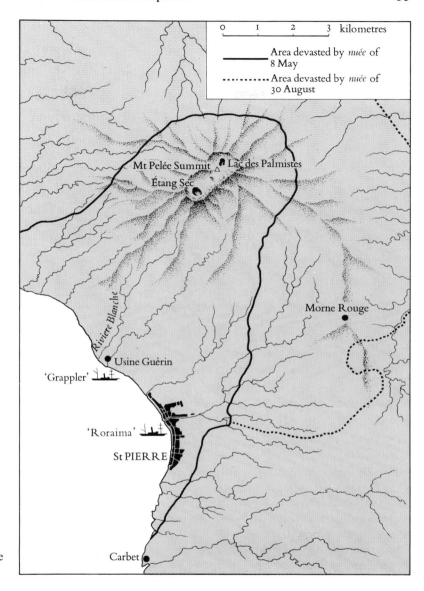

Fig. 4.19 Mt. Pelée, St Pierre and the area devastated by the *nuée ardente* of 8 May 1902.

forcing its way up towards the crater floor had weakened the 100-metre-thick crater rim enclosing the lake, and heated its waters to nearly boiling point. It was thus a scalding torrent that gushed down the valley, travelling at a speed approaching ninety kilometres an hour. As it flowed, it incorporated masses of loose debris, mud, and boulders, some reported to weigh fifty tonnes. This peculiarly unpleasant mixture combined to become a mud-flow or *lahar*, which over-ran everything in its path. Tragically, a

large rum distillery, the Usine Guérin, stood on the flat land in the valley of the Rivière Blanche, surrounded by sugar-cane plantations. Twenty-three workmen died when the distillery was overrun. Nothing remained of it but a tall chimney stack, sticking up like a pillar above a desert of black, seething mud.

Subsequently, the lethal lahar rushed on into the sea at the mouth of the Rivière Blanche, setting up a tsunami powerful enough to capsize the yacht *Prêcheur*, moored off the river mouth.

Fig. 4.20 St Pierre's main street—the Rue Victor Hugo—before the eruption. Many vessels are moored offshore, just as they were on 8 May 1902.

At St Pierre, the 3–4-metre-high tsunami flooded some low-lying parts of the town along the waterfront. Two lesser lahars followed the first.

As the situation deteriorated, the people living near the volcano became increasingly distressed, and many tried to leave St Pierre, heading for Martinique's second town, Fort-de-France. Unsurprisingly, the local authorities did not know how to handle the situation, which was complicated by the fact that an election was due on Sunday 11 May, and the political factions in the town were anxious that nothing should interfere with it. No arrangements to evacuate the city were made, but neither were those who wished to leave prevented from doing so. A 'scientific commission' comprising some of the most learned people on the island was set up, which produced a report designed to set fears at rest. One of the members of this commission and principal of the Lycée, Monsieur Landes, was worried by events, but when interviewed by the local newspaper, made a cautious, non-committal statement, warning that some of the areas nearest the volcano should be evacuated. The nub of Monsieur Landes's thinking was that Mt. Pelée presented no more danger to St Pierre than Vesuvius did to Naples, a distinctly equivocal conclusion! In an editorial in the last issue his paper was ever to publish, the editor of *Les Colonies*, seized on this statement and used it to encourage people to stay in St Pierre: 'Where,' he asked, 'could one be better than at St Pierre? Do those who invade Fort-de-France believe that they will be better off there than here, should the earth begin to quake? This is a foolish error against which the populace should be warned.'

His appeal must have met with some response because although some hundreds of people did leave for Fort-de-France, many hundreds *more* sought refuge in St Pierre itself, flocking in from the surrounding countryside, so that on the morning of 8 May, as many as 28 000 people may have been crowding the town. Mrs Prentiss, the American consul's wife, in her last letter to

America, said that she had the opportunity of leaving St Pierre with her husband on an American schooner, but decided that the situation did not warrant it. She died with her husband on the next day, only a few hours after writing her letter.

Modern studies have shown that until 6 May, the material erupted consisted of old, chilled material blown out by steam explosions. At about 10 a.m. on 6 May, the style of eruption changed, and columns of black ash with incandescent blocks were hurled upwards: hot magma had broken through to the surface. Lava began to pile up at the summit of the volcano, forming an embryonic lava dome (Chapter 7). Small parts of the growing dome collapsed from time to time, sending intermittent hot avalanches cascading down the Rivière Blanche during 7 May.[13]

4.3.1 The events of 8 May 1902

At 07.52 (St Pierre time) on 8 May, the telegraph operator in St Pierre tapped out the single word 'Allez' to his opposite number in Fort-de-France, meaning that he was ready to receive a message. It was the last word to come out of St Pierre. We shall never know exactly the sequence of events in St Pierre on that morning, but we can build up a fairly accurate picture from the physical evidence of the ruins, and the reports of eyewitnesses. We can at least be sure of the time when the town died, because apart from the abrupt halt in telegraph traffic, the big clock on the Military Hospital was found in the smouldering ruins, its hands stopped at 07.52.

The 8th of May was Ascension Day. Many faithful Catholics had assembled in the Cathedral to celebrate the day, and to pray for deliverance from the volcano. At about 07.50, a series of deafening detonations was heard, and a great black cloud was seen to issue out from near the top of Mt. Pelée. Rolling relentlessly down the slopes of the volcano, the cloud spread out into a broad fan which rapidly engulfed St Pierre, and set the town ablaze from end to end. In the next two or three minutes, the population was annihilated. Eyewitness accounts of those few minutes naturally vary in detail, but one of the most reliable descriptions was that of Monsieur Roger

Arnoux, a member of the Astronomical Society of France, who was observing from a vantage point well above and away from the town. He was awakened by a minor earth tremor during the night, but went back to sleep again and experienced nothing else untoward during the night. The following morning, however, at about 8 o'clock:

... while still watching the crater, I noticed a small cloud pass out, followed two seconds later by a considerable cloud. This latter cloud rolled swiftly down towards St Pierre, hugging the ground, but extending upwards at the same time, so that it was almost as high as it was long. The vapours ... were of a violet grey colour, and seemingly very dense, for although endowed with an almost inconceivably powerful ascensive force, they retained to the zenith their rounded summits. Innumerable electrical scintillations played through the chaos of vapours, and at the same time that the ears were deafened by a frightful fracas.

What M. Arnoux described so graphically was a phenomenon new to science. When subsequent examples were observed on Mt. Pelée, they became known as *nuées ardentes*, or glowing clouds. *Nuées* are amongst the most lethal of all the weapons in a volcano's armoury, and are a variety of gravity-driven pyroclastic flow (Fig. 4.21). The densest part of the flow hugs the ground, but above it climbs a towering dark cloud of fine dust particles, which may form a wall two or three kilometres high. These dynamic masses of hot gas and incandescent solid particles present formidable problems to volcanologists (Chapter 12).

An American sailing ship, the *Roraima* arrived at St Pierre early in the morning of Thursday 8 May, anchoring a little way offshore. She could not have come at a worse time; a few hours after she arrived, she was a helpless burning hulk, with most of her passengers and crew dead. Those few who survived reported that they had noticed that the volcano was active even before dropping anchor, many of them had come up on deck to see the spectacle. At about 7.45 a.m. (ship time), they heard a major explosion and a few minutes later a tremendous gust of searing hot gas roared over St Pierre and the ships lying offshore,

Fig. 4.21 A photograph that has not been excelled in almost a century of volcanological study—a *nuée ardente* from Mt. Pelée erupted on 16 December 1902, taken by Albert Lacroix. Following the valley of the Rivière Blanche, the wall of cloud reaches 4000 metres high, while the toe of the flow has just reached the sea.

capsizing the steamship *Grappler*, and rolling the *Roraima* so severely that she lost her masts and smoke-stacks. Everyone on deck was killed by the fiery blast, with the exception of the captain, who lived on for a short time before becoming unconscious and falling overboard. Boiling hot mud and ashes rained down through skylights blown in by the blast, killing many of those below decks, and scalding others.

St Pierre was torn apart by the short-lived *nuée ardente*. Thousands of the casks of rum stored in the town exploded in the heat and caught fire immediately. Burning rum ran down the streets and even into the sea, spreading out as far as the *Roraima* and causing small fires on her. For many hours the conflagration raged through the town, spreading rapidly through buildings wrecked by the blast. Not a single person was left alive to check it (Fig. 4.22 a and b).

4.3.2 *Post-mortem on St Pierre*

One of the first to go ashore at St Pierre after the disaster was the Vicar-General of Martinique, Monsieur Parel, who was at Fort-de-France on the morning of the eighth, having left St Pierre only the afternoon before. He joined a rescue

(a)

(b)

Fig. 4.22 (a) and (b). Two views of St Pierre, taken by S. Poyer a week after its destruction. (a) looks south along the water-front. (b) looks north, towards Mt. Pelée, although the volcano is not visible in the pall of smoke from smouldering fires. Despite the widespread damage, little volcanic ash is evident.

party which set off by boat from Fort-de-France, returning to St Pierre a scant twenty-four hours after he had left. Let him describe the scene in his own words;

Thursday 8 May. Ascension Day. This date should be written in blood . . . When, at about 3 o'clock in the afternoon we rounded the last promontory which separated us from what was once the magnificent panorama of St Pierre, we suddenly perceived at the opposite extremity of the roadstead the Rivière Blanche with its crest of vapour, rushing madly into the sea. Then a little further out blazes a great American packet [the *Roraima*], which arrived on the scene just in time to be overwhelmed by the catastrophe [Fig. 4.23]. Nearer the shore, two other ships are in flames. The coast is strewn with wreckage, with the keels of the overturned boats, all that remains of the twenty to thirty ships which lay at anchor here the day before. All along the quays, for a distance of 200 metres, piles of lumber are burning. Here and there around the city . . . fires can be seen through the smoke. But St Pierre, in the morning throbbing with life, thronged with people, is no more. Its ruins stretch before us, wrapped in their shroud of smoke and ashes, gloomy and silent, a city of the dead. Our eyes seek out the inhabitants fleeing distracted, or returning to look for the dead. Nothing to be seen. No living soul appears in this desert of desolation, encompassed by appalling silence.

It is not clear exactly how many townspeople survived the passage of the *nuée*. It was probably as few as two. The blast that struck St Pierre was remarkable for its sheer force—masonry walls one metre thick were blown down, heavy cannon

Fig. 4.23 The burning hulk of the *Roraima*. Her masts were snapped by violent rolling when the *nuée* engulfed her.

torn from their mounts, and a three-ton statue carried sixteen metres. Angelo Heilprin, an American geologist who visited the scene and wrote a book about the tragedy, described: 'twisted bars of iron, great masses of roof sheeting wrapped like cloth about posts upon which they had been flung, and iron girders looped and festooned as if they had been made of rope'.

Amidst all this evidence of the effects of unlimited destructive power unleashed, there were small pockets in which unexpected things survived—delicate china cups, corked bottles of water, still drinkable, little packets of starch in which the granules were untouched; even a street fountain still splashed cold drinking water in one of the ruined streets. Because the physical damage was so variable in its extent, it was obvious to the early investigators that the almost total mortality was not *solely* the result of the force of the blast, since there should have been survivors in areas where the blast was less severe—there were poignant scenes of families dead in rooms where glass bottles, jugs, bowls, and cutlery stood undamaged on tables where they had been set out (Fig. 4.24).

Evidence that the hot dust of the blast was the cause of death was abundant. Thousands of bodies in the ruins all told the same story, of practically instantaneous death when the scorching hurricane from Mt. Pelée reached them. There had been no time to flee, or even to struggle; hundreds simply died in their tracks. The hot gas did its work swiftly, extinguishing thousands of lives in the space of two or three minutes. Many victims were badly burned after death by the fires that swept the town, but many of those that died in the open showed severe burns on their bodies, even though their clothes were not singed. Although the blast was intensely hot, it can only have lasted a few moments, not long enough to ignite fabrics.

Two survivors told stories which confirm this interpretation. Most famous is the tale of the 'prisoner of St Pierre', a stevedore named Augustus Ciparis, who was incarcerated in the town jail at the time of the disaster. Fortunately for him, his cell was extremely secure, partly below ground level, without even a window—its only aperture was a tiny grating above the door.

Fig. 4.24 A would-be rescuer surveys the poignant scene in a house in St Pierre, where a baby, a small girl, and their mother lie dead, while bottles and jugs remain unscathed on a table.

Ciparis remained locked up in his cell for four days after St Pierre had been laid waste, without food, half-dead from burns and shock, until his cries for help were heard by two islanders picking through the ruins of the town. When he had recovered, Ciparis was able to describe what had happened—and he went on telling the story for the rest of his life, for he joined a travelling circus, and became something of a celebrity.

On the morning of 8 May, Ciparis had been waiting for his breakfast to be brought to him, when it suddenly grew dark. Immediately afterwards, hot air laden with dust began to come through the grating over the door. It was not a strong gust, but was fiercely hot, burning him severely all over his back and legs. He was wearing a shirt and trousers at the time; these were unmarked. The heat did not last for more than a few seconds, and when it had passed, Ciparis was left in an awful solitude, his cries for help going unanswered as the city burned above him. Had he not had a bowl of water in his cell, he too, would have succumbed before his rescuers found him.

So much for the death of St Pierre. Destruction of the city was not the last episode in the eruption, though, and the lethal *nuée* was by no means the last from Mt. Pelée. On 20 May, a second powerful *nuée* swept through St Pierre,

Fig. 4.25 Mt. Pelée's towering spine, extruded 300 metres above the summit lava dome, photographed in June 1903.

flattening many of the ruins left by its predecessor, but taking no lives—there were none left to take. Several other *nuées* swept harmlessly down the valley of the Rivière Blanche during the next couple of months. On 30 August, death came again to Martinique when a powerful *nuée* blasted out from Mt. Pelée and rolled down in a new direction, engulfing the small village of Morne Rouge. Two thousand people died, in circumstances almost identical to those of St Pierre. After this final, fatal episode, the eruption dragged on for many months, lasting well into 1903. Ash which had been sifting down softly on St Pierre since the beginning of May 1902

covered the ruins in a grey pall, burying many of the bodies that still lay in the debris. One should not attribute human qualities to natural phenomena, but the gentleness with which Mt. Pelée buried its victims seemed like an atonement for their awful deaths. A stranger gesture was to follow. In November 1902, a great spine of solidified lava began to rise above the crater of the Étang Sec, forced upwards by the pressure of the magma below. Growing at about ten metres a day, by May 1903 the spine was no less than 310 metres high, rearing above St Pierre like an obelisk, a memorial to the thousands who had died below (Fig. 4.25).

— 4.4 Mt. St Helens, May 1980

At 08.32 a.m. on 18 May 1980 two geologists, Keith and Dorothy Stoffel, were circling the summit of Mt. St Helens volcano in a light aircraft, photographing the snow-covered, inactive crater when, to their horror, the northern side of the mountain began to move. It slid downwards *en masse* initially, then metamorphosed into a huge avalanche. Seconds later, a massive cloud of ash shot upwards, climbing thousands of metres into the air. Thus began the paroxysmal phase of the eruption of Mt. St Helens in Washington State, USA, by far the best-documented event in the annals of volcanology.

Although its violence took scientists by surprise, the outburst was not unexpected. Two years earlier, Dwight Crandell and Donald Mullineaux, volcanologists attached to the United States Geological Survey, had written a report pointing out that Mt. St Helens had the most extensive record of recent activity of any volcano in the continental USA. They predicted that an eruption might take place before the end of this century, and presented a map detailing the areas at risk from mud-flows and ash falls.[14] At 15.47 Pacific Standard Time on 21 March 1980, the value of their meticulous work was borne out when a magnitude 4.2 earthquake took place beneath the volcano, first evidence that it was reawakening.

Seismic activity rapidly increased in frequency,

so plans for evacuation of the area were prepared On 27 March, the first eruption of ash took place. Similar small eruptions soon became commonplace, continuing into April. A characteristic pattern of seismic activity known as 'harmonic tremor' set in, indicating that magma was probably ascending beneath the volcano. Seismic activity slowly waned through the month of April, but by the end of April ground surveys had shown that a large topographic bulge, almost 2 km in diameter, was distorting the northern flanks of the volcano, ballooning upwards and outwards at rates up to 1 metre per day, swelling so obviously that day-to-day changes were perceptible to the eye. By 12 May, the highest point on the bulge was 150 metres above the pre-existing topography. Even discounting volcanic activity, the size of this bulge, and the rate at which it was swelling, presented a clear threat, since it was obvious that it would eventually become mechanically unstable and break away in a large avalanche (Fig. 4.26).

4.4.1 Avalanche

On 18 May the avalanche occurred, initiating a disastrous chain of events. Although the immediate trigger of the avalanche was a magnitude 5 earthquake, the underlying cause was an ascending mass of hot magma a couple of kilometres beneath the volcano. Slumping away of a huge sector of the northern part of the volcano

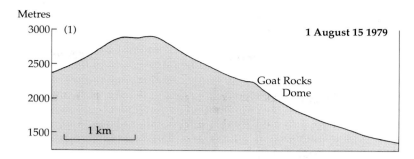

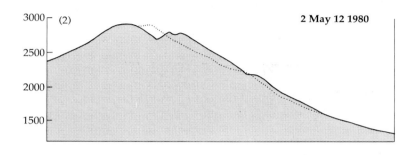

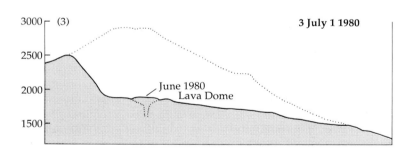

Fig. 4.26 North–south topographic profiles through Mt. St Helens (1) prior to the initiation of activity; (2) showing the growth of the bulge on the north flank by 12 May 1980 (dotted line is original profile); and (3) after the catastrophic collapse. Goat Rocks dome was a conspicuous landmark on the original volcano. After J. G. Moore and W. C. Albee in *The 1980 eruptions of Mt. St Helens, Washington.* (ed. Lipman, P. W. and Mullineaux, D. R. (1981), *US Geol. Surv. Prof. Paper* **1250**, 843 pp.

depressurized the hot rocks in the core of the volcano, unleashing a violently explosive outburst (Fig. 4.27). The eruption, therefore, was a combination of a huge 'ordinary' avalanche and a series of devastating volcanic side-effects. This lethal combination reveals the complexity of volcanic phenomena, and the difficulties of trying to interpret them. Here we follow events in a broadly chronological framework. A comprehensive volume compiled by the US Geological Survey provides a definitive account of the eruption.[15]

In terms of volume, the debris avalanche was by far the largest product of the eruption. A mixture of rock, glacier ice and soil, it swept down the northern flanks of the volcano, rushed into both arms of Spirit Lake (Fig. 4.28), climbed more than a hundred metres up an opposing slope, and surged down the valley of the Toutle River, coming to rest ultimately east of the Camp Baker logging base, more than 20 kilometres from where it entered the valley (Fig. 4.29). Previously, the upper Toutle Valley had been a tranquil, wooded area, much frequented by campers and fishermen; afterwards the valley was filled to a depth of 100 metres by a grim, grey mass with a peculiar hummocky topography produced by compaction of the debris around

Fig. 4.27 Mt. St Helens ten days after the eruption, showing the massive amphitheatre left by the failure of the northern flank. While most of the volume of the original edifice was removed by avalanche, some was reamed out by the later plinian eruption.

larger boulders. Countless small crater-like holes (kettle holes) were caused by melting of lumps of glacier ice. No survivors lived to witness the passage of the avalanche, but estimates based upon the momentum required to climb up opposing slopes suggest that it travelled with a velocity of the order of 75 metres per second.

When the avalanche surged into the Toutle Valley and Spirit Lake, it dammed the river and displaced huge volumes of water from the lake. Catastrophic mud-flows cascaded down the Toutle River, into the Cowlitz River, and ultimately into the Columbia River. Smaller mud-flows swept down the south fork of the Toutle River, and Smith Creek on the eastern flanks of the mountain. In terms of property damage, mud-flows were the most destructive effects of the eruption. Houses, bridges, and roads in the Toutle

Valley were swept away; thousands of mature trees lining the river banks were snapped like twigs and carried downstream, accumulating in a giant log jam in the Cowlitz River, 60 kilometres from the volcano. Huge volumes of sediment were carried into the Columbia River. Dredging a new navigation channel in the river was one of the most costly consequences of the eruption.

Unlike the debris flow, the mud-flows were low-viscosity torrents of sediment and rocks mobilized by water. Although their speed and momentum ensured that they were powerfully destructive, they fortunately caused little loss of life. Two campers who had set up camp only ten metres from the Toutle River were swept away in a mêlée of mud, water, and logs, but survived their ordeal with only bruises and shock to show for it.

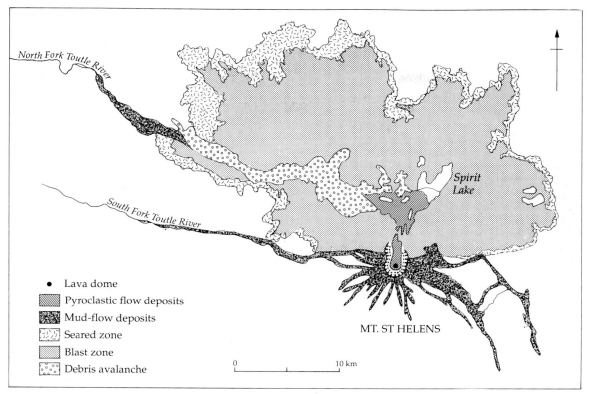

North Fork Toutle River

South Fork Toutle River

Spirit Lake

- Lava dome
- Pyroclastic flow deposits
- Mud-flow deposits
- Seared zone
- Blast zone
- Debris avalanche

MT. ST HELENS

0   10 km

Fig. 4.28 Outline map showing the distribution of the main components of the 18 May 1980 eruption of Mt. St Helens.

4.4.2 Blast

Almost all the fatalities in the eruption were caused by the explosive blast which was detonated moments after the main landslide had started moving. Here the chronology becomes a little complex: the blast of hot gases arrived *ahead* of the debris flow at some points along the Toutle River, where the avalanche appears to have overridden areas of forest *already* knocked flat by the blast. Initially, the blast had a velocity of 90–100 metres per second. It probably accelerated to supersonic velocities (greater than 300 metres per second) as it expanded in the vent area. Since it was travelling so much faster than the avalanche, the blast overtook the avalanche before the latter had travelled more than a kilometre or two.

Millions of trees, many of them full-grown Douglas firs, were flattened by the blast over an area of nearly 600 square kilometres. Most were simply uprooted and stripped as bare as telegraph poles. Some, partially sheltered behind ridges, were snapped like match-sticks. Since wind speeds of as little as 20 metres per second (80 kilometres per hour) can blow trees down, it is scarcely surprising that the blast levelled the forest so comprehensively. After helicoptering over the endless grey, ash-covered wastes of fallen trees, lying oriented on the ground as though a giant comb had been drawn through them, it was difficult even for volcanologists not to feel overawed by the scale of the destruction (Figs. 4.30–4.31).

Compared with the debris avalanche and mud-flow, the crucial feature of the Mt. St Helens blast is that it involved *hot* rocks from the core of the volcano. For the most part, the deposit left by the blast which did so much damage formed a thin layer, less than half a metre thick even at points only a few kilometres from the volcano. It

Fig. 4.29 Oblique aerial photograph taken 4 June 1980, looking westwards along the Toutle River, showing the flat-topped debris avalanche deposit filling the pre-existing valley. The forest in the blast zone has been completely laid flat (right foreground). Avalanche deposit has already been deeply incised by the renascent Toutle River (distance). United States Geological Survey photo.

was much thicker in the floors of valleys, where material unable to stick to the steep valley sides flowed downhill and ponded on the valley floors.

Two weeks after the eruption, the ponded material was still warm to the touch, but it had never been hot enough to carbonize the many fragments of wood within it. In appearance, the deposit resembled the ashes left by a coal fire which has just burned out: many solid lumps in a mass of fluffy, pale-grey ash (Fig. 4.32). Where streams had cut through this deposit, their banks were unstable, and minor collapses were frequent, liberating highly mobile block-and-ash flows with uncannily low viscosity. Although they came from the core of the volcano and were erupted hot, the rocks in the blast deposits were not actually new magma. They must have been already *solid* when they were erupted, because

Fig. 4.30 In the blast zone ten days after the eruption. At this site, mature forest trees were partially screened from the blast by the ridge in the background; they therefore remained rooted in the ground, but their trunks snapped. On the unprotected side of the ridge, trees were uprooted and carried away.

Fig. 4.31 Chilling evidence of the violence of the blast. This centimetre-sized clast was embedded like a bullet in the trunk of a tree.

they did not contain the gas bubbles characteristic of rocks like pumice, produced when an expanding froth of magma is abruptly torn apart by explosions, and the fragments immediately chilled.

Areas affected by the searing blast were sharply defined. Within them, *all* trees were laid flat, either snapped off at ground level or uprooted. At the edge of the blast zone, there was a narrow transitional zone some hundred metres wide in which most trees were standing but had lost their branches; beyond this was a further

Fig. 4.32 A 2–3 metre thick section through the blast deposit ponded in the Coldwater valley. Steam rises lower downstream where hot material is chilled by water. Light-toned warm, dry blast deposit is overlain by a thin layer of wet, grey air-fall deposit.

narrow zone in which trees retained all their foliage, but were scorched brown by the heat of the blast.

About 60 people were killed by the blast. It is not certain how many adventure-seekers had ignored instructions not to enter a 'red zone' around the volcano set up by the local authorities, so the number of fatalities is not known exactly. Studies of the victims' vehicles provide clues to conditions within the blast. In a truck only 6 km from the volcano, temperatures within the cab were high enough for plastic objects to soften slightly, but the effects were only transient—the vehicle's fuel tank did not catch fire. At this location, the blast brought down trees, but was not sufficient to flip over the vehicle. Airborne debris severely battered the side of the vehicle facing the mountain. Three people died in this vehicle, probably killed by flying wood splinters—slivers of wood were driven completely through the seat backs (Fig. 4.33). In another vehicle 12 kilometres north-west of the volcano, the temperature was not hot enough to soften plastics, and the vehicle did not show evidence of heavy battering, probably because it was parked high up on a steep slope overlooking the volcano, out of reach of material picked up by the blast. A *National Geographic* photographer died beneath a metre of ash in this vehicle (Fig. 4.34).

Autopsies carried out on 25 of Mt. St Helens's victims supplied the first modern data on the causes of death in volcanic disasters, and provided insights into the fates of earlier victims, such as those of St Pierre in 1902. Apart from 'trauma' injuries caused by flying rocks or falling trees, two major causes of death predominated: burns affecting large areas of the body, and lung damage resulting from inhalation of hot dust. Some victims died rapidly of simple asphyxiation when the fine dust of the blast cloud clogged their bronchial passages; others who survived short-term asphyxiation died later from acute respiratory distress syndrome, caused by severely damaged lungs and consequent bacterial infection. As the scant two survivors in St Pierre demonstrated, the ratio of fatalities to survivable injuries in volcanic disasters is unfortunately much higher than in other natural catastrophes. Volcanoes leave few walking wounded: a sharp dividing line separates the quick and the dead. Numerous eyewitnesses close to the Mt. St Helens blast zone escaped entirely unscathed, while many of victims rescued alive from within the blast zone perished soon after.[16]

4.4.3 *Plinian eruption column and pyroclastic flows*

All the events described so far took place within the space of a minute or two after the beginning of the eruption at 8.32 a.m. Evisceration of the

Fig. 4.33 Pick-up truck located on margin of the blast zone, 6 km west of the volcano. Flying debris killed the occupants of this truck; only a few hundred metres away, forest trees remained green and unscathed.

Fig. 4.34 Photographer's car, buried more than 1 metre deep in blast and ash-fall deposit. Car was parked on a ridge overlooking the Toutle Valley and Mt. St Helens, obscured by cloud in background 12 km distant. The avalanche deposit is visible behind helicopter; it was channelled along Toutle Valley. The driver did not survive.

volcano by the great avalanche tapped young, fresh magma deep within its core, and moments after the *horizontal* blast, huge quantities of pumice were erupted to form the immense *vertical* plinian eruption column that figured on photographs on the front pages of newspapers all over the world. In less than 10 minutes, the column had climbed to a height of more than 20 km, and persisted for much of the day (Fig. 2.5). Unseen during all the colossal activity, numerous pumiceous pyroclastic flows swept down over the northern flank of the volcano, forming a broad apron known as the 'pumice plain'. Like *nuées ardentes*, pumice flows travel as masses of solid particles in a suspending medium of hot gas; the principal difference is that they

consist mostly of pumice, rather than dense rock (Chapter 10).

A thickness of more than 40 metres of pumice flows accumulated in the area of the Upper Toutle valley, smoothing out the hummocky surface of the pre-existing debris flow into a carpet of yellowish pumice fragments. These deposits were the hottest of all those erupted by Mt. St Helens. Two weeks after the eruption, they were cool enough to allow geologists to walk over them with impunity, but two metres below the surface, temperatures of over 300°C were recorded. At the time of their eruption, the temperature of the flows was probably close to that of the original magma body itself, about 700°C.

Because the pumice flows were hot when they were erupted, some dramatic interactions resulted when they spread over wet ground in the Toutle Valley—dozens of small steam plumes hissed up from the floor of the valley, where steam driven off from wet materials beneath made its way to the surface. A similar, but much bigger eruption, of Mt. Katmai in Alaska in 1912 led to the whole of the Ukak River being flooded with searing hot pumice flow deposits. So many steam plumes were generated there that the first observers to enter the valley called it 'The Valley of Ten Thousand Smokes'. A wry American geologist christened the Toutle Valley 'The Valley of a Thousand Smokes'. Like its larger counterpart, this spectacle faded as the pumice flows cooled.

Around the shores of Spirit Lake, interactions between the pumice flows and water were so vigorous that *secondary* eruptions took place, sometimes so violently that ash was sprayed hundreds of metres into the air. These secondary eruptions provided an instructive demonstration of the mechanism by which pumice flows are generated. Sprays of ash fountained steadily into the air. From time to time, the plumes of ash became too heavy for the escaping steam to support, and they collapsed, surging away from the base of the eruption column, splashing and spreading outwards like a low-viscosity liquid (Fig. 4.35). As soon as the spreading flows lost a critical proportion of their fluidizing steam, they came to an abrupt halt, leaving a bone-dry layer

Fig. 4.35 Vigorous secondary eruptions continuing in pumice flow deposits near the shores of Spirit Lake, ten days after the eruption. Small secondary pyroclastic flows are rolling away radially from the jet-like plume at the centre.

of dusty particles, quite unlike any liquid (Chapter 10).

4.4.4 Plinian ash cloud

Although it was insubstantial compared with the other products of the eruption, the ash cloud erupted by Mt. St Helens had the most far-flung effects. Thousands of eyewitnesses described the extreme rapidity with which the dark cloud boiled upwards into the atmosphere and the visceral fear that they felt when the cloud rolled overhead, cutting out the sunlight and turning day into night. Over a centimetre of ash accumulated at Yakima, 120 km downwind from the volcano. Weather radar and satellite images provided synoptic views of the ash cloud, so that its track over the USA could be monitored. Flights by U2 aircraft showed that at an altitude of about 17 km, the cloud consisted mostly of ash particles, but that at about 20 km it consisted largely of an aerosol of sulphate particles, produced by the condensation of sulphurous gases erupted by the volcano.

Its effects were so widespread that the gigantic ash cloud naturally grabbed more attention from the news media than more crucial aspects of the eruption. Although ash fell as far east as the Great Plains of the United States, the total volume of air-fall ash was rather small; less than half a metre thickness accumulated on the flanks of the volcano itself. During the eruption of Vesuvius in AD 79, more than a metre of ash fell on Pompeii, 10 km distant from the volcano. Thus, while misery was widespread and vehicles and crops ruined over large parts of Washington State, things could have been much, much worse.

Activity did not cease on 18 May. On 25 May and again on 12 June, renewed eruptions sent ash clouds rearing into the sky once more. On 25 May, wind carried the ash north-westwards, depositing over a centimetre of ash at Chehalis, 80 km distant. On 12 June, ash was carried southwards and fell on the city of Portland, the largest urban area to be affected by any of the three ash falls.

4.4.5 After 18 May

Mt. St Helens's topography was drastically changed by the 18 May eruption. It was reduced from a serene, 2949 m high cone to a raw, gaping amphitheatre 1.6 km in diameter, open on its northern side, with the highest point on its rim at 2560 m, and the floor at 1800 m (Fig. 4.27). Most of the 'missing' volume ended up in the 2.8 cubic kilometre debris avalanche deposit in the Toutle Valley. Since the eruption, the crater has been under constant surveillance. In mid-June, 1980 a plug of viscous lava began to be extruded through the floor of the crater. This was the first time during the eruption than any actual lava, as opposed to ash, had been observed. Lava continued to be erupted quietly over a period of

Fig. 4.36 Mt. St Helens amphitheatre and lava dome in 1987. Some fume is rising from the growing dome, which is several hundred metres high. United States Geological Survey/Cascades Volcano Observatory photograph 9704008.

many months, punctuated by occasional minor explosions. A sluggish mass eventually piled up to form a lava dome hundreds of metres high within the confines of the amphitheatre (Fig. 4.36). Such domes are characteristic of eruptions of this sort, where the magma is silicic and viscous. A similar dome grew after the eruption of Bezimianny (Kamchatka) in 1956, where half the side of the volcano was also demolished. Lava domes grow slowly and unspectacularly, but they are none the less potentially dangerous. If a dome grows so steep and high that it is mechanically unstable, it may collapse, sending down an avalanche of hot rock in a lethal *nuée*-like flow over the country beneath. Several fatalities have resulted from *nuées* produced by collapses of the Santiaguito dome, in Guatemala.

Guatemala may seem a long way from Mt. St Helens, and not directly relevant. An important lesson to be learned, though, is one of timing. The Santiaguito dome did not begin growing until 1922, *20 years* after the eruption. It was still growing in 1993. Although the Mt. St Helens dome has been inactive for some years at the time of writing, it will require careful watching for decades to come.

— 4.5 Notes

1. Bond, A. and Sparks, R. S. J. (1976). The Minoan eruption of Santorini, Greece. *J. Geol. Soc. Lond.* **132**, 1–16.
2. Sheets, P. D. and Grayson, D. K. (eds.) (1979). *Volcanic activity and human ecology.* Academic Press, New York.
3. Sigurdsson, H., Carey, S., Cornell, W., and Pescatore, T. (1985). The eruption of Vesuvius in AD 79. *Nat. Geog. Res.* **1**, 332–87.
4. Sigurdsson, H., Cashdollar, S., and Sparks, R. S. J. (1982). The eruption of Vesuvius in AD 79: reconstruction from historical and volcanological evidence. *Am. J. Archaeology* **86**, 39–51.
5. Verbeek, R. D. M. (1895). *Krakatau* [in Dutch] Batavia, 495 pp., 1895; [in French] 1896.
6. Symons, G. J. (ed.) (1888). *The eruption of Krakatau, and subsequent phenomena*, Report of the Krakatau Committee of the Royal Society, Trübner, London, 494 pp.
7. Simkin, T. and Fiske, R. S. (1983). *Krakatau 1883: the volcanic eruption and its effects.* Smithsonian Institution, Washington, DC, 464 pp.
8. Self, S. and Rampino, M. R. (1981) The 1883 eruption of Krakatau. *Nature* **294**, 699–706.
9. Francis, P. W. and Self, S. (1983). The eruption of Krakatau. *Scientific American* **249**, 172–87, 1983.

10. Francis, P. W. (1985). The origin of the Krakatau tsunamis. *J. Volcanol. Geotherm. Res.* **25**, 349–64.

11. Sigurdsson, H., Carey, S., Mandeville, C., and Bronto, S. (1991). Pyroclastic flows of the 1883 Krakatau eruption. *Eos* **72**, 377–81.

12. Sigurdsson, H., Carey, S., and Mandeville, C. (1991). Krakatau. *Nat. Geog. Res.* **7**, 310–27.

13. Chretien, S. and Brousse, R. (1989). Events preceding the great eruption of 8 May, 1902 at Mount Pelée, Martinique. *J. Volcanol. Geotherm. Res.* **38**, 67–75.

14. Crandell, D. R. and Mullineaux, D. R. (1978). Potential hazards from future eruptions of Mt. St. Helens volcano, Washington. *US Geol. Surv. Bull.* **1383-C**, 26 pp.

15. Lipman, P. W. and Mullineaux, D. R. (eds.) (1981). The 1980 eruptions of Mt. St. Helens, Washington. *US Geol. Surv. Prof. Pap.* **1250**, 843 pp.

16. Baxter, P. J. (1990). Medical effects of volcanic eruptions (1990). *Bull. Volcanol.* **52**, 532–44.

5

Magma—the hot stuff

Magma is the volcanologist's raw material, the molten material that is ultimately erupted at the surface in lava flows or pyroclastic eruption columns. *Magma* is not synonymous with *lava*. Magma is an elusive term, difficult to define succinctly, but it is best regarded as fresh, mostly molten rock, still vitalized with the volatiles that it acquired in its source region.

Many changes affect molten rock during its transformation from subterranean magma to surface effusion. Three components, or phases, are usually present in magma. First, a viscous silicate melt; second, a variable proportion of crystals; and third, a volatile or gas phase. Each of these phases influences the way in which the

magma erupts at the surface. When subjected to subtly different eruption mechanisms, a single magma may give rise to startlingly different eruption products. There is a world of difference between obsidian; massy rhyolitic glass that shatters at a blow into razor-sharp splinters, and pumice; rock so light and frothy that it floats on water. Both, however, may be derived from the same magma, and both may be erupted from a single volcano at the same time. Obsidian is erupted quietly in sluggish lava flows, whereas pumice is formed when expansion of gas within the magma causes it to foam into a consistency resembling expanded polystyrene.

5.1 The melt

This is the most complex component. Some of the physics of magmas is still uncertain—it is impossible after all to get into a magma chamber with experimental instruments. Molten silicate rock is physically different from the liquid water that results when ice is melted. Pure water consists only of simple molecules of H_2O, and therefore ices melts (and freezes) at atmospheric pressure at a single temperature: a temperature that is so sharply defined that it forms the starting point of the Celsius temperature scale: 0°C. A molten magma, by contrast, is chemically complex, consisting of silicate molecules in which a

wide range of elements is combined. This complexity has two consequences for volcanology: the melt does not consist of free molecules, but is *polymerized*, and it does not have a single, clear-cut freezing point.

Polymerization describes the way that molecules cluster to form larger complexes by repeated linking of the same molecular groups, retaining an underlying chemical identity. It is a common phenomenon—at one time it was even thought that water molecules might polymerize. An everyday example of a polymer is polyethylene, composed of $CH_2{=}CH_2$ molecules linked to form endless chains with the general structure [$-CH_2-CH_2-$]n, where n is a large number. There are countless other examples of organic polymers, which collectively have transformed daily life—where would we be without all those plastic bags? Although less familiar, silicate polymers are similar. They are important in volcanology because they affect the viscosity of a melt, and hence the way in which it erupts. Polymerization in silicate magmas is due to the strong bonds that exist between the silicon and oxygen atoms which form networks of inter-linked tetrahedra. Silicic magmas contain more silicate tetrahedra, and so are more highly polymerized and therefore more viscous than basaltic ones.

To reduce the viscosity of a silicate melt, the silicon–oxygen bonds must be broken. One way of doing this is to add water, which, by forming OH^- ions, breaks the bonds and causes depoly-merization. Alkaline silicic rocks are also less polymerized than others, so whereas ordinary rhyolites are extremely viscous, their alkaline counterparts may have remarkably low viscos-ities. Conversely, adding carbon dioxide to a melt may help polymerization, and therefore increase its viscosity.

Three factors influence the melting (and freezing) temperature of silicate materials: com-position, pressure, and volatile content. Moving from basaltic compositions towards silicic, there is a marked drop in melting temperature. At high pressures, a silicate of fixed composition will melt at higher temperatures than at lower pressures. And a 'wet' silicate (containing lots of volatiles) will melt (and freeze) at lower temperatures than a dry one. All these variables, coupled with polymerization, mean that when a rock is heated, it does not begin to melt at a fixed temperature and carry on melting at the *same* temperature, like ice, until it is all molten. Rather, for a given set of conditions (composition, pressure, and volatile content), it will begin to soften at a certain temperature, and become progressively more molten as temperature increases until at some higher fixed point it is entirely molten. These temperatures increase with pressure, so on a temperature–pressure graph, two lines may be plotted. One marks the beginning of melting (the

solidus), and the other marks the end (the *liquidus*). Exactly the same applies in reverse when a melt cools (Fig. 5.1).

Fortunately, volcanologists need only con-sider melts at or near the surface, so pressure is not a major variable, and diagrams such as Fig. 5.1 can be safely left to petrologists. For most of the discussions that follow, we can assume that the eruption temperatures of com-mon lava compositions are as shown in Table 5.1.

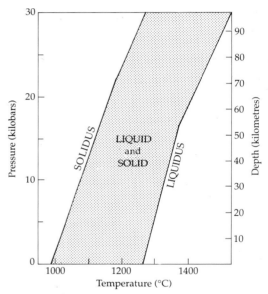

Fig. 5.1 Melting relationships for dry basalt. At temperatures and pressures left of the 'solidus' line, basalt is entirely crystalline, solid rock; within the shaded field it is partially melted, and to the right of the 'liquidus' line it is entirely molten. If water is present, the shapes of the boundaries are markedly different, and melting commences at lower temperatures.

Table 5.1 Eruption temperatures of common volcanic rock types

Rock type	Temperature (°C)
Rhyolite	700–900
Dacite	800–1100
Andesite	950–1200
Basalt	1000–1250

— 5.2 Phenocrysts

Many magmas have begun crystallizing long before they are erupted. Thus, lavas erupted at the surface often contain abundant phenocrysts (Section 1.4). Typically, they are only millimetres across, but exceptionally they may reach several centimetres, forming handsome ornamental rocks. Phenocrystal minerals by their very nature are those that crystallize out at the highest temperatures. In a basalt they are therefore typically olivine and pyroxene, while in silicic rocks they are more commonly feldspar. In basaltic lavas, phenocrysts are often sparse, forming only a few per cent of the total volume. Sometimes, however, so much crystallization takes place that olivine crystals sink downwards through the melt, accumulating at lower levels to form rocks crammed with olivine, known as *picrites*. These lavas, of course, have compositions very different from the original liquids that they crystallized from. (Some of the meteorites from Mars, the nahklites, may have formed by cumulation of crystals in an originally basaltic lava.) In silicic lavas, phenocrysts are often abundant—in dacites they sometimes form more

than 40 per cent of the total volume. In such cases, of course, the magma is not a liquid melt in a meaningful sense, but a pasty mush of crystals.

Study of phenocrysts provides useful volcanological insights. Phenocrysts begin crystallization before eruption, and so they may have complex histories. Plagioclase feldspar phenocrysts often exhibit spectacular compositional zoning. Zoning is described as *normal* when the composition changes from more calcic to more sodic towards the edge of the crystal, *reverse* when the opposite is observed, and *oscillatory* when the composition varies erratically from one to the other. These mineralogical variations can be used to track the evolution of physical conditions within the magma chamber, showing that pressures have varied abruptly over short periods of time, perhaps in response to surface eruptions. Other phenocrysts have mineral compositions at variance with the composition of the lava containing them, suggesting that magmas of different compositions must have mixed together. Phenocrysts derived from an alien source are termed *xenocrysts*.

— 5.3 Volatiles

Wafts of pungent fumes are sure signs that one is approaching a volcanic vent. Sadly, the acrid, lachrymatory gases discourage one from lingering in interesting places, just as they do in street riots. Even if it is not actually erupting, a volcano may release thousands of tonnes of sulphur dioxide every day, far exceeding the worst individual industrial sources of pollution such as copper smelters. Obnoxious smells apart, gases play a dominant role in the eruption of magmas. Determination of the amounts and compositions of gases present in a magma is tricky, not only because of the physical difficulty of sampling the hot raw material but because some gases which are stable within the magma react chemically the moment that they are exposed to the air.

Sulphur dioxide (SO_2) is the most easily recognized volcanic gas, familiar to rioters as tear-gas, but steam and carbon dioxide are

more abundant. Hydrogen, hydrogen chloride, hydrogen fluoride, hydrogen sulphide, carbon monoxide, and several other gases have also been detected. In alkaline lavas, chlorine and fluorine may be abundant. Carbon dioxide is the dominant volatile species in the weird carbonatite lavas of East Africa.

Three different techniques are used to measure gas compositions and abundances. Some scientific knights in armour, panoplied in glittering aluminium, joust with incandescent magma, using long tubes as lances to collect gas samples in evacuated flasks (Fig. 5.2). Although it is fiendishly difficult to obtain samples uncontaminated by atmospheric air this way, many useful data have been assembled. A limitation of the technique is that, by definition, only small, easily approached sources can be sampled, such as chimneys near the source of a basaltic lava. Gas

Fig. 5.2 Aluminium-armoured volcanologist sampling active lava flow on Mt. Etna. The difficulties and hazards of this technique speak for themselves.

data from these sources may not be representative of the more voluminous gas stream released via primary vents.

Second, small glassy blobs of the original melt are often trapped within phenocrysts crystallizing from a magma. Analyses of the volatiles contained in these glasses provide a way of estimating the volatile content of the original magma, but again it is difficult to be certain how representative they are. Finally, at the other end of the size spectrum, concentrations of some gases in an eruption plume can be measured by remote sensing techniques, using instruments such as the Correlation Spectrometer (COSPEC).[1] The COSPEC relies on measuring the absorption of sunlight by sulphur dioxide, using clear sky as a reference standard, and it can be located on the ground, or flown in aircraft below the plume. Absorption is proportional to gas concentration; thus once the concentration has been measured, it is easy to measure release rates from the plume velocity. Although this technique works well for sulphur dioxide, it is much more difficult to measure other gases such as carbon dioxide and water, because these are so abundant normally in the atmosphere.

There is a marked range in volatile content from basalts to rhyolites. Typical ocean ridge basalts often contains less than 0.5 per cent water, whereas a rhyolite may contain 4 to 5 per cent. Just as the carbon dioxide used to give Coke its fizz is more soluble at high pressure than low, so magmatic gases are more soluble at high pressures than low. Thus, gases exsolve from a magma as it is depressurized near the surface, with explosive results. Gases are also more soluble at low temperatures than high, although this has a less marked effect in terms of eruptive behaviour.

Given that a basalt magma may contain less then one per cent by weight of volatiles, these may appear to be unimportant accessories. But the molecular weights of the volatiles such as water are so low relative to the molecular weights of silicate magma components that there is actually a disproportionately large number of volatile *molecules* in the magma. These numerous molecules make their presence felt in a number of ways, out of all proportion to their small mass fraction.[2]

Volatiles diffuse relatively easily through magmas and are of low density; therefore they congregate towards the top of magma chambers, forming volatile-rich caps or cupolas. These volatile-rich layers are necessarily the first to erupt, a fact which explains why the early stages of eruptions are often more violently explosive than the later stages.

— 5.4 Viscosity —

Viscosity is one of the those concepts which can be usefully bandied about conversationally, but which are mine fields of complexity when used quantitatively. In everyday terms, viscosity describes the sluggishness of a fluid; its resistance to being stirred. Notice the use of the term *fluid*, rather then liquid: the same principles apply to liquids, gases, pyroclastic flows, and aerosols. Viscosity is a crucially important parameter in many volcanic processes. It can be defined as *the internal resistance to flow by a substance when a shear stress is applied.* A simple fluid like water, flows in response to the slightest stress; it has a low viscosity. When a glass of water is tilted, the water instantly responds by flowing to find its new level. Honey is much more viscous; when a jar is tilted, it takes far longer to find the new level. In both cases, once flow has started, the rate of flow (or *strain rate*) is directly proportional to the applied stress. Such fluids are termed *Newtonian* fluids, after Sir Isaac who first described them.

More complex substances behave differently. At low stresses, they do not flow at all; they appear to be solid. Once a certain minimum stress has been exceeded, however, they begin to flow and at higher stress levels behave like Newtonian fluids, their rate of flow varying in direct proportion to the applied stress. The initial stress required to make a fluid commence flowing is its *yield strength*. Fluids which exhibit a yield strength are termed *Bingham* substances. Many fluids exhibit behaviour intermediate between Newtonian and Bingham; they do not possess a definite yield strength, and show a non-linear relationship between shear stress and strain rate. They are termed *pseudo-plastic* (Fig. 5.3). Toothpaste is an example of a complex (sometimes thixotropic) fluid; a small squeeze of it retains its shape on the brush, but a large volume deforms under its own weight. While a few exceptional basaltic lavas approach Newtonian behaviour, most have pseudo-plastic, toothpaste-like qualities. For simplicity, however, we can regard most lavas as Bingham substances.

Like many other volcanological parameters, magama viscosity is exceedingly difficult to

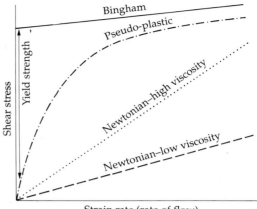

Fig. 5.3 Relationships between shear stress and strain rate (rate of flow) in the main types of fluid flow. Simple fluids like water are Newtonian; they need zero stress to make them flow. For lavas to flow, sufficient stress must be applied to overcome the yield strength. On a volcano, this stress is supplied when lava flows under gravity on a steep enough slope. (After Wolff, J. A. and Wright, J. V. (1981) Rheomorphism of welded tuffs. *J. Volcanol. Geotherm. Res.* **10**, 13–34.)

determine directly. Some measurements have been made on lavas, however, by hardy souls sticking instruments into them. Typical basaltic lavas have viscosities of between 10^2 and 10^3 Pa s (Pascal second; kg m^{-1} s^{-1}; 1 Pa s = 10 poise). To put this in context, pure water has a viscosity of about 10^{-3} Pa s, so these basalts are thousands of times more viscous than water. (Olive oil is about one hundred times more viscous than water.)

5.4.1 *Factors affecting viscosity*

Fluids become more viscous as they get colder— this is why automobile engines require different grades of oil in winter and summer. In magmas, the effect is dramatic—for rhyolitic melts, viscosity increases by more then eight orders of magnitude between 1300°C and 600°C (Fig. 5.4). Basalt shows a similar trend, although not

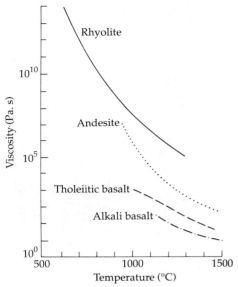

Fig. 5.4 Relationship between viscosity and temperature for the main lava types. From Murase, T. and McBirney, A. R. (1973). Properties of some common igneous rocks and their melts at high temperatures. *Geol. Soc. Amer. Bull.* **84**, 3563–92.

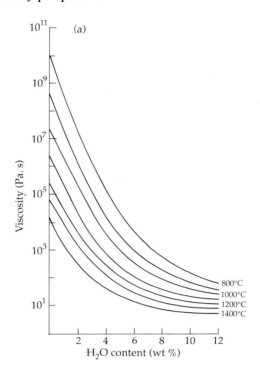

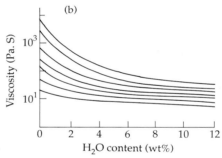

Fig. 5.5 Effect of dissolved water content on the viscosities of the main lava types. (a) is for granitic melts and (b) for basaltic melts. From Murase, T. (1962). Viscosity and related properties of volcanic rocks at 800° to 1400°C. *Hokkaido Univ. Fac. Sci. J., Ser.* **7**, **1**, 487–584.

continued as far—basalts are mostly solid below 1000°C. Fig. 5.4 makes an important additional point: although melts of all compositions become less viscous at higher temperatures, silicic melts are *always* more viscous than basaltic ones at the same temperature.[3]

Dissolved water also has an important effect on magma viscosity because of its ability, mentioned earlier, to render magma less polymerized by breaking silicon–oxygen bonds. This effect is illustrated in Fig. 5.5: at 1000°C, the viscosity of a silicic melt is decreased by many orders of magnitude when the dissolved water content is increased. Basalt shows a similar, but less marked decrease in viscosity. Again, Fig. 5.5 shows that silicic magmas are more viscous than basaltic ones at the same temperature and with the same water content.

Other phases present in a fluid also affect its viscosity. Solid materials such as phenocrysts naturally increase viscosity, but the effects are difficult to quantify because of the wide ranges in sizes and shapes of the crystals. Gas bubbles, the only other likely components, have effects which are even more difficult to untangle, since a well-vesiculated magma is a froth whose flow properties are controlled by factors such as surface tension and the thickness of bubble walls (Fig. 5.6).

Fig. 5.6 Congealed drips and dribbles of basalt lava on the side of a lava channel illustrate convincingly the low viscosity of hot basalt. Surface-tension forces play an important role in shaping these fluid forms. Flow was erupted from Kilauea, Hawaii, in 1974. Stalactites are a few centimetres in length.

— 5.5 Rheology—turbulent v. laminar flow

Much of volcanology is concerned with how large volumes of fluids erupt from holes in the ground, and then flow away afterwards. *Rheology*, the study of flowing materials, is an exceptionally complex discipline, and its mathematics, *fluid dynamics*, involves problems that remain intractable even in these days of supercomputers. So we will adroitly skirt most of these esoteric issues and explore only one: flow regime. It is a matter of common observation that a slow-flowing river does so in an orderly manner. All parts of the water mass flow restfully in the same direction at the same speed without eddies or vortices, except at the edges. This is *laminar* flow. Should the river encounter a steeper slope, its behaviour changes abruptly. It becomes a boisterously disorganized mass of swirling eddies and vortices. Flow lines within the water mass are chaotic. This is *turbulent* flow. A smouldering cigarette demonstrates the difference between laminar and turbulent flow well: initially, the

plume of smoke rises from the tip in a thin, orderly column, but after about twenty centimetres it begins to waver, then rapidly breaks up into an erratic cloud of wisps and whorls.

Whether a fluid flows in a turbulent or laminar mode, with all the implications that ensue, depends on its mass flow rate and viscosity. An English physicist (Osborne Reynolds, 1842–1912) first recognized this in experiments using a slender trail of dye in a pipe full of flowing water. Reynolds's work gave rise to a useful dimensionless parameter, the *Reynolds number*, Re, which relates the flow velocity, U, depth, h, and kinematic viscosity, η, of any moving fluid:

$$Re = Uh/\eta$$

Reynolds's own experimental work, boosted by innumerable later studies, showed that the transition from laminar to turbulent flow takes place at Reynolds numbers greater than about

1000–2000. Silicate melts have relatively high viscosities, and so almost all lavas flow in a laminar mode, except a few exceptionally hot, low-viscosity basalts such as komatiites. (For most lavas, U is small, and η large in the equation.) For some pyroclastic flows, velocities may be so high and viscosities so low that flow may be in the turbulent regime. We return to this issue in Chapters 10 and 11.

Finally, many natural materials, including some lavas, have such high yield strengths that when they flow, almost all shearing is confined to their edge: their centres remain almost undeformed, and the material is merely pushed forward *en masse*. This is what happens when toothpaste is squeezed from a tube, and is termed *plug flow*. Extrusion of spines of solid lava from a lava dome, as happened at Mt. Pelée in 1902, are extreme examples of toothpaste-like extrusion (Fig. 4.25; Figs 5.7–5.8).

Fig. 5.7 Extrusion of viscous lava formed a steep-sided, craggy lava dome on Mt. Pelée dome in November 1930, closely similar to the dome that grew in 1902. Jagged spines often grow on the crests of such domes, pushed upwards in tooth paste-like extrusion of semi-solid lava.

Fig. 5.8 Profile view of a 50-m-high spine that grew on the lava dome of Mt St Augustine volcano in 1986. As here, spines are typically smoothly curved on one side and fractured on the other (cf. Fig. 4.25). Photo courtesy of Lee Siebert, Smithsonian Institution.

— 5.6 Vesiculation

Pumice, the froth of silicic lava that some strange people reportedly use to mortify their flesh in the bathroom, is an example of a highly *vesicular* volcanic rock. Vesicles are small bubbles which are found in lavas of all compositions. They are often only a few millimetres across and thinly scattered, but some lavas are so honeycombed with large bubbles that they resemble Swiss cheese; more holes than solid. Vesicles form when dissolved gases come out of solution from a magma as it is depressurized; just as they do when a bottle of champagne is opened. If the bubbling gas can escape freely, the lava that results will be degassed or 'flat' and the resulting rock essentially free of bubbles. If, on the other hand, gas is still exsolving while the magma is erupted, bubbles will continue to grow, and will be found frozen into the lava when it cools. The ability of exsolving gas to bubble away freely, rather than explosively disrupting the magma, is critical to the understanding of volcanic eruption mechanisms. Two aspects require our attention: the *formation* of vesicles, and their *growth*.

Formation of vesicles in a magma depends on the amounts of dissolved volatiles such as water and carbon dioxide, and the vapour pressure that they exert relative to confining pressure of the magma. Exsolution will commence when the vapour pressure equals or exceeds the confining pressure, just as steam bubbles begin to grow in boiling water when their vapour pressure balances the water pressure. Thus, vesicle formation will be initiated at a greater depth (pressure) in a volatile-rich magma than in a volatile-poor one. One way of initiating vesicle formation, then, is to depressurize a magma, a process physically similar to uncorking a champagne bottle. This is termed *first boiling*.

Vesiculation may also occur as a result of a second, more complex phenomenon. When crystallization takes place in a cooling magma, removal of crystals from the magma concentrates volatiles in the remaining liquid, thus driving up their vapour pressure. Crystallization of minerals also liberates latent heat of fusion, keeping temperatures high, which in turn causes vapour pressures to remain high, ultimately causing bubble formation. When runaway bubble growth is triggered by crystallization, the process is termed *second boiling*. Enormous pressures can be generated in magma by the increase in volume that results, which may have spectacular consequences. For an andesitic magma containing only 2.8 weight per cent water, crystallization and boiling may cause an increase in volume of as much as 50 per cent.[4]

In a study of how vesicles develop in magmas, Stephen Sparks[5] argued that the supersaturation pressure of dissolved volatiles in a magma required for vesicles to be nucleated is quite low, perhaps only a few bars, because of the presence within the magma of molecular species which are surface active and promote vesicle growth. (In pure liquids, such as distilled water, bubble formation is inhibited, so boiling is difficult and often 'bumpy'. Small pieces of porcelain are often added specifically to provide bubble nucleation sites.) Thus, vesicles develop easily in magmas that are only slightly supersaturated with volatiles. Because of the relationships between magma surface tension, bubble radius and supersaturation pressure, *small* vesicles are not stable, and in nature, vesicles less then about 5 microns in diameter are not found.

Once nucleated, growth of vesicles to larger sizes is controlled by the volatile content of the magma, the rate at which volatiles can diffuse through the magma into the bubbles and other intrinsic variables such as the density, viscosity, and surface tension of the magma. Initial formation of vesicles has an interesting side-effect: loss of water from the magma causes its viscosity and yield strength to increase. Near the surface, in the regime where explosive eruptions become possible, the chief extrinsic control on bubble growth rates is rate of decompression of the magma column. The effect of these variables is that bubbles are likely to grow to diameters between 0.1 and 5 cm in basaltic explosive eruptions, but only 0.001 to 0.1 cm in rhyolite eruptions. (It is the lower diffusivity of water in silicic melts and their higher eruptions rates, not the higher magma viscosity, that are chiefly responsible for the small bubble size in rhyolitic pumices, because these give bubbles less opportunity to grow).

In a vigorously vesiculating magma, there comes a time when individual bubbles cannot be considered in isolation; they interact with one another, making the magma more like the foam on top of a beer than a simple fluid. As long as volatiles continue to diffuse from the magma, bubbles will continue to grow until the pressure inside the bubbles is the same as the vapour pressure of the volatile phase in the surrounding magma. Bubbles will not burst into one another, because they are all at similar pressures. Sparks suggests that most bubbles cease growth well before explosive disruption of the magmatic froth, when volume ratio of gas bubbles to liquid is about 4 to 1. At this stage, bubbles cannot expand further because extraction of volatiles increases the viscosity of the remaining liquid, which has to be forced into the thin films between adjacent bubbles (Fig. 5.9).

In a column of magma in the throat of a volcano, the volume change caused by bubbling propels magma higher up the conduit. There is a large volume of bubbles and a huge surface area of liquid forming the bubble walls, so pressure

equilibrium is maintained between the exsolved gases contained in the bubbles and those still in the enclosing liquid. These pressures may reach tens or hundreds of times atmospheric pressure. In a column of basaltic magma, there may be a

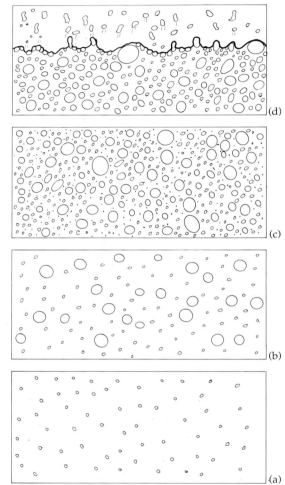

Fig. 5.10 Stages in the vesiculation and fragmentation of magma. In the earliest stage (a), nucleation occurs and bubbles grow freely; in (b) growth continues and new nuclei form; larger bubbles interfere with each other's growth; in (c) the magma froth is saturated with bubbles so bubble growth ceases; in (d) the fragmentation surface moves downwards, propagating downwards by the bursting of bubbles. From Sparks, R. S. J. (1978). The dynamics of bubble formation and growth in magmas: a review and analysis. *J. Volcanol. Geotherm. Res.* **3**, 1–37.

Fig. 5.9 Not an anatomical view of the sinewed chambers of a human heart, but a scanning electron micrograph of a typical specimen of silicic pumice, this one from Tenerife in the Canary Islands. Expansion of exsolving gases stretched and thinned the sinews of glass to create a light but strong froth of rock. Note that the voids are not bubbles; they are interconnected spaces, formed when the magma was explosively disrupted. Photograph is about 1 mm across. See also Fig. 9.23.

free surface to the atmosphere in the throat of the volcano—a lava pond. A huge pressure gradient builds up across this surface. Bubbles near the surface swell, rapidly overcoming the strength of the bubble walls, until they burst, disrupting and ejecting magma. A *disruption surface*, where fragmentation takes place, moves downwards through the magma column, unless bubbles and magma rise quickly enough that it remains at about the same level (Fig. 5.10).

Explosive rhyolitic eruptions are a little more complex, since they are initiated at greater depths. Pressure within the magma has to be built up until the strength of the confining rocks is overcome before a free surface to the atmosphere is opened, permitting downwards propagation of the disruption surface into the magma body. While rocks are extremely strong in compression, they are weak in tension (strengths are about 1.5×10^7 bars). Thus, the strength of the rocks capping a cooling and vesiculating magma body is easily exceeded long before the magma is solid. Once this happens, an explosive eruption results.

5.7 Summary

If nothing else, this brief chapter ought to have demonstrated that magma is complicated stuff. In succeeding chapters, we look at the many ways in which magma erupts. Fundamentally, it is differences in composition and volatile content that are responsible for all variations between extremes of quiet lava effusion and catastrophic explosion. A *very* rough chain of cause and effect can be summed up as follows;

(1) composition dictates melting temperature;

(2) temperature, volatile content, and degree of polymerization dictate viscosity;

(3) viscosity and volatile content dictate ease of vesiculation and degassing;

(4) ease of degassing dictates explosive potential of eruption;

(5) explosiveness of eruption dictates whether lavas or pyroclasts are produced.

Notes

1. Rose, W. I., Stoiber, R. E., and Malinconico, L. L. (1982) Eruptive gas compositions and fluxes of explosive volcanoes: budget of S and Cl emitted from Fuego volcano, Guatemala. In *Andesite: orogenic andesites and related rocks* (ed. R. S. Thorpe), Wiley, London.
2. Fisher, R. V. and Schminke, H.-U. (1984). *Pyroclastic rocks*. Springer-Verlag, Berlin, 472 pp.
3. Williams, H. and McBirney, A. R. (1979). *Volcanology*. Freeman, Cooper and Sons, 395 pp.
4. Burnham, C. W. (1972). The energy of the explosive eruptions. *Earth, Mineral Sci. (Pennsylvania State Univ.)* **41**, 69–70.
5. Sparks, R. S. J. (1978). The dynamics of bubble formation and growth in magmas: a review and analysis. *J. Volcanol. Geotherm. Res.* **3**, 1–37.

6

Types of volcanic activity

Volcanic eruptions are difficult to classify. Some are short, sharp, and easily pigeon-holed, but others drag on for months, varying in character as rapidly and unpredictably as the moods of an opera star. Different things are often going on at different places on the same volcano at the same time. Because eruptions are complex, it is therefore easier to characterize individual phases of activity. As in most scholarly disciplines, volcanological terminology is a mine field into which one treads perilously, stumbling against the concealed trip-wires of earlier usage, and risking explosions of injured academic sensitivities. In this chapter, we review briefly some terminology commonly used in describing eruptions, which provides a useful vocabulary for dealing with the more detailed aspects of later chapters.

— 6.1 Methods of classification

6.1.1 Effusive v. explosive activity

This is a relatively uncontroversial issue. There is a continuous spectrum between *effusive* activity, dominated by passive emission of lavas, and *explosive* activity, dominated by eruption of pyroclastic material. As outlined in the previous chapter, the likelihood of an eruption being either explosive or effusive is dictated by magma properties. On the smallest scale, magmas of any composition may erupt either effusively or explosively, but at the largest scale there is a clear, but not completely watertight, separation by composition. Large-volume basaltic eruptions are almost exclusively effusive; large-volume silicic eruptions are almost exclusively explosive.

6.1.2 Conventional v. hydrovolcanic activity

In military parlance, 'conventional' is the euphemism used to distinguish non-nuclear from nuclear weapons, notwithstanding how barbaric the 'conventional' weapons may be. Here it is used to distinguish between 'conventional' eruptions that do not involve extraneous water, and *hydrovolcanic* eruptions that do. Given that so many volcanoes are located near the sea, it is not surprising that violent interactions between water and hot magma commonly take place, yielding distinctive types of activity and deposits. Even in volcanoes not located near shorelines, percolation of ground water through aquifers often results in similar activity. Recent research suggests that *most* explosive eruptions may involve at least a component of hydrovolcanic activity. Like their nuclear counterparts, 'non-conventional' hydrovolcanic eruptions are exceedingly violent.

6.1.3 Location and frequency

A critical division here is between *central* and *fissure* activity. Central vent eruptions are the run of the mill kind, ejecting lava and pyroclastics from a single hole in the ground, supplied by a pipe-like feeder channel from an underlying magma chamber. As the eruption continues, ejected material piles up around the vent, constructing the accumulation of lavas and pyroclasts ultimately called a volcano. If the eruption ceases after only one episode of activity, the volcano is called *monogenetic*. Basaltic scoria or 'cinder' cones are the commonest examples of monogenetic central volcanoes; they rarely exceed two or three hundred metres in height. If there are repeated episodes of activity from the vent, a bigger *polygenetic* volcano results. Classic symmetrical cones like Mt. Fuji are built up in this way.

Unfortunately, few volcanoes have geochronologically controlled biographies, so there are few data on how long it takes large cones to grow. Mt. St Helens appears to have destroyed and rebuilt itself several times over the last 40 000 years. Most of its present bulk and shape has been acquired during only the last 2500 years.[1] Other large volcanoes, such as those in the central Andes, appear to have had much longer active lives, measurable in hundreds of thousands of years.

Tall, central vent volcanoes like Fuji are simple structures—they appear to have erupted exclusively through their summit craters throughout their history. Other large volcanoes are more complex, and eruptions may take place from the summit vent, or from points on the flanks. Mt. Etna, for example, is on the largest scale a single volcano of roughly conical form with an active summit complex, but its flanks are dotted with innumerable individual scoria cones and lavas, each the result of a distinct historic (or prehistoric) eruption.[2] These are variously termed *flank*, *parasitic*, or *satellite* vents (Figs 6.1–6.2).

Central vent volcanoes are found all over the world, in every plate tectonic setting. *Fissure* eruptions result when magma-filled dykes intersect the surface, so large-scale fissure eruptions naturally occur where the crust is undergoing extension, as in Iceland. As discussed in Chapter 2, Iceland is widening at a rate of centimetre or two per year, permitting intensive dyke intrusion. When a dyke breaks through to the surface, huge volumes of lava may be erupted through fissures in basaltic flood eruptions. In historic times, the largest example of this phenomenon took place between June 1783 and February 1784, when 14 cubic kilometres of lava flooded from the 25 km length of the Laki Fissure in southern Iceland. Environmentally, the long-term results of this eruption were profound; it was Iceland's worst natural disaster (Chapters 7 and 17).

Fissure eruptions are not confined to areas undergoing regional extension (Fig. 6.3). They are common features of many large volcanoes. Often, the commonest episodes of eruptive activity on volcanoes such as Etna are satellite

Fig. 6.1 Scoria cones at an elevation of about 2000 m on the south flank of Mt. Etna. In the foreground is Monti Silvestri, formed by a parasitic eruption of unusually long duration, lasting from July to December 1892.

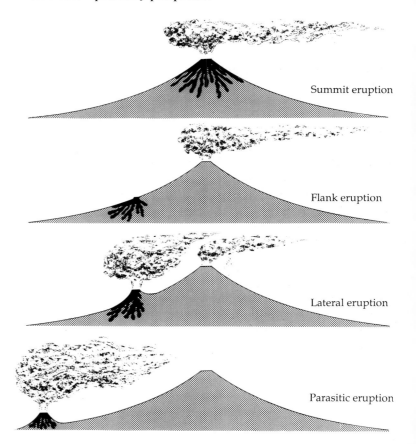

Summit eruption

Flank eruption

Lateral eruption

Parasitic eruption

Fig. 6.2 Some terms commonly used in describing the sites of eruptions relative to a simple conical volcano. Flank eruptions commonly commence with summit activity, but activity shifts to lower levels where most lava emerges. 'Lateral' eruptions are similar, but may involve only minimal summit activity.

Fig. 6.3 Fumes rising above a kilometre-long section of active fissure in the north-east rift zone of Kilauea volcano, Hawaii, signposting the location of a dyke. Pu'u O'o vent in the background, September 1984. (Photo: courtesy P. Mouginis-Mark.)

eruptions low on the flanks, fed by dykes radiating out from the core of the volcano. In the Hawaiian Islands, the great 'hot-spot' volcanoes show even more clearly the effects of dyke propagation in eruptions from satellite vents— Mauna Loa and Kilauea volcanoes are both elongated along major rift zones which are the surface expressions of dyke swarms (Figs 6.4–6.5). For many years, *all* the lava erupted from Kilauea has actually emerged from its East Rift Zone, many kilometres from the summit caldera. In such cases, dyke propagation and extension are related to the *local* tectonic regime, which is an expression of the shape and size of the volcano, and not its plate tectonic setting.

While many eruptions in areas of dyke intrusion such as Iceland or Hawaii start in classic fissure form, yielding marvellous film sequences of 'walls' of fire kilometres in length, this style of activity does not persist. Parts of the dyke may become blocked, or other parts are reamed out, so that activity becomes focused on one or a few sites, where cones are rapidly built up. This was demonstrated by the Great Tolbachik Fissure Eruption (Kamchatka Peninsula, Russia) which started on 6 July 1975 and lasted 450 days. Soviet geologists noted that activity took place along a thirty kilometre length of fissure, commencing at the northern end but focusing on the southern extremity. Four new cones were constructed, the highest 340 m in height; lavas covered more than 40 square kilometres, and more than 2 cubic kilometres of lava and pyroclastics were erupted.[3] An eruption of Izu–Oshima volcano in Japan in 1986 provided another, smaller example of this kind of eruption.

6.1.4 Eruption size

Because volcanic eruptions are so dramatic, their 'bigness' is often wildly misrepresented in popular descriptions. To clarify this, G. P. L. Walker, doyen of pyroclastic volcanologists, suggested that *magnitude* be used in relation to the total volume of erupted material, and *dispersive power* in relation to the area covered by tephra fall-out.

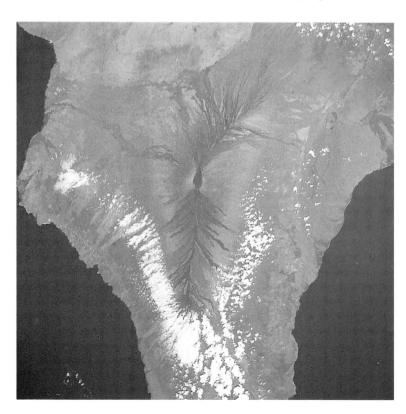

Fig. 6.4 Vertical view from the Space Shuttle of Mauna Loa, Hawaii, showing the elongation of the shield volcano along south-west (bottom) and east rift zones, and the profusion of lavas erupted from them. Many are historic. Dark lavas fill Mokuaweoweo caldera area at summit. Image is about 100 km across.

A third index is *intensity*; the rate at which magma or ash is disgorged.[4,5] Magnitude and intensity are probably the two most useful measures of eruption size; intensity is especially important in pyroclastic eruptions because it controls the height that the eruption column reaches, and therefore the fall-out pattern of the ash. It also influences the degree of fragmentation of the magma, expressed in the proportion of fine ash that is produced (Chapter 9).

Fig. 6.5 Aerial view looking along the south-west rift of Mauna Loa towards the summit, showing numerous small vents along the fissure zone. (Photo: courtesy I. G. Gass.)

— 6.2 Explosivity and deposit type

At the turn of the century, the Italian geologist G. Mercalli (best known for his scale of seismic intensities) initiated a system of volcanic terminology based on the violence of eruptions characteristic of some well-known volcanoes—strombolian, vulcanian et cetera.[6] This necessarily subjective approach was extended to included eruption styles such as hawaiian and peléean, and is still widely used.

It is difficult to get close to an erupting volcano safely, let alone measure anything useful, so attention has turned more recently to erupted *deposits*. This has the bonus that identical techniques can be applied to deposits produced by prehistoric eruptions. In fact, if a prehistoric deposit is well preserved, one can say almost as much about the activity which produced it as about an eruption taking place this century. In Section 6.3, we examine types of activity, and, in Chapter 9, the deposits that result.

6.2.1 Eruption columns

Three distinct elements may be involved in a volcanic eruption column, all or any of which may be present at any one time. First, steady emission of steam forms a white cloud column, like that above a power-station cooling tower. Second, explosions may hurl pyroclastic material up in dark masses, the leading fragments arcing upwards along parabolic paths, trailing smaller fragments behind them, which look from a distance like rockets. In profile, such a blast cloud looks something like a clump of fir trees, so it is sometimes described as *cypressoid*. The term *cock's tail* is also occasionally used (Fig. 6.6). Third, a cloud of fine ash may rise convectively. If there are distinct explosions taking place during an episode of eruptive activity, then each explosion will generate a fresh ash cloud. If explosive activity is sustained, then ash will rise continuously. Ash-laden clouds have a dark, solid appearance, often likened to the convoluted surface of a cauliflower. They resemble the towering, hard-edged cumulo-nimbus clouds that threaten thunderstorms. Often, the three elements (steam, ballistic ejecta, and convecting ash column) are intermingled, so that the result is a mixture of white streaked with grey or black. Beneath and downwind of the eruption column, ash rains down, sometimes falling so thickly that a dark curtain appears to be hanging from the cloud, while lightning flickers in and around it so frequently that the effect is more dramatic than even the most ominous of thunder clouds.

A complicating factor that may have beautiful consequences is the way the eruption column interacts with the atmosphere. By punching abruptly upwards many kilometres, an explosive eruption column naturally perturbs the pressure, temperature, and water vapour regimes of the various levels within the atmosphere. Moist air near the eruption column is suddenly drawn upwards to cooler levels, causing water vapour within it to condense immediately to white cloud. A common consequence is that a delicate, cylindrical veil of cloud suddenly forms around the eruption column; remarkable for its unexpectedness, and delightful for its fragile quality (Fig. 6.7).

Fig. 6.6 Birth of Surtsey, off the coast of Iceland, in 1963. Copious volumes of steam form the white plume at left, while explosive activity hurls dark ejecta upwards and outwards in a prominent cock's tail effect. (Photo S. Thorarinsson.)

Fig. 6.7 Not a volcanic eruption, but a nuclear weapons test in the Pacific. The mushroom cloud generated by the instantaneous explosion of a nuclear weapon provides an extreme illustration of the effects of a powerful, short-lived volcanic outburst punching through the atmosphere. Here, cylindrical sheaths of cloud forming by condensation around the ascending column are beautifully displayed. (Photo: courtesy of Channel 4.)

— 6.3 Types of activity

6.3.1 Hawaiian activity

Hawaiian activity is the mildest of all, taking its name from the Hawaiian Island volcanoes, which have euphonious names such as Kilauea, Mauna Loa, and Mauna Kea. (But many suffer from an excess of vowels—Mokuaweoweo or Halema'uma'u, for example.) Located in the centre of the Pacific, the Hawaiian Island volcanoes are necessarily entirely basaltic, and, as outlined in Chapter 3, their lavas are examples of

ocean island basalts. They are erupted at high temperatures and have low viscosities.

Although Hawaiian basalts contain small proportions of volatiles (less than one per cent by weight), the paucity of volatiles may not be obvious to someone watching escaping gas spraying incandescent lava hundreds of metres into the air for hours on end. Three reasons account for this apparent paradox. One is simply that a great deal of basalt is also erupted at the same time; the second is that prior to an eruption, volatiles accumulate in the upper part of the magma chamber, so the amount of dissolved gas present during early phases of an eruption is quite large. Third the low molecular weight of dissolved gases means that although they form only a small percentage by weight of the magma, they none the less constitute a significant molecular proportion. (In an aerosol can the weight of the propellant gas is a small fraction of the useful content.)

Because Hawaiian basalts have low viscosities and yield strengths (they approach Newtonian fluids more nearly than other lavas), exsolving volatiles can vesiculate within the magma and fragment it without violent explosions. Just as a soda-syphon squirts water, pressure of the rapidly expanding escaping gas in a Hawaiian basalt sprays liquid lava high into the air in *fire fountains*, hallmarks of hawaiian eruptions, and among the most stunning of all natural phenomena (Fig. 6.8). Jets of incandescent liquid rock commonly reach hundreds of metres in height. During the eruption on the Japanese island of Oshima in 1986 they reached *one thousand six hundred metres*.[7] They are not just short spurts, either—some fire fountains play for hours on end. Not unvaryingly, like the monotonous plume in Lake Geneva, but swelling, dying away, and surging up irregularly. At the top of the fountain, smaller individual fragments are carried away by the wind, so that downwind a shifting curtain of glowing droplets showers down.

Fire fountains yield a range of products, depending on their mass eruption rates. At low rates, fountains are said to be 'optically thin'; the clasts of lava within them cool quickly by radiating their heat away to the atmosphere. Most clasts are thus chilled and solid by the time they reach the ground, where they accumulate as cindery *scoria* deposits. At higher mass eruption rates, fountains become optically 'dense', meaning that so much material is present that clasts can no longer radiate freely to the atmosphere; each clast, so to speak, keeps its neighbours hot. In this case, the ejected gobbets of lava that fall back to earth as *spatter* are still hot and plastic enough to weld together to form *spatter cones* and *ramparts*. If the mass eruption rate is fast enough, the falling material scarcely cools at all, so that it blends together on reaching the ground to flow away as *clastogenic* lava. Many lavas formed during hawaiian eruptions commence in this strictly pyroclastic way, rather than welling directly out of the ground. Of course, even the most powerful and dense fire fountains yield a rain of cool, solid particles from their margins, to be borne away by the wind as tephra. Although the total volume of clastogenic lava may be huge, relatively little pyroclastic material results. Some

Fig. 6.8 Fire-fountaining during a typical Hawaiian eruption; Mauna Ulu, 1969. These fountains are of modest proportions, reaching only about one hundred metres; they may attain one kilometre. Lavas from the vent cascade over a cliff in foreground. (US Geological Survey Photo by D. S. Swanson.)

60 million cubic metres of lava were erupted during a magnificent spell of fire-fountaining from the Kilauea Iki vent (Hawaii) in 1959, when fountains reached 580 m high. Scoria piled up to form a cone 50 m high round the vent, and was a metre thick one kilometre downwind from the vent.

6.3.2 Lava lakes

A fascinating aspect of hawaiian activity, although not an essential attribute of it, is the existence of long-lived *lava lakes*. When European missionaries first reached Kilauea caldera in 1823, they found that:

'The southwest and northern parts of the crater were one vast flood of liquid fire, in a state of terrific ebullition . . . fifty-one craters, of varied form and size rose like so many conical islands from the surface of the burning lake. Twenty-two constantly emitted columns of grey smoke or pyramids of brilliant flame and many of them at the same time vomited from their ignited mouths streams of florid lava, which rolled in blazing torrents down their black, indented sides into the boiling mass below'. At night 'the agitated mass of liquid lava, like a flood of metal, raged with tumultuous whirl'.[8]

A nugget for those who relish the curious cross-currents of history: the first European to descend to the floor of the Kilauea caldera was Lord Byron, cousin of the romantic poet and Commander of HMS *Blonde*. He did so in 1824, from a point on the rim where the volcano observatory is now located.

Between 1823 and 1924, a lava lake hundreds of metres across was present almost continually in the Halemaumau crater, nested within Kilauea's summit caldera. It has appeared and disappeared several times since. At the time of writing, Kilauea's lake had been absent for many years, but a smaller lava pond (Kupaianaha) was present at a subsidiary, dyke-fed vent on the north-east rift (Fig. 6.9). At night, a lava lake or pond presents an unearthly sight. Slabs of dark, solidified crust shift about slowly on the surface, while bright lava glows redly through the cracks between slabs. Sudden escapes of gas sometimes squirt glowing jets from the centre of the pond, while surging and sloshing of the lava may throw

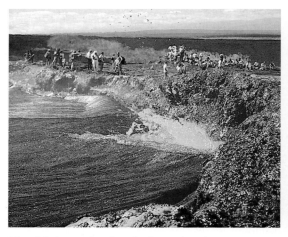

Fig. 6.9 A gaggle of geologists enjoy the Kupaianaha lava pond, Hawaii, 1987, standing on the rim only a metre above the surface of the pond. Minor spattering is taking place. Crusted surface of lake (left) appears solid in black-and-white image, but is actually hot, plastic, and steadily deforming.

up gouts of incandescent spatter around the edges.

Kilauea's lava lake was for long periods so temperate that visitors could come close to its edge to watch. When it became a tourist attraction, the 'Volcano House' hotel was built on the rim of the caldera. Its terrace still provides a welcome place to sip a cool beer and savour the view over the gently fuming caldera, with the gentle slopes of Mauna Loa silhouetted against the sky. In suitable postcard terminology, the lava lake was labelled the 'House of Everlasting Fire', a misnomer, since although adjectives like 'fiery' are often applied to volcanic phenomena (count the number of times in this book!), it is not literally correct, since there is rarely any actual fire, in the sense of anything burning to produce smoke and flames. Almost all the red and yellow colours come from the incandescence of the hot lava, while the 'smoke' consists of steam and pungent fumes escaping from the lava. True flames are rare, because only small quantities of reduced, burnable gases such as hydrogen and hydrogen sulphide are present. Traces of hydrogen escape from some magmas, and can occasionally be seen burning with a weak blue flame. Much more often, lava encroaching vegetated areas sets fire to organic material such as wood.

Sadly for the tourist industry, persistent lava lakes are rare. Most examples are in distinctly inaccessible places. There is a small lake (of odd, alkalic composition) in the summit crater of Mt. Erebus, the 4045-metre-high volcano near McMurdo Sound in Antarctica. There is also a pair of lakes on Erta' Ale volcano, in the scorching wastes of the Danakil Depression, Ethiopia.[9] In the heart of Africa, Nyiragongo has had a persistent lava lake (also of alkali basalt composition) deeply recessed with the crater of the volcano. For a hair-raising account of the great French volcanologist Haroun Tazieff's descent to the lava lake, and the formidable bureaucratic obstacles that he had to overcome before he could get there, his book is compulsive reading.[10]

Hawaiian activity rarely kills anyone. Should a lava lake be abruptly breached, catastrophe may ensue in the flood of lava, as occurred at Nyiragongo in 1977, but this is exceptional. On Hawaii itself, many tourists have lost themselves in the anastomosing lava fields, or died of heart attacks while scrambling around on them, but the number of direct casualities is negligible. Hawaiian legend has it that after battles with King Kamehameha, some of the men of Chief Keoua's army were killed near the edge of Kilauea caldera by a great eruption in 1790. This eruption was unusual for Hawaii because it was hydrovolcanic and therefore explosive. Much of the ejected material consisted of fine ash (the Keanakakoi ash) but during the final stages, large blocks were ejected and it was these that killed the warriors. Footprints said to have been made by the army are still to be seen in the ash.[11]

6.3.3 Strombolian activity

'Strombolian' is an elastic volcanological term. It has been applied indiscriminately to activity ranging from miniscule burps to discharges producing eruption columns more than a kilometre high. Life would be simpler if the term were replaced by others, but because it has been in use since Mercalli's day, it is unlikely to be abandoned. True strombolian activity consists of intermittent, discrete explosive bursts, ejecting pyroclasts a few tens or hundreds of metres into the air. Each burst lasts only few seconds at most, and there may be long pauses between bursts—twenty minutes or more. In strombolian activity, basaltic magma is again involved, but basalt with higher viscosity and yield strength than hawaiian magma. Thus, higher pressures are required to fragment the vesiculating lava column in the throat of the volcano. Most importantly, no sustained eruption column develops: a feature of true strombolian eruptions is that one can often see clear through the haze of fumes above the vent.

Strombolian eruptions are named, of course, after the tiny volcano-island of Stromboli, between Sicily and Italy. Stromboli has been erupting in the same way for millenia, earning a reputation as the 'lighthouse of the Mediterranean'. Explosions occur every few minutes, sometimes more or less rhythmically, often quite irregularly; sometimes causing ear-splitting bangs, sometimes thumps and whooshes. There are often brief, sustained blasts like the din of a jet engine at close quarters. Some explosions blast fragments of hot, plastic lava high into the air, at exit velocities typically of the order of 100 metres per second. Seen from beneath, the larger clasts seem to travel quite slowly, twisting and turning lazily in the air, silhouetted against the sky like great black bats. At night, the glowing lava clasts trace elegant parabolic ballistic paths as they fall back to earth—night-time photographs of strombolian eruptions have sold millions of postcards. Little fine ash is erupted in these explosions, but in more sustained blasts a brownish slug of dusty tephra climbs to a height of a hundred metres or so, before being rapidly dispersed by the wind (Figs 6.10–6.11).

Strombolian activity, then, is a bit noiser than hawaiian, but is still not particularly dangerous—there are two villages on Stromboli, only a few kilometres away from the ever-active vent. After decades of living in the shadow of the volcano, the inhabitants have little cause to worry about whether they will live to see the rosy-fingered dawn rise on another day, or another boat-load of tourists coming across the wine-dark sea to visit the volcano. Strombolian activity is common on volcanoes round the world. Etna has been booming away intermit-

tently in one or other of its summit craters for millennia—its first historic eruption was in 1500 BC. Milton referred to it as 'Thundering Aetna' in *Paradise Lost*. In this century, the lurid spectacle of gouts of glowing tephra being ejected used to make an odd contrast with the winter sports taking place on the smooth snowy slopes below the summit, until the volcano inconveniently obliterated the best skiing slopes by covering them with lava flows. Pacaya, in Guatemala, is even more active than Etna. Its red glare can often be seen at night from Guatemala City, only thirty-two kilometres distant.

Mt. Erebus also has bouts of strombolian activity. It was in this condition when a team of six men from Scott's base camp made an ascent of the volcano in December 1912, led by a geologist, Raymond Priestley. When the party reached the lip of the summit crater safely, they found that there was little to see within the crater apart from swirling clouds of sulphurous steam. Nothing much was happening. They took some photographs, left a record of their ascent in a sealed can in a cairn, and set off back down again. About 150 metres from the top, Priestley realized that the can they had left behind contained their exposed photographic films and what they were carrying with them was the can containing the

Fig. 6.10 Explosive ejection of a slug of pyroclasts from a typical strombolian eruption, Pacaya volcano, Guatemala, 1969.

Fig. 6.11 Night-time view of strombolian activity on Stromboli itself. Two separate vents were active during the time exposure. 'Blips' in the trajectories of some of the ejecta are due to the clasts spinning in flight, exposing larger and smaller surfaces to the camera.

record of their ascent . . . He sent back one of the party to retrieve the film can, but no sooner had this poor unfortunate reached the crater rim than a strombolian blast from the volcano showered him in hot, fine-grained scoria. Fortunately, he survived unharmed.

6.3.4 Vulcanian activity

Between Stromboli and the mainland of Sicily are the Aeolian islands, the islands of the wind. Above one of these islands rises a drab, eroded cone with a deep crater at its centre. This is Vulcano, eponym of all volcanoes (Fig. 6.12). It has been intermittently active throughout history, and may have been more active in Greek and Roman times than it is today. To the Greeks, Vulcano was known as Hiera, the chimney of the forge of Hephaestus, god of fire, but it was the Roman god of fire, Vulcan, whose name stuck. Vulcan was one of the three children of Jupiter

and Juno, and, other divine attributes apart, was blacksmith to the Gods. As part of his professional duties, he forged the breastplate of Hercules, the shield of Achilles and the arrows of Apollo and Diana. And to cap all this mythological irrelevance, the island of Vulcano is now linked by a small spit of land to another small volcano, Vulcanello, which was a separate island until united to Vulcano by an eruption in the sixteenth century. This pair may have comprised the original 'Scylla and Charybdis', beloved of Homer (*The Odyssey*, Book 12, 20–82) and modern politicians, but then so might several other island pairs around the Straits of Messina.

Mythological distinctions apart, Vulcano has also given its name to a style of volcanic activity: *vulcanian*. Vulcanian eruptions are of small magnitude (less than 1 cubic kilometre), but their eruption columns rise much higher than strombolian columns, sometimes reaching 10–20 kilo-

Fig. 6.12 Looking down into the crater of Vulcano. Visible above the steaming fumarole on the rim is the smaller cone of Vulcanello, and beyond that is the island of Lipari.

metres. Thus, their tephra are dispersed over a wider area. A second important difference between vulcanian and strombolian activity is that vulcanian activity is more violently explosive, sometimes destroying parts of the volcanic edifice, and producing highly fragmented ash.

Vulcanian eruptions vary considerably in their noisiness and duration. Some consist of discrete cannon-like explosions at intervals of tens of minutes to hours. Each explosion ejects a slug of ash at velocities of hundreds of metres per second, which then decelerates and rises convectively above the volcano. These violent explosions sometimes form the first phases of a longer eruption, while the volcano is 'clearing its throat'. In these cases, no new material is involved, the pyroclastic ejecta being merely fine fragments of older lava which had previously solidified and blocked the vent. Other vulcanian eruptions continue for long periods, when fresh, viscous magma is constantly supplied. Often, these eruptions seem eerily silent, dense clouds of fine ash billowing steadily from the vent, as noiselessly as smoke from a chimney.

Vulcanian activity requires magmas with high yield strengths and viscosities. Often, rather crystal-rich magmas have suitable properties. Andesitic magmas are commonly involved. Vulcanian activity is therefore commonly associated with growing lava domes, such as those of Mt. Pelée in 1902 and 1929, and of Mt. Lamington in Papua New Guinea in 1951. *Nuées ardentes* are often by-products of vulcanian explosions. Owing to their lethal nature, the *nuées* often attract more attention than the eruptions that generate them.

Between 1963 and 1965 Irazu volcano in Costa Rica was consistently active. Frequent explosions rained ash over a wide area, ruining the coffee crop. Throughout the entire two years, no lavas or *nuées* were erupted, and despite the innumerable explosions, less than two metres thickness of ash accumulated within one kilometre of the volcano. Sakurajima volcano, Japan, which has been erupting since 1955, continues at the time of writing to dump 1–6 kilograms per square metre of ash on the nearby city of Kagoshima every year. Municipal authorities around the world grumble at the problems presented by citizens dropping litter. They should have some sympathy for Kagoshima municipality, which constantly has to shovel away tonnes of ash.

In summary, vulcanian activity is noisy, messy, and produces impressive eruption columns, but its effects are not widespread.

6.3.5 Sub-plinian or Vesuvian activity

Sub-plinian activity is a step up from vulcanian activity. It is associated with higher eruption columns, which give rise to extensive sheets of tephra deposits. Sub-plinian eruption columns are sustained for longish periods, and climb to nearly 30 kilometres. Large volumes of tephra may be erupted in sustained blasts, in the form of new magmatic material, rather than shattered bits of old rock. Because the degree of fragmentation is less than in vulcanian eruptions, clast sizes are generally larger at a given distance from the vent. Dacitic and rhyolitic magmas are usually involved, but sub-plinian deposits of more mafic compositions are known, such as the scoria fall deposit erupted by Sunset Crater, Arizona, in AD 1065.

It should not be supposed that sub-plinian eruptions are necessarily of less consequence than plinian events. Many factors, other than sheer size, dictate how hazardous an eruption will be. A little-known eruption of Vesuvius in 1631 was half the magnitude of the AD 79 plinian event, but none the less took more than 4000 lives: a consequence of the numbers of people crowding the flanks of the volcano. Vesuvius has had many such eruptions in its history, which is often said to follow an irregular cycle; twenty-five to thirty years of repose being followed by an eruptive phase, when a great column of ash rises many kilometres above the city of Naples (Fig. 6.13). Even when casualties are avoided, numerous practical problems ensue, as they did during the most recent eruption in 1944, which coincided with the Allies' long struggle up the Italian Peninsula. Allied airfields around Naples were carpeted with ash, making the runways temporarily unusable, while gritty ash particles found their way into the air intakes of aircraft flying in the area. Understandly, the eruption went largely undocumented, except for a few photographs showing the eruption cloud looming ominously over a city already overhung by barrage balloons. There is much well-founded concern about what will happen when the next eruption takes place.

6.3.6 Plinian activity

With plinian activity, we enter the realm of eruptions that make history. Pliny the Younger's description of Vesuvius AD 79 provides an unsurpassed account of such an eruption (Chapter 4). Although plinian eruptions are fortunately rare (two or three per century) several historic examples have been studied, so their mechanisms are reasonably well understood. Plinian eruptions are driven by powerfully convecting erup-

Fig. 6.13 Vesuvius in eruption on 18 June 1794, depicted in a contemporary engraving. Heavy ash fall is taking place on the downwind (left) side of the column, while lightning flickers here and there. A lava flow descends to the sea on the right side.

tion columns with exit velocities of several hundred metres per second, punching high up into the stratosphere. Some reach up to 45 kilometres altitude. Such high eruption columns naturally cause widespread dispersion of tephra, covering huge areas of ground with an even thickness.

One of the largest eruptions of the twentieth century was the October 1902 plinian eruption of Santa Maria volcano, Guatemala. The year 1902 was an extraordinary one for eruptions in the Caribbean and Central America, with eruptions of historic proportions at Mt. Pelée, Soufrière (San Vincent), and Santa Maria. In the space of only about 18–20 hours about 12 cubic kilometres of dacitic pumice were erupted, forming a powerful column which reached at least 28 km altitude, carefully measured with his ship's sextant by Captain Saunders of the mailboat S.S. *Newport*. Tremendous detonations were audible in Oaxaca, Mexico, and Belize, British Honduras. Air-fall pumice covered an area of more than 1.2 million square kilometres. Thousands of people were killed and much of Guatemala's coffee industry was destroyed.[12]

Plinian tephra deposits are typically composed of bubbly pumice clasts of dacitic to rhyolitic composition, which often appear dazzlingly white in field outcrop. They are found mostly at destructive plate margins where large volumes of silicic magmas can be generated from continental crust. This is only a generalization, however—Hekla volcano in Iceland has a long history of sub-plinian and plinian activity. In Hekla's case, the silicic magma involved accumulates beneath the volcano through the long-term fractionation of the ubiquitous Icelandic basaltic parent magma.

It is also a generalization to suggest that plinian eruptions are silicic. Although rare, plinian eruptions of basaltic composition are known. Most famous of these was a short-lived but intense eruption of Tarawera on the North Island of New Zealand in 1886 (Chapter 9). Thanks to fifty years—a lifetime—of research by R. F. Keam of the University of Auckland, the circumstances of the Tarawera eruption are as minutely detailed as any in history. They are described in a remarkable book which even lists the names and ethnic origins of every one of the 100 victims of the eruption.[13]

6.3.7 Ultraplinian eruptions

Ultraplinian eruptions represent the next increment over plinian eruptions: column heights reach more than 45 kilometres. Fortunately, no such stupendous outburst has taken place in recent historic times, but an idea of the nature of such an event has been gained from several years meticulous work in the Taupo area of the North Island of New Zealand by George Walker and Colin Wilson. They described the Taupo eruption as the most 'powerful plinian . . . event yet documented'.[14-15] This eruption was fairly large in terms of magnitude, since about nine cubic kilometres of pumice were erupted, but its most remarkable aspect is the extreme dispersion of the tephra.

At its maximum, the preserved thickness of the pumice deposit is only 1.8 metres, but it covers a huge area. Fifty kilometres downwind from the volcano, the pumice is still a metre thick, while 100 kilometres distant is is still more than 25 cm thick. Tephra more than 10 cm thick covered an area of more than 15 000 square kilometres. In order for dispersal of the tephra to take place over such a large area, an enormously high eruption column would be required. The Taupo eruption column may have been as much as 50 km high. To generate such a high column requires an exceptional mass eruption rate, of the order of one hundred thousand cubic metres per second.

Radio-carbon dates of carbonized logs show that the Taupo eruption took place about AD 186, well within the period of documented history in the northern hemisphere. Because the Maoris did not reach the islands until about AD 1000, New Zealand was uninhabited at that time. Had the North Island been populated earlier, the eruption would surely have wiped out all settlements. For decades afterwards, much of the North Island would have been a barren desert of pumice, a dismal prospect for would-be colonizers.

— 6.4 Hydrovolcanic eruptions

Here we leave the realm of 'conventional' eruptions. *Hydrovolcanic* is a self-explanatory term: it suggests both water and volcanism. Water can interact with hot volcanic materials in a variety of ways: when a vent opens up under the water of the sea or a lake; when a volcanic vent on dry land intersects water contained in an aquifer (termed *phreatic* water, from the Greek for a well); or simply when a lava or pyroclastic flow moves over water-saturated sediment. In the simplest cases, small amounts of water come into contact with hot volcanic rock—not necessarily molten magma—causing steam explosions which fragment the rock and shower the bits around. These are common, if rather minor, phenomena. Where large amounts of surface or phreatic water interact with magma rather than hot rock, violently explosive *phreatomagmatic* eruptions ensue.

Thermal energy, of course, is what drives all volcanic eruptions. In phreatomagmatic eruptions, some of the thermal energy of the magma is used to heat water and turn it to steam, which fragments the magma as it expands explosively. Thus, a key characteristic of phreatomagmatic eruptions is that they yield exceptionally highly fragmented, fine-grained ash. 'Explosive' interactions may suggest single, instantaneous events, but phreatomagmatic eruptions can be prolonged, self-sustaining phenomena. When hot magma first comes into contact with water, the magma is itself chilled, while a film of steam forms in the contact zone. Huge thermal gradients are set up, and the film is therefore unstable and frequently collapses. Thermal stresses building up in the outer layer of the magma cause it to fragment; when it fragments, a still greater area of hot rock is exposed to water and thermal stresses. Explosive fragmentation thus propagates into the magma body. Similar phenomena are familiar to nuclear engineers morbidly contemplating the 'China syndrome' in runaway reactors, when the core melts its way downwards through the containment: they are called *fuel–coolant interactions* (FCIs) because they result when molten nuclear fuel interacts with coolant.[16]

FCIs are so important that they have been closely analysed in the laboratory, leading to some useful volcanological spin-offs.[17] Experiments show that for the efficient conversion of thermal energy into mechanical energy, the optimal mass mixing ratio of basaltic melt to water lies between 0.1 and 0.3. In this range, rapid vaporization of water and the consequent expansion results in explosive yields that can reach one-quarter to one-third that of an equivalent mass of TNT. Thus, hydrovolcanic explosions are not to be taken lightly. In experiments carried out in the optimal range, finely fragmented particles (less than 50 microns) always resulted. At lower or higher mixing ratios, much larger particles (1–10 mm) were produced.

Water is almost ubiquitous on the Earth's surface, so hydrovolcanic eruptions are common; most eruptions probably involve a hydrovolcanic component, even if it is not immediately obvious. Two kinds of exclusively hydrovolcanic eruption are recognized: *Surtseyan*, which are the 'wet' equivalents of strombolian eruptions, and *phreatoplinian*, the equivalents of sub-plinian and plinian eruptions. Surtseyan eruptions typically involve basaltic magmas; phreatoplinian eruptions more silicic ones. Interestingly, it has been argued that since much of a magma's thermal energy is expended in flashing water to steam, a phreatomagmatic eruption ought not to be as efficient as a conventional one, and that therefore eruption column heights might be less. There is little observational data on this, however.

6.4.1 Surtseyan eruptions

In 1952 a Japanese hydrographic research vessel, the *No. 5 Kaiyo-Maru* was sent to investigate reports of a submarine eruption about 420 km south of Tokyo. She never returned from that mission: an explosion apparently took place while she was directly over the site of the volcano. There were no survivors from the crew of thirty-one, but some floating debris was found later, with fragments of rock particles embedded in it . . .

In July 1990, another Japanese research vessel,

the *Takuyo*, observed submarine activity off the Izu Peninisula, Japan, an area of frequent submarine events. At 18.29 on July 13, small-scale seismicity was detected by a Japan Meteorological Agency seismometer. At sea, an explosion was heard and a 30-second vibration felt at 18.33. One minute later, the JMA seismometer was saturated by seismic events, and remained saturated for the next ten minutes. At 18.40, the crew of the *Takuyo* saw the sea surface dome upward about 500 m from the vessel, then a grey–black plume rose from the same area. Five more plumes, about 30 m high and 100 m across, were observed in the next five minutes. Ejection of each plume was accompanied by violent shaking and vibration of the ship. Seismic activity continued, but no further surface manifestations were seen. Later, bathymetric surveys revealed a new cone in 100 m of water. The cone was about 450 m in diameter, but rose only 10 m above the sea-floor.[18]

What the crew of the *Takuyo* saw were the first stages in the creation of a new volcanic island. In 1963, formation of a new island off the coast of Iceland, Surtsey, provided an instructive example of submarine eruptions, and gave its name to a distinctive style of activity. On the Icelandic mainland, basaltic eruptions are typically effusive: colourful but modest strombolian eruptions building scoria cones, with basaltic lavas flowing quietly away. When such an eruption takes place offshore, things are different. At first, there is little to see other than an area of dirty, discoloured water, with a few dead fish floating around in it. Next, violent explosive activity commences, similar to that described by the *Takuyo* crew. Every few seconds or minutes, powerful blasts propel showers of fragments hundred of metres into the air, forming 'cock's tail' plumes, while a vigorous, persistent steam plume rolls steadily upwards (Fig. 6.6). Such activity may continue for many weeks, until so much ejected material has piled up on the sea-floor that it emerges above water.

Even after the volcano has built itself above sea-level, the style of activity may not change much, since abundant sea-water can still find its way into the vent. Eventually, a visible island may emerge above sea-level, progressively insulating the hot magma from sea-water. Many eruptions do not get beyond this point. In others, copious quantities of hot spatter and small lava flows begin to weld the island into a more solid structure. Explosive activity dies away, and the active vent begins to eject ordinary scoria. In this condition, the eruption has ceased to be hydro-volcanic at all, and has become a conventional strombolian one. Surtsey commenced activity on 4 November 1963. By April 1964 it had progressed to strombolian activity, lavas flowing into the sea through a breach in the cone. Eruptions from the main vent finally ceased in May 1965. Now silent, it has been colonized by sea-birds and scientists, the latter in monitoring the arrival of the first plant species.

A similar eruption, but one with intriguing geopolitical implications, took place in the Mediterranean in 1831, 50 km north-east of the island of Pantelleria. It was decribed by the great geologist Charles Lyell in the very first textbook on geology. On 18 June, an English ship sailing in the area was jolted by 'earthquake shocks' as though it had struck a sandbank. On 10 July, a column of 'water' 800 m in diameter was observed spouting up from the sea to a height of 20 m, shortly to be replaced by dense columns of steam. By 18 July, a small island had appeared, rising a few metres above sea-level, with a central crater from which immense quantities of scoria and vapour were being ejected. By the end of July, the island was over a kilometre in circumference and 30 m high. A couple of weeks later it reached its maximum dimensions of about 5 km in circumference and 60 m high (Fig. 6.14).

A new piece of real-estate appearing in the strategically important western Mediterranean was a mouth-watering prospect to all the nineteenth century European imperialist powers, so the new island was immediately claimed by Britain, Spain, and Italy. Anxious to stake their claim, each proposed different names for the

Fig. 6.14 Strombolian activity is unambiguously depicted in this rare gouache painting of the eruption of Graham Island in the Mediterranean in 1831. An early steam ship investigates on the left.

island. Captain Senhous, R.N., was the first person to land on the lisland, and the name he gave it, Graham island, was adopted by the British. Their rivals promoted Ferdinanda, Hotham, Corrao, Sciacca, Julia, and Nerita. Sadly for these hopes of Mediterranean hegemony, the island rapidly succumbed to wave erosion. By the end of October, it had almost gone. By early in 1832, it had disappeared altogether. As Lyell commented wryly in his book: 'as the isle was visible for only about three months, this is an instance of a wanton multiplication of synonyms which has scarcely ever been outdone, even in the annals of zoology and biology'.

In modern times, the remote submarine volcano Kavachi in the Solomon Islands (SW Pacific) has been the site of at least eight island-forming eruptions since 1939.[19] Each successive island was soon washed away by the sea, but it can only be a matter of time before sufficient debris accumulates for a longer-lived structure to emerge.

Eruptions within crater lakes

Minor variants of subaqueous eruptions are those that take place beneath crater lakes. For a period during the 1970s, this was a normal state of affairs at Poas volcano in Costa Rica, which contains a shallow lake about two hundred metres in diameter. Owing to the steady streaming of sulphur dioxide through the lake water, it was intensely acid, with a pH less than 1. It was also quite hot, about 47°C. Periodically, small eruptions took place beneath the malevolent, greeny-grey waters of the lake, blasting plumes of mud-laden spray into the air like depth charges (Figs. 6.15–6.16). Although entertaining to watch, these eruptions were generally small and harmless—at least to humans. Downwind of the volcano, the luxuriant tropical rain forest has been wiped by sulphurous fumes from the volcano and intermittent showers of acid rain jetted up from the lake. Clothing wetted by lake water disintegrated in a few days, so the effects on vegetation may be easily imaginable.

Because the lake water at Poas buffers the temperature within the crater, fumarolic sulphur deposits accumulate, rather than being sublimated as they would in dry conditions. In some circumstances, when the water level falls, pools of molten sulphur form, some of them several metres across, bubbling vigorously like evil cauldrons. If further explosive eruptions take place, liquid sulphur is sprayed into the air, falling back around the crater to form a bizarre greenish pyroclastic deposit. On some ejected boulders, little dribbles of molten sulphur form stalactites and stalagmites. Altogether, wandering around the grey, ash-draped, malodourous

Fig. 6.15 Poás volcano (Costa Rica) and its crater lake in a peaceful condition in 1978.

Fig. 6.16 Bursting through the lake surface like a depth charge, a small eruption blasts from the lake floor at Poás. Larger eruptions eject showers of acid lake water over large areas, but are hard to photograph.

crater of Poas reminds one of Milton's description of Purgatory:

> A dungeon horrible on all sides round
>   As one great furnace flamed,
> Yet from these flames, no light,
>   but rather darkness visible
>   Served only to discover sights of woe
> Regions of sorrow, doleful shades where peace
>   and rest can never dwell, hope never comes
> That comes to all, but torture without end
>   Still urges, and a firey deluge, fed
> With ever burning sulphur unconsumed
>   *Paradise Lost*

6.4.2 Phreatoplinian eruptions

Rather few phreatoplinian deposits are known; far fewer than surtseyan. In part, this is because the type of activity and the deposits that result have only recently been recognized, and in part it probably also reflects the physical difficulties inherent in mixing water with the large volumes of silicic magma required. No phreatoplinian eruption has been observed in progress, so all that we know of them has been learned from studying old deposits. A well-described example is the Oruanui ash from the Taupo area of North Island, New Zealand. It is one of a series of ash deposits recognized to be unusual because of their very fine grain size (highly fragmented) and

wide dispersion, although the eruption magnitudes were not large.[20–21] Askja volcano in Iceland erupted a similar highly dispersed, fine-grained ash during its eruption in 1875. Both the Oruanui and Askja deposits clearly resulted from high eruption columns. In the case of the Oruanui ash, it is not difficult to see how water was able to interact with magma because the eruption began directly beneath the waters of Lake Taupo, the largest lake on the North Island.

6.4.3 Fire and ice: sub-glacial eruptions

During the last great Ice Age, which ended only 11 000 years ago, almost a third of Earth's land area was covered by ice sheets several hundred metres thick. Inevitably, therefore, many volcanoes erupted beneath ice. Relics of this curious, almost paradoxical expression of volcanic activity are preserved in several high-latitude volcanic regions, but are best known from British Columbia, where they were first recognized[22] and Iceland, where they are best displayed.

Naturally, the first thing that happens in a sub-glacial eruption is that a lot of ice melts, forming copious amounts of water. Thus, in its earliest stages, the eruption is, in effect, a submarine one. Pillow lavas are erupted, piling up to form a steep-sided mass confined by the icy sides of the vault melted in the ice. As the eruption proceeds, masses of chilled, fragmented basalt accumulate. Eventually, the active vent may build itself sufficiently high that fragmentation ceases, so that normal lava flows are able to emerge above the pile of pillow lavas and lava debris, but they remain ponded within the cavity in the ice.

6.4.4 Hlaups and jökulhlaups

Sub-glacial eruptions present unique hazards when torrents of melt-water burst out of their icy dams. Modern sub-glacial eruptions are best known from Iceland, where there are four ice-caps, or *jökuls*, largest of which is the Vatnajökull, situated above the Grimsvotn volcano. Normally, ice-caps are drained by one or more sizeable rivers emerging from their flanks. In a volcanically triggered *hlaup*, a colossal deluge of icy-cold water breaks through the ice-cap, scouring a channel as much as a kilometre wide, and inundating the outwash plain beneath the ice-cap. No over-used adjective can convey an adequate impression of the volume of water liberated in a *jökulhlaup* (glacier-burst). Figures help, though. The Amazon has an estimated flow of something like 200 000 cubic metres per second. *Jökulhlaups* from Grimsvotn have usually lasted 1–2 weeks, with total volumes of 6–7 cubic kilometres, and runoff rates peaking briefly at about 40 000 cubic metres per second. When Katla volcano erupted in October 1918, it triggered floods with a peak of about 100 000 cubic metres per second. Such huge volumes of water have enormous erosive powers, and are capable of transporting millions of tonnes of sediments long distances. A boulder about 400 cubic metres in size was carried fourteen kilometres by the Katla *hlaup* of 1918, and huge icebergs tens of metres across were transported

three kilometres. South-east Iceland's entire coastline has been modified by the deposits left by these volcano-propagated floods. Fortunately they have caused little loss of life because the area is thinly populated.

On the other side of the world, a small *hlaup* was produced during the August 1991 eruption of Mt. Hudson in Chile. Hudson, in latitude 46° S, consists of a beautiful 8-km-diameter caldera, whose rim reaches 2600 m. It resembles the magnificent caldera of Crater Lake in Oregon. However, unlike Crater Lake, Hudson's caldera is filled with glacial ice. Thus, when the August eruption took place, huge amounts of melt-water ponded within the caldera, overflowing eventually in a flood down the valley of the River Huelmules. Blocks of ice up to 5 m in diameter were washed almost as far as the Pacific coast, 35 km distant.

— 6.5 Peléean activity

Peléean activity cannot be as narrowly characterized as the earlier types, and is often linked to vulcanian or plinian activity. It is a dangerous style of activity, and the term is widely employed, thus it merits brief description here.

One feature characterizes peléean activity: the eruption of *nuées ardentes*, or glowing clouds, of the sort that destroyed St Pierre in 1902. Mt. Pelée unleashed many nuées during the eruption of 1902–1903, all of them linked with the growth of a large extrusion of viscous lava on the site of the Etang Sec, where the eruption started. Growth of the lava dome was accompanied by intermittent vulcanian activity, responsible for the loud explosions heard in St Pierre, and for the ash that sifted down on the town before and after the fateful 8 May. On that date, a large ash cloud blasted out sideways from the incipient dome, rather than being propelled vertically upwards as in a normal vulcanian explosion. This was a nuée. A nuée consists of solid fragments—huge boulders mixed indiscriminately with fine dust—avalanching downslope under gravity. Hot escaping gases rise upwards, carrying huge amounts of dust with them, forming a turbulent wall of ash-laden cloud kilometres high, while the

denser part, containing most of the solid material, hugs the ground and rolls rapidly over it at great speed. (Fig. 4.21).

A nuée is at once immensely hot and immensely powerful—the St Pierre nuée has often been called a 'tornadic blast'. Its destructive capabilities are all too clear. It is a little surprising, therefore, that similar phenomena had not been recognized prior to 1902, since they are not uncommon and must have been observed on many occasions in the past. Since 1902, many nuées have been observed at close quarters, many of them linked with lava dome growth, as at Mt Pelée. At 10.40 on the morning of Sunday, 21 January 1952, a violent eruption of Mt. Lamington, Papua New Guinea, generated a nuée which devastated an area of 180 square kilometres, and took the lives of 3000 people. Although local authorities did not even know at the time that Mt. Lamington was a volcano, an experienced volcanologist arrived on the spot within 24 hours, and made detailed daily observations during the following months of the eruption[23] (Section 12.3.2). In 1991, a similar but smaller nuée took 40 lives on Unzen volcano, Japan.

— 6.6 Eruptions in general

Labels such as 'strombolian' or 'plinian' should be applied only to recognizable phases of activity: many eruptions exhibit several different phases of different duration. An eruption of Mt. Etna in 1971, provides an enlightening example of this, although the eruption itself was unremarkable and caused no casualities.

Etna has two summit craters, one of which had been showing mild strombolian activity for many years prior to 1971. When a patch of snow below the summit began to melt at the end of March and the activity in the summit crater began to die away, it was clear that something was afoot. On 5 April, a set of fissures opened at an altitude of about 3000 metres, ejecting showers of incandescent lava fragments. These rapidly acumulated around the fissure, building up a small scoria cone. Later, lava flows were erupted and began to ooze down some of Etna's best ski slopes, engulfing one by one the pylons that carried the skiers' cable-way up the mountain. This was the

first phase of the eruption; it lasted about two weeks (Fig. 6.17–6.18).

Strombolian activity continued for most of that time in the scoria cone built over the fissure, until it was about 30 metres high. The lava flows did not reach the lower, inhabited parts of the mountain and stopped moving after a few days. Aparts from wrecking the skiiing facilities, the worst damage they did was to demolish an impressive volcanological observatory constructed during Mussolini's regime in the 1930s. It is hard to imagine a more appropriate fate! (Fig. 6.19).

On 13 May, a second phase commenced when a linear fissure opened abruptly several kilometres away, low on the eastern flanks of the volcano, at an altitude of only 1800 metres. There was no explosive activity, no eruption cloud; just a crack in the ground opening up, and a lot of pungent blue sulphurous fumes. Soon, lava began to flow freely but quietly from the

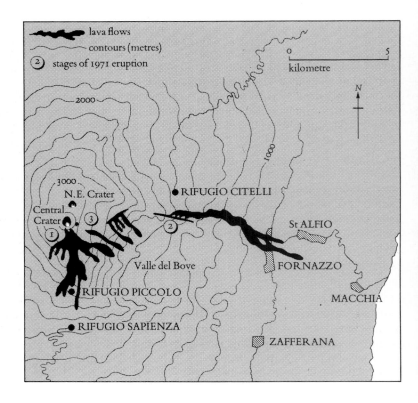

Fig. 6.17 Locality map for the three phases of the Etna 1971 eruption.

Fig. 6.18 Lava flows wrecking Etna's ski slopes during the first phase of the 1971 eruption. A plume of steam marks the active vent at the foot of the central crater. Fumes mark the course of the most recently active flow, at centre right.

Fig. 6.19 Last moments of the Observatorio Volcanologico Etneo in April 1971. Lava has already demolished the left wing; shortly afterwards, the entire structure (built in the 1930s) was engulfed.

fissure, in much larger volumes than in the first phase. This undramatic release of large volumes of lava would, if one had to give it a name, be best described as hawaiian activity, although there was no fire-fountaining. Because the lavas were erupted low down on the volcano, they soon flowed through inhabited areas, overrunning vineyards, bridges, roads, and houses indiscriminately, and threatened the villages of St Alfio,

Fornazzo, and Macchia. In St Alfio, the local priest was talked into exorcising the lava, and with or without divine intervention, the village escaped serious damage.

On 18 May, the third phase was initiated when a deep pit-like hole 150 m across suddenly opened up high on the volcano, at a height of 2980 metres. No lava at all was erupted here, but great, dark, cauliflower ash clouds rolled up from it for several days, laying a thin carpet of ash downwind. This was vulcanian activity.

So, not counting the initial strombolian activity in the summit crater, there were three widely separated sites of activity during the course of the eruption, and the character of activity was different at each: strombolian, hawaiian, and vulcanian. Many of the differences between the phases can be interpreted in terms of the degassing history of the magma. Much of the dissolved gas stored at high levels in the magma escaped during the first, strombolian phase, so that the lavas of the second phase were largely degassed or 'flat', and hence erupted quietly. Although the third, vulcanian phase does not fit so easily into this scheme, high-level degassing followed by lower-level emission of degassed lavas is typical of Etna, and many other volcanoes.

— Notes —

1. Mullineaux, D. R. and Crandell, D. R. (1981). The eruptive history of Mount St. Helens. In *The 1980 Eruptions of Mt. St. Helens, Washington*. U. S. Geol. Surv. Prof. Pap. No. 1250, Washington, pp. 3–16.

2. Chester, D. K., Duncan, A. M., Guest, J. E. and Kilburn, C. R. J. (1985). *Mount Etna: the anatomy of a volcano*. Chapman and Hall, London, 389 pp.

3. Budnikov, V. A., Markhinin, Ye. K., and Ovsyannikov, A. A. (1983). The quantity, distribution, and petrochemical features of pyroclastics of the Great Tolbachik Fissure Eruption, pp. 41–56. In *The great Tolbachik fissure eruption: geological and geophysical data 1975–1976*. (eds S. A. Fedotov and Ye. K. Markhinin), pp. 41–56 [English translation by Cambridge University Press, 1983].

4. Walker, G. P. L. (1980). The Taupo pumice: product of the most powerful known (ultraplinian) eruption. *J. Volcanol. Geotherm. Res.* **8**, 69–94.

5. Carey, S. and Sigurdsson, H. (1989). The intensity of Plinian eruptions. *Bull. Volcanol.* **51**, 28–40.

6. Mercalli, G. (1907). *I vulcani attivi dell Terra*, p. 119.

7. Earthquake Research Institute, Tokyo (1988). *The 1986–87 eruption of Izu–Oshima volcano*. Tokyo, Japan, 61 pp.

8. Dana, J. D. (1891). *Characteristics of volcanoes with contributions of facts and principles from the Hawaiian Islands*. Low, Marston, Searle and Rivington, London, 399 pp.

9. Barberi, F., Cheminee, J.–L., and Varet, J. (1973). Long-lived lava lakes of Erta Ale volcano. *Rev. Geog. Phys. et de Geol. Dyn* **XV**, 347–52.

10. Tazieff, H. (1979). *Nyiragongo: the forbidden volcano*. Cassell, London, 287 pp.

11. McPhie, J., Walker, G. P. L., and Christiansen, R. L. (1990). Phreatomagmatic and phreatic fall and surge deposits at Kilauea volcano, Hawaii, 1790 AD: Keanakakoi Ash Member. *Bull. Volcanol.* **52**, 334–54.

12. Williams, S. N. and Self, S. (1983). The October 1902 Plinian eruption of Santa Maria Volcano, Guatemala. *J. Volcanol. Geotherm. Res.* **16**, 33–56.

13. Keam, R. J. (1988). *Tarawera*. Privately published. Auckland, New Zealand, 472 pp.

14. Walker, G. P. L. (1980). The Taupo pumice: product of the most powerful known (ultraplinian) eruption. *J. Volcanol. Geotherm. Res.* **8**, 69–94.

15. Wilson, C. J. N. and Walker, G. P. L. (1985). The Taupo eruption, New Zealand I. General aspects. *Phil. Trans. Roy. Soc. Lond.* **A314**, 199–228.

16. Peckover, R. S., Buchanan, D. J., and Ashby, D. (1973). Fuel–coolant interactions in submarine volcanism. *Nature* **245**, 307–8.

17. Sheridan, M. F. and Wohletz, K. H. (1983). Hydrovolcanism: basic considerations and review. *J. Volcanol. Geotherm. Res.* **17**, 1–29.

18. Volcanic Events (1990). In *Eos* (*Am. Geophy. Union*) **70**, p. 791.

19. Smithsonian Institution. (1989). *Global Volcanism 1975-85*, Prentice Hall, 655 pp.

20. Self, S. and Sparks, R. S. J. (1978). Characteristics of widespread pyroclastic deposits formed by the interaction of silicic magma and water. *Bull. Volcanol.* **41**, 196–212.

21. Self, S. (1983). Large scale phreatomagmatic silicic volcanism: a case study from New Zealand. *J. Volcanol. Geotherm. Res.* **17**, 433–69.

22. Mathews, W. H. (1947). 'Tuyas', flat topped volcanoes in northern British Colummbia. *Amer. J. Sci.* **245**, 560–70.

23. G. A. M. Taylor, G. C. (1983). The 1951 eruption of Mt. Lamingon, Papua. *Bur. Min. Res. Geol. and Geophys. Bull.* **38** (second edn). Canberra, Australia, 129 pp.

7

Lava flows

Throughout the solar system, from the scorched surface of Mercury to the distant clutter of the asteroid belt, basalt lavas are the dominant expression of volcanism. On Earth, lavas are gently oozing forth from a few volcanoes on dry land at any one time, and more may be emerging below the sea. This book was written in Honolulu between 1989 and 1991. Throughout that time, basalt lava was streaming continuously into the ocean from a vent at Kupaianaha on the neighbouring island of Hawaii. In 1990, lavas became unwelcome visitors at a Visitor Center in the Volcanoes National Park, engulfing it. Later, they burned their way through the entire village of Kalapana. Frequently, the evening news carried harrowing images of houses bursting into flames as flows encircled them, while their owners watched impotently.

Lavas may appear to be merely streams of molten rock, presenting few technical problems for volcanologists. Nothing could be further from the truth. Lavas have a wide range of compositions, from carbonatites through basalts to rhyolites. Composition apart, their physical properties are also influenced by their volatile contents, crystal contents, and cooling histories. Thus, flows of marginally varying compositions may behave quite differently. Only recently has the rheology of lava flows come under close scrutiny, so it remains poorly understood. Even the scale of flows can be misleading. Documentary TV programmes often feature stock film sequences of basalt lavas snaking down the flanks of volcanoes, commonly in Hawaii, where lavas are highly accessible. Although eminently watchable, these sequences convey an impression that lava flows are not particularly big—few of those filmed are more than a few tens of metres wide, and they rarely flow more than a few kilometres. But many flows known from the geological record are of terrifying proportions. In Queensland, Australia, not a place that one often associates with volcanism, there is a series of flows only a few thousand years old, which extend nearly 100 km from their source.[1]

Fourteen million years ago, over 700 cubic kilometres of basalt spewed to the surface in the Columbia River Plateau to form the Roza flow, which travelled more than 300 km westwards from its source along the Columbia River Gorge.[2] One estimate suggests that the lava flow front was 30 m high; extended laterally for over 100 km; flowed forward at five kilometres per hour; and that the eruption rate was so great that the whole volume may have been erupted in a week.[3] The slightly younger Pomona flow travelled even further, reaching over *550* km from its source (Fig. 7.1). Some basalt lavas on the Moon and Mars flowed even further. Thus, while no huge flows have been erupted *recently* on Earth, we should not be lulled into thinking that lavas necessarily have the limited scales that we associate with familiar images of them.

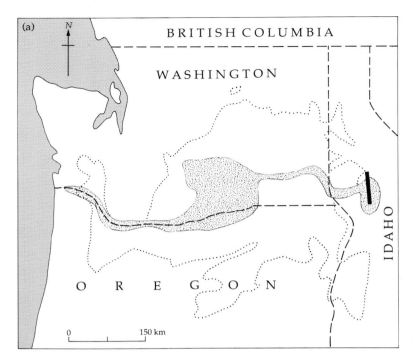

Fig. 7.1 (a) Outcrop of the 14-million-year-old Pomona flow in Washington and Oregon, USA, according to P. R. Hooper.[3] Dotted outline marks outcrop of the Columbia River basalt province, and the short solid line the fissure vent region for the flow.

Fig. 7.1 (b) View across the Snake River to the elegant columns of the Pomona flow, close to its source in the Lewiston Basin. The flow filled an older Snake River canyon. Photo: courtesy of P. R. Hooper.

7.1 A typical basalt lava

Before looking at the physics underlying the flow of lava, let us review briefly what happens to a lava of modest scale, traced from source to toe. For a technical account of an individual flow, Lipman and Banks's description of a 1984 Mauna Loa flow is ideal.[4]

When it first emerges from below ground, a basalt lava is at a temperature of about 1200°C. It glows fiercely reddish yellow; noticeably redder than the colour of molten iron poured from a furnace. At this temperature, the molten rock is at its least viscous, so it flows downslope in a fiery torrent, splashing boisterously over minor obstacles, and cascading over larger obstacles in glowing fire-falls. Flow rates as high as 60 kilometres per hour were observed when the Nyiragongo lava lake was breached in 1977. Although the term 'rivers of fire' inevitably springs to the lips of narrators of volcano films, closer observation shows that the lava flows more like thick cream than water: an expression of its greater viscosity.

Fierce surface radiance quickly cools the flowing lava. Its glow fades to a less intense cherry red, while its viscosity increases to something more like molasses. As the viscosity increases, the lava flows more smoothly, slowing to the speed of a run, and then of a fast walk. Black streaks and blobs of chilled solid lava appear on the surface. (These dark streaks may *appear* black against the brighter background, but are themselves still very hot, and would probably appear red-hot against a dark background.) Initially, these streaks are no more substantial than flecks of foam on a boisterous mountain stream, soon engulfed once more in the main mass. As the lava cools further, more chilled material remains on the surface, agglomerating together to form progressively larger rafts which eventually cover the surface, hot lava itself glowing dully red only through cracks and fissures. At this temperature, the lava is extremely viscous, like sticky treacle. Its stickiness can make it difficult to extract the pole that volcanologists, obeying some irresistible atavistic urge, feel obliged to poke into it, if they can get near enough.

As cooling progresses, a rubbly surface layer of solidified lava blocks accumulates, carried on top of the moving lava like rock on a quarry conveyor belt. Solid lumps of lava also pile up in front of the advancing front of the flow. In the higher, hotter parts, the lava continually advances over this solid material, rolling over it like a caterpillar tractor. In its lower, colder parts the lava may resemble a static heap of slag from the outside, but mobile lava is still arriving deep within it and pushing forward. So the front of the flow is unstable, and there is a constant gentle clatter of small chunks of lava falling forward. Every now and then a larger mass breaks off, toppling forward in a glowing cascade, leaving a sullen red scar, quickly fading, to mark the place on the flow that it fell from. Where such a flow encounters trees, it usually pushes them over, rather than flowing round them.

In this way, the toe of the flow slowly advances, clattering and rattling forward like a shuffling slag heap. Barely perceptible movement of perhaps a few metres an hour continues until the supply dies away at source, and the flow becomes still and silent. In a matter of hours, the surface will be cool enough for geologists to swarm over it without harm, except for scorched boots and singed back sides. Deeper down, the interior mass cools much more slowly, so that for days afterwards the core may still be red hot; its glow visible through cracks and fissures. Rubbly lava is a remarkably effective insulator—in 1938 a couple of Russian geologists hopped on a raft of chilled lava on a moving flow from the Klyuchevskaya volcano in Kamchatka. They were carried along on the gently moving conveyor belt at about one and a half kilometres an hour while they made measurements of the viscosity of the lava, and eventually hopped off again, none the worse for wear, although the surface temperature of their raft was 300°C, and that of the flow interior many hundreds of degrees higher.

7.1.1 The 1783 Skaftár fires (Laki) fissure eruption

Of the innumerable lavas which have been erupted in modern times, the 200-year-old Laki fissure eruption demands mention because of its size and environmental effects. (Icelanders know this event as the Skaftár fires). We follow descriptions by the Icelandic volcanologists S. Thorarinsson[5] and T. Thordarsson. A classic example of a basaltic fissure eruption, the lavas were erupted from a 27-km-long chain of more than 140 craters which cut across an older *moberg* (Section 16.1.6) ridge called Laki; the crater row is called *Lakagigar*, and the lava flow, which entered the Skaftá river valley, the *Skaftareldhraun* (Fig. 7.2).

After a week of precursory seismic activity, the eruption began at 9 a.m. on Whit Sunday 8 June 1783. It lasted eight months, until 7 February 1784. For the first few days, intense hawaiian fire fountains sprayed basaltic magma up to 1400 metres into the air, forming large amounts of Pelé's hair (Section 9.3). Fissures opened initially south of the Laki moberg, lava flowing down the Skaftá River gorge to reach the lowlands 35 km away four days later. On 29 July the crater row north-east of Laki opened up, lava surging along the river-bed of the glacial river Hverfisfljot.

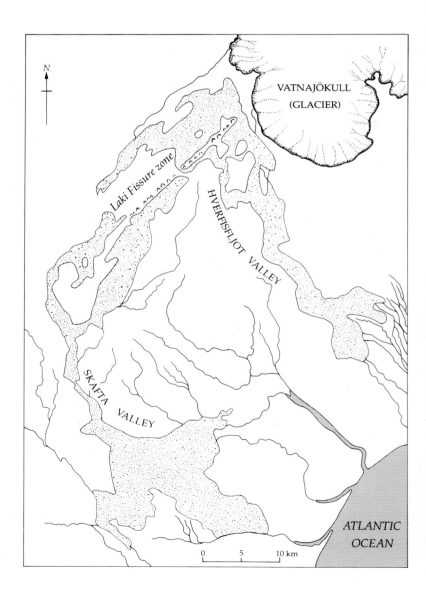

Fig. 7.2 The 27-km-long Lakagigar crater row, south-east Iceland, and lavas from the 1783 eruption filling the Skaftá and Hverfisfljot river valleys.

Several later phases of renewed rifting and copious lava effusion have been identified. At maximum, more than 8000 cubic metres per second were flooding out of the fissures, comparable with the discharge of the River Rhine near its mouth.

In all, 14 cubic kilometres of lava were erupted, burying 565 square kilometres of land, including 2 churches and 14 farmsteads in the neighbouring valleys. Excellent records were made of the lava's progress, notably by the Reverend Jon Steingrimsson, a clergyman based in the hamlet of Prestbakki. His maps show the flow front of the lava advancing at times up to 5 km per day along the Skaftár and Hverfisfljot valleys. Simple overrunning of farm land and buildings was the least serious of the effects of the eruption, which ultimately caused the deaths of almost a quarter of the population of Iceland (Chapter 17).

7.2 Rheology and other considerations

What controls the thickness of a lava flow? How far will a flow reach? Questions like these may seem elementary—as indeed they are—but it is still difficult to answer them satisfactorily. The fluid dynamics of lavas are poorly understood at present, partly because they are inherently difficult to study, but also because the field has attracted less attention than other aspects of volcanology, notably petrology.

As discussed in Chapter 5, magma viscosity varies enormously with temperature, composition, and degree of polymerization. While many measurements of lava viscosity have been made by hardy souls inserting viscometers into active flows, these are single, point source measurements. Measuring temperature and viscosity and their variations across *all* parts of a basalt lava at the same time is much more difficult, probably impossible. Andesite and rhyolitic flows present even more intractable problems. Thus, we can address here only some of the simplest aspects.

7.2.1 Viscosity and flow thicknesses

Superficially, a blob of viscous liquid such as glycerine may seem to behave like lava when it oozes slowly over a flat surface. Over time, however, a blob of glycerine flattens out to a thin film: it is a Newtonian liquid, with zero yield strength. Two forces influence the rate at which a blob spreads: gravity, and viscous resistance. The magnitude of the gravitational force depends, of course, on the mass of the planet our blob of glycerine is oozing over, but viscosity is a property intrinsic to the material. Lavas also possess a yield strength, largely derived from the chilled crust which immediately forms on their surface. Before a blob of lava can spread, its yield strength must be overcome by shear stresses. Thus, lavas never form flat films; they always retain some thickness, which is a measure of their yield strength. Fluid basaltic lavas such as those of Hawaii commonly flow in oozy dribbles as little as a few centimetres thick; rhyolites, by contrast form flows which are never less than many metres thick. Most studies of lava flow rheology have built upon foundations laid by Hulme in a seminal paper published in 1974.[6] (There is also an excellent review by Chester *et al.*[7]). These studies show that the critical thickness, t, that a body of lava resting on a sloping surface must obtain before it will flow is given by a simple equation:

$$t = \frac{\tau}{\rho g \tan \alpha} \qquad (1)$$

where g = acceleration due to gravity, ρ = density, τ = yield strength, and α = slope in degrees (Figs. 7.3–7.4).

Once the flow gets moving, things rapidly get extremely complicated, because of the large number of variables. Apart from temperature and composition, lava viscosity is also a function of the *rate* of shear. Thus, a basalt flow moving on a steep slope where shear rate is high will have a lower viscosity than one moving on a gentle slope, other factors being equal. Disregarding all these complications, to a first approximation, the

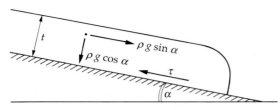

Fig. 7.3 A flow of thickness t is propelled downslope by a force proportional to $\rho g \sin \alpha$, but is resisted by a basal shearing stress. Flow only takes place if the downslope force is larger than the yield strength τ.

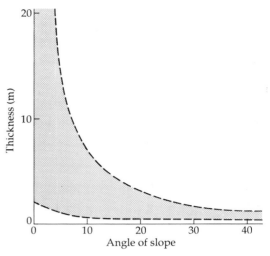

velocity V at which a lava of thickness t will flow down a slope α is given by:

$$V = \frac{\rho g t^2}{B\eta} \sin \alpha \qquad (2)$$

Where η is the viscosity and B is a constant, whose value is around three on a flat surface. (The top line of the equation essentially describes the shear stress generated by the lava.) It does not take a mathematical genius to see that lavas with lower viscosities will flow faster than those with high viscosities.

It follows from their variations in viscosity and temperature that lavas of different compositions will form flows with different morphologies. One useful way of expressing variations in morphology is the *aspect ratio*, a term borrowed from the realm of aviation by G. P. L. Walker. To an aeronautical engineer, aspect ratio is the ratio of

Fig. 7.4 Measured thicknesses of young lavas on Mt. Etna fall within a well-defined field when plotted against the angle of the slope that they descend. Upper and lower bounding limits of the envelope correspond with lava yield strengths of 10^5 and 10^3 Pa. (After Walker, G. P. L. (1967). Thickness and viscosity of Etnean lavas. *Nature* **213**, 484–5.)

the span of a wing to its width; to a volcanologist it is the ratio of the thickness of a flow to the area it covers.[8] Basalt lavas are thin and extensive, whereas rhyolite lavas are fat and compact. These variations can be expressed on a thickness against area plot, where basalt lavas occupy a distinctive field, with aspect ratios less than 1/100 (Fig. 7.5).

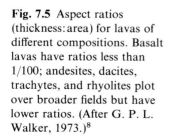

Fig. 7.5 Aspect ratios (thickness:area) for lavas of different compositions. Basalt lavas have ratios less than 1/100; andesites, dacites, trachytes, and rhyolites plot over broader fields but have lower ratios. (After G. P. L. Walker, 1973.)[8]

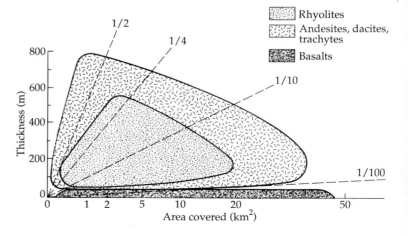

7.2.2 Levées

Because lava flows have yield strengths, they do not slop around untidily all over the place, but are bounded by levées, natural banks at the sides of the flow. Chilled material rapidly builds up these marginal banks, channelling the lava into a self-made canal. Levées are often left standing high and dry, like long, regular walls, after the initial rapid rate of flow decreases (Figs. 7.6–7.7). In detail, levées can form in a number of ways such as overflow and avalanching[9] but more interestingly, the width of a levée in theory provides a measure of the yield strength:

$$W_b = \tau/2g\rho(\tan \alpha)^2 \qquad (3)$$

where W_b is the width of the levée.

7.2.3 Lengths of lava flows

A lava flow begins to erupt from a vent high on the flanks of Mt. Etna or Mauna Loa. How far will it flow? Will it engulf the bustling village of Nicolosi, or incinerate the leafy suburbs of Hilo? Such questions emphasize that lava flows present problems of more than purely academic interest. Many factors influence the distance a flow will travel—composition, slope, rate of cooling, and so on. After collecting a mass of data, G. P. L. Walker suggested in 1973 that, other factors being equal, the *rate of effusion* is critical. On the largest scale, this is clearly correct: basalt lavas achieve far higher effusion rates than silica lavas, and they flow far further; hundreds of kilometres in exceptional circumstances. In flood basalt eruptions, effusion rates of thousands of cubic metres per second are possible, but in a typical andesite eruption such as that of Arenal, Costa Rica, maximum effusion rates do not exceed a few cubic metres per second, and flows rarely extend more than ten kilometres.[10]

Among flows of similar composition, the problem of lava flow length is more complex. While effusion rate remains significant, other factors are important. Pieri and Baloga[11]

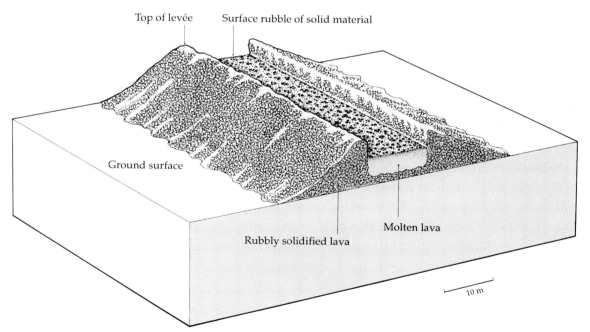

Fig. 7.6 Formation of *levées* in a typical lava. Initially, lava forms a tongue whose thickness is defined by its yield strength, but its edges soon chill, channelling a lava stream which may overflow its banks if effusion rate increases; when effusion rate decreases, lava drains downslope, leaving *levée* walls rising well above the lava surface.

Fig. 7.7 Elegant *levées* on a lava which flowed down from the summit of Teide volcano in Tenerife, Canary Islands. The *levées* are so regular that they look almost artificial.

showed that effusion rate is a better guide to the surface area finally covered by a flow, rather than its length. They argued that since radiation from the surface is the dominant cooling mechanism, the proportion of the lava's surface that exposes glowing, hot lava rather than chilled crust will dictate how quickly the lava cools, and hence how far it travels. This proportion is in turn affected by slope and topographic roughness.

While it may appear obvious that lavas cool mostly by radiation, it is worth emphasizing that radiation is *much* more important at high temperatures than low. At moderate to low temperatures, convection is the chief cooling mechanism. Radiative cooling is summarized in the famous Stefan–Boltzmann equation:

$$E = kT^4$$

where E = radiated thermal power and T = temperature in degrees K.

Because the *fourth* power of T is involved, a small percentage decrease in temperature yields a huge decrease in radiant thermal energy. Like other rocks, lavas are poor conductors of heat. When the surface temperature of a lava is high, there is a tremendously rapid radiative heat loss, which cannot be balanced by thermal conduction from the core. Thus, the crust of an active lava

flow inevitably chills quickly as it moves away from its source.

7.2.4 *Lava flows v. flow fields*

A 'lava flow' is not necessarily a single entity. As anyone who has tried to cross young basaltic lavas will have discovered to their cost, they often form a jumble of overlapping and anastomosing channels, presenting formidable obstacles to movement. When a number of small flows overrun each other and the whole mass cools together, the term *compound* flow is sometimes used; the term *flow field* describes the tangle of flows from a single eruptive episode (Fig. 7.8). Compound flows are usually the products of relatively slow effusion rates, permitting earlier flows to stop and cool a little before being overrun by later ones; at higher rates, lavas flow rapidly away from their sources to form long, *simple* flows.

In a study of flow fields on Mt. Etna, Chris Kilburn and Rosaly Lopez found that cooling and discharge rate along a flow appear to be the main factors controlling the generation of new flows.[12] A lava that cools rapidly relative to its flow rate will sprout many offshoots. As a result, long durations of flow at slow effusion rates will generate flow fields that have large width to length ratios, composed of many branches. Geoff

Fig. 7.8 Vertical air photo of a basaltic andesite lava flow field from the La Poruña scoria cone, Chile. Although a single entity on the large scale, the flow field is made up of innumerable small and large flow lobes and channels, particularly well displayed at top left. Flow field is about 8 km long.

Wadge showed that effusion rates often vary during the course of an eruption, increasing rapidly to a maximum, and then waning gradually.[13] These variations are reflected in the changing character of the resulting flow field. Early on, at peak effusion rates, the field extends in length downslope; later, as the field matures, it ceases to extend much downslope but becomes much broader.

Like the filigree patterns of frost crystals on window panes, branching trees and innumerable other natural phenomena, the outlines of lava flow fields appear to be *fractals*; that is, they appear to have similar shapes when viewed at different scales. Scientists at the University of Hawaii studied photographs of flow fields on the Earth, and satellite images of flows on Venus, Mars, and the Moon, and found that the fractal nature of flow outlines holds over five orders of magnitude in scale, from 0.5 m to 60 km.

—— 7.3 Morphological features of basalt lava flows ——

Aa flows

Two kinds of basalt lavas are familiar to all undergraduate geologists because their Hawaiian names lend themselves to atrocious puns: *aa* and *pahoehoe* (Fig. 7.9). *Aa* (properly written *a'a*) is the commonest type; *aa* surfaces are a jumble of loose, irregularly shaped cindery blocks, often with razor-sharp asperities. Crossing an *aa* flow is a miserable experience. Every step is a balancing act; every slip of the foot ensures deeply slashed boots; a fall means badly lacerated hands and knees.

When revealed in cross-section in cliff or

Fig. 7.9 Sharp contrasts in physical properties of clinkery *aa* lava (left) and smooth, shiny *pahoehoe* (right) are evident here. Both were produced by the same eruption of Mauna Ulu in Hawaii in 1974, and are chemically identical. Photo is about 1 metre across.

quarry, the structure of a typical *aa* flow is apparent. Usually a few metres thick, it consists of two distinct zones, an upper, rubbly part, and a lower, massive part consisting of solid lava which cooled slowly, insulated from the atmosphere by the overlying rubble (Fig. 7.10). Beneath this zone, a reddened, baked layer of pre-existing soil is often present, separated from the massive base by a thin layer of rubbly lava, which was overridden by the advancing flow. Within the massive lava, vesicles are often abundant, sometimes reaching several centimetres in size. Most are spherical, giving the lava a bubbly appearance, but some are elongated upwards, forming *pipe vesicles*. Sometimes, gas *blisters* form large cavities tens of centimetres across (Fig. 7.11). In older lavas, which have been deeply buried and subjected to low-grade metamorphism, a range of new minerals crystallizes within vesicles, forming *amygdules*. Mineral collectors prize especially well-crystallized *zeolites*, which often form magnificent radiating clusters.

7.3.2 Pahoehoe

Pahoehoe lavas are the least viscous of all common lavas—viscosities as low as 10 Pa s have been estimated for some Hawaiian flows, and their yield strengths approach zero. Thus, they form a wide range of fluidal surface struc-

tures. *Pahoehoe* surfaces are less common than *aa*, but can be a delight to walk over when they are smooth and glossy. Sometimes, the smooth surface skin is just that—a thin skin overlying large cavities. In this case, known as *shelly pahoehoe*, to walk over the surface is to invite sudden crashing descents or wrenching stumbles when the thin shell breaks abruptly. Amusing the first time, exasperating the twentieth time, and unprintable after that. Innumerable subtly different structures form on *pahoehoe* lavas. Best known is *ropy pahoehoe*, whose surface is crumpled within channels into closely spaced series of pleats (Fig. 7.12). In Hawaii, aptly named *entrail pahoehoe* is as common; its glossy, bloated intertwined protuberances resembling the viscera of a huge pachydermal animal (Fig. 7.13 a and b). Ropy *pahoehoe* tends to form early in the evolution of a flow field, when lava moves relatively rapidly; slower, more sluggish flows form entrails.

It is enlightening to watch a small Hawaiian *pahoehoe* flow creep slowly forward. Its advance guard is typically a glowing lobe only ten or twenty centimetres thick, which chills to a stop within a metre or two. As it chills and contracts, the hot surface continually spalls off small glassy flakes with a gentle clittering sound. Thermal stresses cause some flakes to pop up many centimetres into the air. Hydrostatic pressure

Fig. 7.10 Quarrying of this nineteenth-century *aa* lava on the south flanks of Mt. Etna (background) reveals the massive rock of the lower part of the flow (useful as building stone); overlain by a couple of metres of loose, rubbly material.

within the lava upstream ensures that the flow continues to advance, so the cooling lobe is soon overrun by another. Often, new lobes are propagated from the junction between one lobe and an overlying one: hydrostatic pressure jacks up the overlying layer; a glowing suture appears along the seam between the two; red-hot lava begins to bulge outwards, and in minutes a fresh lobe is oozing forwards. In this way, a flow front that was initially only one lobe thick rapidly builds itself up into a compound flow several metres thick. Rates of advance can be agonizingly slow—when the village of Kalapana was being destroyed in 1990, home-owners had to endure for weeks the sight of *pahoehoe* lavas creeping towards their houses at only a few metres an hour, uncertain of how much longer they would have before their homes were incinerated, or whether some slight accident of topography might spare them for a few days or weeks.

7.3.3 *Transitions between* pahoehoe *and* aa

Pahoehoe and *aa* lavas are often erupted from the same vent. Visually, the differences between

Fig. 7.11 Not a slice of crusty French bread, but an Etnean basaltic lava full of vesicles and gas blisters. Largest blister is about 20 cm across.

Fig. 7.12 Ropy *pahoehoe* on Mauna Ulu 1974 basalt lava, Hawaii. Pleating of the chilled skin of the cooling flow while still plastic yields a texture like that on custard or congealed, greasy soup. Field of view is about 2 metres across.

pahoehoe and *aa* are striking, but the changes taking place within the lava are subtle. True *pahoehoe* surfaces are developed only in low-viscosity basaltic lavas, whereas *aa* surfaces are found in lavas of a much broader range of compositions. Chemically, *aa* and *pahoehoe* lavas may be identical; the differences between them lie in the physical structure of the silicate melt and the way that it is polymerized. After studying lavas erupted in 1975 on Etna, Pinker-ton and Sparks concluded that effusion rate was an important factor in determining whether a flow becomes *aa* or *pahoehoe*: higher effusion rates favour *aa* formation, slower rates, *pahoehoe*.[14] Scott Rowland and George Walker concluded that *aa* forms when flow rates are greater than about 5–10 cubic metres per second.[15]

During the 1969–74 rift zone eruption of Mauna Ulu in Hawaii, 335 million cubic metres

(a)   (b)

Fig. 7.13 (a) Entrail lava, Mauna Ulu, Hawaii, part of the same flow field as Fig. 7.12. Entrail forms are best developed where the flow descends a steep slope, as here, where it has crossed a *pali* fault scarp. Appendix in foreground is 10 m long, 20–30 cm thick; skyline is about 20 metres distant. (b) Entrail *pahoehoe* is extruded in long, thin lobes. On cooling, a strong plastic outer skin forms, leading to curiosities such as this, where a lobe has been deflected upwards like an inquisitive elephant's trunk. When still hot, the outer skin can be perforated with a probe, allowing a fresh bud of lava to ooze out.

of lava were erupted. Most of this was in the form of *pahoehoe*, but where the flows encountered steeply sloping fault scarps of the coastal *pali* and flowed faster downhill, numerous streams of *aa* developed, presumably in response to abrupt changes in shearing stress. Today, their brownish rubbly surfaces stand out in sharp contrast to the acres of glossy grey entrail *pahoehoe*. Many details of the *pahoehoe* to *aa* transition remain to be understood, but one thing is clear: *aa* lava *never* reverts to the *pahoehoe* form.

7.3.4 Lava tunnels and tubes

Lava tunnels and tubes are common features of basaltic lava flow fields, although by their nature they may be hard to find. Lava tunnels form when the surface of a flow crusts over, while hot lava continues to flow beneath. If the rate of flow is sufficiently fast, the flow may erode its way thermally into underlying lavas, producing tunnels with distinctive 'figure-8' shaped profiles. When the lava supply is cut off at source, or is diverted elsewhere, the lava filling the tube drains away downslope, leaving behind an empty conduit.

Lava tubes are important volcanological phenomena because they extend the distance which a flow field can reach. When flowing through a tube, lava is so well insulated that it loses little heat by conduction to the outside world, and none by radiation. The small volumes of basalt which are persistently erupted from the vent at Kupaianaha in Hawaii could not flow as normal *pahoehoe* over the ground for more than a kilometre before chilling by radiant cooling, but because almost the entire volume flowed through long-established tubes, they could reach the sea coast, more than 10 km distant. One can walk right across such a flow field without realizing that lava is flowing beneath, except when the ground is noticeably warm immediately over an active tube, and where occasional 'skylights' broken through the crust provide glimpses of the glowing lava deep beneath.

'Tube-fed flows' can be found on basaltic volcanoes in most parts of the world, and on other planets. On the giant Martian volcanoes, vast tube systems enabled lavas to flow tens or hundreds of kilometres down slope, emerging low on the flanks of the volcanoes (Chapter 18). Although they may reach generous proportions, up to 30 m wide and 15 m high, few lava tunnels on Earth have been traced continuously for more than a few kilometres. In north Queensland, Australia, a discontinuous set of lava tubes has been followed for nearly one hundred kilometres, mainly by identifying the line of cavities left by collapse of the tunnel ceilings.[16–17]

Exploring lava tunnels can be as captivating as probing limestone caves, because they are sometimes decorated with intricate lava stalactites and stalagmites, resembling dribbles of melted chocolate. Limestone caves are typically dank, wet, and chilly, but are redeemed by their water-worn rocks, which are often smooth enough to be somewhat accommodating of the passage of warm bodies over them. Young lava tunnels, however, are so dry that their cindery surfaces are relentlessly unforgiving to tender flesh. In a recently active flow field, tubes remain hot and smelly with sulphurous gases, so explorers leave with parched mouths tasting like the bottoms of proverbial parrots' cages.

Hawaii boasts many superb young lava tubes on the flanks of Kilauea, some of them explored as recently as 1980. Kazumura cave is as broad as a subway tunnel over much of its length (Fig. 7.14). It winds nearly 12 km over a vertical drop of 261 metres. Apua Cave, developed in the 1973 lavas from Mauna Ulu, can be explored for only a few hundred metres, but is richly decorated with lava formations, including bizarre chocolate-dribble stalagmites up to a metre tall but only a few centimetres in diameter (Fig. 7.15).

7.3.5 *Lava blisters, tumuli, and squeeze ups*

A tube-fed lava flow is effectively confined within a pipe. A hydrostatic head of pressure can therefore build up within a tube system, even on modest slopes. Many curious structures form in response to this pressure. Most common are broad swellings of the lava surface, typically a few metres high and a few tens of metres across. When first formed, *tumuli* are no more than

Fig. 7.14 Kazumura cave, Hawaii, a lava tube five metres wide in many places, and more than ten kilometres in length. Such a lava tube could clearly carry lava at a huge rate of flow. Photo: courtesy of A. C. Waltham.

Fig. 7.15 Tottering stalagmites and stalactites of basaltic lava dribbles, Apua cave, Hawaii. These form only in very hot ambient conditions; where more rapid cooling takes place, as in the open air, lava viscosity is higher, so delicate fluidal features cannot form. Photo: courtesy of A. C. Waltham.

Fig. 7.16 A small lava tumulus, formed in 1990 lavas in Hawaii. Lava has broken through the fissured bulge, to feed a new flow.

smooth, regular domes, but the crust of the swelling usually rents apart, in response to pressure from beneath, leaving deep radial fissures over the highest part of the tumulus. If the head of pressure is great enough, lava may squeeze out to the surface through the fissures to initiate small new surface flows (Fig. 7.16).

7.3.6 Other minor forms

It would be possible to fill a book with descriptions of lava flow phenomena; just such a encyclopaedic book already exists and provides fascinating volcanic browsing.[18] Here are just a few of the more common: *Toothpaste* lava is cool, plastic lava that is extruded in the way that toothpaste is squeezed from its tube. When an active lava breaks through to the surface after a long subterranean passage, a good deal of gas may have come out of solution, and this may cause secondary pyroclastic eruptions at points along the flow. *Spatter cones, rings,* or *hornitos* are often built up over these *adventitious* vents. These are not usually large—a few metres high—but they may be extremely steep; welded spatter

piling up to form prominent pinnacles or chimneys around the vent.

More important landforms arise where flowing lava enters the sea, lakes, or even the damp environments presented by lake beds or snow fields. *Lava deltas* are commonly built on the shores of volcanic islands where lavas enter the sea, such as on the coast of Tenerife, extending the coastline outwards (Fig. 7.17). Often, the interaction between *pahoehoe* lava and water is surprisingly undramatic, and can be as tranquil as pouring coffee into cream. When *aa* lavas enter the sea, they do so explosively, fragmenting the lava and forming piles of debris known as *littoral cones*. On the seaward side, huge quantities of fragmented lava accumulate. This pile of breccia is often strengthened by submarine pillow lavas, and may be covered ultimately by flows advancing from the land, extending the coastline outwards in a lava delta.

Where lavas advance into environments less wet than the sea, such as lake beds or low-lying marshy depressions, ground-water trapped beneath the lava is heated, sometimes enough for it to flash violently into steam, bursting explosi-

Fig. 7.17 A small lava delta on the island of Tenerife, Canary Islands. Owing to the scarcity of flat land at sea-level, property developers seized on the delta, leaving little of it unbuilt on.

vely through the lava to form *pseudocraters* or *rootless vents* at the surface. Pseudocraters are best known from Iceland, especially around Lake Myvatn where dozens of them erupted through the 2500-year-old Laxardalshraun lava as it flowed into the lake basin. Around Myvatn, the pseudocraters are not particularly large, typically 100 m across and 15 m deep, but the Skutustadir crater group has been important in the local agricultural economy. In Iceland, meadow land is scarce amongst all the rugged lavas, so the smooth inner and outer slopes of the pseudocraters provide the best local sources for hay crops. It is said that the hay is fertilized entirely by the dead bodies of the billions of midges that swarm round Myvatn in summer, making the long summer evenings less than blissful . . . .[19]

On a smaller scale, lavas advancing over trapped bodies of water can be hazardous, in unexpected ways. Ten bystanders were killed on

Mt. Etna in 1930 while watching basaltic lava invading a house. Unknown to them, the water supply for the house was a subterranean cistern. When trapped beneath the lava, this heated up until pent-up steam exploded violently through the lava. On Hawaii, otherwise tranquil *pahoehoe* flows present similar hazards when they advance over buried septic tanks in built-up areas. Apart from steam, there is also the risk of explosions of methane gas, and of being showered by unsavoury biogenic products . . . .

7.3.7 Lava moulds

Like other plastic substances, basalt lavas are rather good at preserving the shapes of objects. Trees are the commonest objects to experience high-temperature fossilization, so 'lava forests' are known in many parts of the world. After engulfing the trunk of a tree, *pahoehoe* lava sets it on fire, incinerating the wood, while itself setting solid around the outside. All that is left is a cylindrical hole in the ground, sometimes preserving the pattern of the bark. Less commonly, fast-flowing lava piles up against a tree, chills against and cremates it, but then drains away again, leaving a tall pipe-like 'chimney' of basalt marking the site where the tree once grew (Fig. 7.18).

Trees apart, many other objects, organic and inorganic, may form moulds. A fossil rhinoceros was found in a North American lava, while the cascade of lava draining from Nyiragongo in Central Africa in 1977 formed moulds of hapless elephants unfortunate enough to be overrun. Outside Africa, automobiles are more likely to be overrun than elephants. In Hawaii numerous rusting cars, trucks, and vans protrude above young lavas. When the metal has all rusted away, it will take a skilled palaeontologist to recognize automotive gems such as the mould of a 1976 VW Beetle with customized rear end . . . (Fig. 7.19).

Fig. 7.18 Each of these 1–3 metre high chimneys was originally an *ohia* tree, engulfed by 1974 lavas from Mauna Ulu, Hawaii. Fast-flowing fluid lava chilled around the trees, cremated them, and then drained away, leaving the moulds standing.

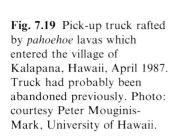

Fig. 7.19 Pick-up truck rafted by *pahoehoe* lavas which entered the village of Kalapana, Hawaii, April 1987. Truck had probably been abandoned previously. Photo: courtesy Peter Mouginis-Mark, University of Hawaii.

7.4 Andesitic lavas

Describing natural phenomena under separate headings has the unwelcome implication that they can be considered as separate entities, each in its own watertight compartment. There are no hard and fast differences between basalt and andesite lavas, except that andesites consistently exhibit higher yield strengths and viscosities than basalts, so that, generally speaking, andesite flows are thicker, and travel slower and less far than basalts. While *aa* surfaces are often encountered in basaltic andesites, *pahoehoe* surfaces are never found on andesite flows. *Block lavas* are typical of andesites. Their surface consists of large, smooth-sided blocks of lava, up to several metres in size, much more agreeable to the touch then cindery *aa*, but still presenting formidable obstacles to movement. Flow fronts of blocky lavas are steep, may be more than 100 metres high, and consist of piles of huge angular blocks, balancing precariously on one another (Fig. 7.20). *Autobrecciated* textures are common in andesite lavas, where blocks cascading from the front of the moving flow are later incorporated

Fig. 7.21 View looking downwards on the same cleft feature as Fig. 7.20, illustrating deep axial fissure.

Fig. 7.20 Margin of a blocky andesite lava flow on the flanks of San Pedro volcano, north Chile, showing a cleft marginal lobe. Here the flow front is more than 30 m high; further upstream, it is more than 100 m high. (Person gives scale.)

Fig. 7.22 Detail of the margin of San Pedro flow, showing smooth-sided, angular blocks banded by regular 'chisel' marks left by splitting of lava during cleft formation.

within it, forming a mass of angular fragments solidly welded together.

Lava tubes are less common in andesites than in basalts, but are not unknown. In large andesite flows, a chilled crust of brittle rock overlies a hotter, central core which carries on moving and deforming. A hydrostatic head commonly builds up within the viscoplastic core of a lava lobe; eventually a point is reached where the thick, chilled crust is fractured, and it splits along a single, elongate fracture along the length of the lobe, forming a *cleft lava* (Fig. 7.21). Further upward pressure forces the two sides of the cleft upwards and outwards, so that they diverge progressively as the fracture propagates deeper and deeper into the lava. Fracture does not take place continuously, but in a series of distinct jerks, leaving long, parallel ridges four or five centimetres apart and a few millimetres high on the walls of the cleft. Each of these *chisel marks* is the result of single discrete fracture (Fig. 7.22). Andesite lavas thus encapsulate the mechanical properties of the lithosphere and mantle: their hot cores deform slowly and plastically to allow flow to take place, while the chilled crust deforms by brittle failure.

7.5 Dacitic lavas

Dacitic lava flows represent a further increment in viscosity and yield strength over andesite flows. They are extremely sluggish, and form thick, steep extrusions. Apart from the usual reasons for increasing viscosity and yield strength (lower temperature, increased polymerization *et cetera*), dacites also exhibit another: they typically have high crystal contents, sometimes up to 50 per cent. Thus, a dacite extrusion may be a crystal-rich mush, rather than a proper melt.

Eruption of viscous dacite magmas may be accompanied by greater or lesser amounts of explosive activity, so that extrusions are often found surrounded by a low cone of ejecta. Commonly, extrusion of dacite lava is the last event in an explosive eruptive cycle, as the last dregs of degassed magma ooze their way reluctantly to the surface. They are so thick that the term 'lava dome' is often preferable to 'lava flow' for many silicic extrusions. It is convenient to discuss lava domes in the context of dacitic compositions, but not all domes are dacitic—lavas with compositions ranging from basaltic andesite to rhyolite may form domes. Steven Blake at the UK's Open University recognized four different kinds of lava dome based on their morphologies:[20] low domes, *coulées*; peléean domes, and upheaved plugs (Fig. 7.23).

7.5.1 *Low domes or* tortas

Low domes are flattish, roughly symmetrical extrusions erupted on level ground. In the central Andes, where such domes are common, they are called locally *tortas*, or cakes (Fig. 7.24). *Tortas* have few external structural features. Their top surfaces are as bouldery as the blockiest of block lavas, but there is little larger-scale topography, which distinguishes them from irregular shaped peléean domes. Blake attributes their more regular topography to a slightly lower yield strength in the lava, which deforms more under its own weight. *Tortas* grow by entirely internal processes; each successive increment of lava from the vent pushes earlier erupted material outwards, stretching and thinning it. A section across such a dome would have a concentric, onion-like structure.

7.5.2 *Coulées*

These are a cross between domes and lava flows: they are thick extrusions erupted on slopes steep enough for shear stresses to exceed the yield strength, thus permitting the flow to ooze down slope. A huge example of such a coulée is the Chao lava in north Chile,[21] which has a volume of about 24 cubic kilometres and was erupted in three distinct flow units; the flow front of the largest unit is over 300 m high (Fig. 7.25). If it could be transported to England, such a flow would be prized as one of the country's major landmarks: it would feature on postcards and would be regularly thronged with ramblers. Huge pressure ridges or *ogives*, best appreciated

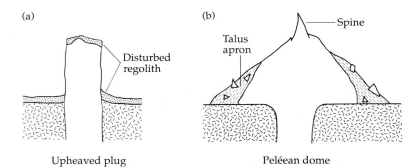

(a) Disturbed regolith

Upheaved plug

(b) Spine

Talus apron

Peléean dome

Fig. 7.23 Cross-sections of the four types of lava dome recognized by Blake, discussed in the text. From Blake, S. (1989). Viscoplastic models of lava domes. In *IAVCEI Proceedings in Volcanology.* Vol. 2. *Lava flows and domes,* pp. 88–126. Springer Verlag, Heidelberg.

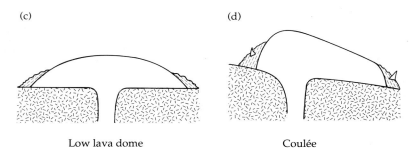

(c)

Low lava dome

(d)

Coulée

Fig. 7.24 Tocorpuri, north Chile, a typical flat-topped dacitic *torta.* Flow front is about 100 m high, and consists of a talus slope topped by a cliff of lava.

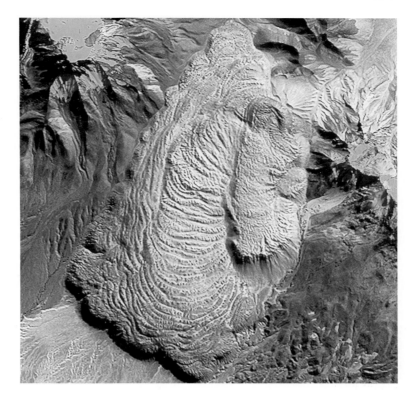

Fig. 7.25 Landsat Thematic Mapper image of the Chao dacitic coulée, North Chile. Two distinct lobes are obvious. Massive ogive ridges wrinkling surface are hundreds of metres apart and tens of metres high. Image is 15 km across.

on air photographs, pleat the surface of the flow. Jon Fink has suggested that such ogives, which may be many metres high and spaced tens of metres apart have similar origins to the 'ropes' on the surface of *pahoehoe* lavas. Their sizes and spacings are controlled by the temperature gradient and relative viscosities of the outer carapace of lava and the hotter inner core.[22] Chao is so young that its internal structure is not exposed, but in older, dissected examples, coulées display *ramp structure*, formed as successive increments of lava are extruded from the vent and squeeze earlier erupted increments outwards. Most of the outward movement of the flow is accommodated in a thin zone of shearing at the base (Fig. 7.26).

7.5.3 Peléean domes

After the catastrophic 1902 eruption of Mt. Pelée in Martinique, a lava dome grew in the vent of the volcano, ultimately thrusting up a spine 300 m high, an ephemeral monument to its victims.

Peléean domes, then, are characterized by craggy topography and lava spines which form jagged battlements along the crest of the dome. Steep collars of debris formed by collapse of unstable pinnacles often encircle the dome. These collapses usually take place as ordinary 'cold' avalanches, such as occur on any steep mountainside. If the collapse is a big one, however, incandescent rocks within the core may be exposed, and a *nuée ardente* liberated (Chapter 12).

A typical peléean dome grew after the eruption of Mt. Lamington, Papua New Guinea, in 1951. Within two months of the catastrophe, the Lamington dome had grown to a height of 500 m; it was subsequently largely destroyed by explosions, but by January 1952 it was once again over 600 m high, with a volume of about 1 cubic kilometre (Fig. 7.27). Numerous spines grew, one individual squeezing upwards at a rate of a metre per day to reach a height of 100 m. Like those of Mt. Pelée, many had a typical 'half-

Fig. 7.26 'Ramp' structure well exposed in a trachytic coulée on the flanks of Teide volcano, Tenerife, Canary Islands.

horn' shape, one side smoothly curved and lineated during extrusion like wire through a die; the other jagged and fractured (cf. Fig. 5.8).

Other peléean domes grow more slowly. In Kamchatka, the Novy dome on Bezymianny volcano has been growing since 1956, while in Guatemala the Santiaguito dome on Santa Maria volcano has been growing since 1922. Santiaguito grows so slowly that its growth is most apparent on photographs taken months apart, but it is none the less constantly active because the piled rocks are mechanically unstable. One can almost hear it growing, since there is a continual rattle of small stones and rocks falling from the higher parts down on to the talus slopes below. Even when the dome is shrouded in mist and invisible, as it often is in the afternoons, the slithering, clattering noise continues, with every now and then a much larger collapse taking place.

7.5.4 Upheaved plugs

Upheaved plugs are masses of rock that are pushed bodily upwards like pistons, and which have yield strengths sufficiently great that they do not deform and spread outwards once above the surface. (Plugs which deform the surface topography without breaking through are termed *cryptodomes*.) Some lava plugs may carry a layer of country rock sediments on top of the piston, leading to some strange geological relationships. Beautifully documented examples of this phenomenon have happened twice on the Usu volcano in Japan, once in 1910 and once in 1943. The first upheaval produced a hill which was christened Meiji Sin-zan, or 'Roof Mountain'; the second, logically enough, was called 'Showa Sin-Zan' or 'New Roof Mountain'. Showa Sin-zan grew during the height of World War II, so there were no official records of its growth, but the astute village postmaster, Masao Mimatsu, kept a unique pictorial record of its progress by drawing a series of profiles on the paper covering of his window (Fig. 7.28). From January 1944 until November 1944 the uplifted area rose steadily, but no new rock was visible the uplifted material was all crater lake sediments, baked hard by the heat of the lava beneath, carried upwards as smoothly as if on an elevator. In November of 1944, a plug of lava eventually punched its way through the elevated dome, still carrying a cap of baked clay, rising until it was about 100 m above the top of the dome, and nearly 300 m above ground level.[23]

7.5.5 Rheological studies of lava domes

Small lava domes are the simplest lava extrusions, so they are relatively easily studied, and the lessons drawn from them can be usefully applied to more complex lava flows. By using stiff mixtures of kaolin slurry extruded on to a flat surface to model lava domes, Steven Blake obtained useful insights into the growth of natural domes and flows. Slow-growing lava domes can be satisfactorily modelled as true Bingham plastic materials, whose viscosity is

Fig. 7.27 Peléan lava dome of Mt. Lamington, photographed on 8 April 1951 by G. A. M. Taylor. Compare the Mt. Pelée dome (Fig. 5.7). Photo: courtesy of Wally Johnson, Bureau of Mineral Resources, Geology and Geophysics, Canberra, Australia.

independent of shear stress, since as a dome increases in size and begins to deform due to its own mass, shear stresses remain only slightly greater than the yield strength.

Blake found that in his laboratory experiments, the height and radius of a dome were related by a simple expression:

$$H = 1.76\,(\tau_0 R/\rho g)^{1/2} \qquad (4)$$

where H = height, R = radius, τ_0 = yield strength, ρ = density, and g = acceleration due to gravity.

It follows that if H and R can be measured, then the yield strength can be determined. Measurements made during the growth of a basaltic andesite lava dome at Soufrière St Vincent in 1979 showed that the height of the lava dome was consistently proportional to the square root of its radius. Empirically, the relationship was found to be:

$$H = 5.75 R^{1/2}$$

This relationship suggests that the yield strength of the Soufrière dome was about 2.6×10^5 Pa, about an order of magnitude greater than would be expected if the whole dome were made of plastic lava. It is not difficult to account for this discrepancy: the carapace of chilled lava surrounding the hot interior provides much of the yield strength of the whole body (Fig. 7.29).

Blake's work with models showed that even simple lava domes grow in rather complex ways. An expanding dome does not simply roll outwards in the caterpillar-track style of some lava flows. Material moves radially outwards, stretching the carapace, such that a single point approaches the margin of the dome linearly, but never quite reaches it—the margin of the dome undergoes simultaneous radial compression and circumferential extension. While one might expect to see radial and circumferential extensional fractures and gashes in an expanding dome, this is not the case—the kaolin models consistently displayed sets of spiral slip planes, spiralling outwards from the centre of the dome both clockwise and anticlockwise, and dividing the surface of the whole dome into rhombic rough-

(a)

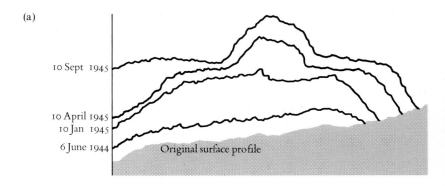

10 Sept 1945

10 April 1945
10 Jan 1945
6 June 1944   Original surface profile

(b)

Fig. 7.28 (a) Masao Mimatsu recorded stages in the growth of Showa Sin-Zan dome, Japan, on a piece of paper covering his window. At the end of 1944, a plug of lava punched through the crust of uplifted sediments, forming the prominent crown. (b) Summit of Showa-Sin Zan today. Small steam plumes still emanate from it. Surrounding area is a spa popular among tourists.

Fig. 7.29 Andesitic dome growing within the crater lake of La Soufrière volcano, San Vincent, Caribbean. Photo was taken in 1972 when dome was 90 metres above lake level and 650 metres in diameter. Lake water was hot and acid. Photo: courtesy of J. Roobol.

ness elements. Such elegant sets of spiral slip planes have not been seen on actual lava domes, perhaps because even the slightest inhomogeneity might disrupt the pattern, but more probably because few domes have been closely studied.

Peléean domes present slightly different problems. These are constructed of such viscous magmas that their yield strengths are very high,

and it is difficult for them to deform under their own weight. They typically have steep, pyramidal profiles, whose shape is influenced by the heaps of talus accumulating around the flanks of the growing dome. The high viscosity of the associated magma means that peléean domes are also more likely to be violently disrupted by explosive activity than low, flat lava domes.

— 7.6 Rhyolitic lavas

Lavas are scarce among rocks of highly silicic compositions, since most silicic magmas fragmented into pyroclasts. Thus, rhyolite lavas are much less abundant than dacites, and few historic rhyolite lava eruptions are known. A small rhyolite obsidian was erupted from Vulcano in the Mediterranean in the eighteenth century, but descriptions of its emplacement are meagre. Only *one* true rhyolite flow has been observed to erupt in modern times. This was during the 1953–7 activity which formed the Tuluman Islands, two new islands in the St Andrew Strait off the north coast of Papua New Guinea.[24] Much of the eruption was submarine, so the effusion rate of the small rhyolitic lava could not be determined.

Rhyolites are pale-grey to brown or buff-coloured rocks with a microcrystalline structure and a rather sugary texture. Morphological, rhyolite lavas form extrusions closely similar to dacites, and are commonly found within calderas formed earlier by major explosive eruptions. Valles caldera in New Mexico displays an especially elegant necklace of post-caldera rhyolite domes all round the ring fracture that formed the 1 million-year-old caldera (Fig. 14.9).

7.6.1 *Obsidian*

At its best, obsidian is jet black, pure glass, free from bubbles and other imperfections. When hammered, it shatters satisfyingly into myriads of razor-sharp splinters. Most fragments have irregular surfaces, but many exhibit a characteristic *conchoidal* fracture: the surface is smooth, but ridged with concentric corrugations, centred on the point of impact of the hammer, creating a pattern reminiscent of the ridges on a cockle

shell. Whereas dacite lavas often contain a high percentage of phenocrysts, rhyolitic obsidians do not contain crystals; they are *aphyric*.

Obsidian was prized in prehistoric times because it provided wickedly sharp tools. Some modern surgeons assert that obsidian flakes provide cutting edges superior to stainless steel. Obsidian was traded extensively throughout prehistoric Europe and Central America because of its value. Archaeologists have been able to trace some of these trade routes for hundreds of kilometres by using geochemical techniques to fingerprint the few sources from which obsidian is found, and then matching these against stone tools found in distant excavations. Apart from its utility, the deep glossy blackness of obsidian makes it attractive for aesthetic reasons, so it has been used in jewellery throughout history. Sad young King Tutankhamen stares at us from his magnificent gold funerary mask through black obsidian eyes. Many other pieces of jewellery found in his tomb included obsidian. Archaeologists suggest that the most probable source of this material was one of the volcanic islands in the Aegean.

In modern Europe, the most famous obsidian locality is the Roche Rosse flow, on the island of Lipari, off the coast of Sicily, erupted after King Tut's demise. In North America, superb obsidian flows were erupted on the flanks of the Newberry caldera in Oregon about 1400 years ago, piling up in a *coulée* which oozed for a couple of kilometres to form the Big Obsidian Flow. In the Medicine Lake Highlands caldera, Sisikyou County, California, another flow formed a coulée several hundred metres thick, which was called, in the inimitable American way, 'Big Glass

Mountain'. A mountain of glass it certainly is. From the air, it looks exactly as if someone had poured a couple of buckets full of molten glass on to the surface, forming a puddle which slowly oozed away. Large, high wrinkle ridges (ogives), concentric with the edges of the flow, resemble the pleats on an elderly elephant's skin (Fig. 7.30).

Glasses are not particularly stable materials, so in the course of time they devitrify, acquiring a microcrystalline structure. In some cases, devitrification begins at centres scattered regularly throughout the whole mass, to produce some attractive effects. Some obsidians, such as those from the Yellowstone caldera, have white spots

about a centimetre across scattered liberally throughout the black glass. This is *snowflake* obsidian, much sought after by rock-hounds for use in jewellery.

Obsidian flows are never glassy all the way through. Much of their volume consists of frothy white pumiceous material interbanded with dark, unvesiculated glass, often on a scale of centimetres. In a masterly study based on the Little Glass Mountain, Jon Fink showed that a distinct stratigraphy is present in an obsidian flow.[25] At the base is a layer of air-fall tephra, explosively erupted during the opening stages of the eruption. This is successively overlain by a basal lava breccia, coarsely vesicular pumice, the

Fig. 7.30 Oblique aerial photograph of an obsidian coulée, Glass Mountain, northern California. Coulée has a volume of about 1 cubic kilometre and may be about 1000 years old. Ogives pleat the surface. Photo: courtesy of Ron Greeley, Arizona State University.

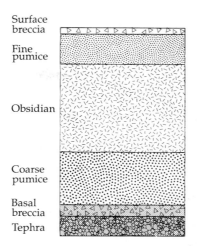

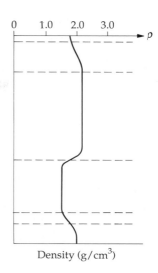

Density (g/cm³)

Fig. 7.31 Zonation within an obsidian lava, as outlined by John Fink, showing the relative densities of the more and less vesiculated layers. From Fink, J. H. (1983). Structure and emplacement of a rhyolitic obsidian flow: Little Glass Mountain, Medicine Lake Highland, northern California. *Geol. Soc. Am. Bull.* **94**, 362–80.

main thickness of obsidian, finely vesicular pumice, and surface breccia (Fig. 7.31).

Each of the different layers has sharply different rheological properties. It is these that give obsidian flows their characteristic surface features. Two aspects are particularly relevant: the large difference in viscosity between the surface carapace compared to the main mass of obsidian, and the variations in density. As the flow advances, it will be subjected to compression or extension as it moves over topographical highs and lows. Because the carapace is so much more rigid than the obsidian core, the deformation causes the carapace to fold or fracture, producing complex surface topography, while the core responds plastically. The wavelength of these folds is dictated by the thickness of the carapace, but may be tens of metres (Fig. 7.30).

Spectacular flow folds are often developed within the low-viscosity obsidian layer itself, expressed in contortions of layers of more and less pumiceous material (Figs. 7.32–7.33). Such folded patterns are characteristic of movement by shearing in many kinds of media, from Newtonian fluids to complex fluids containing layers with different yield strengths. Flow folds can be seen in glaciers, in some kinds of metamorphic rock, and wherever immiscible liquids become intermingled. Iridescent films of oil on top of puddles of water provide good

Fig. 7.32 Flow banding expressed by interleaved bands of light, vesiculated (pumiceous) lava, and dark, glassy obsidian; Roche Rosse, Lipari, Italy.

Fig. 7.33 Small flow folds pleat the flow banding of the Roche Rosse obsidian.

analogues: when the puddle is stirred gently, the swirling, winding folds produced resemble those seen in lavas. Where there are differences in strength between layers, flow folds are not smooth and winding, but are more irregular, and there is often evidence of brittle failure, as layers stretched to breaking point form irregular fragments.

Variations in density yield some curious internal structures. The coarse pumice underlying the main obsidian layer in Fink's examples is less dense than the obsidian, and therefore it is buoyant and a gravitational instability exists. If the flow is large enough, buoyancy forces may be great enough for the low-density pumice to form *diapirs* a few metres or tens of metres in diameter which then rise upwards through the obsidian, often punching through the outer carapace to the surface. Pumice diapirs which have broken through the flow surface are conspicuous because of the textural contrasts with their surroundings (Fig. 7.34).

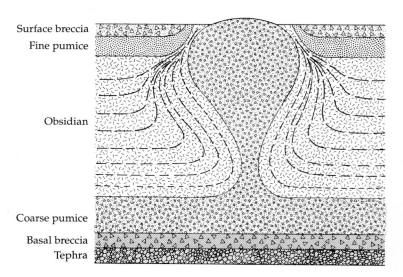

Surface breccia
Fine pumice

Obsidian

Fig. 7.34 Diapirs of buoyant pumice rise through denser layers of overlying obsidian in some flows, producing conspicuous variations in surface textures.

Coarse pumice

Basal breccia
Tephra

— 7.7 Carbonatite lavas

Oldoinyo Lengai in the Eastern Rift of Tanzania is the world's only active carbonatite volcano. Its exotic lavas provide a miniature laboratory of flow processes.[26–27] Effusion rates are only about 0.3 metres per second; most flows are only a few centimetres thick, and travel only a few tens of metres from their source (Fig. 7.35). They are unique because of their low effusion temperatures (575–593°C) and extraordinarily low viscosities (1–5 Pa s), lower than any other terrestrial magma—basalt has a viscosity of about 100 Pa s. Oldoinyo Lengai's lavas are

Fig. 7.35 Active flow of carbonatite lava on the crater floor of Oldoinyo Lengai volcano, Tanzania. Hammer in foreground emphasizes the diminutive scale of these flows. Miniature channels and levées are well displayed on the far side of the half-metre wide active channel. These flows resemble basaltic *pahoehoe*; others resemble *aa*. (Photo: courtesy of Jorg Keller.)

made of the alkali carbonate minerals *nyerereite* $(Na_{0.82}K_{0.18})_2Ca(CO_3)_2$ and *gregoryite* $(Na_{0.78}K_{0.05})Ca_{0.17}(CO_3)$, with a generous content of dissolved carbon dioxide, which exsolves to form abundant vesicles.

— 7.8 Sulphur lava flows

'Lava flows' of sulphur are not strictly magmatic, but merit a brief mention. Many volcanoes, particularly those of andesitic or dacitic composition deposit large amounts of sulphur around fumaroles in their summit regions. Sulphur may accumulate over many hundreds of years, as the body of crystallizing magma beneath the volcano cools and liberates its content of dissolved volatiles. On many volcanoes, so much sulphur is present that it may form more than 30 per cent of the rock mass, constituting an economic resource. In Chile, several of the world's highest mines are located on the summits of 6000-m-high active volcanoes; their fumarolic sulphur is mined to make sulphuric acid and for use as a dressing on vineyards.

If, after a long episode of fumarolic activity, the volcano heats up again, some of the sulphur may melt to form 'lava flows'. Lastarria volcano in Chile has superbly exposed examples, 1–2 metres wide, several hundred metres long and with spendid ropy *pahoehoe* surface structures.[28] Sulphur is exceptionally interesting rheologically, since its viscosity varies hugely with temperature. It melts at *c.* 113°C, and its viscosity decreases initially with increasing temperature, reaching a minimum of about 0.01 Pa s at 160°C, approaching that of water. At 160°C, however, its viscosity dramatically *increases*, shooting up many order of magnitude as the structure of the melt changes. Potentially, this could yield some exotic lava structures, but none has yet been described on Earth because sulphur lavas are uncommon. Some bizarre volcanology may yet remain to be discovered on Io, however, where much of the surface may be covered by sulphur.

— Notes

1. Stephenson, P. J. and Griffin, T. J. (1976). Some long basaltic lava flows in north Queensland. In *Volcanism in Australasia*, (ed. R. W. Johnson), pp. 41–51, Elsevier, Amsterdam.

2. Swanson, D. A., Wright, T. L., and Heltz, R. T. (1975). Linear vent systems and estimated rates of eruption for the Yakima basalt on the Columbia Plateau. *Amer. J. Sci.* **275**, 877–905.

3. Hooper, P. R. (1982). The Columbia River Basalts. *Science* **215**, 1463–8.

4. Lipman, P. W. and Banks, N. G. (1987). Aa flow dynamics, Mauna Loa 1984. In *Volcanism in Hawaii*. US Geol. Surv. Prof. Pap. No. 1350, pp. 1527–67.

5. Thorarinsson, S. (1969). The Lakagigar Eruption of 1783. *Bull. Volcanol.* **33**, 910–27.

6. Hulme, G. (1974). Interpretation of lava flow morphology. *Roy. Astron. Soc. Geophys. J.* **39**, 361–83.

7. Chester, D. K., Duncan, A. M., Guest, J. E., and Kilburn, C. R. J. (1985). The rheological behaviour of basaltic lavas. In *Mount Etna, the anatomy of a volcano*. Chapman and Hall, 404 pp.

8. Walker, G. P. L. (1973). Lengths of lava flows. *Phil. Trans. Roy. Soc. Lond.* **274**, 107–18.

9. Sparks, R. S. J., Pinkerton, H., and Hulme, G. (1976). Classification and formation of lava levées on Mt. Etna, Sicily. *Geology* **4**, 269–71.

10. Wadge, G. (1983). The magma budget of Volcan Arenal, Costa Rica. *J. Volcanol. Geotherm. Res.* **19**, 281–302.

11. Pieri, D. C. and Baloga, S. (1986). Eruption rate, area, and length relationships for some Hawaiian lava flows. *J. Volcanol. Geotherm. Res.* **30**, 29–45.

12. Kilburn, C. R. J. and Lopes, R. M. C. (1988). The growth of lava flow fields on Mt Etna, Sicily. *J. Geophys. Res.* **93**, 14 759–14 772.

13. Wadge, G. (1981). The variation of magma discharge during basaltic eruptions. *J. Volcanol. Geotherm. Res.* **11**, 139–68.

14. Pinkerton, H. and Sparks, R. S. J. (1976). The 1975 sub-terminal lavas, Mount Etna: a case history of the formation of a compound lava field. *J. Volcanol. Geotherm. Res.* **1**, 167–82.

15. Rowland, S. K. and Walker, G. P. L. (1990). Pahoehoe and aa in Hawaii: volumetric flow rate controls the lava structure. *Bull. Volcanol.* **52**, 615–28.

16. Ollier, C. D. and Brown, M. T. (1965). Lava caves of Victoria. *Bull. Volcanol.* **28**, 1–15.

17. Atkinson, A., Griffin, T. J., and Stephenson, P. J. (1975). A major lava tube system from Undara volcano, north Queensland. *Bull. Volcanol.* **39**, 1–28.

18. Green, J. and Short, N. M. (1971). *Volcanic landforms and surface features: a volcanic atlas and glossary.* Springer-Verlag, New York, 522 pp.

19. Thorarinsson, S. (1953). The crater groups of Iceland. *Bull. Volcanol.* **14**, 3–44.

20. Blake, S. (1989). Viscoplastic models of lava domes. *IAVCEI Proceedings in Volcanology.* Vol. 2. *Lava flows and domes.* pp. 88–126. Springer Verlag, Heidelberg.

21. Guest, J. E. and Sanchez, R. A. (1969). A large dacitic lava flow in northern Chile. *Bull. Volcanol.* **33**, 778–90.

22. Fink, J. (1980). Surface folding and viscosity of rhyolite flows. *Geology* **8**, 250–4.

23. Minakami, T., Ishikawa, T., and Yagi, K. (1951). The 1944 eruption of Volcano Usu in Hokkaido, Japan. *Bull. Volcanol.* **11**, 45–160.

24. Reynolds, M. A. and Best, J. G. (1976). Summary of the 1953–57 eruption of Tuluman Volcano, Papua New Guinea. In *Volcanism in Australasia*, (ed. R. W. Johnson), pp. 287–96. Elsevier, Amsterdam.

25. Fink, J. H. (1983). Structure and emplacement of a rhyolitic obsidian flow: Little Glass Mountain, Medicine Lake Highland, northern California. *Geol. Soc. Am. Bull.* **94**, 362–80.

26. Dawson, J. B., Pinkerton, H., Norton, G. E., and Pyle, D. M. (1990). Physico-chemical properties of alkali carbonatite lavas: Data from the 1988 eruption of Oldoinyo Lengai, Tanzania. *Geology* **18**, 260–3.

27. Keller, J. and Krafft, M. (1990). Effusive natrocarbonatite activity of Oldoinyo Lengai, June 1988. *Bull. Volcanol.* **52**, 629–45.

28. Naranjo, J. A. (1985). Sulphur flows at Lastarria volcano in the north Chilean Andes. *Nature*, **313**, 778–80.

8

Bubbles and bangs: the mechanisms of pyroclastic eruptions

No sight on Earth is more inspiring than that of a volcanic eruption column ascending tremendously into the sky. Although photographs convey an impression, they cannot re-create the emotions engendered in those fortunate (or unfortunate) enough to witness one at first hand. If ash fall mantles the landscape over a wide region, the physical effects of a large pyroclastic eruption may be formidable, but even if the dire consequences are limited only to the flanks of the volcano, aerosols may inflame sunsets all over the world for months.

Sadly, it is difficult to learn much about the physics of pyroclastic eruptions simply by standing and watching them. Much pyroclastic research has therefore been done on paper in provincial universities, far from the *frisson* of active volcanism, and also in field studies of ancient deposits. Over the last twenty years, important theoretical advances have been made, notably by George Walker, Lionel Wilson, Stephen Sparks and their co-workers in a seminal series of papers on 'Explosive volcanic eruptions'.[1-5]

In this chapter, a few significant theoretical papers on eruption columns are reviewed. Because some of the details are fairly esoteric, readers more concerned with generalities may prefer to skim through this chapter before going on to chapters dealing with pyroclastic deposits. For the sake of simplicity, these studies deal mostly with *dry* eruptions, whereas in reality large eruptions commonly involve a component of water:magma interaction.

— 8.1 Three kinds of explosive eruption —

An explosive volcanic eruption involves three stages: fragmentation of the magma by bubble growth; blasting of the fragmented mass through the vent to the surface, and subsequent ascent of the eruption column. Vesiculation, the first stage, may be promoted by decompression (*first boiling*) or crystallization (*second boiling*). An explosive eruption will take place when the pressure within a magma exceeds the strength of the surrounding rocks. Expansion of the growing bubbles causes the ascent velocity of magma in the vent to increase; when the volume fraction of bubbles exceeds about $\frac{4}{5}$, the magma will be explosively disrupted (Section 5.6). In this chapter, we examine the two subsequent stages, ejection and column formation.

It is worth emphasizing that explosive eruption columns are driven by the thermal energy stored in the magma. Thermal energy is transferred into the kinetic energy of the eruption column through the expansion of the gases diffusing into growing vesicles. Magma viscosity, dissolved volatile content, and mass eruption rate are therefore the pre-eminent variables influencing the course of an explosive eruption. Lionel Wilson recognized three different types of volcanic explosion: plinian, strombolian, and vulcanian. For each of these, he estimated 'muzzle velocities' for different conditions of pressure and volatile content.[6] To derive ejecta velocities, Wilson started from the 'modified Bernouilli equation':

$$\tfrac{1}{2}U_s^2 = \frac{(P_i - P_s)}{S_c}$$

where U_s is the ejection velocity, P_i is the pressure in the gas reservoir, P_s is the atmospheric pressure, and S_c the magma density.

A second equation is the 'Gun Barrel Equation', whose name hints clearly at its ancestry, and of what is involved in propelling objects at high velocity through narrow vents:

$$P_i = \frac{4wU_s^2}{27gAb}$$

where w is the mass of a projectile accelerated to speed U_s by an initial pressure P_i applied over an area A; g is acceleration due to gravity, and b is a constant.

Unfortunately, these relatively simple equations are only the starting point. Thereafter, things rapidly get horrendously complex. But since neither equation realistically describes natural volcanic processes, we can conveniently forget Bernouilli and gun barrels, and focus on some practical results.

8.1.1 Plinian eruptions

Plinian eruptions are not ordinary explosions, in the sense of discrete bangs. They are sustained jets, which may continue for minutes or hours (cf. Fig. 2.5). Consequently, the time taken for a blob of magma to pass through the vent is much less than the duration of the eruption. In volatile-rich magmas, estimated muzzle velocities are several hundred metres per second. These theoretical estimates compare well with velocities calculated from the sizes and distances of clasts flung out in actual eruptions (Fig. 8.1).

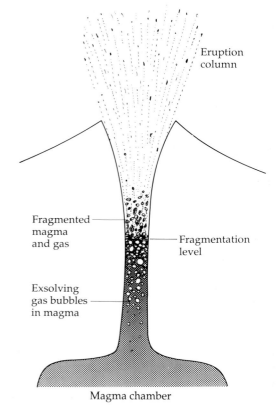

Eruption column

Fragmented magma and gas

Fragmentation level

Exsolving gas bubbles in magma

Magma chamber

Fig. 8.1 In a plinian eruption, exsolution of gases from magma takes place deep within the vent as magma rises continously from its reservoir. Depressurization fragments the magma and drives it steadily from the vent. Depth to the fragmentation level may be more than a kilometre. After Wilson, L. (1980). Relationships between pressure, volatile content and ejecta velocity in three types of volcanic explosion. *J. Volcanol. Geotherm. Res.* **8**, 297–313.

One crucial aspect of plinian eruptions is that magma and exsolving gas remain in close proximity; the gas bubbles cannot move upwards through the magma as they do in effervescent Alka-Seltzer. As bubbles grow, they rise through the magma at about the same rate at which the magma itself is rising through the crust. This can happen in viscous magmas, and in low-viscosity magmas rising so fast that there is no time for much relative bubble movement. Plinian eruptions of rhyolite magmas are examples of the first case; sustained jets of basaltic hawaiian fire fountains may be examples of the second.

8.1.2 Strombolian eruptions

Plinian eruptions involve magma ascent which is rapid relative to the rate of bubble growth. By contrast, strombolian explosions involve lower-viscosity magmas which rise slowly and smoothly, at maybe only a few metres per second. There is thus opportunity for growing bubbles to coalesce, swell to large dimensions, and move upwards through the magma body. Strombolian burps, therefore, are short blasts, separated by periods of less than a tenth of a second to several hours.

Because they are so common, there are many good field observations of strombolian eruptions, including some useful movie film. Tracking of individual glowing clasts on slow motion films of Stromboli itself and of Heimaey, Iceland showed that typical maximum muzzle velocities did not exceed 200 metres per second.[7-8]. Once out of the vent, atmospheric drag sharply decelerates the high-velocity clasts, at up to 50 g; thus, as clasts approach the tops of their trajectories 100–150 m above the vent, they are travelling at only 40–50 metres per second (cf. Figs 6.10–6.11)). Large clasts continue to decelerate until their upwards velocity reaches zero, after which they fall under gravity, tracing elegant parabolas on time exposures. Small particles slow until they reach a minimum velocity; the velocity at which the hot, buoyant gases surrounding them rises convectively.

In strombolian eruptions, the amount of volatiles required to blow magmatic material out of the vent at the velocities observed must be large; perhaps 10 to 30 per cent by weight of the erupted material. This is far more than could be accounted for by gas exsolving directly from the magma—typical basalts contain less than 1 per cent volatiles. Each explosive burp probably therefore represents the bursting of a magma bubble as it reaches the surface (Fig. 8.2). Each glowing clast propelled upwards is a glob of the magma which formed the skin of the exploding bubble. Because vesicles can move and coalesce, and so the expanding gas in the bubble that ejects a particular slug of glowing clasts need not all

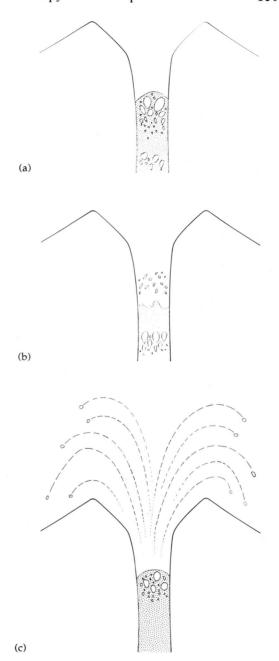

(a)

(b)

(c)

Fig. 8.2 Three stages in a strombolian burp, seconds apart. In (a) coalescing metre-sized bubbles have swollen the surface of magma in the vent, only a few tens of metres below surface. In (b) the magma has been disrupted, and glowing globs are thrown upwards. In (c) the clasts trace the parabolic trajectories familiar from photographs, reaching 100–150 metres above the vent. After Wilson, 1980.

Fig. 8.3 A metre-sized bubble swells on the surface of a lava pond at Mauna Ulu, Hawaii, October 1969. Bursting of lava bubbles yields glowing spatter similar to that of strombolian eruptions. U.S. Geological Survey photo by Jeffrey Judd.

have exsolved from that same slug, but may have accumulated from a larger volume of magma below. Both calculations and observations suggest that bubbles may reach quite large sizes, several metres in diameter, with excess internal pressures of up to 3 bars. Large lava bubbles have often been photographed swelling grotesquely on the surface of lava lakes in Hawaii, bursting in showers of basaltic spatter (Fig. 8.3).

8.1.3 Vulcanian eruptions

Vulcanian explosions are discrete bursts taking place at intervals of minutes to hours. Often, the ejected material is not juvenile, but shattered fragments of a solid lava plug which chilled in the volcanic vent. Vesiculation and growth of bubbles in the *ejected* material itself is not therefore involved. Exsolution taking place in magma at greater depth ultimately generates pressure which exceeds the strength of the overlying rock and shatters it (Fig. 8.4). In some (perhaps most) vulcanian eruptions, pressure may also build up through vaporization of ground-water seeping into the vent. Other vulcanian eruption may be examples of fuel–coolant interactions, a physically different type of process (Section 6.4.).

Velocities of 200 metres per second are predicted from theoretical studies for vulcanian explosions where the volatiles are entirely magmatic. Much faster ejection velocities have been observed in actuality, for example in the 1975

eruption of Ngauruhoe, New Zealand, where velocities of 400 metres per second were measured. Such high velocities would call for an unrealistic 10 per cent magma volatile content. Thus, a significant contribution of ground-water seems inevitable.

In several historic eruptions, large blocks have been hurled long distances, prompting suggestions of inconceivably high ejection velocities.

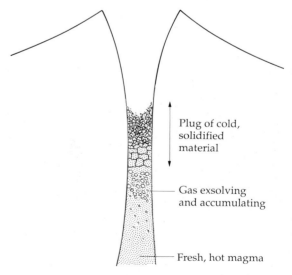

Plug of cold, solidified material

Gas exsolving and accumulating

Fresh, hot magma

Fig. 8.4 Many vulcanian eruptions are thought to involve a plug of solidified material in the throat of a vent, due to build-up of volatile pressure in deeper-seated magma. After Wilson, 1980.

During the 1968 eruption of Arenal, Costa Rica, blocks were propelled up to 5 km from the vent; the larger ones forming huge impact craters, many metres in diameter. Using standard ballistic theory, supersonic ejection velocities of *c*.600 metres per second were invoked to account for them.[9] Such vast velocities are improbable, according to Wilson, since the calculations assume the individual blocks were ejected into still air, where atmospheric drag is a major factor. In reality, a shock wave is propagated through the air, moving upwards ahead of a dense slug of ejected fragments. Thus, drag forces on individual clasts may be minimal at the onset of the blast, and huge initial velocities are unnecessary to explain the distances reached by clasts. None the less, walking around the crater field left by such a blast is apt to make even volcanologists become quietly mediative (Fig. 8.5).

8.1.4 *Importance of magma volatile content to eruption velocity*

Discounting for the moment the effects of magma–water interactions, it is the expansion of hot, compressed gas that powers explosive volcanic eruptions. It follows that the proportion of volatiles present dictates the ejection velocity and behaviour of the ensuing eruption column (Fig. 8.6). Highest ejection velocities are

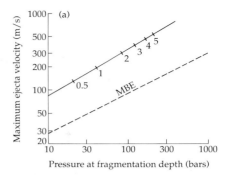

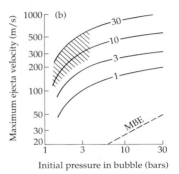

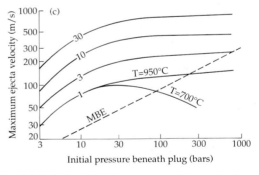

Fig. 8.5 Impact crater several metres in diameter formed by block ejected by an explosion in crater of Poás volcano, Costa Rica, 1978, photographed the following morning. The ejected block probably broke up on impact, and the fragments are buried below surface.

Fig. 8.6 Relationships between maximum ejection velocity, pressure, and volatile content for Wilson's three types of eruption. (a) Plinian eruptions. Gas fractions are shown in weight per cent at the tick marks (MBE is the curve obtained from the Modified Bernouilli Equation). In strombolian eruptions (b) pressure shown is that in a bubble just prior to bursting, for different volatile contents; the shaded area shows the most likely combination of pressure and volatiles. In vulcanian eruptions (c) pressure is that beneath retaining cap at time of failure. From Wilson, L. (1980). Relationships between pressure, volatile content and ejecta velocity in three type of volcanic explosion. *J. Volcanol. Geotherm. Res.* **8**, 297–313.

achieved in plinian eruptions of silicic magmas, where volatile contents of 4–5 per cent may result in velocities of over 500 metres per second. While water is the most probable volatile phase in terrestial plinian eruptions, other volatiles are possible on Venus and Mars, where carbon dioxide may be important. Lower eruption velocities will result when higher molecular weight volatiles like carbon dioxide are involved.

— 8.2 Dynamics and dimensions of eruption columns —

Like Caesar's Gaul, volcanic eruption columns can be divided into three parts. Immediately above the vent is the *gas thrust* region, in which particles are propelled upwards like a shot from a gun. We explored some of the factors that influence muzzle velocities in this region in the preceding section. Above the gas thrust region is a *convective* region, in which hot, buoyant gases carry finer fragments upwards, while larger fragments fall back (Fig. 8.7). We now examine this convective region, and a third, the *umbrella* region, following a useful summary by Stephen Sparks.[10]

Hot-air balloons drifting like brightly coloured bubbles above suburbs and shires on still summer evenings demonstrate that hot gas is buoyant. Within a convecting eruption column, buoyancy forces are so powerful that entrained clasts may even accelerate upwards, a condition sometimes termed *superbuoyancy*. Pilots of hot-air balloons are usually content to drift along a few hundred metres above ground for an hour or so, but a convecting eruption column may rise tens of thousands of metres and be sustained for long periods, perhaps many hours. Small variations in vent radius and mass eruption rate may cause dramatic changes to its behaviour.

8.2.1 Widths of eruption columns

Plumes of smoke or steam widen as they rise. This is a commonplace observation, easily taken for granted. It is far from easy, though, to describe the way that an eruption column widens. In the simplest case, where the atmospheric density decreases linearly upwards, an ascending column with initial width x should widen in a linear fashion, such that:

$$b = x + \gamma h$$

where b = width; h = height); and γ has a value of $c.0.125$.

In real life, this relationship works well for small plumes rising short distances. One can see

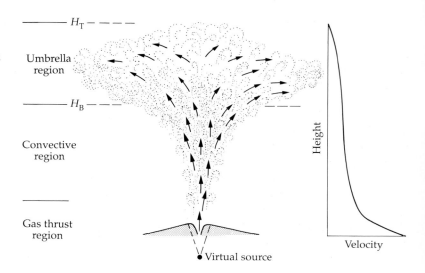

Fig. 8.7 Principal features of a volcanic eruption column, showing schematically the variation of velocity with height. Buoyancy carries plume to height H_B; lateral spreading takes place above H_B; material moving radially outwards. Momentum drives some material upwards to a maximum height H_T. From Sparks, R. S. J. (1986). The dimensions and dynamics of volcanic eruption columns. *Bull. Volcanol.* **48**, 3–15.

H_T

Umbrella region

H_B

Convective region

Gas thrust region

Virtual source

Height

Velocity

it at work sometimes in the plume of smoke rising in calm air from a factory chimney stack. With large volcanic plumes, rising tens of kilometres, not only is the atmospheric density gradient not linear, but air from the surrounding atmosphere is drawn into the column and heated by the hot ash particles, causing the column to expand much further. Measured eruption columns have large γ values, for example 0.17 in the Soufrière 1979 eruption. A major plume 20–30 kilometres high therefore has a profile as in Fig. 8.8, widening rapidly upwards, until the umbrella region is reached.

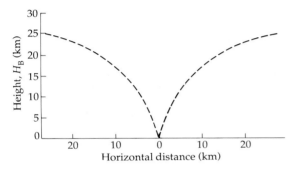

Fig. 8.8 Lateral expansion of a volcanic eruption column with height, calculated for a magma temperature of 1000°C. From Sparks, R. S. J. (1986). The dimensions and dynamics of volcanic eruption columns. *Bull. Volcanol.* **48**, 3–15.

8.2.2 Height of eruption columns

Under normal conditions, temperature decreases upwards in the Earth's atmosphere through the *troposphere* until a level called the *tropopause* is reached. Above this, temperature increases again in the *stratosphere*. The height of the tropopause varies with latitude; it is higher in the tropics (*c*.20 kilometres) and lower at higher latitudes (*c*.10 kilometres). Variations in temperature, of course, are linked to variations in density. An eruption plume will rise until it reaches a level where its density is the same as the surroundings, so it is no longer buoyant. At this level, H_B in Sparks' terminology, it spreads sideways. Some material in the centre rises higher, carried upwards by momentum, not buoyancy. This give rise to a plume whose shape may seem familiar

(Fig. 8.7). It resembles, of course, the mushroom cloud of nuclear explosions, although these are the results of single thermal pulses, rather than sustained convection. Sparks termed the lateral spreading region the 'umbrella' region, a name with less fearful connotations than 'mushroom'.

Decades of study have led to a reasonable grasp of the thermodynamics of hot gases, and therefore it is possible to write an equation describing how high a convective plume will rise:

$$H_T = 5.773 \, (1+n)^{-3/8} \{\sigma Q s(\theta_e - \theta_{ae})\}^{1/4}$$

where H_T is the height climbed; n describes the ratio of the temperature gradient in the column to that in the atmosphere; Q is the volume discharge rate of magma; s is its specific heat, σ is its density and $(\theta_e - \theta_{ae})$ the difference in temperature between the newly erupted magma and the atmosphere at sea-level.

This intimidating equation need not detain us. Its importance is that the height reached by an eruption column can be estimated if the eruption rate is known; the other variables are easier to determine, and some, such as magma density, are routine. Eruption rate is the critical factor because it is a measure of the thermal energy being pumped into the eruption column. Essentially, the height reached varies as the fourth root of the eruption rate; a 16-fold increase in eruption rate is required to double the height reached.

Unfortunately, there is a complicating factor that cannot be ignored: the temperature change with height in the atmosphere is not linear, as the equation assumes: there is an abrupt change at the tropopause. When an eruption column punches its way up through the tropopause into the stratosphere, its buoyancy is reduced because of the decreased temperature contrast. Taking this and other factors into account, Sparks modelled the height reached by eruption columns for different mass eruption rates in tropical and temperate climates, and for different magmatic temperatures. His results are summarized in Fig. 8.9, a diagram that will underpin much of our discussion of eruptions columns.

8.2.3 The umbrella region

Pliny the Younger was the first to describe the

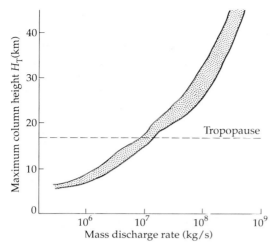

Fig. 8.9 Variation of maximum eruption column height, H_T, with mass eruption rate. Stippled field encloses plausible magmatic temperatures between 1000 and 600°C. Temperature changes at the tropopause cause the slight kinks in the bounding curves. Overall, mass eruption rate is much the most important factor influencing how high an eruption column will reach. After Sparks 1986.[10]

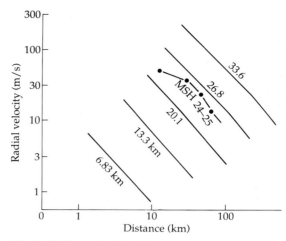

Fig. 8.10 Theoretically derived rates of radial spreading in umbrella region with distance from centre for eruption columns of different maximum heights. Data observed for Mt. St Helens 18 May are shown for comparison. After Sparks 1986.[10]

umbrella region of a large eruption column, although he compared the AD 79 cloud to the spreading canopy of a pine tree. (Did Romans have umbrellas . . . ?) An important aspect of the

umbrella region is the *rate* at which radial spreading take places, and the way that this falls off with distance from the centre of the column (Fig. 8.10). In large eruption columns, radial spreading rates may be extremely high, of the order of 300 metres per second, approaching sonic speeds.

— 8.3 Fall-out and dispersal of tephra from volcanic eruption columns

Pyroclastic deposits provide lots of parameters that volcanologists can measure on the ground, for example the distribution of clasts of different sizes. These parameters provide useful checks of the validity of mathematical models of eruption columns. A theoretical study by Carey and Sparks[11] on the interrelationships between plume parameters and the resulting fall deposits provides a fine introduction to our review of pyroclastic fall deposits in the next chapter.

Their starting point was the eruption column model outlined above, with two further refinements: the assumption that vertical velocity within the column shows a gaussian distribution in a profile across the flow, and the introduction

of *terminal fall velocities* for clasts of different dimensions. Terminal fall velocity is a simple concept; one that can be readily measured in air by the simple means of dropping clasts from a great height and timing their fall. A free-fall parachutist (a large, soft clast) reaches a terminal velocity of about 60 metres per second. Terminal fall velocities can be estimated from this equation:

$$\text{Terminal velocity} = C_d \sqrt{\frac{dg\sigma}{\beta}}$$

where C_d is the drag coefficient; d is clast diameter; g is the acceleration due to gravity; σ is the clast density, and β the atmospheric density.

Owners of expensive motor cars will probably be able to recite a parameter similar to C_d for their own and rival vehicles, namely the aerodynamic drag coefficient. Volcanic pyroclasts are, of course, far removed from the sensuous smoothness of luxury cars—they are irregular in size and shape and have rough, drag-inducing surfaces. Values of C_d of about 1.054 have been found in experiments.[12]

If the terminal fall velocity of a clast is less than the upwards velocity of the column, it will be carried *upwards* to the radially spreading umbrella region. In most eruptions, convective velocities are great enough to carry fist-sized lithic clasts to the top of the column. By modelling the velocity distribution in the column, it is possible to construct *clast support envelopes* for clasts of different sizes, and thus predict the pattern of fall-out that will result. Outside the envelope for a given clast size, fall velocity exceeds ascent velocity, so clasts fall out (Fig. 8.11). A clast can fall out completely only if its descent trajectory carries it clear of the plume core; if in falling it crosses the clast support envelope, it will be carried up again. Thus, most

fall-out takes place from overhanging parts of the umbrella.

Because large clasts have higher terminal fall velocities than small ones of the same density, they naturally fall out nearer the vent. *Isopleths*, lines joining points of the same maximum clast size, can be drawn on a map by plotting the distribution of clast sizes found in a pyroclastic deposit. As eruption column heights increase, so clast dispersal becomes more effective. Thus, the resulting isopleths become more widely spaced on the ground, and enclose larger areas (Figs. 8.12–8.13.)

8.3.1 *The influence of wind on eruption column shapes*

At a superficial level, the influence of wind on an eruption plume is obvious: the plume is deflected downwind. When it comes to predicting the distribution of fall-out from a wind-blown eruption plume, problems emerge. An eruption column may penetrate 40 kilometres through the atmosphere, traversing levels where winds are blowing in sharply different directions at different velocities. Meteorological records provide wind data for modern eruptions, but there are

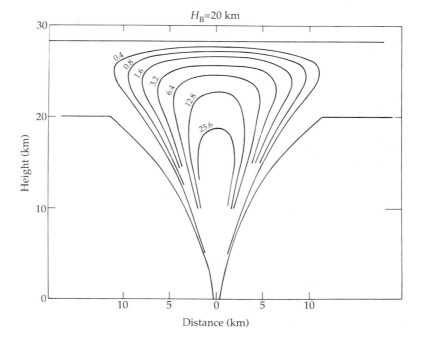

Fig. 8.11 Clast support envelopes for an eruption column in which the neutral density level (H_B) is 20 km, and maximum height H_T is 28 km. Envelopes are contoured for clast diameters in centimetres, for assumed densities of 2500 kilograms per cubic metre. Within each envelope, clasts larger than the contour dimension cannot be supported and fall, but clasts can only fall completely out of the eruption column from the overhanging parts of the umbrella. From Carey, S. and Sparks, R. S. J. (1986). Quantitative models for the fall-out and dispersal of tephra from volcanic eruption columns. *Bull. Volcanol.* **48**, 109–125.

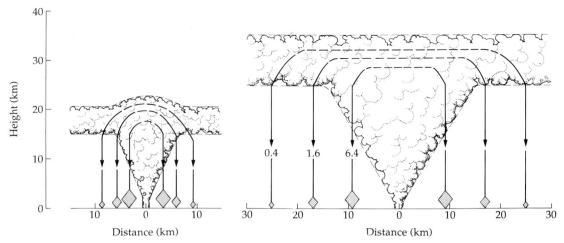

Fig. 8.12 Higher eruption columns disperse clasts of given size much more widely than lower columns. Trajectories for 0.4, 1.6, and 6.4 cm diameter clasts are shown for columns with maximum heights of 21 km (left) and 35 km (right). From Carey, S. and Sparks, R. S. J. (1986). Quantitative models for the fall-out and dispersal of tephra from volcanic eruption columns. *Bull. Volcanol.* **48**, 109–25.

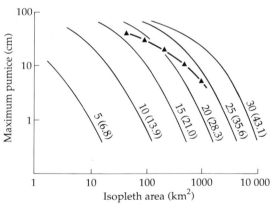

Fig. 8.13 Areas enclosed by isopleths for pumice clasts of given size, calculated for eruption columns of different heights, and for pumice density of 500 kilograms per cubic metre. Curves are labelled with column heights H_B in kilometres; H_T in parenthesis. Triangles show actual field data for the Fogo A pumice deposit (Azores), which has approximately this density. Agreement between theory and observations suggest a column height of about 20 km for the eruption. From Carey, S. and Sparks, R. S. J. Quantitative models for the fall-out and dispersal of tephra from volcanic eruption columns. *Bul. Volcanol.* **48**, 109–25.

none for past eruptions. Fortunately, ascent velocities of large eruption columns are so great that the effects of cross-winds are small, but for smaller columns wind effects can be profound. Carey and Spark's model assumed a 'simple' wind profile with a wind maximum at the tropopause (about 11 kilometres in temperate latitudes), and superimposed these velocity vectors on the motions of clasts in the eruption columns.

On a calm summer evening, the plume of smoke from a bonfire rises vertically, but even a gentle breeze carries it sideways, often causing it to stream horizontally downwind. In powerful volcanic eruption columns, where the plume rises high enough to develop an umbrella region, radial spreading will take place, even *against* a strong wind. One component of radial spreading must be directly upwind, so at some point, the upwind spreading velocity exactly balances the downwind drift velocity. This is the *stagnation point* (Fig. 8.14).

A useful measure of plume shape in a wind is the ratio of the upwind stagnation radius to the horizontal displacement of the plume axis. For

Fig. 8.14 Upwind stagnation point and displacement of plume axis in an eruption column. Arrows within plume show velocity vectors in spreading region; where spreading velocity is balanced by equal and opposite wind velocity (double arrowed vectors) defines stagnation point. The d is displacement of column axis. In cases where the amount of upwind spreading is the same as column displacement, the upwind edge of the column is vertical.

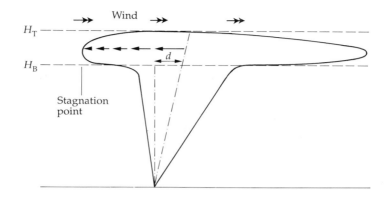

the largest columns, the umbrella is able to spread upwind, forming an asymmetrical mushroom, so the ratio is greater than one. For small columns, the ratio is less than one, and the whole column leans over downwind. For intermediate sized columns, the ratio is close to unity, and the column is nearly vertical, with a straight edge on its upwind side, and an extended downwind overhanging umbrella. (Fig. 8.15). One can sometimes see similar features in towering cumulo-nimbus thunder-heads; they commonly spread out at high altitude to form asymmetrical *anvils*; wispy wedges of ice-crystal cloud.

8.3.2 *Fall-out of clasts from plumes in winds*

Few volcanic eruptions take place in windless conditions. To understand how pyroclasts are

dispersed from an eruption plume, it is essential that the effect of wind be included (Fig. 8.16). Carey and Sparks prepared maps showing the distribution of clast sizes for different combinations of clast density, column height, and wind velocity; in each case the wind velocity is that at the tropopause. Even in strong wind conditions, the largest eruption columns yield roughly circular isopleths. This is because the columns are more than 20–30 kilometres high, reaching well above the tropopause. Considerable upwind spreading is possible within the umbrella region, and although clasts have to fall through levels within the troposphere where winds may be strong, they are still not carried as far downwind as those from lower columns. In these, the umbrella region is near the tropopause, and the

Fig. 8.15 Ratio of distance to stagnation point to axial displacement plotted against column height H_B for wind speeds of 10, 20, and 30 metres per second, showing contrasted shapes of plumes. From Carey, S. and Sparks, R. S. J. (1986) Quantitative models for the fall-out and dispersal of tephra from volcanic eruption columns. *Bull. Volcanol.* **48**, 109–25.

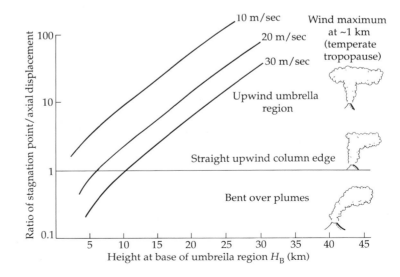

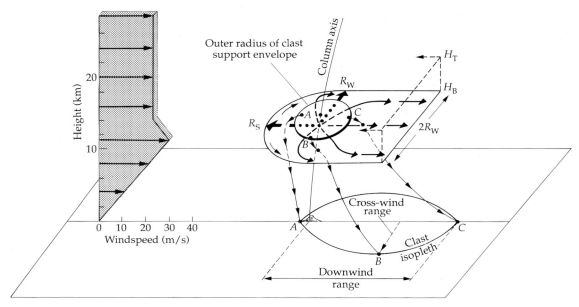

Fig. 8.16 Components required to model distribution of clasts of fixed sizes from a plume. R_S is the upwind stagnation radius; R_W is width of radial expansion. Radial flow is deflected by the cross-wind, so clasts move through a volume defined by heights H_B, H_T, and twice R_W. Trajectories of three clasts define the geometry of any individual isopleth; A is directly upwind; B is cross-wind and C downwind. Clasts A, B, and C are released from outer radius of a clast support envelope which lies between planes H_B and H_T. Note the wind profile, with a maximum value at the tropopause at about 11 km, and a steady value of three-quarters of the tropopause maximum through most of the stratosphere. From Carey, S. and Sparks, R. S. J. (1986). Quantitative models for the fall-out and dispersal of tephra from volcanic eruptions columns. *Bull. Volcanol.* **48**, 109–25.

plume extends a long way downwind. Wind, therefore has most marked effects on small eruptions, which yield attenuated, elliptical isopleths (Fig. 8.17).

Isopleth maps are not merely hypothetical constructs. If by field sampling and measurement one can collect enough data for clast sizes in a real pyroclastic deposit to draw a map, one can infer some critical eruption parameters. The distance of an isopleth from the vent measured along wind direction is a function of both column height and wind speed, and therefore these two factors can be inferred, even for a prehistoric eruption.

Carey and Sparks used the Mt. Helens 1980 eruption to test this technique. Isopleths for pumice clasts 4 and 8 cm in diameter were obtained from field-work, suggesting a column height of 19 kilometres and wind speed of

36 metres per second (Fig. 8.18). Radar studies of the actual eruption column showed a peak height of 19 kilometres at 1700 hrs on the day of the eruption, while a meteorological balloon showed a wind at the tropopause between 27 and 33 metres per second.

Although this technique provides a splendid tool for analysing eruptions, it must be treated cautiously. One serious limitation is that it assumes a consistent situation. In nature, this is unlikely to prevail. Low-level winds are particularly fickle. In the course of a day, they may go through huge changes in both strength and direction, as all sailors know. At stratospheric levels, things are more stable. Thus, the technique is likely to work best for short-lived but powerful eruption columns that reach the stratosphere, rather than long-lasting but weak columns that remain within the troposphere.

Applications: aircraft and ash clouds   This rather formal treatment may suggest that the mechanics of eruption and dispersion of ash clouds are of strictly academic concern. This is far from the case. Understanding eruption columns has made it possible to assess the potential effects of tephra fall-out on communities at risk, thus simplifying the tasks of authorities responsible for planning for emergencies. For example, by modelling the effects of tephra fall, estimates have been made of the likely consequences of renewed pyroclastic activity at Vesuvius, one of the world's most worrying volcanoes.[13]

Apart from these obvious applications, in recent years several disasters have been avoided by the narrowest of margins when passenger aircraft have unwittingly flown into ash clouds. In June 1982 a British Airways 747 jumbo jet carrying 240 passengers encountered a plume from Galunggung, Java, when it was at an altitude of 11 km (36 000 feet). All four of its engines stalled, and its wings and windscreen were abraded. After a terrifying gliding descent of 7.5 km (24 000 feet) the pilot eventually succeeded in restarting the engines, enabling him to scrape into a safe landing in Jakarta. Three weeks later, a Singapore airways jumbo flew into another plume from Galunggung at a height of 9 km. It lost power in three engines, but the pilot was able to restart one, landing safely after an emergency descent of only 2.4 km. In 1990, a KLM 747 flew into the ash from Redoubt in Alaska, again losing all engine power and forward visibility. After a heart-stopping descent, it managed to land safely at Anchorage. Seventeen aircraft encountered ash from the 1991 Pinatubo eruption. All survived, but it can only be a matter of time before disaster strikes.

Hazards to aircraft from ash clouds are especially acute because they may almost literally appear out of the blue, giving little chance of

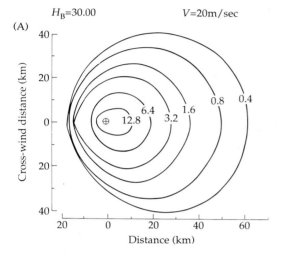

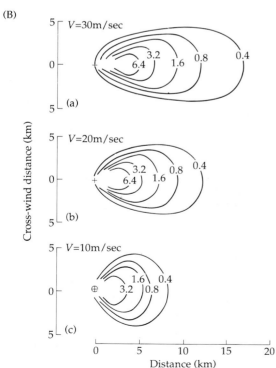

Fig. 8.17 Calculated isopleth shapes for some different eruption conditions. Isopleths are clast diameters in centimetres, and assume densities of 2500 kilograms per cubic metre. (A) is for a column with maximum height 43 km (H_B 30 km) and a wind of 20 metres per second at the tropopause. Isopleths are almost circular. (B) is for an eruption column 14 km high, with wind speeds of 10, 20, and 30 metres per second. There is much more downwind extension of isopleths for the lower column, even at the same wind speed. From Carey, S. and Sparks, R. S. J. Quantitative models for the fall-out and dispersal of tephra from volcanic eruption columns. *Bull. Volcanol.* **48**, 109–25.

Fig. 8.18 A single well-constrained isopleth can be used to deduced important eruption parameters. Theoretical plots for cross-wind range against maximum downwind range are shown here for different wind velocities, with column heights superimposed. An actual data point, the four centimetre pumice isopleth for the 18 May Mt. St Helens eruption is also shown. Column height and wind speed deduced from this plot agree well with actual meteorological observations made on the day, 19 km and 30 metres per second respectively.

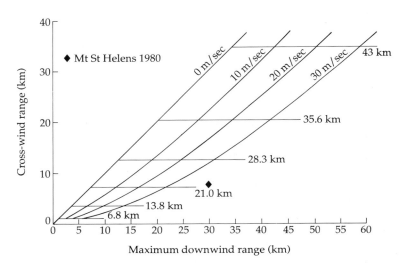

warning pilots. A brief eruption of Lascar volcano, north Chile which took place at 10.39 UT on 16 September 1986 lasted less than five minutes. An eruption column rose to a height of about 15 km, and was rapidly borne downwind. Ground observations and geostationary satellite images showed that the eruption plume formed a discrete extended slug about 2 km thick at elevations between about 10 and 14 km, which passed over the major city of Salta, Argentina, 285 km distant from Lascar, less than 2 hours after the eruption. By 14.12 UT the plume had diffused, having travelled 400 km in 3.5 hours, and covered an area greater than 100 000 square kilometres (Fig. 8.19).

Fortunately, although ash fell out over Salta and its airport, no aircraft incidents were reported. But in situations where an ash cloud drifts downwind at more than 100 kilometres per hour, as at Lascar, there is precious little time to identify the hazard and notify aircraft at risk. It is urgently necessary to refine our understanding of plumes, so that their extent can be predicted, and to develop rapid monitoring techniques so that warnings can be given.

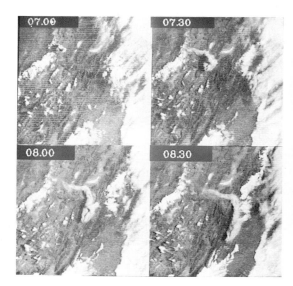

Fig. 8.19 Times series of images from the geostationary GOES weather satellite, acquired every 30 minutes, showing rapid downwind transport of the vulcanian eruption cloud of Lascar volcano, north Chile, 16 September 1986. White mass at right of each image is cloud bank over Argentinian pampas. Each image is about 500 km across.

— 8.4 Mt St Helens, May 1980—an exceptional case study —

All the best remote-sensing technology in the world is 'deep black'; kept under wraps by the military. Only exceptionally do their data leak out to the public. Mt St Helens' 1980 eruption provided such an exception. Images from remote-sensing satellites provided a unique view of the propagation of the initial giant umbrella cloud. An account of this unusual event by Sparks, Moore, and Rice conveniently rounds off our study of eruption columns.[14] Note that this paper dealt with the initial 'blast' cloud, not with the sustained plinian eruption cloud (mentioned in the previous section) that developed later.

It was a landslide on the north flank of Mt. St Helens which triggered the 18 May blast (Chapter 4). The 'blast' travelled at a velocity of about 90 metres per second, rapidly overtaking the avalanche. Initially, it moved *horizontally*, but within about ten minutes had grown *vertically* to form a giant mushroom cloud 25 km high, with stalk diameter of over 20 km and a cap 70 km in diameter. Satellite images show that the centre of the rising cloud was actually 12 km north of the volcano, and that the vertical

velocity of the cloud was approximately 110 metres per second. For roughly the first ten minutes, the cloud climbed vertically, but then it expanded rapidly radially between elevations of 12 and 16 km altitude, at a maximum velocity of about 55 metres per second. It had spread 15 km directly *upwind* of the western edge of the devastated zone less than 20 minutes after the first explosion (Figs. 8.20–8.21). This upwind stagnation point, it will be recalled, marks the point at which upwind spreading is balanced by downwind drift. Meterological data for the spreading altitudes showed winds averaging 22 metres per second, in good agreement with the velocity measured from satellite images of the cloud's spread.

Two facets of this remarkable umbrella cloud must be accounted for: first, the transition from horizontally directed 'blast' to vertically ascending column in a period of only about 1.5 minutes. Second, the extremely rapid ascent of the cloud from ground level to over 20 km in less than 10 minutes.

As outlined earlier, the height reached by an

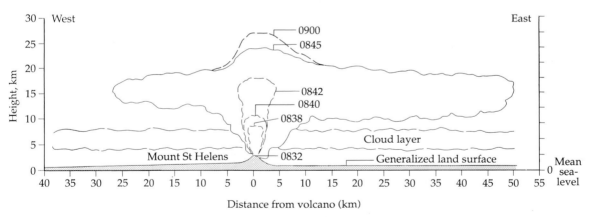

Fig. 8.20 Profiles through the giant umbrella cloud formed during the first half hour after the Mt. St Helens blast at 08.32 on 18 May 1980. Immediately above its source region, the cloud reached 25–30 km altitude (H_T), but the maximum altitude decreased radially away, to about 20 km. The central high point was due to momentum carrying some material higher than the neutral density level. Rapid radial spreading took place after 08.42. From Sarna-Wojcicki, A. M., Shipley, S., Waitt, R. B., Dzurisin, D., and Wood, S. H. (1981). Areal distribution, thickness, mass, volume and grain size of air-fall ash from the six major eruptions of 1980. In *The 1980 eruptions of Mount St. Helens, Washington*, (eds. P. W. Lipman and D. R. Mulineaux), US Geol. Surv. Prof. Paper, No. 1240, pp. 577–600.

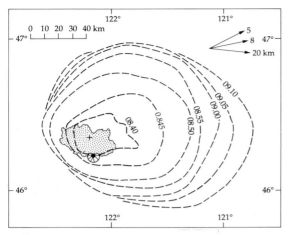

Fig. 8.21 Outline of the giant umbrella cloud at five-minute intervals as observed from geostationary satellites. Cross shows initial centre of cloud ascent; stippled area shows extent of blast deposit; arrows at top right are wind directions at 5, 8, and 20 km altitude, measured at Spokane, Washington. Compare the profile views in Fig. 8.20. From Sparks *et al.* 1986.

eruption column is dictated by the thermal flux, which in conventional eruptions is an expression of the mass eruption rate. In the case of the Mt. St Helens umbrella cloud, the explosion at 08.32 on Sunday morning suddenly propelled a mass of hot rocks, dust, and gas horizontally northwards. All of the thermal energy that later drove the cloud to a height of 25 km was contained in that ejected mass. Field and theoretical studies showed that the blast cloud had a mean temperature of about 400°C, and an energy content of about 10^{17} J. Because it had little vertical thickness and hugged the ground, levelling trees by the million, the blast cloud was clearly dense initially. Only when it had travelled many

kilometres did it become less dense than the surround air, gain buoyancy, and lift off the ground.

Three processes were involved in the decrease in density, causing the rapid transition from horizontal blast cloud to vertical convecting column. First, decompression as the hot gases expanded into the atmosphere; second, fall-out of solid fragments from the cloud; and third, the advancing blast cloud entrained and heated a large volume of ordinary air. Whereas large blocks of hot rock are poor at transferring heat to the atmosphere, dust-sized material is much more efficient. There was plenty of such fine material in the blast. This was probably the most important single factor in generating a powerfully convecting column, capable of punching up through the atmosphere. Initially, the cloud had zero vertical velocity but it rapidly became so powerfully buoyant that it *accelerated* upwards, reaching an upwards velocity of over 100 metres per second soon after it lifted off the ground, comparable with the initial *horizontal* velocity.

It was an exceptional combination of circumstances that enabled the Mt. St Helens umbrella cloud to be closely documented. Generation of the great cloud from a laterally moving blast, rather than a point source vent, may also appear rather exceptional. However, the laws of physics apply in all circumstances, so this spectacular eruptions teaches us that umbrella clouds should develop whenever large mass flux eruptions take place, irrespective of their sources. Specifically, we should expect umbrella clouds in eruption columns higher than 20 km and with volume eruption rates greater than 5×10^4 cubic metres per second. Whenever large pyroclastic flows are erupted, they too may generate huge umbrella clouds (Chapter 10).

Notes

1. Walker, G. P. L., Wilson, L., and Bowell, E. L. G. (1971). Explosive volcanic eruptions, I. The rate of fall of pyroclasts. *Geophys. J. Roy. Astron. Soc.* **22**, 377–83.
2. Wilson, L. (1972). Explosive volcanic eruptions, II. The atmospheric trajectories of pyroclasts.

Geophys. J. Roy. Astron. Soc. **30**, 381–92.
3. Wilson, L. (1976). Explosive volcanic eruptions, III. Plinian eruption columns. *Geophys. J. Roy. Astron. Soc.* **45**, 543–6.
4. Wilson, L., Sparks, R. S. J., and Walker, G. P. L. (1980). Explosive volcanic eruptions, IV. The

control of magma properties and conduit geometry on eruption column behaviour. *Geophys. J. Roy. Astron. Soc.* **63**, 117–48.

5. Sparks, R. S. J. and Wilson, L. (1982). Explosive volcanic eruptions, V. Observation of plume dynamics during the 1979 Soufrière eruption, St. Vincent. *Geophys. J. Roy. Astrom. Soc.* **69**, 551–70.

6. Wilson, L. (1980). Relationships between pressure, volatile content and ejecta velocity in three type of volcanic explosion. *J. Volcanol. Geotherm. Res.* **8**, 297–313.

7. Chouet, B., Hamisevicz, N. and McGetchin, T. R. (1974). Photoballistics of volcanic jet activity at Stromboli, Italy. *J. Geophys. Res.* **79**, 4961–76.

8. Blackburn, E. A., Wilson, L., and Sparks, R. S. J. (1976). Mechanisms and dynamics of strombolian activity. *J. Geol. Soc. Lond.* **132**, 429–440.

9. Fudali, R. F. and Melson, W. G. (1972). Ejecta velocities, magma chamber pressure and kinetic energy associated with the 1968 eruption of Arenal volcano. *Bull. Volcanol.* **35**, 383–401.

10. Sparks, R. S. J. (1986). The dimensions and dynamics of volcanic eruption columns. *Bull. Volcanol.* **48**, 3–15.

11. Carey, S. and Sparks, R. S. J. (1986). Quantitative models for the fall-out and dispersal of tephra from volcanic eruptions columns. *Bull. Volcanol.* **48**, 109–25.

12. Wilson, L. and Huang, T. C. (1979). The influence of shape on the atmospheric settling velocity of volcanic ash particles. *Earth Planet. Sci. Lett.* **44**, 311–24.

13. Macedonio, G., Pareschi, M. T., and Santacroce, R. (1990). Renewal of explosive activity at Vesuvius: models for expected tephra fall-out. *J. Volcanol. Geotherm. Res.*, **40**, 327–42.

14. Sparks, R. S. J., Moore, J. G., and Rice, C. J. (1986). The initial giant umbrella cloud of the May 18th, 1980 explosive eruption of Mount St. Helens. *J. Volcanol. Geotherm, Res.* **28**, 257–74.

9

Pyroclastic fall deposits

9.1 Introduction

For many years, pyroclastic rocks were the poor relations of lavas, which were the only 'respectable' volcanic rocks. It is hard to give an ash deposit a satisfying swipe with a hammer to get a hand specimen of *proper* rock, so geologists often ignored messy pyroclastic rocks. Many geochemists claimed to have 'sampled' a volcano, when all they had done was to sample its lavas. This despite the fact that many volcanoes erupt far greater volumes of pyroclastic rocks than lava. In the last two decades, more attention has been paid to pyroclastic rocks as a result of ground-breaking work by volcanologists such as G. P. L. Walker in Britain, R. L. Smith in the USA, and their associates. So much innovative work has been done that the period has been a stimulating one, punctuated by heated discussions by volcanologists in the field, perched precariously on slopes of loose ash, their boots full of gritty pumice, and their notebooks full of scrawled diagrams designed to convince unbelievers. Today, the study of pyroclastic rocks is its own special field, with its own excellent textbook.[1] Many problems have been solved, but some important ones remain unsolved.

To bring some order to this large subject, pyroclastic rocks are dealt with in four groups. This chapter deals with the simplest group, pyroclastic fall deposits. Chapter 10 deals with pyroclastic flows, the most voluminous of all volcanic deposits. More controversial phenomena, pyroclastic surges and *nuées ardentes*, are discussed in Chapters 11 and 12.

9.1.1 Describing pyroclastic fall deposits

A pyroclastic fall deposit is conceptually simple —it is what falls to earth after an eruption column has propelled material skyward. A fresh fall deposit is also simple to look at—it mantles the landscape evenly, like a grey snowfall; a snowfall that does not melt.

After a snowfall, the first thing everyone does is to measure its *thickness*. Similarly, after an eruption, it is natural for people to estimate the thickness of the layer that has turned their neighbourhood a pale shade of grey. With modern deposits, this is straightforward, except where the deposit is thin. This is more important than it may sound—the film of ash on a car roof may indeed be thin (and difficult to measure), but given that such thin deposits may cover many thousands of square kilometres, their total *volumes* may be large.

In prehistoric examples, deposits less than one or two centimetres thick are rarely preserved, and the vagaries of weathering and erosion are such that it is usually necessary to scout round to find the *maximum* thickness preserved in any one area. Once enough data have been collected, it is possible to compile an *isopach* map. (Isopachs are contour lines joining points of equal deposit thickness.)

Another obvious parameter to measure at any location is the *grain size*. A terminological point: although news media (and volcanologists speaking loosely) often use the term *ash* to describe pyroclastic materials when no grain-size connotation is intended; it is better to use the more general term 'tephra' (Table 9.1).

In any natural deposit, a wide range of clast sizes is generally present. For many purposes, it is the *maximum* clast size that is important. In the field this can be determined by picking out by eye the five or ten largest visible clasts at any one site and measuring them. Where fine-grained deposits are concerned, sieving is necessary, either in the field or in the laboratory. Once sufficient grain-size data have been obtained, an *isopleth* map can be drawn.

Sieving is also essential to determine another valuable parameter, the degree of *sorting*, or proportions of clast sizes present at any location. Most air-fall deposits are well sorted, as one would expect given the wind—winnowing that takes place as they fall to earth. A well-sorted deposit produces a sharp peak on a size/frequency graph, whereas a poorly sorted one produces a much broader distribution (Fig. 9.1).

Apart from these crucial dimensional parameters (thickness, clast size, sorting) three other parameters are useful in describing pyroclastic deposits: first, the *vesicularity* of the clasts can be an important guide to the eruption process. Bruce Houghton and Colin Wilson[2] devised a *vesicularity index*, basically the volume fraction of vesicles in a clast, measured by comparing the density of the clast in air and water. They found that most 'dry' magmatic eruptions yield clasts with vesicularity indices uniformly between 70 and 80 per cent, regardless of magma viscosity, so that the vesicularity of basaltic strombolian scoria is broadly similar to that of a rhyolitic plinian pumice. Where magma–water interactions are involved, the timing of fragmentation

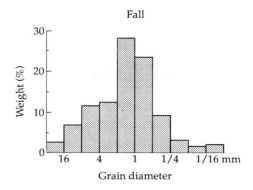

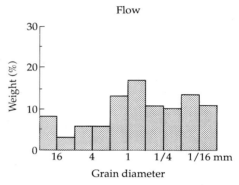

Fig. 9.1 Sorting in pyroclastic deposits. Fall deposits are well sorted. At any one locality, most grains fall within a narrow size range, so the histogram shows a strong central peak, here at about 1 mm diameter. Pyroclastic flow deposits are much more poorly sorted, so their histograms lack clear peaks. These data are for the plinian fall and related pyroclastic flow forming the Upper Bandelier Tuff, New Mexico. (After Cass, R. A. F. and Wright, J. V., 1987.)

relative to vesiculation is variable. Thus, variable degrees of vesiculation may result.

In violent eruptions where the magma is highly fragmented, crystals originally present in the magmas as phenocrysts may become disaggregated from their frothy, often glassy, host rock because of their differences in strength. Since they are also denser than the glass and occupy a restricted range of sizes, crystals may have distinctive distribution patterns. Thus, *crystal concentration* is sometimes measured in pyroclastic deposits. A simple way of determining this is to estimate the volume proportions of crystals in

Table 9.1 Size ranges for tephra

Name	Size range
Ashes	Less than four millimetres
Lapilli*	Between four and 32 millimetres
Blocks	Greater than thirty-two millimetres

* *Lapilli* (Italian, 'little stones')

intact lumps of pumice, and compare this value with the proportion found on sieving the ash in question.

Finally, most pyroclastic deposits include a proportion of *lithics*, fragments of rocky material plucked from the walls of the vent, which may be of entirely non-volcanic origin. Measurements of the size, proportion, and nature of such lithic clasts can be as valuable as those of the magmatic clasts. In the Vesuvius AD 79 plinian deposit, the sequence of different lithic types derived from the subvolcanic basement provides a record of the opening of the vent during the eruption (Section 9.7.1).

9.2 Classifying eruptions and their pyroclastic deposits

'By their fruits ye shall know them', the Bible has it (Matthew 7: 20). This hallowed epigram applies in volcanology—pyroclastic deposits identify the type of eruptive activity giving rise to them. George Walker was the first volcanologist to distinguish between eruptions by sieving their deposits.[3] His method was later refined by David Pyle.[4] In essence, Pyle's method uses a graph which plots the decrease in thickness of a tephra deposit with distance away from the vent against the decrease in fragment size.

Two parameters are involved. One is a measure of the maximum clast size in the deposit at any given location: the *clast size half-distance* (b_c) is the distance over which the size of the largest clast in the deposit decreases by half. It is a measure of the eruption column height and therefore the intensity of an eruption, and can be derived from isopleth maps. Regardless of the initial size distribution of particles within the eruption column, b_c should be the same for any given eruption column height. The second parameter is the *thickness half-distance*, b_t the distance over which the measured thickness decreases by half. This parameter is obtained from isopach maps. b_t is a measure of the dispersive power of the eruption, and its value depends both on the column height *and* on the size distribution of particles in the eruption column. (b_c and b_t are best measured from circular isopleths and isopachs. If these are elliptical, a more elaborate technique is needed.)

Different types of pyroclastic deposits can be distinguished on Fig. 9.2. Along the bottom is the thickness half-distance, b_t. The vertical axis is a little more complex; the ratio of b_c to b_t. Plotting this ratio against b_t is a convenient way to smooth out differences introduced by variations in *magnitude*—large eruptions will always tend to have larger values of b_t than smaller ones, all other things being equal. Several different fields are labelled on the diagram; these encompass the most important types of terrestrial volcanic activity.

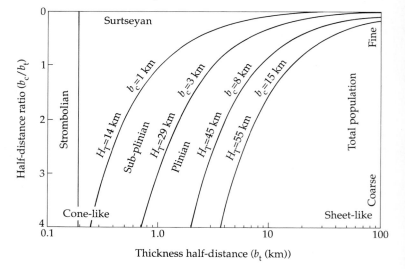

Fig. 9.2 Pyle's diagram for classifying pyroclastic fall deposits from measurements of maximum clast size and deposit thickness. The figure is contoured for clast half-distance b_c and eruption column height H_T. Construction of the diagram is explained in text. From Pyle, D. M. (1989). The thickness, volume and grain size of tephra fall deposits. *Bull. Volcanol.* **51**, 1–15.

— 9.3 Hawaiian deposits —

Hawaiian eruptions are clean and tidy—they produce small volumes of tephra, of which rather little is fine ash. None the less, they yield some intriguing pyroclastic deposits. As discussed in Chapter 8, when low-viscosity basaltic magmas are erupted, gas and liquid phases separate easily, so that the magma is not explosively fragmented. In a standard hawaiian eruption, gas-rich magmas spray liquid lava into the air in fire fountains, often hundreds of metres high. Fountain structure and the character of the resulting deposits are determined by the velocity profile and the maximum spread angle of the fountain; a measure of whether the plume is a narrow jet or a broad spray[5] (Fig. 9.3). Clasts are often ejected at such high emission rates and in such narrow jets that the eruption column is optically dense. Little radiative cooling of the clasts can take place before they fall back to earth, so they remain liquid, coalescing with others at the base of the fountain to form clastogenic lavas.

When the emission rate is lower, clasts hit the ground too cool to flow, but globs of lava (up to a metre or more in diameter) weld together to form steep-sided spatter cones around the vent. Such cones are rarely more than a few tens of metres high. Some globs are flung higher, tracing longer, lazier parabolas back to earth, twisting slowly, until they hit the ground with a thump, splodging out to form flattish masses. Because they resem-ble the brown steaming splodges observed near grazing cows, these are often known as *cow-pat bombs*.

Smaller particles, of course, are carried further away from the vent by the wind (Fig. 9.4). They too are distinctive. Because the molten lava has such low viscosity (sometimes 100 Pa s or less), surface tension controls the shapes of liquid droplets in the fire fountain. Being small, these chill quickly, falling to the ground as small black glistening spheres, dumb-bells or elongated tear-drop shapes, known as *Pele's tears*, after Pele, the Hawaiian volcano goddess (not to be confused with Mt. Pelée in the Caribbean). Walker proposed the term *achnelith* (Greek, 'spray stones') for pyroclastic particles shaped by sur-face tension.

When a piece of ordinary glass is heated over a flame, it softens until the heated part eventually gathers itself into a blob, droops downwards and begins to fall. However, the molecules in the glass are polymerized, so the blob does not detach itself completely like a water droplet: as it falls it draws out behind it a thin, flexible thread of glass which seems capable of stretching indefinitely. Pele's tears behave in the same way: when they separate from larger fragments, they draw off behind them a hair-like tail of glass, sometimes more than a metre long. Similarly, when blobs of fluid are shredded during spattering events at Hawaiian lava lakes, long, thin filaments of glass

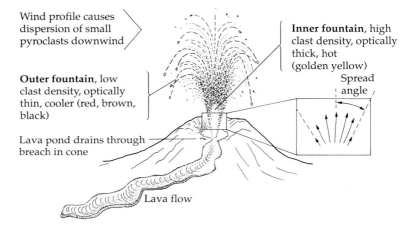

Fig. 9.3 Components of a basaltic pyroclastic eruption, illustrating the structure of a fire-fountain. Lengths of arrows on the inset indicate schematically the velocity profile. From Head, J. W. and Wilson, L. (1989). Basaltic pyroclastic eruptions: influence of gas-release patterns and volume fluxes on fountain structure, and the formation of cinder cones, spatter cones, rootless flows, lava ponds and lava flows. *J. Volcanol. Geotherm. Res.* **37**, 261–71.

Wind profile causes dispersion of small pyroclasts downwind

Outer fountain, low clast density, optically thin, cooler (red, brown, black)

Lava pond drains through breach in cone

Inner fountain, high clast density, optically thick, hot (golden yellow)

Spread angle

Lava flow

Fig. 9.4 A small fire-fountain spurts incandescent lava from the Mauna Ulu vent, Hawaii, 20 October 1969. A shower of fine particles drifts downwind from the main jet, which is inclined at an oblique angle. US Geological Survey photograph by D. A. Swanson.

are drawn out as the fragments separate. If the filaments should snap off, they may drift for many kilometres on the wind, coming to rest far from the vent. *Pele's hair* consists of thin, delicate strands of glass, which are soft and flexible, and have a light golden-brown colour (Fig. 9.5). It is rumoured that some discriminating birds use this hair to build their nests—but there are no reports of the phoenix nesting in Hawaii, or in Masaya, Nicaragua, or Erta'Ale, Ethiopia, where Pele's hair is also common.

Another hawaiian pyroclastic curiosity is *reticulite*, a basaltic glass foam (Fig. 9.6). More than a metre thickness of reticulite accumulated around Kilauea caldera after a nineteenth-century eruption. Reticulite is an attractive gold-brown glassy froth with a bubble structure of interlocking polygons so delicate that it was called 'thread lace scoria' by James Dana, who made the first volcanological study of the Hawaiian Islands. So large a fraction of the total volume consists of void spaces that the density of reticulite in air is extraordinarily low—less than 300 kilograms per cubic metre. (Recall that the density of solid basalt is 2700 kilograms per cubic metre.) Paradoxically, these ultra-light rocks

Fig. 9.5 Pele's hair produced during the 1969 Mauna Ulu eruption. The strand is more than 30 cm long. US Geological Survey photograph by D. A. Swanson.

(a)   (b)

Fig. 9.6 (a) Abraham Lincoln's profile on a 1976 one cent coin, which is resting on a reticulite fragment from the Mauna Ulu eruption. It is hard to photograph the lacy network of transparent vesicles constituting the glassy froth. (b) James Dana, the great American mineralogist, drew the delicate cellular texture of 'thread-lace' scoria (reticulite) in his 1890 textbook on volcanology.

sink in water, unlike pumice, because their voids are interconnected, allowing water to enter the whole mass. Another consequence of the huge ratio of air to basaltic glass is that reticulite is quite weak—it can be crushed between the fingers.

Pele's hair and tears are absorbing pyroclastic curiosities, but their attractiveness should not detract from the crucial aspect of hawaiian eruptions: although huge volumes of magma may be sprayed into the air as fire fountains, little ash accumulates to form tephra deposits. Both in the vent and in the eruption column, the incandescent material remains so plastic that it does not fragment easily.

9.4 Strombolian deposits

Strombolian eruptions involve magmas significantly more viscous than hawaiian ones. Thus, achneliths, Pele's hair, and related phenomena are not found, and clastogenic lava flows are minor. Basaltic *scoria* is much the most recognizable product of strombolian activity. Cindery in appearance, scoria fragments have a light, frothy texture, the lava being highly vesiculated. Individual clasts are sharp-edged and angular, the fragmented surfaces cutting across vesicles: vesiculation clearly preceded fragmentation. Normally, basaltic scoria is a drab greyish-black colour, but when fresh it may be stunningly iridescent, shining in peacock-blue colours.

Steam percolating through a scoria cone often oxidizes the iron in the rock, giving the scoria a deep reddish-brown colour, which makes it attractive for ornamental pavements and other uses.

Because the lava is easily fragmented by escaping gas, only a small proportion of the finest grained material is produced, so a typical scoria deposit is not dusty—one can sit on it without getting grimy. Maximum clast sizes decrease rapidly away from the vent, so b_c is small. Erupted volumes are rather small, of the order of 0.01 cubic kilometres of magma, and the eruption column rarely reaches more than a few hundred metres above the vent. Thus deposits thin rapidly away from the vent, and therefore plot on the left margin of the Pyle diagram, in the field of 'cone-forming' deposits (Figs. 9.7–9.9).

Basaltic scoria is used extensively for road-making, and so scoria cones are often quarried, displaying their internal structures. A regular layering is present, each layer the result of successive pulses in the eruption. Little fine-grained material is present within the cone, but there are many large blocks and bombs. Primary layering is sometimes accentuated by rolling, the coarser lumps rolling downslope over earlier-deposited material. Individual layers in the cone sometimes show impressive reverse grading, partly due to increasing vigour in the eruption, but partly also due to larger clasts rolling down slope over finer material which remains *in situ* (Fig. 9.10). *Ball-milling* in the vent sometimes produces clasts which are as well rounded as

Fig. 9.7 A rain of basaltic tephra forms a dark curtain during the eruption of Surtsey, Iceland in 1963. When this photograph was taken, Surtsey had built itself sufficiently above sea-level that activity was predominantly strombolian, rather than surtseyan. Compare with Fig. 6.6. Because Stromboli volcano itself does not exhibit this kind of activity, a better name for it would be *microplinian*. Photo: S. Thorarinsson.

Fig. 9.8 Welded strombolian spatter armours the crest of this small scoria cone, one of many parasitic vents on the flanks of Mt. Etna, Italy.

Fig. 9.9 Roadside section through a scoria cone, Tenerife. Regularly sloping layering shows how the cone was built up through successive pulses in the eruption. Many large blocks and bombs are prominent. White deposits are recent efflorescences of salts.

Fig. 9.10 Rhythmic layering in a scoria cone in Costa Rica, showing splendid reverse grading, due to increasing intensity of eruption during a single pulse, but accentuated by rolling downslope of larger clasts. Hammer gives scale; layers are about 1 metre in thickness.

marbles. They roll exceedingly well, and accumulate around the lower flanks of the cone (Fig. 9.11).

Finer ash may extend several kilometres downwind from the vent: there is an abrupt topographic break between steep-sided cone and flat sheet of air-fall tephra. Some prehistoric deposits in the Azores provide useful examples of scoria fall deposits. Isopach and isopleth maps for the Galiarte deposit show the deposit rapidly thinning and getting finer-grained away from the vent;[6] at a distance of 3 km the deposit is less than 10 cm thick, and maximum clast size is less than 1 cm (Fig. 9.12). Like other fall deposits, distal strombolian tephra are well sorted (Fig. 9.13). While layering is a common feature within scoria cones, it is not often well developed in fall deposits away from the vent.

A scoria fall deposit from Helgafell buried the town of Heimaey, Iceland in 1973. During the space of a few hours, a one-metre thickness of tephra accumulated in parts of the town only a few hundred metres from the vent. Many house roofs collapsed under the weight of tephra, but in other parts of the town, a kilometre away from the vent, tephra accumulation was so much less that little damage was done, demonstrating in a practical way the same properties of the deposit as the isopach maps for the Azores (Fig. 9.14).

While the vast majority of strombolian fall deposits are basaltic, it is possible for more silicic eruptions to produce deposits with similar

Fig. 9.11 Suite of clasts showing impressive ball-milling, from the same scoria cone as Fig. 9.10. Clasts are dense, unvesiculated lava, not scoria.

characteristics. To be sure, 'scoria cones' of silicic composition are not nearly as common as basaltic ones, but they do exist: Monte Nuovo, a trachytic cone, grew to a height of over 100 metres on the shore of the Bay of Naples in a few days in 1538, causing consternation to the local savants who were not accustomed to such outbursts. *Pumice cones* are also fairly common,

often surrounding extrusions of dacitic or rhyolitic lavas.

9.4.1 Bombs

Strombolian eruptions produce a number of colourfully named kinds of bombs. When long strands of viscous lava are flung from the vent, twisting in the air, *rope* or *ribbon* bombs more

(a)
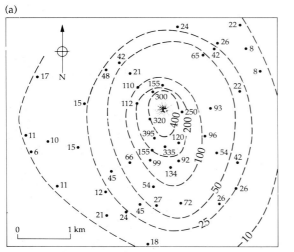
—10— Thickness (cm)

(b)
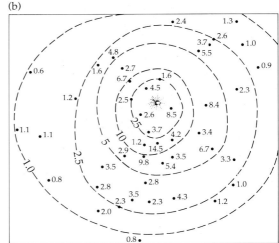
—5— Average maximum size (cm)

Fig. 9.12 (a) Isopach and (b) isopleth maps for tephra deposit around the Galiarte scoria cone, on Terciera island, Azores, an example of a typical strombolian deposit. From Self, S. (1976). The recent volcanology of Terciera, Azores. *J. Geol. Soc. Lond.* **132**, 645–66.

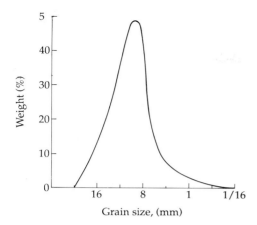

Fig. 9.13 Histogram for a sample of strombolian scoria collected close to the Serra Gorda cone, Sao Miguel, Azores. After Booth *et al.* (1978). A quantitative study of five thousand years of volcanism on Sāo Miguel, Azores. *Phil. Trans. R. Soc. Lond.* (A) **288**, 271–319.

than a metre long may result. When more equant lumps of lava are ejected, the outermost skin of plastic lava is pushed backwards towards the trailing edge of the bomb by the rush of passing air. If the bomb spins, the skin is drawn out into a twisted flange, producing a *spindle* bomb. Such bombs range from tiny specimens only a centimetre across to boulders more than a metre in diameter, and are common products of strombolian activity (Fig. 9.15). Some bombs form round a kernel of chilled lithic material, sometimes non-volcanic. These are *cored* bombs (Fig. 9.16).

Fig. 9.15 A metre-sized spindle bomb from the La Poruña scoria cone, north Chile.

Fig. 9.14 Remarkably, this house did not collapse, despite being buried under 2–3 metres of strombolian scoria from the 1973 eruption of Heimaey. Its steeply sloping roof, designed to shed snow, clearly worked equally well for tephra. Photo: S. Thorarinsson.

Fig. 9.16 Cored bomb from a Mexican scoria cone, with core of lighter-toned andesite.

When bombs fall out on to the slopes of a cone, they may embed themselves in the loose tephra, still glowing and fuming, or they may bounce off and roll downwards at high speed, developing a rapid spin. As momentum builds, they hurtle down slope in leaps and bounds, making a loud whirring noise, travelling sometimes several hundred metres before ending up in a rattling shower of small stones at the bottom. At each bounce, their corners are knocked off, so at the bottom of the slope they are found as peculiarly smooth, rounded lumps: *cannon-ball* bombs.

— 9.5 Vulcanian deposits

'Vulcanian' is another elastic volcanological term. It is easier to say what vulcanian eruptions are *not*, rather than what they are: they are not very big eruptions of not very vesiculated magma. Eruptive activity which has been called 'vulcanian' ranges from solitary, cannon-like explosions that blast a slug of ash from the vent, to sustained convecting plumes that may conti-nue for hours. Eruption columns are rarely high, typically 5–10 km. Their deposits are character-istically rather fine grained (although not so fine as in surtseyan deposits) and often contain much non-magmatic material, ripped from the throat of the volcano. High degrees of magma fragmen-tation suggest involvement of water in phreato-magmatic styles of eruption, but it is far from clear that this is always the case. Most deposits are less than a cubic kilometre in volume, a volume which is often the total accumulation of months or years of individual eruptions. Modest dispersal of the tephra, and the preponderance of fine-grained material, ensure that vulcanian deposits plot well up on the Pyle diagram.

Vulcanian eruptions typically involve magmas

(a)

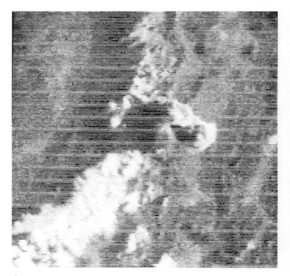

(b)

Fig. 9.17 (a) Eruption column from a single vulcanian explosion of Lascar, north Chile, 16 September 1986. Explosion occurred about three minutes prior to photograph; cloud is already many kilometres above vent, ultimately reaching about 15 km. (b) GOES satellite image of the same cloud at the same time. Image is about 100 km across.

of andesitic or dacitic composition, and so they are common on circum-Pacific volcanoes. A notorious example was that of Irazu, Costa Rica, which erupted on innumerable occasions from March 1963 to 1964, but still deposited a total of less than two metres of ash close to the vent. None the less, coffee crops in the surrounding area suffered severely. The 16 September 1986, eruption of Lascar volcano in north Chile mentioned in Chapter 8 was a short vulcanian blast, which propelled an ash cloud to a height of about 15 km in a few minutes. Fine ash fell out on the city of Salta, Argentina, 285 km downwind

Fig. 9.18 Impact crater on flanks of Lascar volcano, 3.9 km from the active vent, formed by a vulcanian eruption on 20 February 1990. Fragments of the bomb (dark) are visible within the crater.

within a couple of hours of the eruption (Figs. 9.17–9.18).

A vulcanian eruption of Ngauruhoe, New Zealand, on 19 February 1975 was closely observed and photographed:[7] the initial ejection velocity of the erupted slug was measured at more than 400 metres per second; supersonic, so that a shock wave should also have been propagated. Some 500 m above the vent, the velocity had decreased to less than 50 metres per second. As argued in Section 8.1.3, such a high ejection velocity is best interpreted in terms of abrupt failure of a strong cap within the vent, perhaps one made of congealed lava—an accumulation of pyroclastic debris would not be strong enough. To generate the overpressures necessary to fracture the cap and trigger the explosion, high magma volatile contents are required, higher than can be accounted for by simple exsolution of magmatic volatiles. Thus, a component of ground-water interacting with the magma may be involved with some violent vulcanian eruptions. (According to a Russian volcanologist, P. I. Tokarev,[8] the specific energy content of a volcanic system containing 3 per cent water is 5.1×10^4 J kg^{-1}, comparable with some chemical explosives. TNT, for example, has an energy content of 4.2×10^6 J kg^{-1}).

But many vulcanian eruptions are not exceptionally violent, and furthermore, some are sustained over long periods. Many volcanolo-

Fig. 9.19 West rim of the crater of Vulcano. Light-toned rocks formed the cone which existed prior to the last major eruption, 1888–1890. Overlying dark material represents total accumulation of vulcanian tephra from the eruption.

gists have watched vents from which a stream of fine ash and gas boils silently and ceaselessly, forming a plume which climbs a few kilometres into the air, drifting off downwind in a rolling grey-brown plume; compulsive to watch and photograph, but misery for the people living in towns and villages nearby.

Close to the vent, large ballistic bombs and blocks are commonly found in the deposits ejected by vulcanian eruptions. Probably the best known of all vulcanian deposits is that produced by the eruption of Vulcano itself in the 1880s. Nowhere more than a few metres thick, the deposit mantles the slope of the pre-existing cone (Fig. 9.19). Littering the flanks of the volcano are thousands of large bombs, many of them more than a metre in diameter. Most eye-catching of these are *bread-crust bombs*: rounded or angular lumps with a smooth, glassy crust broken by deep cracks and fissures, exposing the frothy, vesicular core of the bomb, and reminiscent of a well-baked, crusty loaf of bread (a sight sadly denied to North Americans). Bread-crust bombs are formed when lumps of viscous, gas-rich lava

Fig. 9.20 Classic example of a bread-crust bomb, ejected during the 1888–90 eruptive phase of Vulcano.

are ejected from the vent: the outer crust chills quickly to form the glassy crust, but the interior remains hot, and continues to vesiculate, frothing up the interior. As in a decently baked loaf, the internal expansion causes the brittle outer crust to crack (Fig. 9.20).

— 9.6 Sub-plinian deposits —

Surprisingly few sub-plinian deposits have been studied in detail, probably because they lack the glamour of their larger counterparts (Fig. 9.21). Sunset Crater, Arizona, produced an instructive example during its eruption of AD 1065, an eruption which impinged heavily on the local Sinagua Indian tribe, burying many of their pit houses. Sunset Crater itself is actually a classic 300-m-high monogenetic basaltic scoria cone, of the sort usually associated with strombolian scoria fall deposits, but its eruption was vigorous enough to produce a sustained convecting eruption column, much higher than a typical 'strombolian' column. About 0.45 cubic kilometres (dense rock equivalent) of magma were erupted.[9]

A more recent example was the 1986 eruption of Izu-Oshima in Japan. At 16.15 on

21 November, a fissure opened at the north-eastern foot of Mihara Yama, the central cone within the caldera of Oshima. Fire-fountaining started over the whole length of the fissure, gradually extending southward, until the active fissure was about 1 km long. By 17.10, fire-fountains were spraying incandescent tephra no less than 1600 m high; a truly remarkable phenomenon. Owing to the enormous emission rate, clastogenic lavas formed around the base of the fountains. The convecting part of the column reached 16 km in altitude, from which a fall-out deposit of about 1.6×10^6 cubic metres of andesitic scoria accumulated within the space of a few hours.[10] By 12.00 on 22 November, the most active phase of the eruption was over.

Fig. 9.21 Eruptions of Vesuvius were common in the eighteenth and nineteenth century, but there has been none since 1944. This 26 April 1872 photograph shows a typical sub-plinian eruption, which continued over several months. Compare the 1794 eruption illustrated in Fig. 6.13. (Photo by G. Sommer.)

—— 9.7 Plinian deposits ————————————————

Many of the greatest eruptions in history have yielded plinian air-fall deposits. Plinian eruption columns are characterized by their great heights—several have been more than 30 km high. Whereas strombolian eruptions build steep-sided cones, plinian eruptions are *sheet* formers: they yield extensive sheets of tephra which mantle the topography uniformly (Figs. 9.22–9.23). Although the deposits thicken towards the vent, the vent itself may be a negative, rather than a positive, topographical feature—a hole in the ground. Many of the world's great calderas formed during plinian eruptions. It is difficult to pin-point precisely the source vents of

many prehistoric plinian deposits, because they are so self-effacing, although their approximate positions can be inferred.

Numerous plinian deposits have been studied all over the world. In 1991, while this book was being written, powerful eruptions of Pinatubo volcano in the Philippines on 15 and 16 June sent ash columns more than 30 km into the atmosphere, raining tephra down over vast areas. Civil authorities ordered the evacuation of tens of thousands of civilians, while the US military evacuated Clark Field air force base and Subic naval base. The dust from Pinatubo has yet to settle, literally as well as figuratively, and so our

Fig. 9.22 This road section near the city of Arequipa, Peru, exposes a major plinian air-fall deposit, demonstrating the regular thickness with which fall deposits mantle topography. Little is known about the eruption that yielded this deposit.

Fig. 9.23 Close-up of a plinian fall deposit, showing the characteristic angular, irregularly shaped clasts and well-sorted character, with little fine grained ash. Tenerife, Canary Islands.

choice of case study is obvious: the AD 79 Vesuvius deposit (Chapter 4). Not only because it is the 'type example', renowned both in literary and geological history, but also because it has been investigated by many workers. Rather than attempting to review all of these excellent studies, we follow here a paper by Sigurdsson *et al.*, which provides an excellent chronological overview, and a valuable introduction to the earlier literature on the eruption.[11]

9.7.1 The Vesuvius AD 79 eruption

Casual mention of the 'AD 79 ash' suggests that there is a single tephra layer around Vesuvius. There are, however, several distinct horizons within the AD 79 deposits (and there are several older deposits as well). Resting directly on the soil of Roman times is the first, a thin layer of fine ash from the explosions that commenced the eruption. This ash was dispersed mostly east of the volcano. Overlying it is the main fall deposit, which has two conspicuous components: a lower, 'white' layer, which grades without break into an upper 'grey' layer. The change in colour denotes a difference in composition: the white layer contains 10–15 per cent more silica than the grey, and its pumice clasts are slightly less dense than the grey (Fig. 9.24). Few plinian pumice deposits

Fig. 9.24 Outcrop of the AD 79 plinian pumice fall deposit near Pompeii. In this black-and-white photo, the colour change is not conspicuous, but is discernible in the scraped (centre) section of the deposit, below level of geologist's head.

show such a plainly visible change, but the AD 79
Vesuvius deposit is an excellent example of a
common volcanological phenomenon: *compositional zonation*. In the plinian fall deposit, the
white-to-grey zonation observed is the mirror
image of the situation in the magma chamber: the
more highly evolved, silicic magma lay at the top,
and was erupted first. As the eruption proceeded,
it tapped successively deeper levels in the magma
chamber, until eventually the grey, more mafic
pumice began to accumulate on top of the earlier
erupted silicic pumice.

In Pompeii itself, about 130 cm of white
pumice are overlain by 120 cm of grey. Sigurdsson and his colleagues mapped the thicknesses of
both grey and white layers around Vesuvius and
the variations in maximum pumice and lithic
clast size in both components. These variations
are summarized in a series of isopach and
isopleth maps, from which the physics of the
eruption column can be inferred (Figs.
9.25–9.26).

Isopach maps Unfortunately, much of the
tephra fell into the sea in the Bay of Naples and
beyond the Sorrento Peninsula, so there are not
as many data as one would like. Analysis of the
isopach maps (Fig. 9.25) gives an estimate of the
volumes of air-fall tephra erupted: 6.4 cubic
kilometres of grey, and 2.5 of white. Taking
account of the densities of the pumice in the two
components, this suggests volumes of unvesiculated magma erupted of 2.6 and 1 cubic kilometres respectively, or 3.6 cubic kilometres in all.

Isopleth maps At Pompeii, the largest pumice
clasts to fall were about five centimetres across—
certainly large enough to warrant tying a pillow
on one's head, as Pliny the Elder's party did at
Stabiae. Comparison of the isopleths of the white
and grey pumices is revealing (Fig. 9.26 a and b).
Consider, for example, the 15 cm isopleths: the
15 cm isopleth for the white pumice plots much
nearer the volcano than that for the grey. This
shows that clasts of a given dimension were
falling out much closer to the vent during
eruption of the white pumice than the grey, which
in turn shows that the column height *increased* as
the eruption progressed. Using the principles

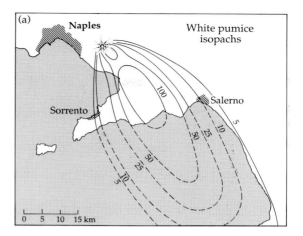

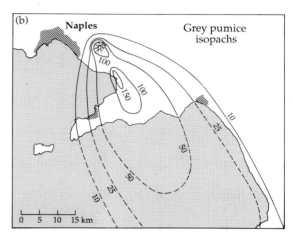

Fig. 9.25 Isopach maps of thickness of the AD 79
Vesuvius plinian fall deposit around the volcano,
showing pronounced downwind elongation.
Contours are for thickness in centimetres; (a) is for
the lowermost white pumice; (b) for the grey
pumice. Shape of contours offshore is conjectural.
After Sigurdsson *et al.* 1985).

outlined in Chapter 8, Sigurdsson *et al.* suggest
that the eruption column height was approximately 27 km during the fall-out of the white
pumice, but increased to about 33 km later.

As the afternoon of 24 August wore on, the
magnitude of the disaster overwhelming them
must have become acutely obvious to the people
living around the volcano. Objectively, the
increasing size of lithic clasts in the deposit
documents for us the increasing vigour of the

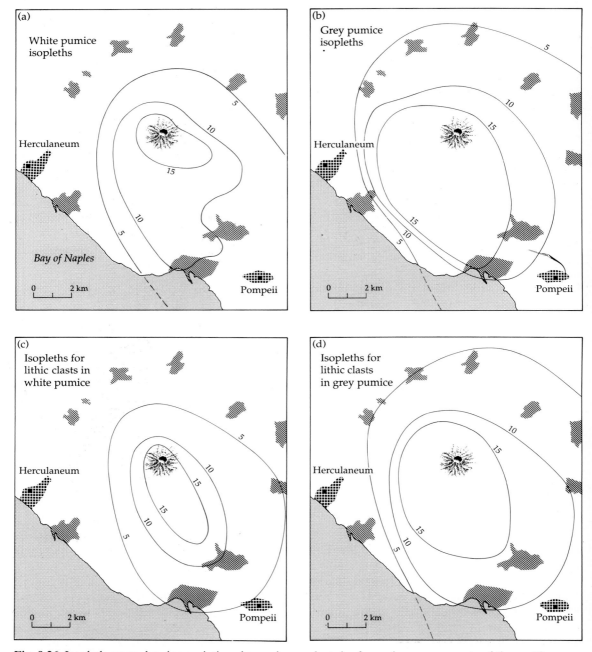

Fig. 9.26 Isopleth maps showing variations in maximum clast size for various components of the AD 79 Vesuvius plinian fall deposit. Outlined areas are modern towns ringing the volcano. (a) Maximum white pumice clast size, (b) maximum grey pumice clast size, (c) maximum lithic clast size in white pumice, (d) maximum lithic clast size in grey pumice. After Sigurdsson et al. 1985.[11]

eruption which they experienced at first hand. There are only about 10–15 per cent of lithics in the tephra, but they are useful because they are denser than pumice clasts. Pumice clasts are also angular and rough, thus they are aerodynamically 'dirty' and more easily affected by small variations in atmospheric parameters than compact lithic clasts. The five-centimetre isopleth for lithics in the white pumice almost coincides with the *10 cm* isopleth for the grey pumice (Fig. 9.26 c and d; Fig. 9.27).

The isopleths are all elongated south-eastwards, so it follows that a stiff north-westerly wind was blowing during the eruption. This tallies nicely with Pliny's account—his uncle could not put off in his boat from Stabiae, because of brisk onshore winds. By applying the

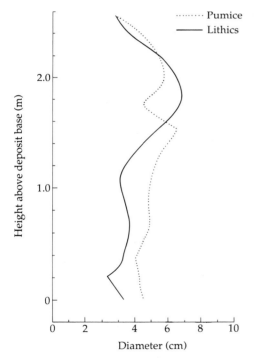

Fig. 9.27 By sampling the AD 79 plinian deposit at roughly 20 cm intervals, Sigurdsson *et al.* were able to document the increasing size in maximum pumice and lithic clasts during the progress of the eruption. Data for lithic clasts show the increase much more clearly than pumice. Section studied was at the Necropolis in Pompeii. After Sigurdsson *et al.* 1985.[11]

techniques outlined in Chapter 8 to the lithic isopleths near the volcano, a wind speed of about 25 metres per second is indicated. This value, of course, is for the wind at the tropopause. At sea-level, where Pliny the Elder was endeavouring to escape, the wind speed would have been much less.

Incidentally, the value of careful studies of even the most unpromising material is demonstrated by the *compositions* of the lithics: those from the lower, white pumice are dominantly clasts of volcanic rocks plucked from shallow levels beneath the volcano, perhaps only a kilometre or so. In the grey pumice, there is a marked concentration of limestone fragments, derived from much deeper beneath the volcano, probably 5–6 km.

9.7.2 A chronological summary of the eruption

Sigurdsson *et al.* compiled a chronology of the events of August AD 79 by comparing historical data with the results of theoretical modelling. By combining this work with that of many other volcanologists on the mechanics of the eruption, valuable insights into other pyroclastic eruptions emerge. Several important assumptions are involved in establishing the chronology, not least that activity was initiated at about 1 p.m. on 24 August, the 'seventh hour' mentioned in Pliny's account.

Prior to AD 79, a reservoir of magma had accumulated beneath the volcano. It is not possible to say when this magma chamber began to form, but it had a volume of at least 3.6 cubic kilometres; was about 3 km below surface, and was compositionally stratified, with volatile-rich alkalic magma (55 per cent SiO_2 and almost 10 per cent K_2O) overlying slightly denser, more mafic magma. This magma body was less dense than the surrounding rocks, and so it tended to migrate buoyantly upwards in the volcanic edifice. At some point, hot magma interacted with ground-water seeping downwards through the volcano, initiating the first event, the minor phreato-magmatic eruption which showered fine-grained grey tephra over the eastern flanks of the volcano. This probably took place during the night or on the morning of 24 August.

While undoubtedly alarming to those showered by ash, this eruption was inconsequential. It might have gone unnoticed by those living as far distant as the Plinys in Misenum. None the less, this minor event probably triggered the huge eruption which followed, by opening a vent to the surface, allowing explosive decompression of the main magma body. Exsolution and expansion of the dissolved volatiles caused violent fragmentation of the magma, propelling it upwards through the vent at an accelerating rate. Exit velocities were probably of the order of 400 metres per second. This developing column probably first attracted the attention of the Plinys in Misenum at about 1 p.m. on the 24th.

The gas thrust region may have been as much as 3 km high, while convection carried incandescent gas and pumice clasts up to 27 km. To sustain such a high eruption column required a mass eruption rate of the order of 1×10^8 kilograms per second during the first phase of the eruption, while the white pumice was being erupted. For the first seven or eight hours, white pumice fell, accumulating at rates of about 12–15 cm an hour in Pompeii, directly downwind. Few roofs can support such enormously increasing weights, so roofs probably began to cave in during the afternoon of 24 August. Some victims were probably crushed by the weight of tiles and pumice. During the afternoon, a few may also have been killed by falling lithic clasts, much denser and more dangerous than pumice clasts. Some of these were falling at 50 metres per second when they hit the ground.

During this phase, the vent radius was probably of the order of 100 m. As the eruption continued, inevitable widening of the vent permitted still higher mass eruption rates. By the evening of the 24th, the column height had increased. Progressively deeper levels within the magma chamber were tapped, until after about seven hours the more mafic grey pumice was reached. This was ejected at about 1.5×10^8 kilograms per second, and carried by convection to maximum heights of around 33 km. Grey tephra accumulated rapidly. Although extensive damage was done to buildings by the weight of tephra, most people in Pompeii had probably survived up until this time.

Increasing vent radius and mass eruption rates

cannot be sustained indefinitely. There comes a point when so much magma is being erupted so quickly that the eruption column density becomes too great for stable convection to persist. When this condition prevails, *column collapse* takes place, generating pyroclastic flows and surges, which are far more lethal than tephra fall. As far as Vesuvius is concerned, the salient point is that the most destructive episodes did not occur until many hours into the eruption. After a few hours of grey tephra accumulation, devastating pyroclastic surges were generated, the first sweeping into the city of Herculaneum (which had escaped the heaviest tephra fall) at about 1 a.m. on 25 August. Others reached Pompeii in the following hours. Most of the victims of the eruption had survived long, terrifying hours of heavy tephra fall: they died only when the style of the eruption changed. We return to the absorbing story of the surges in Chapter 11.

9.7.3 *Plinian deposits of basaltic composition*

Most plinian pumice fall deposits are of silicic compositions, like the Vesuvius pumice. But silicic composition is not a *requirement* for a plinian eruption. From the principles outlined in Chapter 8, it might seem reasonable that whenever the appropriate combinations of volatile content, vent radius, and mass eruption rate are met, then a plinian eruption should result. Life is not so simple, however. Basaltic magmas normally contain less volatiles than silicic ones (less than 1 per cent compared with up to 4 per cent), but a basaltic magma body may none the less build up enough volatiles to sustain enormous mass eruption rates. Instead of forming a convecting plinian column, however, this may simply lead to fire fountains of exceptional height. To generate true basaltic plinian eruptions requires an *additional* factor: there must be sufficient fine ash to stoke up the column and drive it convectively to the stratosphere. Fine, dusty ash is better at conveying heat to the atmosphere than a spray of liquid basaltic magma. Thus, some degree of water–magma intraction is required to fragment the magma.

As outlined in Section 6.3.6, the most destructive episode in New Zealand's recent volcanic history was a basaltic plinian eruption. In the

Fig. 9.28 View looking westwards along the Tarawera rift, showing dark-toned deposits from the 1886 eruption mantling the topography. IAVCEI rambling party on left gives scale.

space of a few hours in the early morning of 10 June 1886, about 2 cubic kilometres of basaltic scoria were blasted from a 7-km-long fissure that opened up on Mt. Tarawera, North Island (Fig. 9.28).[12] Ragged, rather poorly vesiculated basaltic scoria fell out over a huge area. Clasts up to a centimetre in size fell at distances of 30 km. Estimates of the mass eruption rate during the paroxysmal eruption were 1.8×10^8 kilograms per second, while the eruption column height was estimated to be 28 km. (Unfortunately, because it was dark, there were no direct observations of the column height.) Evidence for the involvement of water in the eruption was messily abundant—much tephra fell as mud. Prior to the eruption, a world-class display of geysers and sinter terraces had existed at Rotomahana, at the south-easternmost end of the fissure—it was the Yellowstone of its day. A major reservoir of geothermal fluids at Rotomahana may have been breached during the eruption, powering the plinian eruption. Sadly, the gorgeously coloured terraces at Rotomahana were completely obliterated.

Other examples of basaltic plinian deposits have been recognized around the world, for example in Nicaragua,[13] but they are small by comparison with the largest silicic deposits.

9.7.4 Welded plinian air-fall deposits

Welding of basaltic spatter is a ubiquitous feature

of hawaiian and strombolian eruptions, normally forming spatter cones, but generating clastogenic lavas if eruption rates are high enough. Welding in silicic deposits is less common, because clasts travel so far through the air that they have time to cool before they hit the

Fig. 9.29 Densely welded plinian air-fall deposit, Socompa volcano, north Chile. Large dacite clast near knife was originally a lithic clast in the plinian fall; ghostly shadows of original pumice clasts can be seen around and beneath it.

ground. However, if mass eruption rates are high enough, welding can still be achieved, so many examples of welded plinian fall deposits are known. They are easily confused with glassy lavas, but can usually be recognized by spotting the transition from the unwelded base, cooled against the ground, through to the glassy central part of the deposit. Welded fall deposits also have quite different morphological characteristics from lava flows, since they mantle the topography with even thicknesses over large areas (Fig. 9.29).

— 9.8 Ultraplinian deposits

Ultraplinian pyroclastic deposits are exceptional. In principle, they are similar to plinian deposits, but are produced by higher eruption columns and are thus dispersed over wider areas. As outlined in Chapter 6, the AD 186 Taupo pumice of New Zealand was described by its discoverer, George Walker, as 'the product of the most powerful known (ultraplinian) eruption'.[14] The Taupo plinian pumice is not much to look at, however: it is only 1.8 metres thick at maximum. What makes it unique is that the Taupo pumice retains this thickness over a huge area. Its maximum thickness occurs 20 km from the vent; 100 km from the volcano the deposit is still *more than 25* cm thick. Isopleth data show similar remarkable values for the clast size distributions. Some 100 km from the vent, pumice clasts as big as 3 cm are still found—a statistic whose significance is easier to appreciate by comparison with Vesuvius AD 79, where 100 km from the vent, maximum pumice clasts are smaller than 1 cm. This extraordinary dispersal and clast size distribution call for an exceptionally high eruption column. Walker considers that it may have been 50 km high, almost twice that of Vesuvius AD 79.

— 9.9 Surtseyan deposits

Since most volcanoes are located near, or actually beneath, the oceans, magma–water interactions inevitably take place often. Submarine volcanism and the effects of unlimited supplies of water are deferred to Chapter 15. Here, we examine only the pyroclastic deposits generated when eruptions take place at or near sea-level, on coasts, and the many wet environments on dry land where hot magma and cold water can meet—in lakes, river beds, and even subsurface shallow aquifers. The common characteristic of hydrovolcanic (or phreatomagmatic) eruptions is the high degree of fragmentation that results. Since the maximum clast size at a given range is smaller in a hydrovolcanic eruption than a conventional one of similar magnitude, b_c/b_t ratios are low and hydrovolcanic deposits therefore plot along the top of the Pyle diagram.

9.9.1 Surtseyan pyroclastic deposits

Surtseyan eruptions are the hydrovolcanic equivalents of strombolian eruptions on dry land (Section 6.4.1). Surtseyan pyroclastic deposits are easily recognizable. Their most important characteristic is their overall fine grain size, caused by thorough fragmentation of the magma. Few surtseyan deposits extend far from the vent. But although the *average* grain size is small, the *range* in clast size is great, and thus they are much more poorly sorted than strombolian deposits. (Grain-size data from these deposits are often hard to obtain, because the fine, glassy fragments quickly alter to *palagonite*, forming a dense, unsievable mass). They are also unlike strombolian deposits in that they are often finely laminated, indicating that the eruption consisted of numerous brief explosive bursts. Eyewitness accounts and film records of the Surtsey eruption demonstrate that this is exactly what happened. Explosions took place every few seconds to tens of seconds, blasting out 'cock's tail' jets of bombs and tephra hundreds of metres high.

Surtseyan deposits are most distinctive near

Fig. 9.30 Fine laminations and splendid bomb sag in proximal surtseyan deposits, Koko Head, Oahu, Hawaii. White clast at top left is coral fragment. Key gives scale.

the vent. There, they typically contain large blocks in a fine ash matrix. Such blocks can range from metre-sized bombs of scoriaceous basalt to small non-juvenile lithic clasts. In the Koko crater on the island of Oahu, Hawaii, many of the blocks are made of gleaming white coral, ripped from the sea-bed during the eruption. They stand out in stark contrast to the greys and browns of the rest of the deposit. Large ejected blocks, whether basalt or coral, have lots of momentum when they fall from a height of hundreds of metres. Thus, when the coral blocks at Koko crater smashed into the muddy tephra on the slopes of the cone, they dug deep craters, sometimes several metres in diameter. When seen in vertical section, the craters are revealed as *bomb sags* (Fig. 9.30). Often the bomb is still present within its crater; other sections reveal only the sagging layering. Where a tuff cone is exposed along a bedding surface, bombs and their craters are often exhumed, so that one can wander over a surface evocative of a World War I battleground.

— 9.10 Phreatoplinian deposits

Phreatoplinian deposits are to surtseyan deposits what plinian deposits are to strombolian: they result when large volumes of silicic magma are explosively fragmented by contact with water. Good examples are known from Iceland (e.g. Askja 1875), New Zealand, and elsewhere. These all resulted from eruptions taking place beneath lakes within pre-existing calderas. Only the Askja deposit was formed in recent historic times. Pain-staking mapping of Ordovician rocks in the English Lake District has shown that the Whorneyside Tuff may be an exceptionally large, 400-million-year-old example of a phreatoplinian deposit.[15]

In terms of their volume and dispersal, phreatoplinian deposits are closely similar to plinian deposits, and must have fallen from eruption columns of comparable height. Their characteristic property is their extremely fine grain size,

even close to the vent. At the distance from the vent, where a plinian deposit might have a mean grain size of 16 mm, the phreatoplinian equivalent typically would have a mean grain size of only 0.25 mm.

9.10.1 Accretionary lapilli

Accretionary lapilli are characteristic of phreatomagmatic deposits, although they are not diagnostic of them. These small spheres of extremely fine ash, which may reach more than centimetre in diameter, are commonly found in tephra deposits formed by eruptions in which water has been involved. Often, they consist of an outermost shell of harder ash surrounding a more friable core; sometimes they have a concentric, onion-like structure. In some places, they are scarce and barely distinguishable from the matrix of the deposit; in others they are so abundant and conspicuous that the deposit appears to be crammed with small brown marbles. It is usually difficult to extract the lapilli without breaking them; sometimes, the outermost shell is so tough that they are robust enough to be used as marbles (Fig. 9.31).

Although it is clear that accretionary lapilli are associated with 'wet' eruptions, it is not clear exactly how they form. It used to be thought that they resulted from raindrops flushing through dense tephra clouds, fine ash accreting around raindrops. While this is one possible mechanism, it is clear that there are others, since lapilli are found in a range of deposits, not only air falls. They are common, for example, in base surge and other pyroclastic flow deposits (q.v.).

9.10.2 Distal air-fall ashes

It has been implicit in all the preceding discussions that the grain size of pyroclastic fall deposits decreases progressively away from the source. For most purposes this is a reasonable assumption, but in detail it is not completely accurate. Where very small grain sizes are involved, some different physical phenomena such as electrostatic forces come into play. These can cause small grains to clump together to form larger particles. As a result, some pyroclastic fall deposits show slightly bimodal grain-size distributions, with a population of apparently large clasts resulting from fine particles clumping together. Clumping can also lead to premature fall-out of material from eruption clouds, confusing isopach maps and leading to *secondary thickening*. This was well observed at Mt. St Helens, where the 1980 air-fall deposit thinned

Fig. 9.31 Accretionary lapilli 1–2 centimetres in diameter eroding out of a phreatomagmatic deposit, Wairakei Formation (Oruanui) north shores of Lake Taupo, New Zealand.

exponentially to 1 cm at 180 km distance, but then *increased* in thickness again, reaching 4 cm at 300 km.

Strong winds can also yield unusual distributions of ashfall. During the 1991 eruption of Mt. Hudson in Chile, which ejected several cubic kilometres of ash to a height of 20 km, westerly winds were blowing vigorously, at speeds of up to 140 km per hour. Patagonian towns and communities 500 kilometres distant were affected by ash within twenty four hours, experiencing 36 hours of darkness. Ash fall appears to have been most severe in the area of Los Antiguos, on the Chile–Argentina frontier, rather than in the thinly populated mountainous areas around the volcano itself.

Notes

1. Fisher, R. V. and Schmincke, H. U. (1984). *Pyroclastic rocks.* Springer-Verlag, Berlin, 472 pp.
2. Houghton, B. F. and Wilson, C. J. N. (1989). A vesicularity index for pyroclastic deposits. *Bull. Volcanol.* **51**, 451–62.
3. Walker, G. P. L. (1973). Explosive volcanic eruptions—a new classification scheme. *Geol. Rundsch.* **62**, 431–46.
4. Pyle, D. M. (1989). The thickness, volume and grain size of tephra fall deposits. *Bull. Volcanol.* **51**, 1–15.
5. Head, J. W. and Wilson, L. (1989). Basaltic pyroclastic eruptions: influence of gas-release patterns and volume fluxes on fountain structure, and the formation of cinder cones, spatter cones, rootless flows, lava ponds and lava flows. *J. Volcanol. Geotherm. Res.* **37**, 261–71.
6. Self, S. (1976). The recent volcanology of Terciera, Azores. *J. Geol. Soc. Lond.* **132**, 645–66.

7. Self, S., Wilson, L., and Nairn, I. A. (1979). Vulcanian eruption mechanisms. *Nature* **277**, 440–3.

8. Tokarev, P. I. (1984). Volcanic explosions. On the concept of 'volcanic explosion'. *Volc. Seism.* **5**, 315–22.

9. Amos, R. C., Self, S., and Crowe, B. (1981). Pyroclastic activity of Sunset Crater: evidence for a large magnitude, high dispersal strombolian eruption. *EOS* **62**, p. 1085.

10. Earthquake Research Institute, Tokyo (1988). *The 1986–87 eruption of Izu–Oshima volcano.* Tokyo, Japan, 61 pp.

11. Sigurdsson, H., Carey, S., Cornell, W., and Pescatore, T. (1985). The eruption of Vesuvius in AD 79. *Nat. Geog. Res.* **1**(3), 332–87.

12. Walker, G. P. L., Self, S., and Wilson, L. (1984). Tarawera, 1886, New Zealand—a basaltic plinian fissure eruption. *J. Volcanol. Geotherm. Res.* **21**, 61–78.

13. Williams, S. N. (1983). Plinian air-fall deposits of basaltic composition. *Geology* **11**, 211–14.

14. Walker, G. P. L. (1980). The Taupo pumice: product of the most powerful known (ultraplinian) eruption? *J. Volcanol. Geotherm. Res.* **8**, 69–94.

15. Branney, M. J. (1991). Eruption and depositional facies of the Whorneyside tuff formation, English Lake district: An exceptionally large magnitude phreatoplinian eruption. *Geol. Soc. Amer. Bull.* **103**, 886–97.

10

Pyroclastic flows: ignimbrites

'Awesome' is a word devalued by over-use in describing volcanic phenomena. Pyroclastic flows, nevertheless, are truly awesome in the authentic sense of inspiring reverential fear. A single pyroclastic flow may sweep over 100 kilometres from its source at more than 100 kilometres per hour, depositing a layer of pumice many metres thick over thousands of square kilometres. An area the size of Belgium could be obliterated in a few minutes. More than a thousand cubic kilometres of material may be erupted in a few hours. Calderas tens of kilometres in diameter are formed through subsidence of magma chamber roofs. It is difficult to conceive of the social consequences that would ensue if such an eruption were to take place today.

Owing to the magnitude of the threat that they present, large pyroclastic flows ought to concern society as a whole, not just volcanologists. Pyroclastic flows have some of the attributes of dinosaurs: they are huge and dangerous. But they have little of the fascination for the public that dinosaurs have, because *large* pyroclastic flows are rare, and have not earned the atten-

tion of the media. Unlike dinosaurs, however, pyroclastic flows are not extinct. While small pyroclastic flows are frequent, the last eruption of massive pyroclastic flows took place at Toba, Indonesia, about 75 000 years ago, when more than 2000 cubic kilometres of material were erupted. Smaller eruptions are correspondingly more frequent—about 35 000 years ago, the Campanian tuff (80 cubic kilometres) over-ran much of the area around modern Naples.

Lava flows and ash falls are relatively straightforward phenomena. Pyroclastic flows, by contrast, are opaque, literally and metaphorically. Many small flows have been observed, but all that can actually be *seen* are billowing curtains of fine ash which veil the flows deep within. And the physics which control the movement of pyroclastic flows is equally impenetrable. However, the last decade has seen great progress in understanding pyroclastic flows. As George Walker has put it, our understanding of pyroclastic flows has been 'a success story for volcanology'.

— 10.1 Terminology

Pyroclastic flow terminology can be frustrating. Pyroclastic flows range between two end-member types: those that involve vesiculated, low-density pumice, and those that involve unvesiculated, dense lava clasts (Figs 10.1–10.2).[1] Pumice flows leave deposits called *ignimbrites*. Flows of unvesiculated, dense clasts are

commonly called *nuées ardentes* (glowing clouds), while the deposits they form are *block-and-ash* deposits (Chapter 12). Pyroclastic *surges* are flows of lower density than either pumice flows or block and ash flows. But surges, pumice flows and block and ash flows are all varieties of pyroclastic flow.

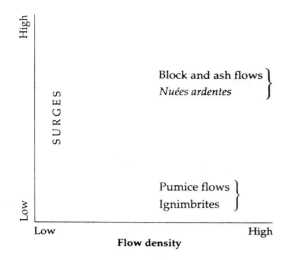

Fig. 10.1 Principal types of pyroclastic flows, distinguished in terms of clast and flow density. Surges are essentially dilute kinds of pyroclastic flow. There are no hard and fast boundaries to the three fields.

One must distinguish between *flows* themselves, and the *deposits* they produce. Flows are transient, dynamic phenomena; deposits are what they leave behind. Volcanologists rarely witness large eruptions at first hand, so they have only static masses of pyroclasts to work with. From these, they attempt to infer the nature of the flows. In practice, flows are often spoken of synonymously with their deposits. One often hears, for example, of a deposit being referred to as a *nuée ardente*. Strictly, this is absurd, since a *nuée* is a cloud. Anything less like a cloud than the rocky rubble of a block and ash deposit is hard to imagine!

Pyroclastic flows originate in more than one way. Pumiceous pyroclastic flows are associated with convecting eruptions columns, so discussion of these here forms a logical link with the previous chapter. Mechanisms leading to the formation of the smaller (but no less dangerous) varieties are discussed in later chapters.

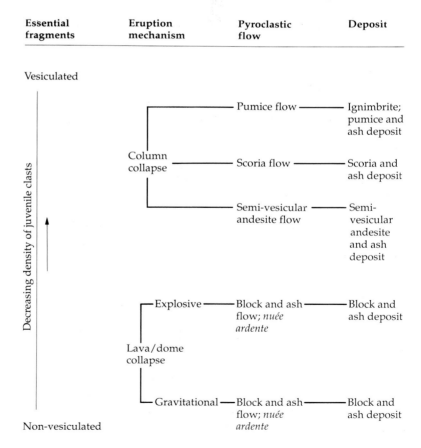

Fig. 10.2 Genetic classification of the main varieties of dense pyroclastic flows. After Wright, J. V. Smith, A. L., and Self, S. (1980). A working terminology of pyroclastic deposits. *J. Volcanol. Geotherm. Res.* **8**, 315–36.

10.2 Formation of pyroclastic flows by eruption column collapse

In Chapter 8, three important components of pyroclastic eruption columns were introduced: a lowermost gas thrust region, an intermediate convecting region, and an upper umbrella region. In simple terms the formation of pyroclastic flows from eruption columns is a matter of density: an eruption column can only rise convectively if it is less dense than the surrounding atmosphere. If it is more dense, once the momentum imparted by the gas thrust runs out, it can only collapse downwards under gravity. In essence, this is all that pumice flows are: masses of pumice that plummet downwards from dense eruption columns, spreading outwards around the volcano. As they fall, the potential energy they gained in the eruption column is transformed into kinetic energy, driving them at high velocity over the ground (Fig. 10.3).

10.2.1 Stability of eruption columns

Lionel Wilson, Stephen Sparks, and George Walker first tackled the problems of collapsing eruption columns.[2-4] They analysed eruption columns as turbulent jets, in which convection of hot magmatic gases is aided by heat transfer between small, hot pyroclasts and entrained air.

They showed that two independent parameters are important in controlling the mass eruption rate, and hence the density and stability of an eruption column. They are the volatile content of the magma and the vent radius.

Vesiculation of magmatic volatiles drives all pyroclastic eruptions: without volatiles a magma would be a flat as warm British beer. It would ooze away from the vent as a humble lava flow. More volatiles mean more fizz. Other things being equal, this yields higher eruption velocities and therefore mass eruption rates. For a given vent radius, the calculated eruption velocity increases sharply as volatile content increases. Typical silicic magmas have 2–3 per cent volatiles; thus according to the model, they should achieve velocities of 300–400 hundred metres per second when erupted through vents 200 m in diameter (Fig. 10.4a). If the volatile content in the model is held fixed, and the vent radius increased, the result is an increase in eruption velocity (Fig. 10.4b). This is intuitively reasonable: it is easier to blast gas and magma up through a wide pipe than a narrow one. But eruption velocity translates directly into mass eruption rate, and into *thermal flux*, which is the

Fig. 10.3 Pyroclastic flows spread radially outwards in dense clouds of fine ash from an eruption column a few tens of metres in height, Mt. St Helens May 1980. While the pumice deposits were still hot, interaction with water-generated secondary eruption columns such as the one shown. Activity continued for many days. Cf. Fig. 4.35.

(a)

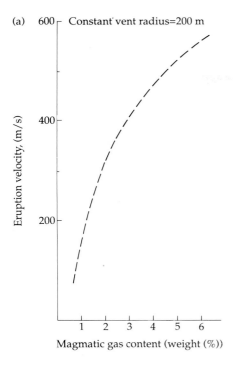

(b)

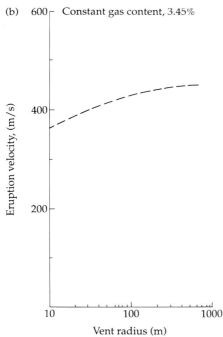

Fig. 10.4 Effect on eruption column velocity of (a) increasing magmatic gas content and (b) increasing vent radius. Radius scale is logarithmic in (b). After Wilson *et al.* 1980.[2]

chief control on eruption column height. Thus, increasing magmatic volatile content and vent radius both cause, initially at least, increasing column height (Fig. 10.5).

So far so good. Things get interesting at higher mass eruption rates, though. As mass eruption rate increases with increasing vent radius, a point is ultimately reached where convection cannot be sustained. In this condition, the column contains so much pyroclastic material that it is no longer buoyant; it is denser than the surrounding atmosphere. So it collapses (Fig. 10.5b). 'Collapse' carries an implication that merrily convecting eruption column abruptly gives up the ghost and spreads itself over the ground. This is not the case. The abrupt descent of the graph in Fig. 10.5b has no *time* implication; rather, it marks a transition from one regime to another. At a certain vent radius, stable convection can be sustained, as a slightly greater vent radius, convection cannot be sustained, so pyroclastic flows result. Although it may sound odd, 'sustained collapse' is possible. All that it means is that after ballistic ejection in a sustained fountain from the vent, pyroclastic material continuously falls back and spreads laterally as pyroclastic flows. This phenomenon may have been observed during the 1877 eruption of Cotopaxi, Ecuador, when an observer described activity at the summit vent as resembling 'a pan of rice boiling over'.

Chuck Yeager and others with the right stuff would appreciate the concept of stability regimes. In probing the 'envelopes' of their high-performance aircraft as test pilots, they were exploring the boundaries of stable flight in the regime of speed and angle of attack; outside the envelope stable flight is not possible (*I've tried A! I've tried B! I've tried C! . . .*). Our 'envelope' is summarized in Fig. 10.6; an important diagram, which underpins the understanding of column stability. In essence, it shows that an eruption column may occupy only one of two regimes in terms of vent radius and volatile content: either *stable*; forming a convecting plinian eruption column, and consequent tephra fall deposit; or *unstable*; collapsing to form pyroclastic flows.

Two implications of the envelope should be noted. Vent radius will probably increase during

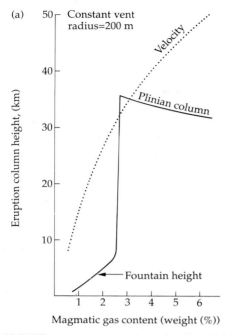

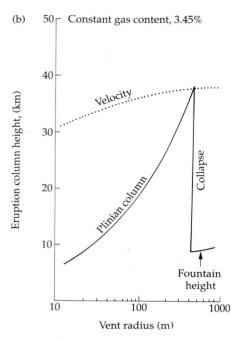

Fig. 10.5 Effect on eruption column height of (a) increasing magma volatile content and (b) increasing vent radius. Column height is a function of eruption velocity. In (a) a high convecting plinian eruption column develops as soon as volatile content is increased above about 2.5 wt per cent; in (b) steady increase in plinian eruption column height with vent radius leads to collapse as soon as vent radius exceeds about 400 m. Collapse also takes place in (a) if volatile content is *decreased* during course of eruption. After Wilson *et al.* 1980.

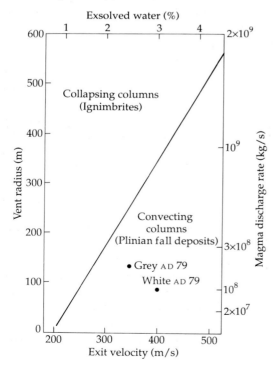

Fig. 10.6 Stability fields for convecting and collapsing columns, in terms of vent radius and magmatic volatile content. Magma discharge rate (right side) is largely a function of vent radius (left side) while exit velocity (bottom) is largely a function of volatile content (top). Data points for the Vesuvius AD 79 white and grey plinian pumice columns are also shown—the column which gave rise to the grey plinian pumice fall plots much nearer the stability boundary than the white. After Wilson *et al.* 1980.

an eruption by erosion of the conduit (Section 9.7.2). This implies an upward movement on Fig. 10.6, pushing the envelope. If continued, then at some critical radius, the envelope will be broken, the previously stable convecting column will become unstable; collapse will ensue, and pyroclastic flows will be formed.

Similarly, as an eruption progresses, progressively deeper levels in the magma chamber, poorer in volatiles, will be tapped. This implies movement from right to left on Fig. 10.6: again

from the regime of stable convection to collapse and pyroclastic flow formation.

Both these implications help to explain a common field observation: large pyroclastic *flow* deposits are commonly immediately underlain by plinian *fall* deposits. This relationship is readily interpretable when the physics of eruptions is considered: convecting eruption columns that give rise to fall deposits become unstable through vent erosion and collapse to form pyroclastic flows.

10.3 Emplacement of pyroclastic flows

Pyroclastic flows are made of solid lumps of pumice, dust, and gas. They do not remotely resemble lava flows, and are certainly not liquids. But pyroclastic flows demonstrate some remarkably fluid properties. Their mobility has been rightly described as 'spectacular'.[5] According to Tad Ui, the 6000 year Koya flow travelled for more than 60 km from its source in Kagoshima (southern Japan), 10 km of this over open water, and left a deposit which was only 2 m thick. The AD 186 Taupo ignimbrite, New Zealand covered no less than 20 000 square kilometres. It had a volume of 30 cubic kilometres, and swept out radially from its source to distances of about 80 km, rushing over hills in its path more than 1000 m high, reaching a velocity of more than 200 metres per second.[6] Other pyroclastic flows have travelled further (more than 100 km) but none is known to have climbed higher. How, then, do pyroclastic flows made of solid materials attain such spectacular mobility?

Part of the answer is that they are highly energetic. In collapsing from an unstable eruption column, a pyroclastic flow transforms potential energy into kinetic energy which propels it laterally. The further it falls vertically from the column, the more kinetic energy it will acquire, and the further and faster it will travel horizontally. An ordinary large rock avalanche can travel more than 30 km horizontally if it falls through a vertical distance of only 3 km. Since the purely ballistic component of an eruption column can carry material to a height of several kilometres, there is no shortage of energy. Thus,

part of the explanation for the apparently exceptional mobility of pyroclastic flows is simply that they are exceptionally energetic.

Another factor is the way in which the flows travel: a moving pyroclastic flow has properties more like those of a liquid than a mass of solid fragments. Pyroclastic flows are *fluidized*, endowing them with much lower viscosities than ordinary bodies of rock fragments.

10.3.1 Fluidization

Fluidization is central to many chemical engineering processes, but is best known through the use of fluidized bed combustion to improve efficiency in coal-fired power stations. A *fluidized bed* is not some new Californian water bed, but a layer of dry, granular material, up through which a stream of gas is blown. It is often demonstrated to engineering students with dry powder in a glass-fronted apparatus. As the gas flow is increased, the particles begin to shiver slightly, then dance around, until finally the whole mass of particles appears to be boiling, all of them in a state of constant motion. In this state, a light object will bob around on the surface, floating on the powder just as it would on liquid—lecturers in chemical engineering love to use a plastic duck. If the duck is pushed under the surface, it immediately pops up again, as it would in the bath tub at home. In this condition, the dry powder is behaving like a liquid, and is said to be fluidized. When fluidized, the dynamic mixture of gas and particles has a lower density than the original powder alone. Note that the particles do not actually 'float' on the gas stream, like ping-pong balls at a fair ground shooting gallery; they remain in intermittent contact with one another, and only part of their weight is taken by the gas stream. In technical terms, fluidization can be defined as occurring when inter-particle friction disappears, and the angle of repose become zero—in other words, when the bubbling powder in the apparatus has a horizontal surface, like a liquid, and will no longer stand up in heaps.

So much for plastic ducks and chemical engineering. Fluidization of volcanic pyro-clastic flows is much more complex, not least because the flows themselves are not laboratory materials. Laboratory powders are chosen for uniformity in size, shape, and density, whereas a pyroclastic flow, like a fall deposit, consists of pumice fragments of all sizes and shapes, plus a variable proportion of denser lithic clasts and separated crystals. It is thus difficult to apply the lessons learned from chemical engineering directly to volcanic flows. But two milestone papers by Steven Sparks[7] and Colin Wilson[8] have elucidated many issues; Sparks's through grain-size studies of actual deposits, and Wilson's through laboratory studies of pyroclastic materials.

Several key points emerge from these studies. First, because pyroclastic materials show such a range in grain size, the flow as a whole is not fluidized: only the finest particles are small enough to be fluidized. Thus, a fluidized pyroclastic flow is best defined as *a dispersion of large clasts in a medium of fluidized fines*. Like water in mud-flows, fluidized fines in a pyroclastic flow act as a lubricant.

It is self-evident that it is easier to fluidize small grains than large ones—a large clast requires a high gas velocity to support it. As Fig. 10.7 shows, typical clasts one millimetre in diameter are fluidized at gas velocities on only 0.1 metres per second, but one centimetre clasts require velocities of about five metres per second. Fig. 10.7 also makes a second important point: because the finest particles in a flow are fluidized at such low gas velocities, some degree of fluidization is easy to produce. Grain-size distributions in natural pyroclastic flows are such that about 40 per cent of clasts are small enough to be

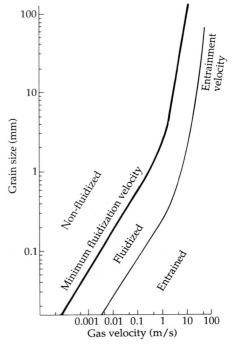

Fig. 10.7 Gas velocities required to fluidize particles of different grain sizes. At gas velocities and grain sizes left of the heavy line, no fluidization takes place; between the two lines is the fluidization field; at higher velocities, particles are entrained—carried away bodily. The graph is constructed from laboratory data for spheres of density 1000 kilograms per cubic metre in a carbon dioxide atmosphere, so cannot be directly extrapolated to natural pyroclastic flows. From Sparks, R. S. J. (1976). Grain size variations in ignimbrites and implications for the transport of pyroclastic flows. *Sedimentology* **23**, 147–88.

fluidized at reasonable gas speeds; the remaining larger clasts are transported within the fluidized medium so formed.

Through his laboratory studies, Colin Wilson discovered another reason why natural materials are unlikely to be completely fluidized: when a mixture of grain sizes and densities is fluidized, *segregation structures* begin to form, as clasts of different sizes and densities jostle apart from one another. Segregation structures provide direct routes through the body of the material, along which gas readily escapes. In the laboratory, vertical pipes filled with coarse particles with large void spaces between them are the most obvious gas escape structures developed: they have easily recognizable counterparts in real deposits (Fig. 10.8).

Segregation by density has another effect readily visible in the field: coarse, low-density pumice clasts float upwards through the flow, congregating at the top, while denser lithic clasts settle towards the bottom. In the field, this phenomenon is manifested in *pumice concentration zones*; rafts of large pumice clasts, often tens of centimetres across, which form eye-catching horizons within deposits, and help to distinguish the deposits of one flow unit from another (Fig. 10.9). Measurement of the density of the various clasts provides a measures of the density

Fig. 10.9 Pumice concentration zone at top of the 4.5-million-year-old Real Grande ignimbrite, Cerro Galan caldera, north-west Argentina. Geologists are standing at the level where half-metre sized pumice clasts have congregated. Fine-grained base of overlying unit is level with their heads.

Fig. 10.8 Spectacular examples of segregation structures in an ignimbrite. These lithic-filled pipes rise many metres through the Real Grande ignimbrite, Cerro Galan, north-west Argentina. Largest lithic clasts in the pipes reach twenty centimetres; most of the finer-grained material has been elutriated—blown away. Steve Sparks for scale.

of the flow itself: the highest-density clasts that have floated to the top give the density of the flow. A wide range of flow densities is possible—field examples of pumice flotation zones with densities ranging from 0.6×10^3 to more than 1×10^3 kilograms per cubic metre are known.

Flow density is also an expression of how *expanded* or inflated the flow is in its fluidized form. Wilson showed that typical pyroclastic flows are not highly expanded: a flow that is 100 m thick while fluidized will deflate to form a deposit about 75 m thick when it comes rest. This conclusion is important because earlier workers had argued that the only way to explain extraordinary observations of pyroclastic flow deposits sitting on surfaces hundreds of metres above valley floors was that the flows must have been extremely highly expanded; low-density clouds of gas and ash hundreds of metres thick (Fig. 10.10). It should not be supposed, though, that wherever one sees an ignimbrite deposit in the field, one is seeing something approximating to the original maximum thickness of the flow—most outcrops probably form from the thinner,

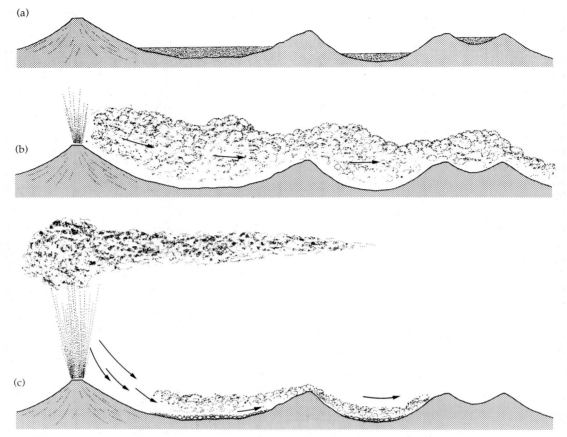

Fig. 10.10 (a) Pyroclastic flow deposits separated from their source volcano by a high topographic ridge and preserved at high elevations (right) presented problems to early students. (b) Initial attempts to explain these relationships invoked highly expanded, low-density pyroclastic flows of enormous thickness, perhaps several hundred metres. (c) Pyroclastic flows can surmount topographic features several hundreds metres high because they acquire enormous amount of kinetic energy in collapsing from high eruption columns. Although topped by clouds of dust, flows themselves are dense and ground-hugging. Only modest amounts of expansion are likely—deposit thickness is probably about three-quarters of flow thickness. Not to scale.

waning parts of the flow, behind the thickest, advancing part.

10.3.2 Rheology of pyroclastic flows

Field and laboratory studies tell us a good deal about the properties of fluidized flows. But one factor remains difficult to determine: their *rheology*. In conventional fluids, such as lava flows, rheology is controlled by viscosity and yield strength. A perfectly fluidized particulate material will have zero yield strength. But since pyroclastic flows are only partially fluidized, their yield strengths are not zero. One way of

estimating yield strength is to consider the size of the largest lithic clast that can be supported: the fact that large lithic clasts sink through pumiceous flows indicates that their yield strengths are low. Viscosity can be measured directly in laboratory fluidization apparatus, but it is difficult to extrapolate such measurements to real flows. By calculating rates at which lithic clasts apparently sank through flows in his field area in Italy, Sparks estimated viscosities of 1–100 Pa s. (Recall that basalt lavas typically have viscosities of 100–1000 Pa s.)

Armed with estimates of flow density, yield

strength, and viscosity, it is possible to say something about how pyroclastic flows achieve such apparently remarkable mobility. It might be supposed that fluidized flows which travel huge distances at great speed and over-run tall obstacles *must* be turbulent. This may be true close to source regions, where extreme velocities obtain, but further away, laminar or plug flow regimes are more probable. The Taupo flow was exceptionally fast-moving, but even this flow was probably turbulent only within 20 km of the vent. In Mexico, the Acatlan ignimbrite provides further evidence for laminar flow, since strong compositional zonation is preserved throughout one flow unit. Had the flow been turbulent, vertical mixing would have smeared out the original compositional zonation.[9] In North America, ignimbrites erupted from Crater Lake, Oregon, are also spectacularly zoned (Fig. 10.11).

Fluidization of pyroclastic flows assumes a stream of gas passing upwards through the flow. But where does the gas come from? Two important sources are usually invoked: gas exsolving from juvenile clasts, and air ingested during movement of a flow. Exsolving gas escaping from hot pumice clasts seems an intuitively obvious source and has long been invoked as a principal source of fluidization, since flows consist almost entirely of hot clasts. Recent work, however, suggests that pumice loses most of its gas *early*, probably in the eruption column, so it may not be an important source during lateral flow.

As it collapses down from the eruption column and hurtles forward over the ground, a flow necessarily displaces a large volume of air. How much of it is ingested is difficult to determine, but this remains a probable source of fluidizing gas. Other possible sources are from ground and surface water volatilized by the flow, and from vegetation carbonized by it, but neither of these is likely to be of more than secondary importance.

Finally, this discussion of fluidization has been thoroughly dry and detached. In the field, the *fact* of fluidization of pyroclastic flows is remarkably real. Step out of a helicopter on to a warm, freshly deposited pyroclastic flow and stir it up with a stick—the apparently inert deposit magically

Fig. 10.11 A striking example of a compositionally zoned ignimbrite; the 7000-year-old Mazama ignimbrite, erupted during the formation of the Crater Lake caldera, Oregon, USA. Darker, upper part of the deposit is andesitic (56–62 per cent SiO_2); lower, lighter part, erupted first, is rhyodacitic (70–72 per cent SiO_2).

changes as trapped hot gas fountains upwards, carrying a plume of fine dust with it; the surface shivers like quicksand, and then settles downwards into a less-expanded form. On the right sort of slope, small secondary flows can be started by judicious digging. Their startling mobility is then revealed as they sweep silently and dustily downslope (Fig. 10.12).

10.3.3. Beyond fluidization: momentum transfer and rocket fuel

All the laboratory and field evidence shows that fluidization is crucially involved in the emplacement of pyroclastic flows. However, it is not in

Fig. 10.12 Miniature ignimbrites, produced by slumping from much thicker, still-hot primary deposits exposed in a gully; Mt. St Helens, May 1980.

itself sufficient to account for the remarkable distances reached by some flows. As outlined in Chapter 13, large 'cold' rock avalanches travel comparable distances, and in these there is no internal source of fluidizing gas, and little likelihood that they can entrain sufficient air to achieve significant fluidization.

Large rock avalanches have little in common with pyroclastic flows. But the one feature they share illuminates the problem of pyroclastic flow mobility: both are endowed intially with huge amounts of momentum. Pyroclastic flows obtain their momentum in collapsing from eruption columns; avalanches in slumping away from a mountain side. Studies of avalanches suggest that they conserve stunningly high velocities by forward transfer of momentum from the rear of the flow. Instead of the whole avalanche mass starting off rapidly and progressively slowing down as a single entity, the front of the avalanche maintains much the same velocity throughout its journey, coming to a very abrupt halt.

Enormous frictional losses are, of course, involved in moving a mass of rock large distances horizontally. Because these frictional losses have to be overcome, and momentum has to be conserved, the only way the front of the avalanche can reach the end of its run with much the same velocity it started with, is if matching amounts of material are travelling a lot slower, or stop altogether, transferring their momentum to the front of the flow. Thus, a large, rapidly moving avalanche is 'fuelled' by transfer of momentum, a progressively decreasing mass of material at the front rushing on headlong while deposition takes place in the rear. The head of a moving avalanche thus resembles the spacecraft born by a rocket: the rocket moves forward through the momentum imparted by the burned fuel, which constitutes by far the largest fraction of the mass of the assembly.

Pyroclastic flows are nothing more than rather exotic avalanches. Their extraordinary mobility becomes easier to understand if one thinks of them as being propelled by the same rocket fuel that drives ordinary avalanches.

— 10.4 Ignimbrites: deposits of pumiceous pyroclastic flows –

Ignimbrite is a name compounded from two Latin words to mean something like 'fire cloud rock', a felicitous name which reflects the origin of these rocks. They are found all over the world, dominating substantial parts of the landscape in countries as diverse as Mexico, New Zealand,

and Japan. Despite their abundance, ignimbrites remained obscure for decades. They were recognized as a distinct rock type only as recently as 1935, when the New Zealand geologist P. Marshall christened them.[10] Several reasons account for their initial obscurity. Partly, it was because an 'ignimbrite' can be anything from a loose, sandy ash to a solid, glassy rock, difficult to distinguish from lava; partly because an ignimbrite can range in scale from a few hundred metres in length to tens of kilometres; and partly because no *large* ignimbrite-forming eruption has ever been observed. Much of the pioneering field-work in understanding ignimbrites was carried out by members of the United States Geological Survey, notably R. L. Smith, working on the magnificent rocks of New Mexico and Colorado. In his influential publications, Smith used the term *ash-flow tuffs* for what would now be called ignimbrites. This descriptive term is still widely used.[11]

10.4.1 The Valley of Ten Thousand Smokes: an example of an ignimbrite-forming eruption

It is fortunate that large ignimbrite eruptions are rare, because a single flow could devastate a huge area in a few hours. Before going any further, therefore, it is appropriate to examine the largest ignimbrite-forming eruption of modern times, which took place in an uninhabited area of Alaska in 1912.

Mount Katmai is located on the horn-like peninsula leading westwards from the mainland of Alaska into the Aleutian Islands. On the evening of 31 May 1912, seismic tremors were felt at Katmai village, 30 km south of the volcano, and on 4 and 5 June severe shocks were felt up to 200 km away by the thinly scattered population. On the morning of 6 June, explosions were heard 240 km away, and an ash cloud was observed just after noon. During the afternoon of 6 June, the eruption reached its climax. There were at least two huge explosions, audible for many hundreds of kilometres, and a plinian eruption column showered air-fall ash over a large area. Ash began to accumulate at Kaflia Bay, about fifty-six kilometres from the volcano, and by 5 p.m. ash was falling on the port of Kodiak, 170 km away. At Kodiak, three distinct episodes of ash fall were recorded over the next three days (Fig. 10.13). A native of Kaflia, Bay, Ivan Orloff, wrote a letter to his wife on 9 June in which he conveyed starkly

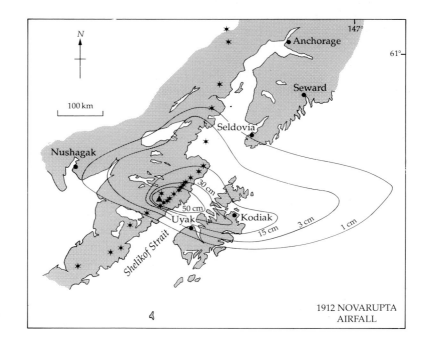

Fig. 10.13 Isopach map for the Novarupta plinian fall deposit associated with the 1912 eruption of the Valley of Ten Thousand Smokes ignimbrite. Kodiak Island and part of the Upper Alaskan Peninsula are shown. Small stars are other volcanoes. Triangle shows site of Novarupta vent. From Hildreth, W. (1983). The compositionally zoned eruption of 1912 in the Valley of Ten Thousand Smokes, Katmai National Park, Alaska. *J. Volcanol. Geotherm. Res.* **18**, 1–56.

the nightmarish quality of the long-drawn-out event:

We are waiting death at any moment. A mountain has burst near here . . . We are covered with ashes, in some places ten feet and six feet deep. All this began June 6. Night and day we light lanterns. We cannot see the daylight. We have no water, the rivers are just ashes mixed with water. Here are darkness and hell, thunder and noise. I do not know whether it is day or night. The earth is trembling, it lightens [*sic*] every minute. It is terrible. We are praying.

At the time, all that was known of the eruption was that several major explosions had occurred, presumably on the Katmai volcano, and that a huge quantity of ash had fallen. It lay 60 centimetres thick in the streets of Kodiak. No one had seen or experienced anything more than this, so no one had the slightest idea of what had happened to the volcano itself. Although the scientific world was aware that this had been a big eruption, it was not until 1916 that an expedition sponsored by the National Geographic Society of America was sent to investigate. Naturally, the first object of the expedition was to examine Mt. Katmai itself, a volcano that had never before been photographed, let alone studied. Some wonderful surprises awaited the explorers.

In place of the original 2300 m high cone, there was now a deep, roughly circular crater two kilometres in diameter and 600 metres deep, containing a turquoise-blue lake. A small island, formed of platy dacite lava, rose about the surface of the steaming hot lake. North-west of the volcano, where there had previously been a valley along which the Ukak river had flowed, the exploring party found a flat plain, four kilometres across at its widest, from which countless jets of steam were playing. As Robert Griggs, leader of the expedition put it: 'It was as though all the steam engines in the world, assembled together, had popped their safety valves at once, and were letting off steam in concert'.[12] Suitably enough, he named this the 'Valley of Ten Thousand Smokes'. It must have been an wonderful spectacle. Sadly, the steam jets died away after a few years, and now the valley is still. But what had caused the remarkable transformation?

Griggs' party found that for a length of about twenty-two kilometres the original Ukak Valley had been filled with loose pyroclastic material, which they described as a 'sand flow'. They inferred that the 'sand' must have been incandescent at the time of eruption, and that the thousands of steam vents were produced by the vaporizing of water trapped in the sediments of the river bed beneath the cooling mass. The party spent some time camped in their valley and found good use for the steam—they cooked their meals over the jets. (Not such a good idea as it sounds, since the steam was quite acid. Their cooking gear was rapidly corroded, and holes appeared in their saucepans, kettles, and billy cans.)

Since Griggs' expedition in 1916, several other have visited the area, which continues to be a major focus of research. Wes Hildreth of the US Geological Survey has made a comprehensive recent study.[13] An early realization was that although the upper part of Mt. Katmai was demolished, the pyroclastic eruption was not centred on Mt. Katmai itself, but on a series of vents eight kilometres west of it, near the head of the Ukak Valley (Fig. 10.14). It was from these vents that the 'sand flow' (ignimbrite) was erupted. Some 10–15 cubic kilometres of ignimbrite filled the old Ukak Valley tens of metres deep, reaching a thickness of 250 m near the vent. Hildreth's stratigraphical work showed that plinian fall deposits corresponding to the three successive episodes of ash fall noted at Kodiak could be identified in the field. Eruption of the ignimbrite overlapped with, but outlasted, the eruption of the first of these, terminating within about 20 hours of the initial outbreak. He estimated that a total volume of 20–25 cubic kilometres of tephra were erupted over a period of about sixty hours. Taking both air-fall deposits and ignimbrites into account and converting to dense rock, a total volume of about 15 cubic kilometres of magma was erupted during this period. Given the brevity of the climactic phase of the eruption, this implies an *average* mass eruption rate of no less than 1.9×10^8 kilograms per second, but the rate would have been much greater during the three periods of high column activity.

Not suprisingly, rapid eruption of this large

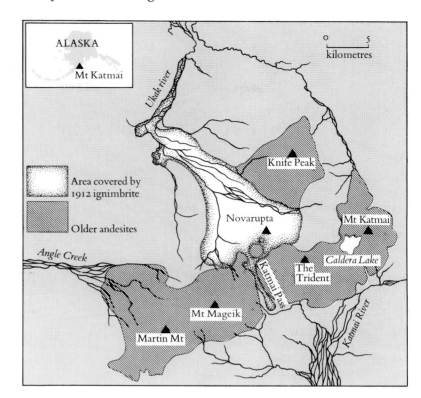

Fig. 10.14 Spatial relations of Valley of Ten Thousand Smokes ignimbrite, Mt. Katmai caldera and Novarupta vent. Area is one of complex topography. Only selected peaks are shown. The ignimbrite was largely confined to the valley of original Ukak River. (After G. H. Curtiss.)

volume of magma had considerable topographic consequences. Apart from causing the collapse of the top of Mt. Katmia via a poorly understood hydraulic connection, a pumice-filled depression 2 km across formed over the vent; a sort of subdued caldera. After the catastrophic phase of the eruption, possibly several months later, a small dome of viscous rhyolitic lava was extruded through the vent, partly filling the depression (Fig. 10.15). This dome was named *Novarupta*. It represented the last gasp of the eruption.

10.4.2 *Morphology of ignimbrites*

Since the eruption of a *large* ignimbrite has never been observed, it is difficult to be certain about what one would look like immediately after it was formed. However, there have been many small ignimbrite eruptions, so it is possible to deduce a good deal about their larger cousins from them, and from the eroded remains of larger bodies.

A simple but important point first: because ignimbrites are the deposits of gravity-driven flows, they are controlled by topography, and are thus preferentially channelled along valleys and other depressions. This is an essential means of discriminating between pyroclastic flow and fall deposits: fall deposits mantle topography uniformly, like snow, except on slopes so steep that they avalanche off. By contrast, ignimbrites usually form flat-topped deposits, confined to the floors of older valleys, as in the Ukak River. Such deposits may be tens or even hundreds of metres thick. In cases where the ignimbrite is exceptionally energetic (such as Taupo), it is capable of sweeping over irregular topography, leaving a thin skin of deposit even over topographically high points. These contrasting types have been called *valley pond deposits* and *ignimbrite veneers*[14] (Fig. 10.16).

Where small ignimbrites are not confined by valleys and are free to spread out over open ground, they form branching, multi-lobed deposits whose general outlines are not dissimilar to those of lavas on similar topography. Central channels bounded by subdued marginal *levées*

Fig. 10.15 At the head of the Valley of Ten Thousand Smokes, Alaska is a gentle caldera-like depression, largely filled by the Novarupta lava dome. View looks to the north-west, with Broken Mountain in the background. US Geological Survey photo.

Fig. 10.16 Valley pond ignimbrites and ignimbrite veneers. Valley ponds may be tens of metres thick; veneers 0.5–3 metres. Veneers are products of only the most energetic pyroclastic flows, which overrun topography, rather then being confined by it. They are preserved only as thin skins over ridges.

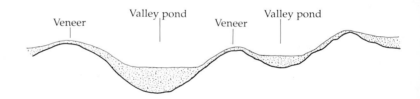

are sometimes present, though these are not so well developed as in lavas. Flow fronts and margins are well defined by accumulations of coarse pumice. The smallest ignimbrites have flow fronts less than a metre high. Large ignimbrites probably have similar morphologies when fresh: the 1.3-million-year-old Purico ignimbrite in Chile, for example, seems to have terminated in a multi-lobed boundary with flow fronts 20–30 m high (Fig. 10.17).

10.4.3 The standard ignimbrite

Volcanologists usually see ignimbrites exposed in vertical sections in quarries or canyons. Sometimes, they may be confronted with towering cliffs of ignimbrite hundreds of metres high (Fig. 10.18). Whatever the scale, close examin-

ation of a vertical section usually reveals that it is not a homogeneous mass. Near-horizontal boundaries which may be extremely obvious or extremely subtle divide the ignimbrite up into *ignimbrite flow units*. Each unit is the product of a single pyroclastic flow. Some large ignimbrites may be composed of only one or two flow units; others of scores. Where there is evidence in the form of jointing or welding (Section 10.4.4) that successive flow units were emplaced so quickly that they all cooled together, the assemblage is termed a *compound cooling unit*.

After diligent field-work followed by innumerable dusty sieve analyses, Sparks *et al.*[15] showed that ignimbrite flow units can be divided into distinctive components. Fig. 10.19, which summarizes these components, has become an icon

Fig. 10.17 Landsat Thematic Mapper image of western flanks of Cerro Purico, north Chile. Purico is a 1.3-million-year-old 5000 metre high ignimbrite shield. At centre, digitate lobes of ignimbrite may correspond approximately with original outlines of the 20 to 30-metre-high flow fronts. Image is 25 km across.

Fig. 10.18 Ignimbrites totalling hundreds of metres in thickness are exposed in cliffs on the flanks of the Cerro Galan caldera, north-west Argentina. Individual flow units are tens of metres thick. Note prominent horizontal boundary between units. Figure at bottom gives scale.

cherished by a generation of zealous graduate students.

Layer 1 is the most controversial part of the ignimbrite canon: the unit generally forms only a tiny fraction of the thickness of the whole deposit, is often only a few centimetres thick and commonly missing altogether. It is fine grained, rich in crystals, sometimes finely laminated and sometimes shows cross-bedded structures. It is often described as a *pyroclastic surge deposit* (Chapter 11). Occasionally, coarse, lithic rich, fines-depleted deposits are found in the same stratigraphic position. These have been termed *ground-layer* deposits.

Layer 2 forms much the most important part

of an ignimbrite, and in essence defines an individual flow unit. It may range in thickness from a few centimetres to tens of metres. *Layer 2a* is a fine-grained basal layer, which rests directly on the ground surface when layer 1 is absent. It is rarely more than a metre thick, and is typically 20 to 30 cm thick (Fig. 10.20). Both pumice and lithic clasts in 2a show reverse grading, the sizes of both increasing upwards until they have the same value as layer 2b, at which point the basal layer is no longer distinguishable. The boundary between the two layers is therefore always transitional.

Layer 2b, much the thickest part of the ignimbrite, consists of a poorly sorted mixture of

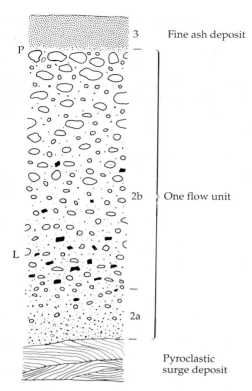

Fig. 10.19 Section through the 'standard' ignimbrite, showing one flow unit with layers 2a and 2b; underlying pyroclastic surge deposit (layer 1) and fine ash deposit (layer 3). Note concentrations of coarse pumice towards top of 2b, and of lithics towards base of 2b. P = pumice; L = lithic clasts. After Sparks *et al.* 1973.

Fig. 10.20 Layer 2a, the fine-grained base of an ignimbrite in Tenerife, Canary Islands. Layer 2a is about 60 cm thick (coin gives scale). It is homogeneous and lacks lithics, while there is a characteristic band of lithics at base of overlying layer 2b.

pumice clasts and dusty ash with a variable content of lithics and crystals (Fig. 10.20). Many, but not all, layers 2b show well-developed grading, with coarse pumice concentrated upwards—*reverse grading*—while lithic clasts show *normal grading*, and coarsen downwards. Pumice concentration zones are sometimes spectacularly developed, forming conspicuous marker horizons defining the tops of flow units (Fig. 10.9; Fig. 10.21; Fig. 10.22).

Layer 3 deposits are rarely preserved, but where they are found, are thin, extremely fine-grained ash deposits that mantle the topography, unlike layers 1 and 2, and can be traced into areas where these layers are not found. They are composed of highly fragmented pumice (dusty

Fig. 10.21 A quarry exposure of an ignimbrite in Tenerife. Layer 2a is visible level with the geologists' heads; above that is the main thickness of layer 2b. Black holes are moulds of tree trunks swept up by the flow and subsequently carbonized.

Fig. 10.22 Fist-sized pumice clasts and small, sparse lithics near the top of layer 2b in a unit of the Real Grande ignimbrite, Cerro Galan, north-west Argentina.

glass shards) and sparse crystals. Some layer 3 deposits contain abundant accretionary lapilli.

Leaving aside the vexatious issue of pyroclastic surges and layer 1 deposits for the time being, the other components recognized in real ignimbrites tally well with our understanding of the physical processes controlling pyroclastic flows: layer 2a, the fine-grained basal layer, is the result of intense shearing taking place between the moving flow and the underlying ground surface. Increasing grain size upwards in this layer is the result of mechanical, grain-dispersive forces rather than fluidization. In layer 2b, fluidization plays the dominant role in segregating clasts by density, resulting in pumice clasts floating upwards and lithic clasts sinking downwards. Fluidization 'fossils' are often found in layer 2b which illustrate the process almost as clearly as if it were still continuing: vertical pipes, filled with fines-depleted coarse pumice snaking up through the deposit (Fig. 10.8). In some localities, carbonized logs, with small pipes above them, also show how gases released by their combustion contributed to fluidization.

Layer 3 is the end result of all the fluidizing gas

blowing upwards through the ignimbrite: the finest ash particles are carried clear away from the ignimbrite, winnowed out and borne upwards by hot convecting gases, to fall out far from the ignimbrite as a fine-grained air-fall ash deposit. Photographs of small pyroclastic flows being emplaced, such as those of Mt. St Helens show this process clearly: the small pyroclastic flows scurrying downslope are themselves invisible, concealed by the curtains of fine ash that roll upwards above them, mimicking the giant convecting cauliflower clouds of the main eruption column.

Layer 3 deposits are often usefully termed *co-ignimbrite ashes*,[16] since they form consistent complements to ignimbrites. Their complementarity can be expressed in terms of crystal content. During elutriation of fines from an ignimbrite, relatively dense crystals, such as quartz and plagioclase, originally present as phenocrysts in the pumice, remain behind while the lighter, dusty pumice matrix is swept away, leaving *crystal-concentrated* ignimbrites and forming *vitric-enriched* layer 3 deposits. Sparks and Walker suggested that co-ignimbrite ash falls may represent loss of up to 35 weight per cent of the ignimbrite glass fraction. More recent studies have shown that wholesale fines depletion may take place, including both glass and crystals.

Co-ignimbrite ashes are emphatically not merely minor volcanological curiosities. They may have large volumes, and, as we shall see, some of the largest of all tephra fall deposits are thought to be of co-ignimbrite origin.

Ignimbrites shown important lateral, as well as vertical, variations. Not suprisingly, they are much coarser near their sources than further away, but near-source deposits are so complex that they have only recently come under detailed study. A sure guide to the proximity of the ignimbrite source area is the presence of coarse *lag breccias*,[17] consisting of lithic clasts up to a metre or more in diameter—no mere pebbles these, but boulder-sized rocks. Such lag breccias are distinct from the lithics that form part of 'normal' ignimbrites: they form separate lenses or horizons within which the breccias are clast-supported and there is little ignimbrite matrix (Fig. 10.23). They probably resulted from strong

Fig. 10.23 Well-developed lithic lag breccia on flanks of the Cerro Galan caldera. Lithic clasts here are up to 30 centimetres in diameter, and show some rounding. Breccia layer forms a prominent marker horizon locally.

gas fluidization. *Lag-fall breccias*, as their name implies, are coarse deposits that fell from the eruption column close to the vent, and became incorporated within the moving pyroclastic flows.

10.4.4 Tambora 1815: a case study in ignimbrites and co-ignimbrite ashes

The eruption of Tambora in April 1815 was the largest volcanic event in modern history. It would demand our attention for that reason alone, but it is instructive for two broader reasons: First, its world-wide atmospheric effects have been of paramount importance in understanding the effects of volcanism on climate. These are explored in Chapter 17. Second, from a

volcanological point of view, Tambora illuminates the crucial role of co-ignimbrite ashes in major eruptions. Sigurdsson and Carey's volcanological account is summarized here;[18] Stothers has compiled a fascinating collection of contemporary observations.[19]

Background   Tambora volcano forms the Sanggar Peninsula on Sumbawa Island in the Sunda arc segment of the Indonesian Archipelago. At sea-level, the volcano has a diameter of 60 km and rises to a present elevation of 2650 metres. Prior to 1815, Tambora may have been 4000 m high, but today the summit of the volcano is occupied by a great caldera 6 km in diameter and 1100 m deep. An ephemeral lake is present on the caldera floor and numerous fumaroles are active along the lower caldera walls (Fig. 10.24).

About 150 cubic kilometres of tephra were ejected during the great eruption, causing ash fall-out at least as far as 1300 km from source, and plunging a region up to 60 km west of the volcano into darkness for up to three days. Explosions were audible 2600 km away. As many as 90 000 people may have died on Sumbawa and the nearby island of Lombok as a result of both direct and indirect consequences of the eruption, though there are few detailed records. Such records as exist were made by a handful of British residents, sea captains and army officers scattered in tiny European enclaves amongst the islands. Their accounts were rapidly collected and published in 1817 by the colourful Sir Stamford Raffles, lieutenant governor of Java at the time.

Evolution of the eruption   Historical records suggest that the earliest events leading up to the paroxysmal eruption probably took place in the period 1812–1815, when the volcano was mildly active. Sigurdsson and Carey showed that the main eruption consisted of two distinct phases; an initial phase of at least four tephra fall episodes, and a subsequent phase of major pyroclastic flow eruption (Fig. 10.25).

Events began with a large explosion on 5 April. Isopleth data suggest that the eruption column reached a maximum altitude of 33 km, slightly higher than that of the AD 79 plinian eruption

Fig. 10.24 View across the Tambora 1815 caldera from the south wall to the north rim showing the one-kilometre-high walls and the small lake at bottom. No photograph can do justice to the dimensions of this immense abyss. Photo: courtesy of S. Carey, University of Rhode Island.

Stratigraphy of 1815 Tambora deposits

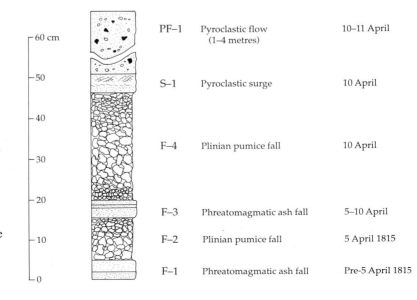

Fig. 10.25 Stratigraphy of the 1815 Tambora deposits, with chronology established by Sigurdsson and Carey. From Sigurdsson, H. and Carey, S. (1989). Plinian and co-ignimbrite tephra fall from the 1815 eruption of Tambora volcano. *Bull. Volcanol.* **51**, 243–70.

PF–1	Pyroclastic flow (1–4 metres)	10–11 April
S–1	Pyroclastic surge	10 April
F–4	Plinian pumice fall	10 April
F–3	Phreatomagmatic ash fall	5–10 April
F–2	Plinian pumice fall	5 April 1815
F–1	Phreatomagmatic ash fall	Pre-5 April 1815

column, and that it was propelled by a magma discharge rate of about 1.1×10^8 kilograms per second. About 12 cubic kilometres of highly vesicular grey-green trachyandesite pumice were erupted during this plinian event, which lasted only about two hours. Between 5 April and 10 April, Tambora lapsed back into state of low-level activity, during which several smaller eruptions produced a thin tephra layer. On 10 April, a much larger plinian eruption broke out, this time driving the eruption column to a height of 44 km. An estimated eruption rate of 2.8×10^8 kilograms per second was required to drive the column so high, making it the most energetic eruption of modern times, surpassing even the Bezymianny eruption of 1956, which reached a height of 38 km. Although exceptionally intense, this eruption was also short-lived, lasting perhaps 3 hours. Only 3 cubic kilometres of tephra were deposited.

This outburst heralded the beginning of the world-shaking events of 10 April 1815. Explosions were heard as far away as Benkoelan on Sumatra, 1775 km distant, and earthquakes were felt at Surabaya on Java, 600 km distant. In one of the few accounts of the eruption the Rajah of Sanggar reported an intensification of activity at about 7 p.m. on 10 April, followed by a rain of pumice on Sanggar east of the volcano at approximately 8 p.m. Tephra fall continued until about 10 p.m., when the village was ravaged by winds which uprooted trees and buildings. At this time the whole volcano was reported to appear as a flowing mass of 'liquid fire'.

Sigurdsson and Carey suggested that this event marked the critical transition from the plinian stage of the eruption to the pyroclastic flow stage, arguing that vent widening caused by the massive eruption rate led to eruption column collapse. Early stages in formation of the huge summit caldera may also have accelerated the mass eruption rate, overloading the eruption column and causing it to become unstable (Fig. 10.26).

Ignimbrite eruption and co-ignimbrite ash formation At least seven massive ignimbrite flows were erupted during the second phase of the eruption, flowing radially all round the volcano,

Fig. 10.26 Landsat MSS image of Sanggar Peninsula, Sumbawa Island, Indonesia, showing 6-km-diameter caldera formed by the great eruption of Tambora, 1815.

entering the sea and extending the shoreline. Today, the 1815 ignimbrites are exposed in sea cliffs all round the Sanggar Peninsula (Figs 10.27–10.29). About 2.6 cubic kilometres of magma were erupted to form the deposits now preserved on dry land. Much more entered the sea.

Self *et al.*[20] used available contemporary records of ash fall on neighbouring islands and core samples collected from the sea floor during an oceanographic expedition in 1929–30 to compile an isopach map of the distal ash. The deposit thickness varies remarkably little with distance, underlining the tremendous dispersal of the ash; a one-centimetre thickness of ash covered more than 500 000 km^2 (Fig. 10.30).

Sigurdsson and Carey concluded that much of this far-flung ash was of co-ignimbrite origin. They suggested that as pyroclastic flows entered the sea, violent secondary explosions took place, leading to elutriation of huge quantities of fine material. In coastal exposures at sea-level, they found that 32 per cent of the fine fraction had been removed from flows during passage into the ocean, and that a further 8 per cent of glass had been elutriated (relative to magmatic propor-

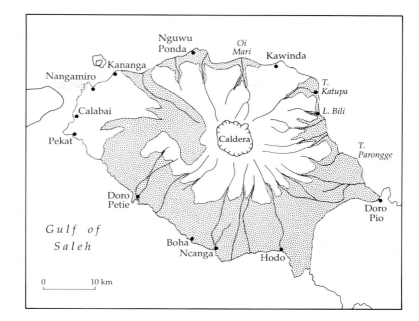

Fig. 10.27 Distribution of ignimbrites from 1815 eruption around Tambora. From Sigurdsson, H. and Carey, S. Plinian and co-ignimbrite tephra fall from the 1815 eruption of Tambora volcano. *Bull. Volcanol.* **51**, 243–70.

Fig. 10.28 Outcrop of the Tambora 1815 ignimbrites on the coast of the Sanggar Peninsula. Photo: courtesy of S. Carey, University of Rhode Island.

tions), indicating removal of no less than 40 per cent of the pyroclastic flow material as co-ignimbrite ash. In distal areas, as much as 80 per cent of the tephra fall deposit is probably of co-ignimbrite origin.

A total of about 50 cubic kilometres of magma (dense rock equivalent) was erupted as pyro-clastic flows, of which 20 cubic kilometres were co-ignimbrite ash and 30 cubic kilometres were fines-depleted flow deposits, most of which now lie beneath the ocean surrounding Tambora. Judging from the period of darkness at Maedora Island, 500 km north-west of the volcano, co-ignimbrite ash fall out continued for about three days. The mass eruption rate of the ignimbrites was probably about 5×10^8 kilograms per second, or about twice the peak rate reached during the plinian stage of the eruption.

All accounts agree that the great 1815 event was overwhelmingly an ignimbrite eruption. Plinian and associated fall deposits formed only 4 per cent of the total erupted mass. Most of the enormous ash-fall deposit that sifted down on surrounding areas was of co-ignimbrite origin. A satellite view of the volcano on 10 April 1815 would have shown a central eruption column rising from the huge caldera, surrounded by almost complete ring 40 km in diameter of massive *secondary* eruption columns, rising where pyroclastic flows launched themselves violently into the sea.

Fig. 10.29 Detail of Tambora 1815 surge deposits overlying an older breccia deposit on the east coast of the volcano. Photo: courtesy of S. Carey, University of Rhode Island.

Fig. 10.30 Isopach map, contoured in centimetres, of the tephra fall from the Tambora 1815 eruption. After Self *et al.* (1984). Volcanological study of the great Tambora eruption of 1815. *Geology* **12**, 659–63.

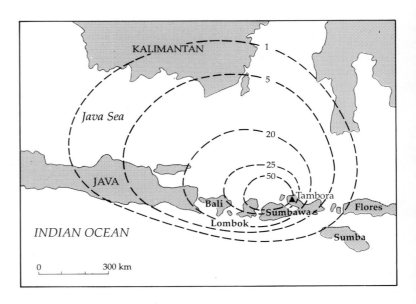

— 10.5 Welding in ignimbrites —

Ignimbrites are composed of hot pumice clasts. Pumice is frothy volcanic glass. Hot glass is soft. These facts add up to account for an important property of ignimbrites which often overprints the subtle features described earlier: they show various degrees of *welding*. Welding is the sintering together and flattening of hot clasts to form a denser, more coherent rock. It may be so extensive that the ignimbrite is a dense black mass of obsidian-like glass, like a glassy lava. In many cases, glassy ignimbrites have indeed been mistaken for lavas, and mapped as such.

Such wholly glassy welded ignimbrites are relatively rare, but many ignimbrites show varying degrees of *partial welding*. Laboratory experiments show that welding of rhyolitic glasses begins at temperatures of 600–750°C, but this depends on a variety of parameters, such as composition, viscosity, volatile content, and pressure. As noted earlier, alkalic magmas have lower viscosities than others, thus the highest degrees of welding are found in alkalic ignimbrites such as those of the East African Rift Valley.

Variations arising from welding can cause startling changes in the appearance of an ignimbrite over small distances, and were one reason why it took so long for ignimbrites to be properly understood. One volcanologist is always associated with the unravelling of these complex relationships: R. L. Smith.[21–22] In a milestone paper with G. S. Ross, Smith described the characteristics that enabled geologists to identify ignimbrites.[23] He also showed that three zones exist in a typical welded ignimbrite: a lowermost, unwelded zone, a middle, densely welded zone, and an overlying partly welded zone that grades upward into unwelded ignimbrite. The lowermost unwelded zone is straightforward: here the ignimbrite chilled against the ground it was in contact with, and was unable to weld (Fig. 10.31).

In order to produce a densely welded zone, an ignimbrite has to remain at high temperature and pressure long enough for welding to take place. Thus, densely welded ignimbrites are typically found towards the bottom of thick deposits ponded in valleys. In a 100-m-thick ignimbrite, often less than 10 metres of the lowermost part shows dense welding, most of the thickness being only partly welded. An important point that Smith demonstrated was that when a sequence of flow units is emplaced in rapid succession, they may cool together as a compound cooling unit, in which the various zones of welding are overprinted on the flow unit boundaries. Such cooling units are often obvious from a distance, if vertical cooling joints in the face of the outcrop cut across horizontal flow unit boundaries, defined by grain-size variations and pumice concentration zones.

A paradox for volcanologists is that many ignimbrites show little or no sign of welding, even though they are extremely thick. Why are not *all*

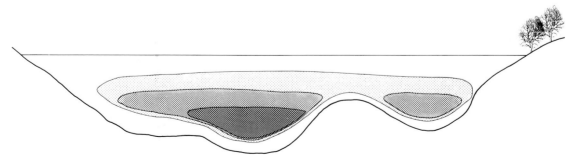

Fig. 10.31 Variations in degree of welding in an ignimbrite, seen in cross-section. Density of shading indicates degree of welding. Material in contact with ground and at top of deposit is unwelded. Dense, glassy welding is usually found only in the thickest part of deposit.

thick ignimbrites welded? There is no simple answer to this, but column height and eruption rate are probably involved: an ignimbrite that is deposited from a high eruption column will probably have more time to cool during collapse and transport than one deposited from a low column.

The most densely welded ignimbrites consist of nothing but glass and crystals. More commonly, it is only the pumice clasts that are conspicuously glassy: instead of pale-coloured frothy lumps, they are squashed into black, glassy flattened pancakes. These, seen in cross-section on a broken piece of ignimbrite, look a little like candle flames, and are known by their Italian name of *fiamme*. (Fig. 10.32). Similar textures can be seen on a microscopic scale, with tiny glass shards flattened and moulded over one another. This *eutaxitic* texture cannot be produced in other rocks, so it is a valuable criterion for identifying ancient ignimbrites, even those hundreds of millions of years old (Fig. 10.33). Extreme degrees of flattening can be produced by squashing of shards at the base of thick ignimbrites, whereas higher up, where pressure is less, the flattening is proportionally less.

Occasionally, densely welded ignimbrites remain thick and hot enough to flow downslope

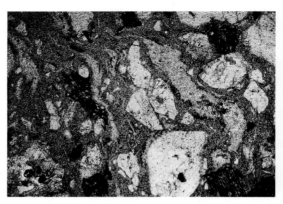

Fig. 10.33 Photomicrograph of eutaxitic texture in a densely welded but devitrified ignimbrite; the 20-million-year-old Kari Kari ignimbrite; Bolivia. *Fiamme* are about 1 cm across.

to form *rheomorphic ignimbrites*. These can be exceedingly difficult to distinguish from lavas, because *fiamme* are obliterated during flow. But in transitional areas, extraordinarily attenuated *fiamme* can be seen, drawn out like toffee strands as the hot glassy material oozed downslope. Rheomorphic ignimbrites are most common in silicic rocks of alkalic compositions, since these glasses have lower viscosities.

10.5.1 Vapour phase alteration and devitrification in ignimbrites

As the cooks in Grigg's party observed in the Valley of Ten Thousand Smokes, a cooling ignimbrite is an aggressive chemical environment. Apart from welding, other manifestations of the thermal energy of an ignimbrite are *vapour phase alteration* and *devitrification*. In essence, these processes take place as the ignimbrite cooks in its own juices. Vapour phase crystallization, as its name implies, results from intensely hot gases passing up through the body of the ignimbrite. Some of the gas may be juvenile, exsolved from pumice, and some may be from heated groundwater. Quite significant lithological changes result. Large pumice clasts loose their familiar spongy texture and become greyish masses of crumbly material, while the matrix becomes a hard, compact mass, often brilliantly white. It is difficult to see more than this with the naked eye, but microscopic and X-ray analyses show that

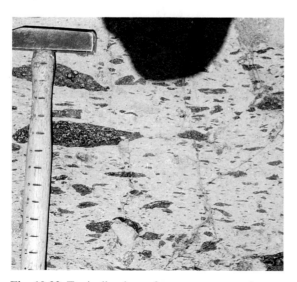

Fig. 10.32 Typically glassy *fiamme*, compressed relics of original pumice clasts, in an ignimbrite. Photo: courtesy of I. G. Gass.

Fig. 10.34 Working face of an quarry developed in an ignimbrite, Naples, Italy. The 10 000-year-old Neapolitan Yellow Tuff is an economical and widely used building material. Rock saws slice rapidly into the face to produce slabs of uniform thickness convenient for building.

Fig. 10.35 Portico of a public building in Arequipa, Peru, built of dazzling white *sillar*. A volcano forms the centre of the city's coat of arms.

the pumice has recrystallized. Tiny new crystals of tridymite, cristobalite (high-temperature varieties of quartz), and alkali feldspar fill many of the void spaces in pumice and matrix that existed previously. This late-stage alteration and crystallization acts as a cement which binds previously loose material together, forming a compact, lightweight rock.

In many parts of the world, such rocks form outstanding building materials not only because of their light weight and excellent insulating properties, but also because although light and strong, they are soft enough to be easily sawn into useful blocks (Fig. 10.34). A Peruvian term, *sillar* is often used to describe these exemplary rocks. Arequipa, second largest city in Peru, is constructed largely of a crisp white sillar, giving the city a pleasing purity and creating many cool but light shady spaces (Fig. 10.35). Mineralogically, *sillars* are complex rocks that have been little studied in detail, but they are distinctive because they are among the most musical of rocks. They ring with a pleasant bonging tone when lightly hammered. Those volcanologists who are not tone deaf can play a modest tune on a series of blocks of varying sizes.

Devitrification is another consequence of the slow cooling of thick ignimbrites. It is usually seen in densely welded glassy ignimbrites, where the glass becomes replaced by microcrystalline cristobalite and alkali feldspar. An identical process takes place in other glassy rocks such as obsidian lavas. Devitrification is almost inevitable as glasses age, so it is commonly observed in ignimbrites.

Notes

1. Wright, J. V., Smith, A. L., and Self, S. (1980). A working terminology of pyroclastic deposits. *J. Volcanol. Geotherm. Res.* **8**, 315–36.

2. Wilson, L., Sparks, R. S. J. and Walker, G. P. L. (1980). Explosive volcanic eruptions, IV. The control of magma properties and conduit geometry on eruption column behaviour. *Geophys. J. Royal Astron. Soc.* **63**, 117–48.

3. Sparks, R. S. J. and Wilson, L. (1976). A model for the formation of ignimbrite by gravitational column collapse. *J. Geol. Soc. Lond.* **132**, 441–51.

4. Sparks, R. S. J., Wilson, L., and Hulme, G. (1978). Theoretical modelling of the generation, movement and emplacement of pyroclastic flows by column collapse. *J. Geophys. Res.* **83**, 1727–39.

5. Miller, T. P. and Smith, R. L. (1977). Spectacular mobility of ash flows around Anuachak and Fisher calderas, Alaska. *Geology*, **5**, 173–6.

6. Wilson, C. J. N. (1985). The Taupo eruption, New Zealand II. The Taupo ignimbrite. *Phil. Trans. Roy. Soc. Lond.* A **314**, 229–310.

7. Sparks, R. S. J. (1976). Grain size variations in ignimbrites and implications for the transport of pyroclastic flows. *Sedimentology* **23**, 147–88.

8. Wilson, C. J. N. (1984). The role of fluidization in the emplacement of pyroclastic flows, 2: Experimental results and their interpretation. *J. Volcanol. Geotherm. Res.* **20**, 55–84.

9. Wright, J. V. and Walker, G. P. L. (1981). Eruption, transport and deposition of ignimbrite: a case study from Mexico, *J. Volcanol. Geotherm. Res.* **9**, 111–31.

10. Marshall, P. (1935). Acid rocks of the Taupo–Rotorua volcanic district. *Trans. Roy. Soc. New Zealand*, **64**, 323–66.

11. Ross, G. S. and Smith, R. L. (1961) Ash flow tuffs, their origin, geological relations and identification. *US Geological Surv. Prof. Pap.* **366**, 1–77.

12. Griggs, R. F. (1922). *The Valley of Ten Thousand Smokes*. National Geographic Society. Washington, DC, 340 pp.

13. Hildreth, W. (1983). The compositionally zoned eruption of 1912 in the Valley of Ten Thousand Smokes, Katmai National Park, Alaska. *J. Volcanol. Geotherm. Res.* **18**, 1–56.

14. Wilson, C. J. N. and Walker, G. P. L. (1985). The Taupo eruption, New Zealand, (I). General aspects. *Phil. Trans. Roy. Soc. Lond.* (A) **314**, 199–228.

15. Sparks, R. S. J., Self, S., and Walker, G. P. L. (1973). Products of ignimbrite eruptions. *Geology*, **1**, 115–18.

16. Sparks, R. S. J. and Walker, G. P. L. (1977). The significance of vitric-enriched air fall ashes associated with crystal enriched ignimbrites. *J. Volcanol. Geotherm. Res.* **2**, 329–41.

17. Druitt, T. H. and Sparks, R. S. J. (1981). A proximal ignimbrite breccia facies on Santorini, Greece. *J. Volcanol. Geotherm. Res.* **13**, 147–71.

18. Sigurdsson, H. and Carey, S. (1989). Plinian and co-ignimbrite tephra fall from the 1815 eruption of Tambora volcano. *Bull. Volcanol.* **51**, 243–70.

19. Stothers, R. B. (1984). The great Tambora eruption of 1815 and its aftermath. *Science* **224**, 1191–98.

20. Self, S., Rampino, M. R., Newton, M. S., and Wolff, J. A. (1989) Volcanological study of the great Tambora eruption of 1815. *Geology*, **12**, 659–63.

21. Smith, R. L. (1960). Ash flows. *Geol. Soc. Amer. Bull.* **71**, 795–842.

22. Smith, R. L. (1960). Zones and zonal variations in welded ash flows. *US Geol. Surv. Prof. Paper* **354-F**, 149–59.

23. Ross, G. S. and Smith, R. L. (1961). *Ash flow tuffs, their origin, geological relations and identification.* US Geol. Surv. Prof. Pap. No. 366, 1–77.

11

Pyroclastic flows: surges

Next time you see the mushroom cloud of a nuclear explosion climbing skywards (hopefully on the TV screen, and not across the street), look at the base of the mushroom's stalk. A ring of cloud will roll outwards away from it. In often-shown film sequences of a test at Bikini Atoll in 1946 the rolling cloud engulfs battleships deliberately moored near ground zero to study the effects of the explosion. Ring-shaped *base surges*, as the radially directed clouds are termed, formed within 10 seconds of the initial explosion, and moved up to 4 kilometres outwards at initial velocities of 50 metres per second, decreasing to 20 metres per second after the first 1.4 km. Certain kinds of volcanic eruptions, particularly those involving magma–water interactions, produce comparable ground-hugging surges. Volcanologists became aware of this phenomenon through the work of J. G. Moore;[1] A. C. Waters and R. V. Fisher (who studied surges both at Capelhinos in the Azores and Taal in the Philippines[2]); and H. U. Schminke, who examined the Laacher See deposits in Germany.[3] (Fisher also had the unusual experience of witnessing some of the weapons tests at first hand.)

Magnificent examples of base surges were observed during the early stages of the 1957–8 eruption of Capelhinos on the western extremity of Fayal Island in the Azores (Fig. 11.1). That eruption also became famous because a new island was constructed, which ultimately fused with the mainland, adding about a square kilometre to Fayal. In the process, it left an existing lighthouse stranded pathetically far

Fig. 11.1 A 270-m-diameter base surge rolls out radially from the base of the eruption column formed during the early stages of the 1957–8 eruption of Capelhinos, Fayal Island, Azores. (Photo: courtesy of R. V. Fisher.)

from the coast. It was, however, Moore's work on the 1965 eruption of Taal volcano, fifty-six kilometres south of Manila in the Philippines, that focused studies on base surges. Taal is a dangerous volcano, with a history of lethal eruptions stretching back to 1572, when Spanish colonists arrived. Taal volcano forms an island in a lake of the same name. Both the island and the shores of the lake are thickly populated because of population pressures, notwithstanding the known volcanic hazard. Fifty-four years after its previous outburst, which took the lives of more than 1000 people, Taal burst abruptly into activity on 28 September 1965. The eruption was as brief as it was devastating. Fortunately, a small observatory on Taal gave warning of danger when the lake-water temperature reached an abnormal level, leading to a partial evacuation of the area around the volcano. None the less, 190 people died in the ensuing eruption.

During a forty-eight hour period, powerful horizontal blasts swept out radially from the base of the eruption column, whipping across the lake on to the shore. Within a kilometre of the vent, trees were uprooted or broken off; further away they were stripped bare of leaves and the sides of their trunks facing the volcano were deeply scoured. Up to ten centimetres of wood were eroded away by the hurricane-force ash-laden surges, while the bark was unaffected on the sides facing away from the volcano. Damage extended up to about 6 km from the vent. Deposits from the surges accumulated to a thickness of over a metre around the shores of the lake. Significantly, although tremendously violent, these base surges were comparatively cool: none of the trees in the affected area showed signs of burning or charring. In the outer part of the devastated area, thick deposits of mud formed on trees and other objects, showing that the temperature could not have exceeded that of boiling water (Fig. 11.2).

Since the events at Taal in 1965, similar base surges have been observed in other eruptions, for example during the eruption of East Ukinrek in the Aleutians in 1977. Hundreds of people were killed by surges during a brief eruption of El Chichón in 1982. And innumerable deposits attributed to 'pyroclastic surges' have been recognized in the deposits of prehistoric and historic eruptions.

So what are pyroclastic surges? This brings us to a controversial area in modern volcanology. It is clear that base surges themselves form only one example of a range of related phenomena. But it is difficult to disentangle from the welter of literature exactly what that range encompasses. One might suppose the 1980 Mt. St Helens eruption to be well understood, since it was the most closely studied in history. Far from it. Some volcanologists have called the 'blast' which felled 500 square kilometres of forest a surge; others have called it a 'blast surge' and still others a pyroclastic flow akin to a nuée ardente. Part of the confusion that has arisen is almost wilful, as some protagonists studiously ignore the work of others. In attempting a simple overview, this brief chapter side-steps controversial issues, which often generate more heat than light.

Pyroclastic surges are varieties of pyroclastic flows. They have three distinctive attributes, manifested in the deposits they leave:

1. In comparison with other pyroclastic flows, they are low density, dilute phenomena. One consequence of this is that they are not so constrained by topography as denser flows such as nuées ardentes.

2. Because of their high velocity and low density, surges are turbulent, whereas most other pyroclastic flows exhibit laminar flow.

3. Because of their low density, surges have less momentum and thus travel less far.

R. V. Fisher has suggested that an important difference between surges and 'ordinary' pyroclastic flows is that surges are not fluidized in the way that flows are, and that surge deposits are the results of two transport processes. Part of the pyroclastic material forms a 'bed load' which is swept along the surface of the ground; the remainder is transported in turbulent suspension. A desert sandstorm, or *simoom*, provides a sort of analogy. Streams of sand are swept over the surface of the desert, stinging the legs of men and camels unfortunate enough to be exposed, while overhead a pall of fine dust darkens the sky, turning the sun a dull red. Of course, when one looks at actual examples of surges, one can find evidence that different mechanisms seem applicable in individual cases. . . . However, rather

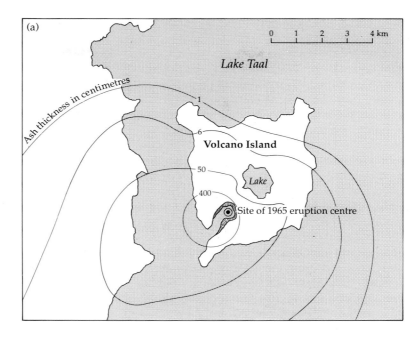

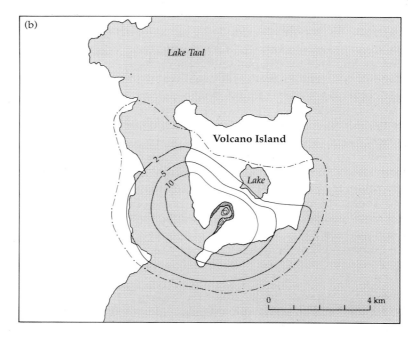

Fig. 11.2 (a) Isopach map, contoured in centimetres, showing total thickness of *all* deposits from the 1965 Taal eruption (Philippines). (b) Isopach map, contoured in centimetres showing thickness of the surge component alone of the 1965 deposits. This unusual map shows the thickness of deposits clinging to *vertical* surfaces. Dotted line marks limit of sand-blasting of objects. (After J. G. Moore, 1967).

than throwing in the towel here, let us examine some of the main types of surge. Three different kinds have been recognized, partly on the basis of their deposits, but also, importantly, on the basis of their context; that is, the nature of the eruption and the other pyroclastic deposits it yielded.

— 11.1 Base surges

Base surges are the most explicitly distinct kind
of surge. As at Taal, they are consistently
associated with hydrovolcanic explosions, and
develop from the collapse of overloaded vertical
eruption columns, in much the same way as
'ordinary' pyroclastic flows. Base surges differ,
however, in that they blast out in turbulent flow
from the base of the column. In this, they
resemble the proximal parts of high-velocity
pyroclastic flows which have not deflated to form
dense, ground-hugging laminar flows. More
importantly, many (but not all) base surges
contain a great deal of condensed steam, so they
are often wet and sticky. Thus, muddy material
newly deposited from a base surge is capable of
clinging to steeply sloping surfaces and of
slumping off them in a plastic manner. This is
quite unlike the behaviour of 'ordinary' bone-
dry, granular pyroclastic flows, or of other kinds
of surge.

Base surges are best known in small basaltic
eruptions, but they may result whenever water
and magma interact violently. When rhyodacitic
magma destroyed the Minoan settlement on the
island of Santorini in the late Bronze Age, the
first event was a conventional plinian pumice
eruption, which buried the town of Akrotiri in
air-fall ash. Immediately after this, immensely
powerful base surges ravaged the island, leaving
deposits up to 12 m thick near the vent.[4]
Interestingly, these base surges were emplaced at
temperatures of about 200–300°C, so they were
dry. Today, the deposits are exposed in some of
the most magnificently sited quarries in the
world, perched 300 m high on cliffs overlooking
the astonishingly azure waters of the Aegean.

11.1.1 Deposits of base surges

Base surge deposits result from the violent
interaction between water and magma, and so
they resemble surtseyan fall deposits. Like other
pyroclastic flows, they are poorly sorted, comprised of highly fragmented material and include
large proportions of non-juvenile lithic material.
'Bomb sags' containing large ejected blocks are
common, but such bombs presumably followed
ordinary ballistic trajectories (Fig. 11.3). Base

Fig. 11.3 Twin bomb sags in proximal surge
deposits, Koko Head, Oahu, Hawaii.
Serendipitously, two 20-cm coral clasts have
followed identical trajectories. Deposits are well
laminated; large-scale cross-bedding is also present
at the locality.

surge deposits often form much of the thickness
of the walls of tuff rings and *maars* (Ch. 17). But it
is important to stress that such thick deposits are
usually the results of innumerable separate
surges. Unlike a nuclear weapons test, a hydrovolcanic eruption may continue for weeks, each
separate explosion depositing its own surge and
fall deposits.

In bulk, base surge deposits are wedge-shaped,
thick near the vent and thinning rapidly away
from it. They rarely reach more than 5–6
kilometres from the vent. Three different facies
have been recognized. Near the vent, easily
recognizable structures are present in the form of
dunes and cross-bedding, typical of the structures found in water-lain sediments deposited in
turbulent environments (Fig. 11.4). These struc-

Fig. 11.4 Dune-bedded surge deposits exposed in a road cut, Punalica, Ecuador. Dunes have wavelengths of several metres.

tures form at right angles to the surge direction, and are asymmetrical; that is, it is possible to infer the direction of movement from them. They immediately distinguish surge deposits from the surtseyan fall deposits with which they are often intimately associated. Cross-bedded layers provide convincing evidence that surges are turbulent. Striking evidence that turbulent surges can be powerfully erosive is provided by U-shaped erosion channels carved by later surges into earlier deposits.

At intermediate distances from the vent, more massive, structureless deposits are found. Further out still, these pass into fine-grained, planar-bedded deposits which closely resemble air-fall deposits and are difficult to distinguish from them. Unlike true fall deposits, thinly bedded surges show thickness variations over topographical features, thinning over highs and thickening in lows. Accretionary lapilli are common in these beds, evidence of their origin in a wet pyroclastic environment (Fig. 11.5).

Fig. 11.5 Bed of accretionary lapilli in surge deposits, Koko Head, Oahu, Hawaii. These lapilli have cores of basaltic soria armoured with fine ash.

— 11.2 Ground surges and their deposits

Once we leave the well-defined class of base surges, we enter more difficult ground, where there are few observations of processes actually happening and much inference from old, rather equivocal deposits.

Ground surge deposits are found immediately underlying ignimbrites: they form 'Layer 1' of the basic ignimbrite unit, although they are not invariably present. A typical deposit is thin, usually much less than a metre thick, even if the overlying flow is tens of metres thick. It is finely laminated, often shows pinch-and-swell variations in thickness and sometimes shows cross-beds. Sieving reveals that the deposits have a grain size like that of coarse sand (median diameter around 2 mm), and are better sorted than other pyroclastic flows, though not so well sorted as air-fall deposits. The most characteristic feature of ground surge deposits is that they are enriched in lithics and especially in crystals. *Crystal enrichment* in some surge deposits is so pronounced that they look and feel like coarse sugar; facets of individual crystals glittering as they catch the light (Fig. 11.6).

Crystal and lithic enrichment implies density segregation—the finer, less dense pumice dust has been elutriated away. What combination of circumstances can produce this, and the distinctive bed forms of ground surges beneath an ignimbrite? Much ink (and hot air) has been spent on this topic, but the problem is not really resolved.

Ground surge deposits are intimately associ-

Fig. 11.6 Abraham Lincoln on a 1977 one cent coin provides scale for a crystal-rich pyroclastic surge deposit, which underlies the Upper Bandelier Tuff (ignimbrite), Valles caldera, New Mexico. Crystals are mostly plagioclase and quartz. Grain size is that of coarse sand.

ated with ignimbrites, even at points far distant from their vents, so it is likely that they were derived from the flow itself, rather than from the eruption column. Surge-like deposits have been found up to *100 km* from the putative vent area of the 18-million-year-old Peach Springs Tuff, which is exposed over huge tracts of California, Nevada, Arizona, and Colorado.[5] A widely held view is that such surge deposits form at or in front of the advancing head of the flow. Eyewitness observations of some pyroclastic flows speak of forward 'jetting' of ash clouds ahead of the main part of the flow. One account by Frank Perret

Fig. 11.7 A generic pyroclastic flow, showing usually recognized components of head, body, and tail. Head is most highly fluidized and turbulent; body moves mostly in laminar mode. Most deposition takes place from body, not head. Here, forward-springing jets are shown, which may be involved in formation of some ground surge deposits.

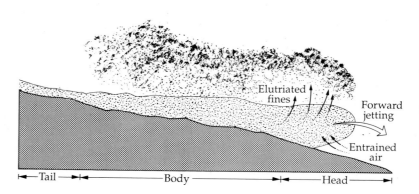

(*vide* Section 12.3) describes 'a flowing mass of incandescent material advancing with an indescribably curious rolling and puffing movement, which at the immediate front takes the form of forward-spring jets, suggesting charging lions'.

These 'jets' may be caused by the advancing flow ingesting air as it rushes forward. Strong fluidization of the head of the flow may result, and explosive heating of the air may cause some material to be hurled forwards as a low-density turbulent surge. Almost immediately, the surge deposits are overrun by the denser main body of the flow, moving in a more sedate (but still catastrophic) laminar flow regime (Fig. 11.7).

— 11.3 Ash cloud surges and their deposits

Ash cloud surge deposits are physically similar to ground surge deposits—they are finely laminated, sometimes cross-bedded, and rich in crystals and lithics. They differ only in their relationship to denser pyroclastic flow deposits such as ignimbrites and nuées: instead of being found below ignimbrites, ash cloud surge deposits may be found *within* them, on *top* of them, and as their *lateral equivalents*. Important ash cloud surge deposits are also found associated with air-fall deposits, as at Pompeii.

Lens-like surge layers are often found within large ignimbrites where there is clear evidence that the flow had encountered a topographic obstacle—flowed over a cliff, for example, or even a steep slope. Large-scale examples of this were found within the Mt. St Helens pyroclastic flow deposits, high on the flanks of the volcano. Such deposits probably resulted from a change from laminar to turbulent flow regimes as the flow speeded up on encountering a steep descent, perhaps ingesting additional air in the process. A similar situation exists where surges appear to be lateral equivalents of denser pyroclastic flows. In large flows, the dense, main parts are constrained to valley bottoms. But around their edges, the density is less, and the 'ash cloud' has a sufficiently low concentration of solids to sweep in a turbulent flow regime over the topography. In this case, as with ground surges, the dense pyroclastic flow, not the surge itself, is the primary phenomenon. But whatever the case, if one could see an actual 'ash cloud surge' being emplaced, it would probably look exactly like a nuée ardente. Only the deposits are different.

Surge deposits found overlying ignimbrites, or associated with fall deposits, can best be interpreted in terms of variations in stability of convecting eruption column regimes. It was shown in Section 10.2.1 that eruption column behaviour has two regimes; one of convection, forming plinian ash falls, and the other of collapse, forming pyroclastic flows. Surges may form when the condition of the eruption column is close to the envelope separating the two regimes. Such a column would be unstable, liable to switch abruptly from one regime to the other. Ash-cloud surges might result from abortive switches; brief excursions from straightforward convection into the regime of pyroclastic flows. This is probably what took place at Pompeii early in the morning of 25 August AD 79.

11.3.1 Pyroclastic surges and flows of the Vesuvius AD 79 eruption

We left Pompeii in Section 9.7.2 in the cimmerian darkness of the night of 24/25 August. At that time, most buildings had been wrecked when their roofs had caved in under the weight of falling ash, but most of the inhabitants who had not fled were probably still alive. As we pick up the chronology of the eruption established by Sigurdsson and his colleagues, recall that the timings are derived both from calculations of the rate of ash accumulation, and from clues in Pliny the Younger's written account of the fate of his uncle.

Ash had been raining down from the stably convecting eruption column for almost twelve hours when the first brief perturbation caused an ash cloud surge to sweep down over the south and west flanks of the volcano. This small surge (S1) travelled less than eight kilometres: it did not reach Pompeii, but it did reach Herculaneum, wreaking terrible havoc there (Fig. 11.8). Travelling at about 30 metres per second, it rushed

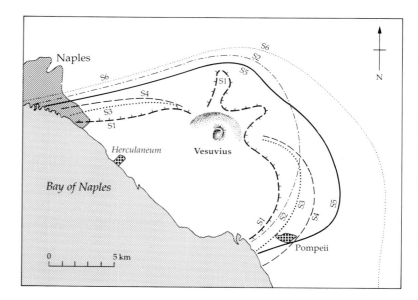

Fig. 11.8 Approximate limits reached by the six surges formed during the AD 79 eruption of Vesuvius. Surges reached into area now occupied by modern Naples. (After Sigurdsson, H., Carey, S., Cornell, W., and Pescatore, T. (1985). The eruption of Vesuvius in AD 79. *Nat. Geog. Res.* **1**, 332–87, 1985.)

down from the volcano in less than four minutes, blasting through the town with sufficient force to topple the colonnade of a public building, and to transport heavy masonary several metres. Temperatures in the surge cloud were high enough to carbonize dry timber, but not green wood.

Herculaneum lay upwind of the volcano, and therefore little ash had fallen in the town, but the residents can have had little doubt of the magnitude of the eruption taking place only a few kilometres away. Many attempted to escape, congregating on the beach and sheltering there in some arched chambers used for storing boats and other fishing gear. It was dark, so they cannot have seen the surge that swept over them out of the night, but they did have time to huddle together for mutual comfort before their lives were snuffed out. Some were probably killed by flying debris propelled by the surge. Most were asphyxiated. Their remains tell a poignant human story (Fig. 11.9).

A small pyroclastic flow followed soon after the surge, but was topographically confined to a valley south of the town, doing relatively little damage. Although of little consequence in itself, this flow underlines the intimate relationships between pyroclastic falls, flows, and surges. The lethal surge was triggered by temporary instability of the eruption column, perhaps a momen-

Fig. 11.9 Skeletons of some of the victims of the Vesuvius AD 79 surges, found in arched chambers on the original sea-front at Herculaneum. The individuals in the foreground were preserved in their last embrace. Photo: courtesy of S. Carey, University of Rhode Island.

tary overloading, which generated a small, dense pyroclastic flow. Stable convection resumed after this brief interruption, and ash fall continued for about an hour. At about 2 a.m. in the morning of the 25th, a second surge (S2) was triggered. This was three times bigger than the first, and travelled much more quickly, perhaps 100–200 metres per second. It caused severe damage in Herculaneum, demolishing buildings left standing by the first surge, but did not reach quite as far as Pompeii. A second pyroclastic flow was also generated, but went mostly north of Herculaneum.

Ash fall continued for several hours after the second surge. Erosion and widening of the vent was taking place during this time, expressed in the increasing proportion of lithics in the fall deposit. Widening of the vent, it will be recalled, is one of the factors contributing to instability of the eruption column. At about 6.30 a.m., a third surge (S3) was triggered, which also failed to reach Pompeii. A major pyroclastic flow followed, which entered Herculaneum and completely buried it. Only the remains of the theatre protruded above the pumice.

An hour later, at about 7.30, a fourth surge (S4) swept down from Vesuvius (Fig. 11.10). This one did reach Pompeii, sweeping over the northern wall of a city already buried under 2.4 metres of pumice. Many people were apparently still moving around on the surface of the fall deposit, perhaps searching for friends, or perhaps

Fig. 11.10 Plinian fall and pyroclastic surge deposits from the AD 79 eruption, exposed at an archaeological site at Boscoreale. Plinian deposits form the thick layers behind ladder, while surge horizons form thin horizontal benches. In the foreground is a plaster cast of a tree buried by the eruption. Photo: courtesy of S. Carey, University of Rhode Island.

merely trying to survive, when the surge struck. Two thousand people, or about 10 per cent of the city's population, may have died in the few moments it took for the surge to pass. Their remains were preserved as hollow moulds in the surge deposit (Chapter 4). A fifth and larger surge (S5) followed a few minutes later, and a subsequent pyroclastic flow covered many areas on the southern flanks of the volcano.

At approximately 8 a.m. on the 25 August, the sixth and largest surge (S6) swept down from the volcano. Pliny the Elder was in Stabiae on that morning, pinned there by unfavourable winds which prevented him from sailing away. Although two metres of ash had accumulated there overnight, he and his companions were in relatively good shape up until that time. Two centimetres of surge deposit overlie the air-fall ash in Stabiae, not much in itself, but clear evidence of an unpleasant episode. Sigurdsson and his colleagues read this line in Pliny the Younger's account as a direct record of the passage of the edge of the surge: 'Then the flames and smell of sulphur which gave warning of the approaching fire drove the others to take flight'.

Pliny the Elder was not a fit man, much overweight. His companions survived, so it seems probable that he died of heart failure in the choking, dusty surge cloud.

In his excellent book on Vesuvius,[6] John Phillips wrote in 1869 that the reference quoted above was an instance of Pliny the Younger's account being 'defective', because 'flames and sulphurous vapours could hardly be actually present at Stabiae, ten miles from the centre of the eruption'. Phillips, of course, knew nothing about the horizontal movement of surges. It is reassuring to know that Pliny's writing should be borne out, so long after the fact.

On the other side of the Bay of Naples, that reliable young man and his mother may also have seen a surge cloud spilling outwards from the volcano 30 km distant and spreading implacably over the water:

'. . . soon afterwards the cloud seemed to descend, and cover the whole ocean; as it certainly did the island of Capri and the promontory of Misenum . . . the ashes now began to fall upon us, though in no great quantity.

I turned my head, and observed behind us a thick smoke, which came *rolling after us like a torrent. . . .'* (My italics).

Although no surge deposit is preserved at the present day at Misenum, Pliny the Younger's account hints strongly that he and his mother were enveloped in the outer fringes of a powerful surge that had spread out from Vesuvius.

11.4 El Chichón 1982—a case study in pyroclastic surges

Between 29 March and 4 April 1982, El Chichón in southern Mexico was transformed from obscurity to one of the most renowned volcanoes in the annals of volcanology. Prior to 1982, only one geologist, a certain Dr Mullerried, had visited it. He made a reconnaissance study in 1930. In his paper, he recorded pointedly that he had not received full field expenses for his work. Nothing has changed . . .

El Chichón's 1982 eruption is best known for its atmospheric effects (Chapter 17), but it also produced the most lethal pyroclastic surges in modern history. Nine villages were destroyed and 2000 people were killed. Sigurdsson *et al.*'s 1987 work on the deposits left by these tragic but instructive surges is summarized here.[7]

11.4.1 Sequence of events

An unexpected plinian eruption lasting six hours broke out on El Chichón on 29 March. It blasted a new vent in the prehistoric summit and deposited 1.4 cubic kilometres of air-fall tephra. During this plinian phase the mass eruption rate was estimated to be 3.5×10^7 kilograms per second, almost twice as great as that of the plinian phase of the Mt. St Helens 18 May eruption. After five days of inconsequential explosive activity, a second major eruptive phase commenced on 4 April. It lasted four and a half hours. Early in this phase, a devastating surge was generated, sweeping radially outwards, but extending mostly southwards, reaching 8 km from the crater (Fig. 11.11). A plinian eruption column developed immediately *after* the surge event. This second plinian eruption column reached 24 km, and had a very high eruption rate, about 6×10^7 kilograms per second. It was followed by a second major surge, which spread radially from the crater to cover about 100 square kilometres. A few hours later, a third major plinian eruption column was generated, driven by a mass eruption rate of about 4×10^7 kilograms per second to a height of 22 km, and sustained for about seven hours. A third small surge terminated the eruption sequence, covering an area of only about 40 square kilometres.

During the whole eruption, only about 1 cubic kilometre of magmatic material was erupted, in the form of trachyandesitic tephra. Of this only about 12 per cent went into the surges, the remainder forming air-fall tephra. Although subordinate in volume to the air-fall deposits, it was the pyroclastic surges that took the lives of all the victims in the surrounding villages.

11.4.2 Deposits

Both the major El Chichón surges were powerfully erosive near the vent. The first scoured away air-fall ash deposits from the preceding plinian phase, and sand-blasted tree stumps protruding above ground. Within 3 km of the crater, the first surge was highly erosive on ridges and scarps, in places cutting right down through soil into bedrock, while the second surge may have eroded away up to 50 per cent of the underlying pumice deposits. Bending over of the stumps of trees and bushes in the downrange direction demonstrated the hurricane-like violence of the surges.

Both surge deposits consist of pumice with variable proportions of lithic clasts; the lowermost part of the first surge also contains abundant charcoal from carbonized vegetation ripped up during its passage. In contrast to the muddy surge deposits of Taal and elsewhere, presence of the charcoal demonstrates that the temperature within the surge was too high for steam to condense to water.

Most volcanologists recognize surge deposits primarily through their dune-bedded characteristics. 'Sand' waves were outstandingly developed at El Chichón. They provide valuable insights into the way that material in surges is trans-

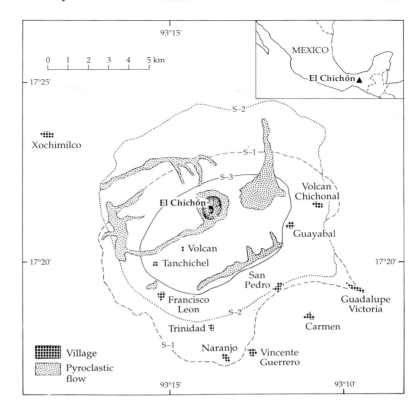

Fig. 11.11 Locality map for El Chichón, showing extent of the three surges, and the villages affected. From Sigurdsson, H., Carey, S. N., and Fisher, R. V. (1987). The 1982 eruption of El Chichón volcano, Mexico (3). Physical properties of pyroclastic surges. *Bull. Volcanol.* **49**, 467–88.

ported. Like sand dunes, 'sand waves' have two slopes, a *stoss* side, facing up stream, and a *lee* side, facing downstream. In the Chichón surges, the original stoss slopes were often eroded away, but individual laminae on the lee slopes could often be traced for many metres. At any one locality, successive dune crests migrate either uprange or downrange, forming patterns of *climbing dunes*. Traced away from the vent, the wavelength of the sand waves steadily decreased, ranging from tens of metres near the vent to only a few metres at a range of 6 km. As in other surge deposits, there was a marked loss of fine ash by winnowing. Up to 25 per cent of the original mass of the surge cloud may have been lost by the time it came to rest.

11.4.3 *Flow mechanisms*

Theories for the emplacement of the Chichón surges have to account for a number of pieces of observational evidence:

(1) the deposits are dune bedded, with stoss side erosion and lee side deposition;

(2) the lee side laminae are reversely graded and show rounding of pumices;

(3) the surges were hot enough to carbonize wood;

(4) in some places the surge deposits show transitions into denser pyroclastic flow deposits—ignimbrites.

Sigurdsson *et al.* concluded that the Chichón surges were not fluidized flows such as those which deposit ignimbrites. They argued that in flows composed of pumiceous materials such as those of El Chichón, elutriation of materials from the flow during fluidization at the necessary high gas-flow rates would have led to a higher degree of sorting than that actually observed. And because the pumices in the surge deposits show a high degree of rounding, they must have been carried along over the ground surface by a highly expanded turbulent mixed flow either in a *traction carpet* in which particles roll or slide along, or by *saltation*, (jumping, bounding, and bouncing along), with the pumice clasts bumping and grinding against each other often enough to round off their corners.

Within a highly turbulent flow, individual particles may whizz around at velocities much higher than the mean velocity at which the flow advances over the surface. This is an expression of the energy content of the flow, which is ultimately of thermal origin. Even in the most distal parts of the Chichón surges, where the deposits are only a few centimetres thick, temperatures were high enough to carbonize vegetation. Thus, there was plenty of thermal energy available from the hot pumice clasts to heat entrained air and drive violent turbulent eddies. As in the case of the giant cloud of Mt. St Helens, heating and entrainment of air, coupled with sedimentation of large clasts from the lower part of the flow, led to a reduction of density of the flow. At El Chichón, the surges reached this critical density at the margins of the devastated area, where they 'lifted off' to form a buoyant cloud from which fine grained air-fall ash sedimented. This fine ash-fall was volumetrically insignificant at El Chichón, and has not been studied.

11.5 Summary

Given all the controversy over surges, it is worth asking again: what are pyroclastic surges?

Compared with the less contentious kinds of pyroclastic flows that deposit ignimbrites, surges are less dense, more highly expanded; travel faster but less far, are turbulent rather than laminar, are not confined by topography, and as a consequence form thin but extensive deposits. Surge *deposits* are generally small in volume, show evidence of fines depletion by elutriation, exhibit cross-bedded 'sand wave' or dune structures, and rarely extend more than a few kilometres from the vent. While it seems logical that there ought to be a continuum between dense pyroclastic flows and less dense surges, in fact there seems to be a fairly marked polarization between the two—there are not many transitional varieties.

Surges like those of El Chichón consist of two distinct regimes: an upper layer and a lower boundary layer. Swirling, turbulent flow predominates in the upper layer, which forms much the largest part of the surge at any one time. Material constantly falls out from it into the boundary layer, which is characterized by shearing and strong vertical velocity and particle concentration gradients. Particles move in the boundary layer by sliding, rolling, and traction during the later stages of the passage of the surge—in the earliest stages, turbulence is so extreme that erosion rather than deposition takes place. Bed-load transport in the boundary layer led to the formation of the prominent 'sand' waves at El Chichón. Ribbons of pyroclastic material streamed over the ground surface in an essentially non-turbulent manner, as in violent sand storms. During growth of the 'sand' waves, turbulence on the upstream (stoss) side of growing dunes led to stoss-side erosion.

Notes

1. Moore, J. G. (1967). Base surge in recent volcanic eruptions. *Bull. Volcanol.* **30**, 337–63.
2. Waters, A. C. and Fisher, R. V. (1971). Base surges and their deposits: Capelhinos and Taal volcanoes. *J. Geophys. Res.* **76**, 5595–5614.
3. Fisher, R. V. and Schminke, H. U. (1984). *Pyroclastic rocks.* Springer-Verlag, Berlin, 472 pp.
4. Bond, A. and Sparks, R. S. J. (1976). The Minoan eruption of Santorini, Greece. *J. Geol. Soc. Lond.* **132**, 1–16.
5. Valentine, G. A., Buesch, D. C., and Fisher, R. V. (1989). Basal layered deposits of the Peach Springs Tuff, northwestern Arizona, U.S.A. *Bull. Volcanol.* **51**, 395–414.
6. Phillips, J. (1869). *Vesuvius.* Clarendon Press, Oxford, 355 pp.
7. Sigurdsson, H., Carey, S. N., and Fisher, R. V. (1987). The 1982 eruption of El Chichon volcano, Mexico (3). Physical properties of pyroclastic surges. *Bull. Volcanol.* **49**, 467–88.

12

Pyroclastic flows: *nuées ardentes*

Nuées ardentes are the most infamous of all volcanic phenomena—it was a *nuée* that eliminated the population of St Pierre at a stroke on 8 May 1902. In its earliest usage, the term *nuée ardente* (literally 'glowing cloud') may have been applied to an ordinary cloud, lit from below by the ruddy glow of a lava flow; an entirely different phenomenon from what we consider here. Unfortunately the term was applied to the pyroclastic flows from Mt. Pelée, captured the imagination of geologists, and has been used confusingly ever since. 'Glowing avalanche' would be a more precise term, but has never gained wide acceptance. As defined in Section 10.1, a nuée ardente is a block-and-ash flow; a pyroclastic flow whose magmatic component is dense rock, in contrast to the vesiculated pumice that forms ignimbrites. Poorly vesiculated andesitic and dacitic magmas usually provide the juvenile material, but there are also some scoriaceous basaltic varieties.

— 12.1 *Nuées ardentes:* glowing avalanches ——

12.1.1 Mt. Pelée 9 July 1902

The year 1902 was a remarkable one for volcanology. On 8 May, St Pierre in Martinique was wiped out by a nuée from Mt. Pelée. On the *previous day*, 7 May, a similar eruption of the Soufrière volcano had ravaged the island of St Vincent, 150 km distant, causing about 2000 deaths. As St Vincent was a British colony, the Royal Society in London sent a Scientific Commission there to study the eruption, consisting of a distinguished traveller, Tempest Anderson, and a geologist, John Smith Flett. After their work on St Vincent, they visited St Pierre to make a comparative study. While there, they witnessed the eruption of a nuée from Mt. Pelée on 9 July 1902. There is a vivid description of it in their monograph on Soufrière and Mt. Pelée, one of the most riveting of all the great volcanological texts.[1]

Their descriptions of two nuées closely spaced in time illuminate many later events. They were in a small boat, moored about 3 km south of St Pierre when:

In the rapidly falling twilight we sat on deck watching the activity of the volcano, and calculating the chances of an ascent next morning, when our attention was suddenly attracted to a cloud which was not exactly like any of the steam 'cauliflowers' we had hitherto seen. It was globular, with a bulging, nodular surface; at first glance not unlike an ordinary steam jet, but darker in colour, being dark slate, approaching black . . . for a little time we stood watching it, and slowly we realised that the cloud was not at rest, but was rolling straight down the hill, gradually increasing in size as it came nearer and nearer . . . it seemed that the farther the cloud travelled the faster it came, and when we took our eyes off it for a second and then looked back it was nearer and still nearer than before. There was no room for doubt any longer. It was a 'black cloud', a dust cloud, and was making directly for us. So with one accord we prepared to get out of its path. We helped the sailors to raise the anchor and, setting the head sails, we slipped away before the wind. By the time the mainsail was hoisted we had time to look back, but now there was a startling change. The cloud had cleared the slopes of the hill. It was immensely larger, but still rounded, globular, with boiling, pillowy surface, pitch black, and through it little streaks of lightning scintillated. It had now reached the north side of the bay, and along the base, where the black mass rested on the water, there was a line of sparkling lightnings that played incessantly.

This appears to have been a fairly small nuée. About thirty minutes later, however:

. . . a red hot avalanche arose from the cleft in the hillside, and poured over the mountain slopes right down to the sea. It was dull red, and in it were brighter streaks, which we thought were large stones, as they seemed to give off tails of yellow sparks. They bowled along, apparently rebounding where they struck the surface of the ground, but never rising high in the air. The main mass of the avalanche was darker red, and its surface was billowy like a cascade in a mountain brook. Its velocity was tremendous. . . . Its similarity to an Alpine snow avalanche was complete in all respects, except the temperature of the respective masses. The red glow faded in a minute or two, and in its place we now saw, rushing forward over the sea, a great rounded, boiling cloud, black, and filled with lightnings . . . it was a fear inspiring sight, coming straight over the water directly for us, where we lay with the sail flapping as the boat rolled gently on the waves of the sea.

The cloud was black, dense, solid, and opaque, absolutely impenetrable, like a mass of ink, . . . covered with innumerable minor excrescences, rounded, and filled with terrific energy. They shot out, swelled and multiplied until the whole surface seemed boiling; one had hardly time to form before another sprung up at its side . . . their effect was that the cloud drove onward without expanding laterally to any great extent.

The display of lightning in the cloud was marvellous. In rapid flashes, so short that they often seemed mere points, and in larger, branching crooked lines, it continually flickered and scintillated through the whole vast mass. It was often greenish, perhaps when seen through some slight depth of the dust cloud, at other times yellowish, and always rapid, short lived . . .

Nearer and nearer it came to where our little boat lay becalmed, right in the path of its murderous violence. We sat and gazed, mute with astonishment and wonder, overwhelmed by the magnificence of the spectacle, which we had heard so much about, and had never hoped to see. In our minds there was little room for terror, so absorbed were we in the terrible grandeur of the scene. But our sailors were in a frenzy of fear, they seized the oars and rowed for their lives, howling with dread every time they looked over their shoulders at the rushing cloud behind us. Their exertions did little good, as the boat was too heavy to row, and fear gave way to despair.

Fortunately for them, and for posterity, the nuée was running out of steam, and although it did later drift over their boat, its energy was spent, and they were unharmed.

12.1.2   Three types of nuée ardente

Three different mechanisms for generating nuées are often recognized (Fig. 12.1). *Merapi*-type nuées, named after the Indonesian volcano they characterize, are associated with gravitational collapse of lava flows and domes: *Peléean* type nuées are triggered by explosive events on growing lava domes, whereas *Soufrière* type are not related to lava domes at all, but to eruption column collapse. Needless to say, these distinctions are rather blurry.

12.1.3 Frank 1903—a cold avalanche

Sir Isaac Newtown is alleged to have said: 'If I have fallen further, it is because I was standing on the shoulders of giants'. This mischievous corruption of one of science's most felicitous aphorisms is the key to understanding the mobility of nuées ardentes. We begin by examining briefly an entirely non-volcanic phenomenon, to emphasize that all moving masses of debris have much in common, and that while volcanic glowing avalanches may be shockingly lethal, there is nothing mysterious about them.

On 29 April 1903, at 4.10 a.m. the small coal-mining town of Frank in Alberta, Canada, was overwhelmed by a rock avalanche. One-half of the town was wiped out, with severe loss of life, but many residents survived to give eyewitness accounts of what happened. Frank is situated in the valley of the Crow's Nest pass in the Rocky

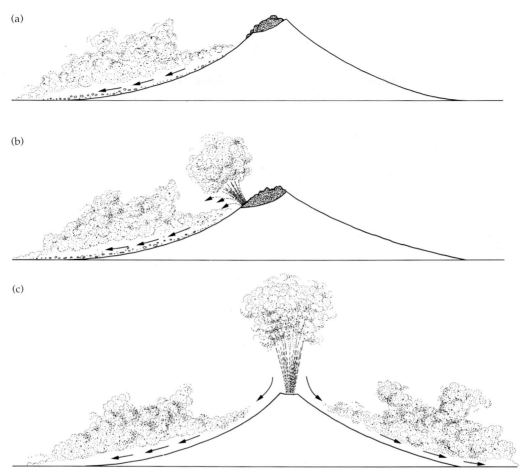

Fig. 12.1 Three common mechanisms for generating *nuées ardentes*. (a) Simple gravitational collapse of a growing lava dome or flow on a volcano (merapi type). (b) Explosive disruption of growing lava dome (peléean type). (c) Collapse from eruption column (soufrière type).

Mountains, with peaks rising on either side of it. Coal-mining in the area had made the valley sides unstable so the shock of a mining subsidence may have triggered the avalanche. A mass of limestone 150 metres thick fell away from the face of Turtle Mountain, crashed about 640 metres vertically down one side of the valley, rushed four kilometres across the flat floor of the valley, *and climbed about 140 metres up on the other side.* Two-and-a-half square kilometres of the valley floor were covered in rock debris to a depth of about twenty metres (Fig. 12.2). The entire event lasted less than 100 seconds.

Significantly, although the avalanche travelled at high speed, about 160 kilometres per hour, much of the material within it was transported without being knocked about much, so that moss and soil were found clinging to some boulders, along with unscathed tree-trunks. Press reports spoke of domestic equipment, houses, and even people being carried for tens of metres without damage, although some of these reports strain credulity. Many accounts agree, however, that the avalanche was immediately preceded by a violent air blast. This matches exactly with what happened at St Pierre in 1902. An eyewitness on a ship lying safely offshore stated that the ash cloud was *preceded* by 'a great rush of wind, which

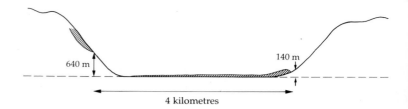

Fig. 12.2 Horizontal and vertical distances travelled by the 1903 Frank landslide, Canada. (Topography to scale, but thickness of deposit exaggerated.)

640 m

140 m

4 kilometres

immediately agitated the sea and tossed the shipping to and fro'.

Four important points emerge from descriptions of the Frank avalanche:

1. Gravity was the *only* force at work.

2. The way in which the avalanche swept downslope, travelled horizontally and then climbed upslope again shows that even in a cold, non-volcanic environment, potential energy can be converted efficiently to kinetic energy and back again, and momentum is conserved.

3. This high degree of mobility may have been helped by fluidization brought about by the air trapped within the falling mass—the violent air blast ahead of the avalanche was simply air displaced by the flow, and some must have been ingested by it. An important point here is the *extent* to which fluidization is involved. As with pumice flows, only the finer-grained material could be fluidized.

4. The evidence of minimal abrasion of components in the avalanche shows that at least some part of it travelled in a laminar mode: as a coherent 'plug' which moved over a highly sheared base.

— 12.2 Merapi-type *nuées* —

Merapi-type nuées are essentially hot avalanches; pyroclastic flows formed by gravitational collapse from lava flows and domes. Clearly, the higher a dome is located on a volcano, and the steeper the slopes, the greater the hazard. Eruptions of Mount Merapi in Indonesia in 1942–3 provided the eponymous examples of the phenomenon. Subsequently, there have been many others. Growth of the Santiaguito dome on Santa Maria volcano, Guatemala since 1922 has spawned many nuées. Most were small, but there were larger events in 1929, 1973, and 1989–90.[2]

Many volcanoes are high mountains. They are, therefore, just as prone to 'cold' avalanches as other mountains—perhaps even more prone, for reasons discussed in Chapter 13. Such 'cold' volcanic avalanches do not differ significantly from the Frank avalanche. Hot avalanches take place when gravity makes itself felt on fresh lava, whether a lava dome is present or not. In the simplest case, an ordinary basaltic lava flowing down a volcano may encounter a slope too steep for it to descend smoothly: the front part crumbles, breaks off, then cascades downslope in a shower of glowing fragments. In *aa* flows, chunks of plastic lava commonly gather themselves into *accretionary lava balls*, which roll along in the flow, getting bigger and bigger, like snowballs. On steep slopes, where the balls can pick up rolling momentum, they sometimes detach themselves completely from the flow, and roll away ahead of it like giant marbles (Fig. 12.3).

Much bigger avalanches take place when viscous lavas, usually andesites or dacites, are erupted high up on volcanoes. Since they are too sluggish to flow, they pile up to form mechanically unstable *coulées* and domes which from time to time break up and crash downslope. In some cases, explosions of volatiles within the cooling lava mass may propagate its collapse, but often

Fig. 12.3 A scatter of large accretionary lava balls, several metres in diameter, which rolled downslope from a lava flow on the flanks of Teide volcano, Tenerife, Canary Islands.

the lavas are effectively gas free, so there is no question of explosive activity triggering the collapse: gravity does it all.

Apart from being composed of hot, possibly incandescent material, 'hot' avalanches behave much like cold ones, and travel comparable distances. Several high volcanoes of the Andes of northern Chile provide enlightening examples.[3] On San Pedro volcano, a thick mass of andesites has been piling up over the last few thousand years, oozing from a vent near the 6100-m-high summit of the mountain. When the last lavas were erupted, they found themselves overhang-

ing a steep slope, which falls away in a long sweep through 3000 metres vertically and more than 20 km horizontally towards the valley of the Rio Loa. As each successive flow oozed out on to the steep slope, the front end reached a point of instability, and broke off, sending a mass of glowing lava hurtling downslope. As more lava was extruded, the process was repeated, covering the lower flanks of the volcano with a carpet of avalanche debris, in which deposits from innumerable successive individual avalanche events can be identified (Figs. 12.4–12.5).

Each avalanche deposit is bounded by a well-

Fig. 12.4 Recent eruptions of the 6100 m high San Pedro volcano, north Chile, constructed a pile of andesitic–dacitic lavas almost 2000 m high. Avalanching from unstable parts of the extrusions formed an extensive apron of hot avalanche deposits (foreground) which extends for 20 km.

Fig. 12.5 Lateral margin of upper slopes of avalanche apron, San Pedro, showing trains of large boulders. Tents give scale.

(a)

(b)

Fig. 12.6 (a) One of myriads of prismatic jointed blocks in San Pedro hot avalanche deposits. Block is *c*.3 m in diameter (penknife gives scale); (b) small-scale prismatic jointing on periphery of block.

defined snout and margins several metres high. Massive boulders are strewn over the surface, some of them so large that they resemble craggy castles. One feature positively demonstrates that these deposits were hot when emplaced: many of the boulders are almost spherical, but their outer surfaces are broken up by joint planes radiating outwards from the centre of the block, dividing it into pyramidal prismatic segments, a style of fracture known as *prismatic jointing*. Closer examination shows that the outer skin of prismatic jointed blocks is glassy, and that the joint spacing gets progressively finer outwards (Fig. 12.6). The larger joints penetrate right to the heart of the blocks, so the blocks are extremely fragile, and could not possibly have survived transport in the avalanche. Thus, the joints resulted from rapid chilling of the blocks *after* the avalanche came to rest. At present, many of San Pedro's block lie in prismatic pieces like giant

executive puzzles, having fallen apart simply as the result of weathering.

12.2.1 Unzen 1991

After a devastating 'cold' collapse in 1792 (Section 13.1.4) Unzen volcano, Japan, remained inactive until 20 May 1991, when a lava dome began growing on the Fugen-dake cone. By 23 May, the dome had reached 100 m in diameter and 44 m in height, and had just begun to emerge above the rim of the crater in which it was extruded. Material began to detach from the margins of the dome and crash down the steep slopes of the cone. Blocks up to 5 m in diameter were seen breaking off from the dome. A small hot avalanche on 24 May generated a minor block and ash flow, leading to the evacuation of 3000 people from the town of Kamikoba, only 4 km from the volcano's summit. Other flows followed on the succeeding days, some of them coming within 500 metres of Kamikoba.

Unzen's growing lava dome and the glowing avalanches that it spawned naturally attracted much attention from press and public. Millions of Japanese TV viewers saw small avalanches cascading from the growing lava dome. At 4.00 p.m. on 3 June 1991, a much larger mass of lava (about half a million cubic metres) collapsed from the dome. Incandescent ash and boulders raced 3.2 kilometres down the flanks of the volcano at speeds reported to be 100 km per hour, descending almost 1000 metres vertically before sweeping into the outskirts of Kamikoba.

Forty-two people died, many of them journalists who had ventured close to the volcano for their stories and photographs. For volcanologists, the poignant aspect of this otherwise unexceptional eruption was that it took the lives of three cherished colleagues: Maurice and Katya Krafft, French volcano film-makers, and Harry Glicken, an American expert on debris avalanches who had narrowly escaped death at Mt. St Helens ten years earlier. They died attempting to record for science the phenomenon that took their lives.

Many other block and ash flows were generated after 3 June as Unzen's dome continued to grow. Ten thousand people were evacuated from its lower slopes. Many houses were destroyed in Kamikoba. But they can be rebuilt.

12.3 Peléean-type *nuées*

Although much was learned about nuées as a result of the calamitous 1902 Mt. Pelée eruption through the work of scientists such as Anderson, Flett, and Lacroix, more was learned when Mt. Pelée reawakened in 1929. This episode provided a unique opportunity to study nuées at first hand, and, so to speak, in their ancestral home.

At this point, an intriguing and indefatigable character arrives on the scene: Frank Perret, an American electrical engineer who had worked with Edison. In 1902, he suffered a breakdown in health. When he heard about the dreadful fate of St Pierre, he decided to make a complete change and to devote his life to the study of volcanoes. He spent the next 30 years observing volcanic eruptions wherever he could—in Italy, Japan, Hawaii, Tenerife, and elsewhere. He happened to be in nearby Puerto Rico when he heard news of the 1929 eruption of Mt. Pelée, so he immediately dropped everything and rushed to Martinique, equipped with only what he stood up in, plus a home-made earth contact microphone (to detect tremors) and 'a folding pocket Kodak'. Soon after arriving, he built himself a small observatory near the volcano, and then spent the next three years observing Mt. Pelée (Fig. 12.7). He saw hundreds of nuées at close quarters. Almost too close, for on one occasion he was engulfed in

the dust clouds of a small nuée. Fortunately, he survived unscathed, and reported that even with his 'trained sense of smell' he could detect no trace of hydrogen sulphide or sulphur dioxide, although he did experience 'burning of the air passages'. Perret's observations were a milestone in studies of pyroclastic flows.

Growing lava domes were essential features of both the 1902 and 1929 eruptions of Mt. Pelée In 1902 a lava dome grew on the site of the Étang Sec on Mt. Pelée. Renewed explosive activity in this same dome produced some of the most violent of the 1929 nuées, but over the course of months, the old dome was largely demolished, or 'eviscerated' in Perret's graphic terms, by successive explosions. As the eruption continued, a new dome grew, eventually rising higher than the old. As it grew, nuées became less violent. Many of the later ones were probably simple hot avalanches, formed by gravitational collapse of unstable parts of the new dome. Almost all the earlier nuées that Perret observed were initiated by powerful explosions, directed either upwards or laterally (Fig. 12.8).

Perret considered that the fragmentary ejected material was itself highly gas charged, and that gas was continually escaping in large volumes from the solid material during the passage of a

Fig. 12.7 Mt Pelée's 1929 dome, seen from Frank Perret's observatory. Most of the nuées he observed travelled down the valley of the Rivière Blanche, the white scarred area on left.

Fig. 12.8 Four minutes after its emission on 7 January 1930, a nuée ardente rushing down the slopes of Mt. Pelée, Martinique was photographed by Frank Perret. A closely similar nuée overwhelmed St Pierre in 1902, but that one may have been even bigger, and spread further laterally. Compare with the photograph by Lacroix in 1902 (Fig. 4.21).

nuée. It was to this continuous discharge of gas that Perret attributed the peculiar mobility of nuées. This is one aspect on which modern volcanologists might disagree with Perret. Clasts in nuée ardente deposits are generaly poorly vesiculated, so there is little scope for much gas exsolution. Furthermore, one only has to stand on the top of the Mt. Pelée dome (something that Perret may never have had the chance to do) and look down the vertiginous south-west flank to St Pierre to appreciate the crucial role that gravity must have played—the slope falls clear away through 1400 m of descent. As the Frank avalanche showed, a little gravity goes a long way.

Notwithstanding this point of interpretation, Perret's descriptions of nuées are as graphic as those of Anderson and Flett:

The first thing seen but rarely heard except in the more highly explosive early outburst is a more or less obliquely advancing mass, expanding at a rate so rapid that it should, seemingly, fill the entire heavens in a moment or two, but suddenly the cloud ceases its upwards expansion, spreading out horizontally, or even downwards on the mountain slope, at the same time developing upwards in cauliflower convolutions of dust and ash. These convolutions grow out of a flowing mass of incandescent material advancing with an indescribably curious rolling and puffing movement which at the immediate front takes the form of forward springing jets, suggesting charging lions . . . the horizontal movement is . . . extraordinarily mobile and practically frictionless because each lava particle is separated from its neighbour by a cushion of compressed gas. For this reason too its onward rush is almost noiseless.[4]

Sadly, most of us will never have the opportunity to observe a real nuée at first hand. Fortunately, a realistic model can be made by stirring up the water in a clear pond with steep, muddy sides. Inevitably, clouds of sediment will come rolling up, obscuring everything. With luck, though, some convoluted clouds of sediment will roll off in long trains directly downslope, hugging the bottom, with opaque curtains of sediment-laden clouds rising above them. Similar results can be obtained using gloopy, immiscible disinfectant liquid when cleaning the toilet bowl (who said domestic chores were mundane?). These non-volcanic models are miniature replicas of a submarine geological phenomenon, known as *turbidity* or *density currents*. Owing to the huge differences in viscosity and density between water and air, they cannot be directly related to volcanic phenomena. But the principles are similar, and they are fun.

One instinctively expects any powerfully destructive force to be immensely noisy. So the silence of nuees that Perret described may be a surprise. This eerie quiet may be due to two causes. First, the only area in which any noise at all can be generated is in the extreme base of the flow, where boulders are being battered around. Second, any noise that is generated will be muffled by the envelope of dense, dust-laden gas in the lower part of the *nuée*.

12.3.1 *The problem of St Pierre*

Much confusion has arisen over the years over exactly what happened to St Pierre: why was the mortality so complete? How could the town have been so comprehensively wrecked by a volcanic phenomenon so puny that left only a few centimetres of ash behind? Even now, almost one hundred years later, there is still much controversy. One reason for the confusion was the

extent of the fires started by the nuée—it was difficult to decouple the results of the two. Another reason, often overlooked, is that more than one nuée tore through the town in 1902. Had there been anyone left alive, the nuée which struck St Pierre on 20 May might have proved just as lethal as that of 8 May. Anderson and Flett observed another on 9 July, and there were others. Clearly, what the later investigators saw was the accumulated damage done by several nuées.

An eloquent reminder of the violence of the blast that ruined St Pierre is preserved in the ruins of the mental hospital there. At the beginning of the century, it was thought appropriate to confine alcoholic patients in immense steel chairs. One of the chairs can be still seen today, bent over as though it were made of plastic. Best not to think about the fate of the patient. Appalling though such reminders are, they place the violence of the nuée in perspective. Tornadoes frequently wreck houses and even devastate small towns in the central USA: these tornadoes are nothing more than spiralling blasts of ordinary air. And as many sailors will know, the katabatic winds that rush down from mountainsides as *williwaws* can also be of terrifying violence. They consist of nothing but cold, dense air.

In the extent of the damage done, and the nature of the deposit left by the nuée, the situation in St Pierre resembled the 'blast zone' of Mt. St Helens. St Pierre may indeed have been ravaged by a 'nuée ardente', but certainly not by a 'glowing avalanche' *per se*—if it had been, the town would simply have disappeared beneath a dense mass of incandescent rubble.

A nuée can be thought of as having three components: first, concealed at the base of the billowing mass of moving cloud so well described by Anderson and Flett, and responsible for it all, is the core component, the dense avalanche of fast-moving incandescent debris. This forms a ground-hugging pyroclastic torrent consisting of everything from fine dush to lava boulders more than a metre in diameter. Necessarily, this dense avalanche is channelled along topographical depressions. Next, there is a lower-density component which is the lateral equivalent of the avalanche, and is not topographically confined. This is similar to the pyroclastic surges described earlier, and the deposits are sometimes described as surge deposits. Third, there is the towering wall of cloud itself, which rises convectively many kilometres into the air, forming much the most conspicuous part of the whole affair. This is similar to the fine ash cloud winnowed from a pumice flow, and gives rise to thin deposits of fine ash, equivalent to those of ignimbrite 'layer 3' (Fig. 12.9).

One school of thought argues that the lethal 8 May nuée was of exactly this tripartite type; and that the main glowing avalanche cascaded down the valley of the Rivière Blanche, where there is a thick accumulation of coarse nuée deposits. (They are coarse breccias forming layers up to ten metres thick, extremely poorly sorted, internally unbedded, contain blocks up to 5 m in diameter, and are true block and ash deposits.) Such dense material clearly did *not* pass through St Pierre.

But the nuée was big enough that its surge-like lateral margins rolled over St Pierre, depositing some fine ash in what has sometimes been loosely (and confusingly) called an *ash cloud hurricane.*

Unfortunately, it is not easy to make stratigraphical correlations between the thin deposits in St Pierre, and the thick ones in the Rivière Blanche. Rampant growth of tropical vegetation since 1902 (and 1929) has meant that there are few good natural exposures, and most studies have been made in excavations, such as one in the old Fort cemetery. For those interested in details, a complete volume of the *Journal of Volcanology and Geothermal Research* was devoted to Mt. Pelée,[5] and Fisher *et al.* have also discussed the various facies of the deposits.[6]

Impressed by the thinness of the deposits in St Pierre, another school of thought argues that there was never any glowing avalanche on 8 May. They suggest that the phenomenon was actually a 'blast' similar to that of Mt. St Helens

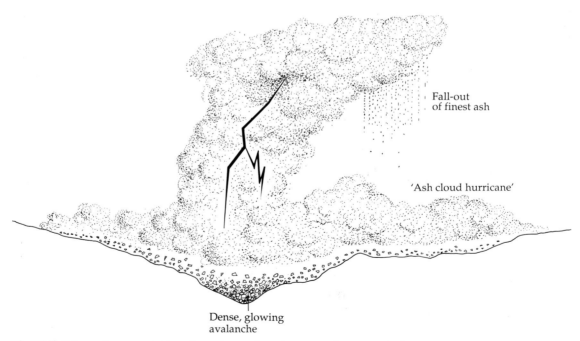

Fall-out of finest ash

'Ash cloud hurricane'

Dense, glowing avalanche

Fig. 12.9 Schematic cross-section through a nuée ardente, coming towards the viewer. Densest part is a true glowing avalanche, confined to valley bottom, in which largest clasts may be metres in diameter. Marginal parts are less dense, spread over topography; may be turbulent, and deposit thinner, finer materials. Convecting rising cloud of fine dust may reach several kilometres, and is the most visible component, although it gives rise only to very fine ash fall-out. Formidable lightning displays may occur.

18 May, namely a thin, horizontally extensive, exceptionally fast moving, ground-hugging pyroclastic flow. We may never know for sure.

12.3.2 Mt. Lamington, Papua New Guinea, 1951

G. A. M. Taylor's work on Mt. Lamington combined valour sustained in the face of serious hazard over a period of several months, with meticulous scientific observations. For it, he was awarded the George Cross, Britain's highest decoration for civilian gallantry. His study, another milestone in volcanology, is essential reading, not only for its intrinsic importance, but because of the light it casts on the Mt. St Helens eruption.[7] Comparison with Mt. St Helens is striking in another sense: Taylor's study was a *tour de force* by a single individual, whereas dozens of scientists of the United States Geological Survey joined forces in the investigations of Mt. St Helens and publication of the resulting volume.

In his first paragraph, Taylor stressed the similarity between Mt. Lamington and Mt. Pelée:

When Mt. Lamington erupted in 1951 a long dormant volcano sprang suddenly into life and produced a paroxysmal outburst of disastrous proportions. Its resemblance to Mont Pelée was most marked, both in the form of its cone and the pattern of its activity.

An age determination of 13 000 years on a piece of carbonized wood from an earlier nuée eruption showed that Lamington had erupted violently previously, but the local authorities did not even realize that the mountain *was* a volcano. (Parenthetically, the eminent pre-war Dutch geologist, van Bemmelen, had recognized this fact, but the knowledge was lost.)

Precursory activity to the catastrophic explosion consisted of landslides in the crater area, earthquake swarms, and emission of gas and ash which increased daily. At 10.40 on Sunday morning, 21 January 1951, a paroxysmal eruption burst from the crater, spawning a nuée ardente which devastated an area of about 174 square kilometres. Almost 3000 people died, many of them in the plantations and missions of Higaturu and Sangara on the north flank of the volcano. Subsequent powerful eruptions continued intermittently until early March, when the large lava dome which had been growing in the crater since late January was destroyed. After the March eruption, explosive activity subsided, and extrusion of the dome continued quietly; by January 1952 the dome was almost 600 metres high.

By good fortune, the climactic phase of the eruption was photographed from an aircraft on a scheduled flight from Port Moresby to Rabaul, about 40 km from the volcano. According to Captain Jacobson, a dark mass of ash shot up from the crater and rose within two minutes to about 12 km (Fig. 12.10). The base of the column expanded rapidly laterally as if 'the whole of the countryside were erupting', in a manner strikingly reminiscent of the initial giant umbrella cloud at Mt. St Helens. Prudently, Jacobson descended at speed to clear the area at 290 kilometres per hour, but was alarmed to find the mushroom head of the cloud looming *above* the aircraft twenty minutes later, at a height of about 15 km.

On the ground, the topography of the volcano was an important factor in controlling the extent of the radially expanding nuée, which was restricted to 12 km on the northern flanks and 8 km on the southern flanks (Fig. 12.11). A conspicuous effect of the passage of the nuées was abrasion: the soil surface in areas close to the crater was grooved and scoured. Buildings were destroyed where the nuée swept through inhabited areas, and people died in similar circumstances to those in St Pierre. Structures were damaged by high-velocity 'ash hurricanes' which deposited only a few centimetres of fine ash, while the denser, bouldery components of the nuées were confined to valleys. In Higaturu, about 9 km from the crater, the District Commissioner's house was pushed bodily northwards five metres and damaged on its southern side by flying debris. A jeep ended up suspended in mid-air from two truncated trees (Fig. 12.12). Plastic objects recovered from nearby houses suggested that temperatures in the nuée were of the order of 200°C and lasted only 1.5 minutes; long enough to cause dreadful burns to some victims, but not

(a)

(b)

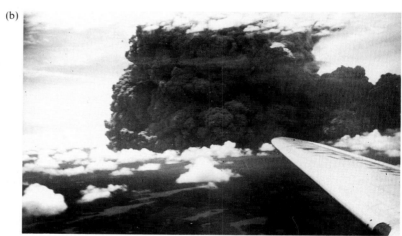

Fig. 12.10 (a) Oblique aerial photograph showing climatic eruption column from Mt. Lamington, Papua New Guinea, which generated the lethal radial nuées of 21 January 1951. (b) Same cloud a few minutes later, showing rapid lateral spread, especially on right. Photos: courtesy of W. R. Johnson, Bureau of Mineral Resources, Geology, and Geophysics, Canberra, Australia.

long enough, for example, to explode a corked bottle of highly volatile ether.

Lamington's lessons Taylor's account of Mt. Lamington is valuable for two reasons. First, while major dome growth took place after 21 January (Fig. 7.27), and while many nuées were subsequently erupted during the construction of this dome, the first, catastrophic nuée of 21 January was apparently *not* triggered by explosion within the dome, but by radial collapse from a powerful eruption column that was propelled upwards to a height of 15 km. While dome growth had commenced at Mt. Pelée prior to the 8 May 1902 catastrophe, it seems likely that there also the dome was too small to have played a

major role in the eruption of the catastropic nuée.

Second, Taylor studied carefully the literature on Mt. Pelée, particularly the work of the French geologist Alfred Lacroix, who made some of the earliest technical studies of nuées. Lacroix was impressed by the extreme violence of some of the nuées of 1902, and by the relatively mild nature of the others. He recognized these two types as having distinct origins: *nuées peléeans d'explosions dirigées*' he thought were blasted laterally like cannon shots, while '*nuées ardentes d'explosions vulcaniennes*' he considered to be formed from vertical eruption columns. Much confusion stemmed from this analysis: confusion which still prevailed at the time of the Mt. St Helens eruption and its 'laterally directed blast'.

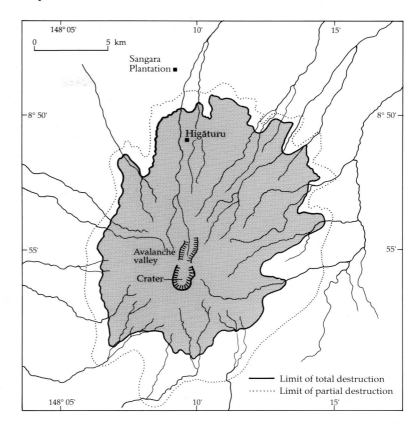

Fig. 12.11 Sketch map of Mt. Lamington volcano, showing areas affected by nuées of 21 January 1951. The pre-existing topography of the volcano concentrated most of the pyroclastic flows towards the north, along the avalanche valley, resulting in a much wider zone of devastation on the north flanks than south. After G. A. M. Taylor.

Taylor witnessed many nuées from Lamington, some at close quarters (Fig. 12.13). An important conclusion he drew from his hair-raising observations was the importance of *gravity*. While explosive eruptions—Taylor called them 'shallow pocket' eruptions—taking place in and around the base of a lava dome can have significant lateral 'blast' effects, the most important force driving 'blasts' is not explosion of expanding gas, but simply gravity. In almost

Fig. 12.12 Remains of two jeeps in the village of Higaturu, destroyed by the Lamington nuée of 21 January 1951. One is barely recognizable at the roadside; the other dangles from trees. Although extremely violent, the nuée left only a thin deposit on the road where the man is standing. Photo: courtesy of W. R. Johnson, Bureau of Mineral Resources, Geology, and Geophysics, Canberra, Australia.

(a)   (b)

(c)   (d)

Fig. 12.13 a–d Sequence of photographs taken by G. A. M. Taylor within a few minutes, showing a nuée from Mt. Lamington on 5 March 1951 rushing towards the photographer. The mountain is obscured by the ash cloud. The nuée travelled along the valley to the left of the camera position, thus sparing the photographer. The convecting cloud becomes more disorganized with height, less tightly convoluted. Some ash fall is visible in (d). Photos: courtesy of W. R. Johnson, Bureau of Mineral Resources, Geology, and Geophysics, Canberra, Australia.

all nuées, the laterally moving mass of rock and gas acquires its kinetic energy from potential energy, either in collapsing from an eruption column, or in simply avalanching down the steep flanks of a volcano. Thermal energy contained within hot magma may propel an initial explosion and the ascent of a vulcanian eruption column, but after the first moments does not contribute much to the *lateral* motion of nuées. Although an initial explosion may be of terrifying violence, its blast necessarily expands and decays outwards. A gravity-driven pyroclastic flow, however, may even accelerate and become more formidable as it continues downslope. Thus, while *nuées peléeans d'explosions dirigées* sound appealing, they are probably of limited significance.

There is one additional complication to this scenario. Once a lava dome has been extruded, its magma has been largely degassed. But as the lava cools and crystallizes, gases may exsolve from it, and become trapped beneath the solid carapace. If concentrations become high enough, explosions may take place, forming craters on the surface. And if the lava should fracture as it creeps down a steep slope, fracturing may trigger a shallow blast, thus generating a pyroclastic flow which crashes downslope under gravity.[8]

— 12.4 Soufrière-type *nuées*—Mayon 1968 —

Lava domes, even if only nascent ones, have been associated with the most notorious nuées, such as those of Mt. Pelée and Lamington. But nuées can also be produced in the absence of lava domes, simply by collapse from an eruption column. The 1902 eruption of Soufrière St Vincent that Anderson and Flett went to study was of this type, and so probably was the first Lamington nuée. More recent examples include well-documented studies of the 1974 eruption of Fuego, Guatemala[9] and the 1968 eruption of Mayon.[10]

Mount Mayon in the Philippines may be the most perfectly symmetrical volcanic cone in the world (Fig. 12.14). It rises in smooth, unbroken curves from sea-level to 2462 metres, its concave slopes steepening steadily upwards towards a summit crater only 200 m in diameter. Mayon has had a long history of destructive eruptions since the first recorded one in 1616. Forty eruptions over the succeeding centuries were responsible for the deaths of more than 1500 people.

A new episode of activity began on 20 April 1968, when a reddish glow was observed in the summit crater. On 21 April, a small ash cloud was seen, and a series of explosions ensued during the next few days. Blocks were hurled up to 600 m in the air, and ash-laden vulcanian eruption columns rose vertically about ten kilometres. From the base of these vertical columns, nuées ardentes emerged, rushing down the steep flanks of the volcano, channelled by existing deep ravines and gullies (Fig. 12.15). Two American volcanologists, J. G. Moore and W. G. Melson, were luckily on hand at the time and were able not only to fly around the volcano while the eruption was in progress, but also to examine the deposits left by the nuées soon after they formed. This is how they described the appearance of the nuées at night:

Formation of nuées ardentes was clearly seen during the night of 1–2 May under favourable cloud conditions. Explosions occurred every few minutes, hurling incandescent blocks and finer ejecta as high as 600 metres above the summit. This material fell back and produced a glowing collar on the surface around the summit crater. As this material flowed down the slope, it produced a rapidly-moving pinkish mass with bright moving particles speckling its surface. From a distance of 12 kilometres, these glowing blocks could be clearly seen through binoculars; many of them must have been several tens of metres across. Commonly, the downward moving blocks changed direction abruptly and simultaneously divided into several brighter particles, apparently as a result of bouncing or striking ravine walls, breaking into smaller pieces, and exposing the hotter, brighter interiors. The downward flowing blocks rapidly cooled and faded in colour, but incandescence could commonly be seen at an elevation of 1,000 metres down from the summit. However, the continuous mantle of glowing material generally

Fig. 12.14 Mt. Mayon, Philippines, in peaceful mood. Legaspi is town in the foreground. US Geological Survey photo by G. A. Macdonald.

Fig. 12.15 Mt. Mayon, Philippines, in a violent mood during the eruption of 1968. Dense clouds of dust and ash largely obscure the volcano, but beneath the main vertical eruption column convoluted dust clouds from nuées ardente rushing down radial valleys are visible. Photo: courtesy of W. G. Melson, Smithsonian Institute.

divided about 500 metres from the summit, and glowing material was confined to narrow tongues, probably in ravines, below that elevation.

During the day, the nuees appeared as rapidly moving, tongue-shaped, light-grey clouds, which expanded outwards and upwards, and became darker grey in colour, ultimately forming great cauliflower cloud-masses which often became so extensive that they obscured the whole volcano. Only about five minutes elapsed from the time of the first emission of a nuée until it disappeared into a disorganized mass of expanding cloud.

From their visual observations of the nuées, Moore and Melson concluded that they were the result of the avalanching of material which had initially been ejected vertically upwards from the crater, before falling back to earth. On the ground, they found that the deposits of Mayon nuées were several metres thick, had well-defined flow fronts and margins, and consisted of loose masses of ash and blocks of all sizes. One block twenty-five metres across was found low down on the flanks of the volcano. Many of the blocks were still hot when examined in the field. Some were extremely fragile, with a 'bread-crust' textured surface, showing that the lava had puffed up due to internal vesiculation. They were found intact, and so it follows that vesiculation, expan-

sion, and bread-crust cracking must have taken place very late, after the nuée had come to rest. This evidence for vesiculation indicates that the Mayon nuées involved a more gas-rich magma than those of Mt. Pelée or Lamington, but it is still not necessary to invoke magmatic gas as an essential fluidizing agent—Mayon is a steep, high volcano, with plenty of scope for gravity to work.

Surrounding each nuée deposit was a clearly defined 'seared zone' which ranged from a few metres in width to over two kilometres (Fig. 12.16). Plant leaves were shrivelled and browned on the outer parts of these zones; on the inner parts all the vegetation facing towards the crater was charred. Many trees were completely stripped of foliage on the same side. All of the animals in the seared zone were killed outright. A farmer, unfortunately caught within one zone, died shortly afterwards as the result of severe burns; before dying in hospital he is reported to have said that he was caught in 'a blast of hot gas'. He shared the same fate as the unfortunate inhabitants of St Pierre and Higaturu. Other witnesses suggested that each nuée from Mayon was preceded by a 'tornadic blast' of cold air, which may have travelled at up to 200 km per hour. Estimates of the speeds of the nuées themselves are rather variable, but one which was filmed appeared to have been moving at

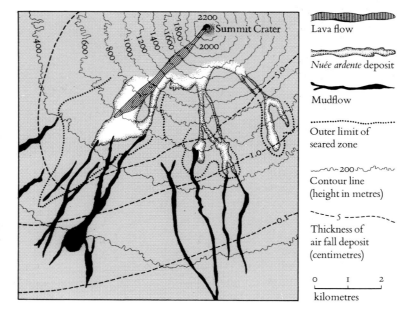

Fig. 12.16 Moore and Melson's map of Mt. Mayon after the 1969 eruption, showing distribution of nuée deposits, mud-flows, and seared zone. After J. G. Moore and W. G. Melson (1969).

something between fifty and a hundred kilometres per hour.

Finally, it is worth noting that the immense clouds of dusty ash generated by the eruption appear to have triggered torrential rainstorms which fell within minutes of the passage of the nuées. Rain-water from these swept away the fresh, loose tephra from the slopes of the volcano to form destructive mud-flows which flooded down many kilometres beyond the lowest points reached by the nuées themselves. These mud-flows added to the misery caused by the eruption, damaging roads, a railway line, pipelines, and farm-land.

— 12.5 Deposits of *nuées ardentes* ——————

It is well to recall here that nuées ardentes can also be called block and ash flows, and that the deposits laid down are *block-and-ash deposits*. This is a usefully descriptive term, although *block-and-dust* deposits is even better, since examining fresh deposits is often tiresomely dusty work.

Block-and-ash deposits consist of clasts of juvenile magma in an ash matrix. They have well-defined flow fronts and margins, and often exhibit linear trains of boulders on their upper surfaces. Transverse ridges of boulders are also sometimes seen where the flow has extended over flat-lying ground. In general, however, block-and-ash flow deposits are usually confined along pre-existing valleys. Where ponded in valleys, the deposits have flat tops, even where individual units are quite thin. This raises an important point: although a block-and-ash deposit may represent the accumulated result of many separate nuées, the *total* volume is rarely large, and is usually much less than 1 cubic kilometre. Although the flows that give rise to them are deadly, block-and-ash deposits are therefore *small* on the scale of volcanic phenomena (Fig. 12.17).

In typical block-and-ash flows, the juvenile magma is either andesitic or dacitic and is poorly vesiculated. Where vesiculation is pronounced, the term *scoria flow* is more appropriate. Perhaps the most impressive feature of block and ash deposits is the sheer size of the clasts they contain—boulders well over a metre in diameter are common. A second striking feature is *reverse grading*—the larger clasts are concentrated upwards. Block-and-ash deposits contain only lithics and no pumice, and so there is no segregation by density of the sort that encountered in ignimbrites, where pumices and lithics show reverse and normal grading respectively in response to fluidization.

The presence of huge lava blocks at the top of a block-and-ash deposit shows that the moving flow had a high yield strength. But why should such large blocks come to the surface? They clearly did not 'float' upwards like large, low-density pumices in an ignimbrite. Another, purely mechanical, process is at work here. When

Fig. 12.17 A typical block-and-ash deposit, laid down by a nuée erupted from Arenal volcano, Costa Rica in 1968. The deposit here is less than a metre thick. Dr Thorpe is examining a dense andesite clast, whose diameter is greater than the deposit thickness. The flat surface of the deposit is indurated and reddened; below the finer-grained material is still unconsolidated. Seventy people were killed by the nuée which left this deposit when it destroyed a nearby village.

a bucket of mixed sand and gravel is shaken for a minute or two, the large gravel pebbles soon come to the surface, even though they have the same, or even higher, density as the sand. When a packet of breakfast muesli is shaken, the same thing happens—the nuts all come to the surface. This simple observation is difficult to explain simply. Part of the answer is that when transient void spaces are opened up by shaking, smaller particles can fall downwards into them, while larger particles can not.

Purely mechanical reverse grading of this sort can be produced in 'cold' equally as well as hot moving flows. Two pointers are often helpful in confirming that an ancient block and ash deposit was actually hot when emplaced. Thick block and ash deposits accumulate in river valleys around a volcano, so they remain hot for long periods. Violent secondary explosions take place when rainstorms send water coursing down the valleys, sapping away at the friable deposits, and exposing hot material in cliff faces. Hot steam rising through such fresh deposits from beneath may cause rapid alteration, oxidizing the iron in it, particularly near the surface. Thus, a deposit that was emplaced hot may show reddening upwards, evidence of it having been 'cooked' by hot steam.

Even better evidence of hot emplacement of a block-and-ash deposit is the presence of prismatic jointed blocks, showing that individual blocks cooled, chilled, and cracked after the deposit had been emplaced.

Block-and-ash deposits are deposits of high-density flows. But as discussed earlier, the margins of the moving flows may be much lower density, and more surge like. Consequently, the marginal deposits are thinner, finer-grained, and sometimes thinly laminated. Cross-bedding structures are sometime present. Finally, the most conspicuous part of a *nuée*, the part that dominates photographs, is the towering wall of cloud which rises above the dense pyroclastic torrent. The cloud is composed of the finest grain-sized material elutriated out of the flow; where it falls out, a thin film of exceedingly fine-grained air-fall ash deposit results. Such delicate deposits are rarely preserved and are hard to find, but they are integral parts of the full suite of block-and-ash deposits.

— Notes

1. Anderson, T. and Flett, J. S. (1903). Report on the eruptions of the Soufriere in St. Vincent, in 1902 and on a visit to Montagne Pelée, in Martinique—Part 1. *Philos. Trans. Roy. Soc. Lond.* A**200**, 353–553.

2. Stoiber, R. E. and Rose, W. I. (1969). Recent volcanic and fumarolic activity at Santiaguito volcano, Guatemala. *Bull. Volcanol.* **33**, 475–502.

3. Francis, P. W., Roobol, M. J., Coward, M. P., Cobbold, P. R., and Walker, G. P. L. (1974). The San Pedro and San Pablo volcanoes of Northern Chile and their hot avalanche deposits. *Geol. Rundschau.* **63**, 357–88.

4. Perret, F. A. (1937). The eruption of Mt. Pelée, 1929–32. *Carnegie Inst. Publ.* **458**. 126 pp.

5. Boudon, G. and Gourgaud, A. (eds.) (1989). Mount Pelée. *J. Volcanol. Geotherm. Res.* **38**, 200 pp.

6. Fisher, R. V., Smith, A. L., and Roobol, M. J. (1980). Destruction of St. Pierre, Martinique, by ash-cloud surges, May 8 and 20, 1902. *Geology* **8**, 472–6.

7. Taylor, G. A. M.: GC (1983). The 1951 Eruption of Mount Lamington, Papua. *Bur. Min. Res. Geol. Geophys. Bull.* **38**, 2nd edn., Canberra, Australia, 129 pp.

8. Fink, J. (1991). Scientific correspondence. *Nature* **351**, p. 611.

9. Davies, D. K., Quearry, M. W., and Bonis, S. B. (1978). Glowing avalanches from the 1974 eruption of the volcano Fuego, Guatemala. *Geol. Soc. Amer. Bull.* **89**, 369–84.

10. Moore, J. G. and Melson, W. G. (1969). Nuées ardentes of the 1968 eruption of Mayon volcano, Philippines. *Bull. Volcanol.* **33**, 600–20.

13

Debris flows: magic carpets and muck

Considered as mountains, volcanoes are exceptional in that they *grow*. Each time that they erupt lavas or pyroclasts, they add cubits to their stature. Ordinary (non-volcanic) mountains are formed by tectonic processes, which uplift blocks of the crust, and then by erosional processes which sculpt the blocks into peaks and valleys by *removing* rock. One might deduce, therefore, that volcanoes ought to be the highest mountains on any planet. On Mars, volcanoes actually are the highest mountains, but the Earth's highest volcano, Nevado Ojos del Salado, is two kilometres *lower* than Mt. Everest, whose summit is carved from marine sediments. Two factors account for this. One involves plate tectonics: terrestrial volcanoes are constructed on moving lithospheric plates. Thus, they erupt for geologically short periods of time before being pulled away from their mantle heat sources and simmering into silence. The second factor concerns relative rates of eruption and erosion. In the Earth's thick, wet atmosphere, the higher that a volcano grows, the more rapidly erosion works to destroy it. Consequently, it is difficult for a volcano to build an edifice more than a few thousand metres high above sea-level.

We explore some of these issues further in Chapter 16, but here we look at one process that limits the heights of volcanoes: gravity-driven debris flows. It should not be supposed that we have forsaken the exhilarating sights, sounds, and smells of volcanic violence for the tedium of senescent volcanoes mouldering away. Far from it. Massive debris avalanches form key parts of some major eruptions. Flank failure may remove one-quarter of a huge volcanic cone in less time than it takes to read this page.

Debris flows are gravity-driven fragmental flows which do *not* include young magmatic material as essential components, although magmatic events may be intimately linked to cone collapse. The transported materials may consist exclusively of volcanic rocks, but these may be old and cold. A broad spectrum of related phenomena exists. At one extreme of the spectrum are dry *avalanches*; at the other, wet *mudflows*. Inevitably, there is no clear dividing line between the two.

— 13.1 Dry rock avalanches

Nuées ardentes—glowing avalanches—and ordinary rock avalanches form another spectrum of related phenomena. Apart from the semantic issue of how much hot material they may contain, rock avalanches differ in an important respect from hot avalanches: hot avalanches are small, but ordinary avalanches may be *huge*, involving failure of complete mountainsides. Although volcanoes have collapsed many times in history, it was Mt. St Helens in May 1980 which made volcanologists aware that an entire flank of a volcano could fail, forming a massive debris avalanche and triggering a violently explosive eruption. Since Mt. St Helens, investigators such as Lee Siebert,[1] Tad Ui,[2] and my colleagues[3-6] have identified more than one hundred other major avalanche deposits on volcanoes around the world, ranging in age from a few years to tens of thousands of years, and in size from less than one cubic kilometre to more than 20. It is now plain that occasional massive collapses are normal events in the life cycles of high volcanoes. It follows that debris avalanche deposits are normal components of volcanoes, like lavas and pyroclastic flows.

There are three kinds of collapse event: those which involve a magmatic eruption; those which involve non-magmatic explosions, and those which are 'cold'.[7]

Mt. St Helens was an example of the first type. A body of dacitic magma had intruded the volcano, forming a cryptodome (Section 4.4. and 7.5.4). Injection of the magma raised a prominent bulge that destabilized the mountain's flank. Magma also interacted with ground-water to produce superheated steam, which remained trapped within the volcano. When a small earthquake finally caused the unstable bulge to fail, the steam escaped explosively. Collapse and explosion eviscerated the volcano, triggering a full-blown magmatic eruption. Although Mt. St Helens is fresh in the memory, this class of events has been called *Bezymianny*-type, celebrating a similar, but less well-known eruption in Kamchatka in 1956. Bezymianny produced an avalanche of modest dimensions, only 0.8 cubic kilometres, but its blast devastated an area of more than 500 square kilometres, similar to Mt. St Helens.

Bandai volcano, Japan, gave its name to the second type. Failure of the north flank of the Ko-Bandai peak took place in 1888, yielding an avalanche with a volume of 1.5 cubic kilometres. Although minor explosive activity propelled an eruption column four kilometres high, no new magma had intruded into the volcano—the explosions were purely phreatic steam blasts (Fig. 13.1).

The third type involves no volcanic activity at all: the volcano simply fails in a 'cold' avalanche. A recent example was Ontake volcano, Japan, in 1984, when a magnitude 6.8 earthquake triggered slope failure.

Mt. St Helens is the only volcano to have collapsed in full view of modern media. But how does one recognize a volcano which experienced collapse unobserved, in prehistoric times? Sometimes, the task is straightforward: the vast, horseshoe shaped amphitheatre left by the avalanche may remain prominent for millennia (Fig. 13.2). Two factors complicate the issue, though. Many volcanoes have similar amphitheatres, caused by

Fig. 13.1 Bandai-san volcano, Japan, collapsed on 15 July 1888. This contemporary lithograph shows the debris-filled amphitheatre (with steaming vents) and the hummocky topography of the avalanche deposit. Unlike Mt. St Helens, Bandai-san had not been destabilized by new magma. Many villages were buried by the avalanche, and several lakes were formed where rivers were dammed by the deposit. Some of these lakes later burst, flooding downstream villages. A total of 461 people were killed.

(a)

(b)

Fig. 13.2 (a) Mt. St Helens ten days after its 18 May eruption. Its north-facing amphitheatre has yet to be occupied by lava, although activity continues. (b) By August 1981, a dacitic lava dome was well established within the amphitheatre. Despite occasional explosive disruptions, the dome is now a permanent feature.

slow block-rotation, landsliding, and erosion, rather than catastrophic failure. And if there has been subsequent activity, an avalanche scar may be healed rapidly by effusions of lava. At Mt. St Helens, the 2.8 cubic kilometre avalanche and concomitant eruption left an amphitheatre almost two kilometres wide and 600 metres deep. Only ten years later, a lava dome already occupied about one-third of the floor of the amphitheatre. At Bezymianny, thirty years of lava dome growth has almost filled the amphitheatre. At Parinacota in Chile, which collapsed 13 000 years ago, no trace of an avalanche scar remains (Fig. 13.3).

In such instances, evidence of collapse is best preserved in avalanche deposits. Identifying these is not easy, because they are often camouflaged by vegetation and erosion, so they can be easily mistaken for quite different deposits, such as glacial moraines. Nevertheless, the physiognomy of an avalanche deposit often shows seductively through the veil of time. Its most obvious feature is *hummocky terrain*: thousands of small hills and closed depressions covering large areas at the base of the volcano (Fig. 13.4). Two hundred and fifty square kilometres of hummocky landscape on the slopes of Galunggung volcano (Java) define the famous 'Ten Thousand Hills of Tasikmalaja'. Long before their origin was understood, the distinctiveness of the terrain was familiar to the local people, obliged to shape their farms, fields, and lives

Fig. 13.3 Thirteen thousand years ago, 6000-m-high Parinacota volcano, north Chile, experienced flank failure. Subsequent lava eruptions have smoothed away all evidence of the collapse scar on the cone, although a debris avalanche deposit extends up to 20 km from the volcano, forming hummocks in middle distance. A younger basaltic andesite lava occupies valley in foreground.

Fig. 13.4 More than 10 000 years ago a large debris avalanche from Tata Sabaya volcano, Bolivia (top left), spilled out 20 km across the Salar de Coipasa, producing an excellent example of the hummocky topography characteristic of avalanche deposits. White deposits between hummocks are gypsum and other evaporites. The image is about 35 km across. Landsat Thematic Mapper image.

around a myriad topographical inconveniences.

13.1.1 Socompa volcano, north Chile: case study of collapse

Of all the world's volcanoes which have collapsed, Socompa is paramount. Whereas the Mt. St Helens amphitheatre occupies about 30° of the volcano's circumference, the avalanche at Socompa carved out a 70° wedge, displacing a volume of material ten times greater than at Mt. St Helens. About 500 square kilometres of the hyper-arid Atacama Desert were covered by the avalanche deposit. The area is so dry that the deposit is magnificently preserved, providing a natural laboratory for studying avalanche deposits (Figs. 13.5–13.6).

At present, Socompa is 6051 m high. Prior to collapse, its altitude was probably about 6300 m.

On its western side, the volcano soars 3000 m above the Atacama Desert, but on its eastern side, it rises only about 2000 m above the high tableland of the Argentinian *puna*. This asymmetry between western and eastern base levels is an important clue to the cause of the massive avalanche. East of Socompa, the *puna* block has been uplifted along a series of north–south major faults which underlie the volcano's flank. About 7200 years ago the jolt of an earthquake on one of the faults triggered failure of Socompa's western flank, disembowelling the volcano (Fig. 13.7).

But Socompa must have been shaken by scores of powerful earthquakes during the many millennia of its history. What made it collapse so definitively 7200 years ago? Again, Mt. St Helens suggests an answer. Field evidence shows that fresh magma had been intruded into Socompa

Fig. 13.5 Socompa volcano, north Chile and its debris avalanche deposit. The primary avalanche travelled 35 km in a north-westerly direction from the volcano; material forming the secondary flow travelled more than 40 km; its final motion being in a direction at right angles to the primary one. Image is 60 km across. Landsat Thematic Mapper image, courtesy of the Lunar and Planetary Institute, Houston.

prior to collapse. Lava may even have been squeezing out high up on the cone. Conspicuous within the avalanche deposit, but forming volumetrically an insignificant proportion of it, are numerous bread-crust blocks, up to 20 m in diameter. Their fragile, glassy outer crusts and frothy, vesiculated interiors are unambiguous evidence for the presence of hot magma within the volcano at the instant of its collapse (Fig. 13.8). Thus, massive failure of the western flank was triggered by the lethal combination of an earthquake acting on the mechanical instability of a volcano swollen by new magma.

When the avalanche occurred, it took a huge bite out of the mountain. But whereas the debris at Mt. St Helens came exclusively from the volcanic edifice itself, the avalanching material at Socompa included rocks from the underlying, non-volcanic basement. Hence, the avalanche deposit consists not only of young lavas, but also of gravels, lake sediments, ignimbrites several million years old, and much else besides. Clearly, the failure surface along which the avalanche started moving sliced deep down through the volcano and into the basement.

Dynamically, the Socompa avalanche was complex. In its earliest stages, its motion can best be described as *sliding*. Enormous blocks of rock, kilometres in size, became detached from the mountain and accelerated downslope. Although some jostling and backward rotation took place, the blocks remained coherent. During this phase,

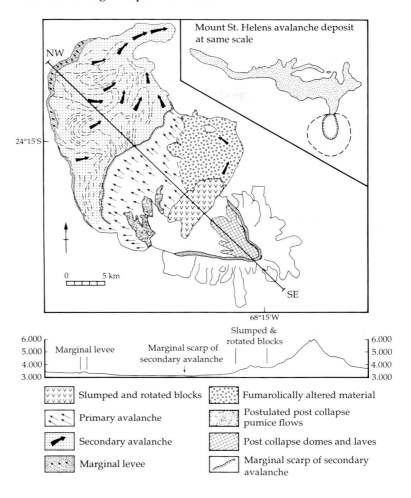

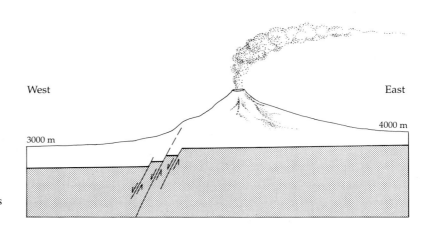

Fig. 13.6 Sketch map and section illustrating the principal components of the Socompa avalanche deposit discussed in text.

Fig. 13.7 A major fault zone in the subvolcanic basement beneath Socompa shaped its pre-collapse topography, causing marked asymmetry, and also triggered failure of its western flank.

Fig. 13.8 Fresh bread-crust block, 2 m diameter in the Socompa avalanche deposit. It provides conclusive evidence for the existence of juvenile magma in Socompa (background) when it collapsed.

it would have been possible to ride safety on top of a sliding block, if one had had the nerve. At Socompa, some of these *toreva* blocks are preserved at the mouth of the amphitheatre, several kilometres from where they started. Initially, it is difficult to credit that these blocks have slid several kilometres, since the largest of

them is three kilometres long and half a kilometre high (Fig. 13.9).

Other blocks slid further downslope, accelerating all the while, until they were moving at velocities between 100 and 200 kilometres per hour. At these speeds, the combination of inertial forces and shearing stresses set up within the moving mass was sufficient to disrupt large blocks into smaller masses, most of them less than a few metres in size. In this condition, motion of the avalanche was more akin to *flow* than sliding, but none the less, subtle original stratigraphical relations are preserved within the deposit, even in the most distant points reached by the avalanche, more than 35 km from the source.

When the first wave of material reached the foot of Socompa, it was plummeting northwestwards, travelling so fast over the north-eastwards sloping terrain that its trajectory was almost straight. (A ball rolled slowly across a sloping surface follows a curved trajectory, until it rolls directly downslope.) At its fastest, the avalanche was probably travelling at nearly 300 kilometres per hour. Decelerating continuously, the avalanche eventually encountered gently rising ground, ran out of steam, and came to a stop 35 km from the peak of the volcano. Its furthest extent is marked by a steep wall of debris 30–40 metres high, forming a marginal *levée*

Fig. 13.9 Oblique aerial view of largest *toreva* block at Socompa; it is more than 3 km long and 400 m high. Numerous secondary fractures cut its upper surface. Disrupted outline of a lava flow can be seen on upper surface, coming towards camera.

bounding the western and northern limits of the avalanche deposit. Marching for tens of kilometres across the barren desert, this enormous wall is as marvellous a monument to the scale of geological forces as the Great Wall of China is to puny human exertion (Fig. 13.10).

Not all of the twenty cubic kilometres of material in the moving avalanche came to rest at the same time. Only a small proportion 'froze' in place to form the primary avalanche deposit and the great marginal *levée*. Most of the material came to rest transiently behind the *levées*, and then moved off down the north-eastwards-facing slope, in a direction almost at right angles to the primary avalanche. Thus, much of the material in the avalanche followed a billiard-ball-like trajectory, changing direction abruptly to form a *secondary* avalanche, which overrode material deposited moments earlier by the *primary* avalanche.

13.1.2 *Debris avalanche deposits*

Unlike other volcanic deposits such as ignimbrites, avalanche deposits are easier to identify from their large-scale properties than from the minutiae of clast composition or sorting. Individual blocks within an avalanche are often so huge, that it may be difficult to realize that they are parts of an avalanche at all—the largest slide blocks at Socompa are more than a kilometre in size, and are made of apparently undisturbed lavas. While kilometre-sized blocks are unusual, blocks tens of metres in size are common (Fig. 13.11). In such cases, spotting that they are components of an avalanche is only possible if the blocks are unusual in some way: perhaps because they are composed of rocks which do not belong in the local context, or because blocks of strongly contrasted properties have been forced into unnatural proximity, as at Socompa, where lake sediments and lava blocks are intermingled.

In places where large blocks are not present, avalanche deposits resemble angular breccias. Large clasts are supported by a matrix of fine material which often reveals evidence of intense shearing. Although a wide range of clast types may be present at any one location, their compositions are not random. Some homogenization takes place within the moving avalanche, but it is far from complete. In one of the first studies of a large avalanche deposit, R. L. Shreve[8] showed that broad stratigraphical relationships in the source rocks were preserved in the Blackhawk deposit, California, and the same has been observed at Mt. St Helens and Socompa. At Mt. St Helens, mapping of the avalanche deposit showed that it was formed by three successive great 'bites' out of the cone, as *retrogressive slope failure* occurred[9] (Fig. 13.12).

At Socompa, it is plain that little lateral mixing

Fig. 13.10 Oblique aerial view of southern margin of Socompa avalanche deposit. Field of view extends about 10 km towards the west; railway line winding along margin conveys an impression of scale.

Fig. 13.11 Boulder twenty metres in diameter in proximal facies of Socompa debris avalanche deposit. Prismatic jointing suggests that it was hot when emplaced.

took place, as the debris is organized in distinct trains of different composition. Different varieties of debris—glassy dacite or reddish oxidized lavas, for example—can be traced back along linear trajectories to individual sectors of the original volcanic cone. Some vertical mixing took place, because blocks of basement ignimbrite are juxtaposed with fresh lavas from the cone itself, but complete homogenization did not take place. Even in the most distal parts, individual sheets of breccia with distinctive lithologies are so coherent that they can be picked out on satellite images.

Finally, two facets of clast size distribution in avalanche deposits are noteworthy. First, the lowest parts of an avalanche deposit are the finest grained. Given the intense shearing along the base of the moving mass, this is scarcely surprising. Second, huge boulders are concentrated on the surface of the deposit. This prosaic statement takes on a startling reality at Socompa, where a jumble of millions of boulders extends unbroken

Fig. 13.12 Retrogressive slope failure at Mt. St Helens led to the formation of three slide blocks, I–III. Prior to collapse, a cryptodome of new magma was intruded beneath the sixteenth to seventeenth century summit dome (a). In (b), about 20 seconds after movement began, two slide blocks have detached, and explosions have begun on head-wall of block where cryptodome is exposed. Moments later, a third slide block became detached. Modified after J. G. Moore and W. C. Albee (1981). In Lipman, P. W. and Mullineaux, D. R. (ed.). *the 1980 eruptions of Mt. St Helens, Washington.* US Geol. Surv. Prof. Pap. No. 1250, 843 pp.

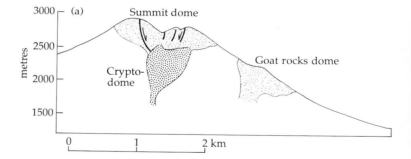

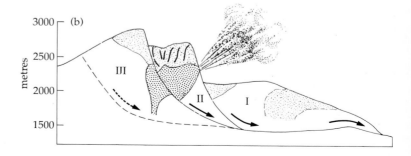

to the horizon. Alice might have contemplated a similar scene on a beach in Wonderland, if the grains of sand were metres across, many times bigger than Alice. Segregation of the largest of boulders on the surface probably took place in the moving avalanche due to the muesli effect described in block and ash deposits (Section 12.5).

13.1.3 *Flow mechanisms of debris avalanches*

When an 'ordinary' block of rock slides down a textbook inclined plane, it travels roughly 166 metres horizontally for every 100 metres it descends vertically—the H/L ratio is said to be 0.6 (Fig. 13.13). Larger avalanches are more mobile than small ones—once their volume exceeds about a million cubic metres, their H/L ratio begins to decrease. Some extremely large avalanches may travel as much as 1000 metres horizontally for every 100 metres descended ($H/L = 0.1$). How do they do it? This question of apparently anomalous mobility brings us back to the problem that we first explored in relation to the spectacular mobility of ignimbrites. In that case, fluidization by air ingested by the flow during its collapse from the eruption column was proposed as an explanation for their mobility.

Does this apply also to debris avalanches? Or are there things that avalanches can teach us that

we can apply to ignimbrites? The notion of an avalanche sweeping down the flanks of a volcano at tremendous speed may suggest disorganized, turbulent flow. Flow is indeed complex, but it is neither chaotic nor turbulent: it is laminar. Little deformation takes place within the body of the flow, except within the layer in contact with the ground, which is highly sheared.

This style of movement is essentially plug flow (Section 5.5). Theoretical studies show that in flows of this sort (which resemble glaciers) strain rate is proportional to stress to the *eighth* power, and shearing is concentrated at the base of the moving mass—one-half of the total shear is concentrated in only the lowest eight per cent of moving debris.[10] This accounts for the common observation that subtle stratigraphical relationships are preserved within avalanche deposits. At Socompa, for example, light-coloured dacite overlay dark andesite in the original volcano, and white ignimbrite underlay both. Identical relationships are preserved within the avalanche deposit. To be sure, each layer has been heavily sheared—whereas they were originally tens or hundreds of metres thick, in the avalanche they are reduced to thin screens a few metres thick. But the important point is that the original relationships are preserved, and that the avalanche was far from turbulent.

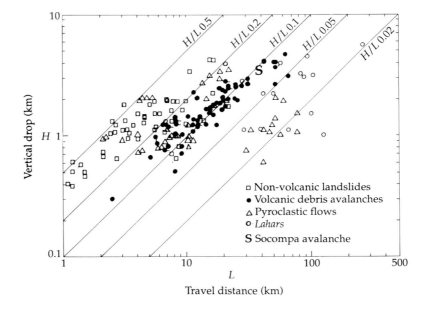

Fig. 13.13 Vertical height descended (H) against horizontal length travelled (L) for non-volcanic landslides, volcanic debris avalanches, pyroclastic flows, and *lahars*. Precise positions of individual points are moot, because data for centres of mass of displaced bodies should be plotted, but are often difficult to determine. After Siebert, L. (1984). Large volcanic debris avalanches: characteristics of source areas, deposits and associated eruptions. *J. Volcanol. Geotherm. Res.* **22**, 163–97.

sliding over highly sheared bases. Their extraordinary mobility is explained by a form of fluidization within the basal layer. Whether described as acoustic, dynamic or mechanical fluidization, this results from the infinite number of collisions taking place between clasts in the basal layer. The properties of this mass of colliding clasts resemble those of molecules in a hot compressed gas. Thus, although the details remain far from clear, it becomes possible to understand the parallels between pyroclastic flows and avalanches, and to see that the causes for their mobility have to be sought in the complexities of the fluidization process.

Whatever the fluidization process, in both cases huge frictional losses take place in travelling tens of kilometres. In order for a moving avalanche or pyroclastic flow to continue to advance at high velocity, these frictional losses have to be compensated for by mass loss: momentum is continually transferred forward to the head of the flow from material deposited and left behind in the rear. Sacrificial transfer of momentum is the rocket fuel that propels both avalanches and pyroclastic flows.

13.1.4 Tsunamis generated by avalanches—the ultimate volcanic hazards?

Some of the preceding discussion may have seemed academic. But it is important to understand volcanic avalanches because their hazard potential is extreme. When an avalanche crashes into an ocean or even a lake, the tsunamis that it generates make most other kinds of volcanic hazard pale into insignificance: volcanic avalanche-generated tsunamis have caused some of the worst natural disasters in historic times.

Collapse of the Rakata cone on Krakatau during the 1883 eruption illuminates the preeminent danger of tsunamis: because they can travel immense distances across oceans, they can kill at startling distances from the source volcano. When the northern half of Rakata slumped into the Sunda Strait in 1883, several hundred kilometres of the coastlines of Java and Sumatra were ravaged by tsunamis which killed 30 000 people. Death and disaster were so widespread that the contemporary investigators found it

difficult to piece together the sequence of events. Some of the horror comes through even in the measured prose of the Royal Society report:

> The times of arrival of the waves at different places on the shores of the Strait (of Sunda) are but vaguely noted, and this is especially the case with the great wave after 10 o'clock on the twenty seventh (of August). Terror and dismay reigned everywhere and darkness settled over the land. At Anjer . . . where this wave must have come, no one was left to see it, the few survivors having fled to the hills.

The 'great wave' was the largest in a series. It may have been 40 m high near its source, but was about 15 metres high at the shoreline. In places where the wave washed up into constricted channels, it rose much higher. At Telok Betong, the sea reached within two metres of the Dutch Governor's residence, 24 m above ordinary sea-level. Where the country was low lying, the wave swept headlong inland. Near Telok Betong, a man-of-war (the *Berouw*) was carried nearly three kilometres inland, to be left stranded high and dry ten metres above sea-level. Lighthouses were swept away and other landmarks obliterated along much of the Sunda Straits. As far away as Ceylon, the tsunamis were big enough to leave small boats in the harbour temporarily stranded, and then refloat them, while at the Cape of Good Hope they were easily measurable with tide gauges. Small signals were even detected in the Bay of Biscay, 17 255 kilometres distant. (These may not have been true tsunamis, but air–sea coupled waves, caused by the response of the ocean's surface to the transient pressure pulse transmitted through the atmosphere by the huge explosions.)

In 6 October of the same year, tsunamis generated by a collapse on Mt. St Augustine volcano in Alaska ravaged the small community of English Bay, on the coast of Cook Inlet, 85 kilometres from the volcano.[16–18] Fortunately, the tide was out when the tsunami struck, and damage was minimized (Fig. 13.16).

It is not necessary for an avalanche to be large for it to generate devastating tsunamis. On 21 May 1792, an earthquake-triggered collapse of a dome on Unzen volcano, Kyushu, Japan, generated an avalanche with a volume of only 0.3

Fig. 13.16 Hummocks and foreground pond are part of the 1883 Burr Point debris avalanche deposit, Augustine volcano, Alaska (background). When it entered the sea, this avalanche set up tsunamis which swept the coasts of Cook Inlet (Fig. 13.17). Photo courtesy of Lee Siebert, Smithsonian Institutions.

cubic kilometres, which travelled only 6.4 km from its origin. None the less, the resulting tsunami swept 77 km of coastline along the Shimabara Peninsula, drowning 9528 people. It then propagated 15 km across the Ariake Sea to wash the coastlines of the Higo and Amasuka provinces, killing 4996 people there.

Given the terrible potential of small avalanches, the effects of a really large tusunami defy the imagination. Yet it has long been suggested that some of the highest sea cliffs in the world, bounding the north shores of the island of Molokai in the Hawaiian Islands were formed by huge failures biting into the flanks of the extinct shield volcano.[19] If so, the tsunamis generated may have wrought prehistoric havoc all round the Pacific. Bathymetric maps of the ocean floor off Molokai show that large parts of the island have indeed slid under water, but it is less clear whether there they did so at high speed, as tsunami-generating avalanches, or slowly, as creeping landslides. On the adjacent island of Lanai, J. G. Moore and W. G. Moore have described what they believe to be the deposit left by a giant wave caused by just such a gigantic avalanche.[20]

Forecasting tsunamis   On 1 April 1946 an earthquake of magnitude 7.5 took place in the Aleutian Islands. Six hours later, tsunamis generated by this earthquake struck the Hawaiian Islands. More than 150 people were killed. As a consequence of this disaster, an early-warning system was set up around the Pacific. Underlying this system is that fact that tsunamis behave like any other waves. Like the ripples set up when a pebble is thrown into a pond, tsunamis have consistent wavelengths, and once initiated, move outwards from the source in a regular train of crests and troughs. The ocean behaves like a large pond, except that the time, or period, between successive tsunami crests may be as much as half an hour, while that for ripples on a pond is less than a second. Since tsunamis are regular waveforms, it is possible to predict how fast they will travel and therefore their arrival times at points distant from their source. A simple equation gives the velocity of any water wave, in pond or ocean:

$$v = \sqrt{(gD)}$$

where v is the velocity, g is acceleration due to gravity, and D the water depth.

At 18.56 GMT on 21 May 1960 one of the most powerful earthquakes of the century struck Chile. It 'set the world ringing', meaning that it made the entire Earth vibrate, like a vast spherical gong, and triggered major tsunamis. By 22.04, reports of the tsunamis in Chile had

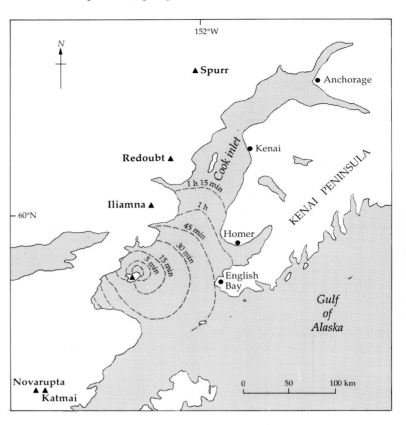

Fig. 13.17 Computed tsunami travel times in Cook Inlet for waves propagated by avalanches from Mt. St Augustine, Alaska. A wave arriving at Homer would have an amplitude of about 1 metre. Calculations were based on the avalanche of 6 October 1883, which was travelling at about 50 metres per second when it entered the sea on the north shore of the island. Many prehistoric volcanic avalanches have been at least an order of magnitude larger. After Kienle, J. *et al.* (1987). Tsunamis generated by eruptions from Mt. St Augustine volcano, Alkaska. *Science*, **236**, 1442–7.

reached Hawaii. A warning was issued by the authorities that the people living round the coasts of the islands could expect trouble. A few hours later, geophysicists had calculated arrival times of the tsunamis and their predictions had been published. At 09.58 the following day, fifteen hours and two minutes after the earthquake, tsunamis reached Hawaii, within a minute of the predicted time. In the town of Hilo, the effects were particularly serious. The third wave was highest, reaching ten metres, and it swept well inland, causing 62 deaths and extensive damage. Those killed had chosen to ignore the warnings given, some succumbing to an excess of curiosity, but others because earlier warnings had not be followed by serious tsunamis. Twenty-two hours after the earthquake, 180 people were killed in Japan, where warnings had also been published. It is easier to understand the deaths in Japan than those in Hawaii: Japan is nearly twice as far from Chile as Hawaii (17 000 km) and few could have imagined that the tsunamis could be lethal at such extreme range.

This should be a lesson for the future (Fig. 13.17).

13.2 Mud-flows and *lahars*

At 9.15 on the morning of Friday 21 October 1966, about 140 000 cubic metres of coal-mining waste slumped abruptly from a tip-heap built on the valley sides above the Welsh village of Aberfan. A black tongue of coal dust, shale, and water flowed down the valley slopes into the village, engulfing successively two farm cottages, a canal, a railway embankment, a school, and eighteen houses before it finally came to rest. One hundred and forty-four people were killed, 114 of them children inside the Pantglas Junior School. This singularly distressing event demonstrated how easily large volumes of fragmental, rocky material can flow, and flow rapidly, when enough water is present to lubricate the mass. At Aberfan, the tip heaps were thoroughly sodden after heavy rain and were also supposed to have been built up above a surface stream. This was not in itself dangerous, but the tips had been piled so steeply above the solid rock of the hillside that they became mechanically unstable. A great mass of the waste material broke away, turning in an instant from a solid pile into a fast-flowing slurry of rock and water.

Similar lethal slurries are all too common on volcanoes (Fig. 13.18). In tropical areas with heavy rainfall, such as Indonesia and the Philippines, tephra that accumulates on the slopes of active volcanoes may become saturated during the monsoon. A lot of it will be washed away by ordinary erosion processes, but some of it will rush downslope in Aberfan-like flows. The speed of such flows varies with the steepness of the slope and the proportions of solid and liquid in the slurry, but it can be very fast indeed, of the order of ninety kilometres per hour. Indonesia has suffered dreadfully from volcanic mud-flows. Consequently, the Indonesian word for them, *lahars*, is now used widely by volcanologists.

Volcanic eruptions themselves may trigger mud-flows. Major ash-producing events often propagate torrential rainstorms, since the ash particles act as nuclei, around which water collects to form raindrops, and the resulting deluges may initiate 'cold' mud-flows like that of Aberfan. A distressing complication during the 1991 eruption of Pinatubo in the Philippines was that the climactic phase of the eruption coincided with the passage of typhoon Yunya. Huge volumes of new-fallen ash mantled the slopes of the volcano, clogging pre-existing drainages. Torrential tropical rainstorms swept away large amounts of ash, generating some of the worst mud-flows in memory. All around the volcano, thousands of houses were washed away or buried.

A month after the main eruption, mud-flows still presented serious hazards. On 9 July 1991, an intense cloudburst on the volcano triggered a minor mud-flow, which surged down the Bucao River. It was studied by a team specially set up to monitor mud-flows. They reported that they could *hear* it coming long before they saw it—'a strong noise of rushing water and rolling boulders'. They noted that the flow was moving at about 12 km per hour, and that boulders up to 1.5 metres in diameter were being carried along. Most alarming of all, the mudflow undercut and eroded away the confining banks of the river in front of their eyes, at a rate of as much as 1 metre per minute on bends.

A second witness stated that the roaring brown

Fig. 13.18 This small mud-flow swept down the flanks of the inactive volcano Chigliapichina, north Chile, following a rare rainstorm in 1969. After emplacement, the mud set hard, blocking a major road. Broad *levées* and a central channel are well displayed.

torrent which churned down the river-bed and quickly overflowed its banks 'sounded like an airplane'. His house was swept away, and his four-acre farm buried beneath mud. By November 1991, about 100 people had been killed by mud-flows, and at least 230 square kilometres of agricultural land paved over. About 7 cubic kilometres of tephra were erupted from Pinatubo. Much of that material remained piled up on the flanks of the volcano. Inevitably, damaging mud-flows will continue to surge down the slopes of the volcano for several years.

13.2.1 Lahars *from crater lakes*

Some of the worst mud-flows in history have resulted when eruptions have disrupted crater lakes, expelling water and debris at the same time. A volcanic crater lake is often a devil's brew of highly acid, sulphurous water and mud. If, in addition, the lake water is hot, the mud-flow which rushes down the gullies around a volcano will be a peculiarly unpleasant, lethal mixture of scalding, acid water, and mud; a combination even more deadly than that which poured into Aberfan. It was a hot flow of this kind which burst from the basin of the Étang Sec of Mt. Pelée and swept over the Usine Guérin, three days before the St Pierre disaster of May 1902.

Kelut volcano on Java is notorious for its mud-flows, and for the attempts to prevent them. Kelut has a deep crater lake, which has been blown out by several eruptions, only to re-form again later. After a particularly disastrous episode in 1919 when 5000 people were killed by *lahars*, the Dutch colonial authorities took action. They dug a series of horizontal tunnels through the crater wall to drain the lake water, lowering its level by about fifty metres (Fig. 13.19). For nearly thirty years, the mud-flow peril seemed to have been averted. In 1951, however, another eruption took place which did *not* cause any mud-flows, but *did* ruin the drainage system. It also reamed out the crater, making the lake deeper than ever. Only seven people were killed during this eruption. By ill chance, two of them were officials from the local volcanological observatory who had come to check up on the volcano, but were caught by an

unexpected blast while actually inside one of the drainage tunnels.

The original tunnels had been rendered useless, and so an ineffectual attempt was made to drain the lake via a single tunnel, through which it was hoped that the trapped water would escape by seepage. In 1964, the newly deepened crater lake was estimated to contain 40 million cubic metres of water. It presented such a serious threat that two Indonesian scientists published a paper pointing out the inadequacies of the seepage drainage system, and the necessity of digging a new tunnel system.[21] They forecast a new eruption within five years. It came in 1966, only two years after their prediction. Many hundreds of people were killed by *lahars*. After this needless tragedy, new drainage tunnels were dug . . .

Kelut volcano erupted once more on 10 February 1990. During the previous two months, the lake water temperature had risen several degrees, reaching a maximum of 41°C and its acidity had increased. Seismic data also indicated that an eruption was pending, so 60 000 people were evacuated from the densely populated area around the volcano. A brief but violent eruption ensued, sending pyroclastic flows 8 kilometres down a steep-walled valley on the east flank of the volcano. As before, the lake was again emptied, but this time, thanks to timely evacuation, few people died. There were 32 casualties, most of them killed when the roofs of their houses collapsed under the heavy weight of accumulated tephra.

Another tragic example of *lahars* propagated from a crater lake was the Tangiwai disaster in New Zealand in 1953. A small eruption of Ruapehu volcano caused a breach in the walls of its summit crater which contained a crater lake. Water gushed out from the crater, lowering the lake level by ten metres, and the sudden flood generated a mud-flow which swept down the flanks of the volcano. It was not a large mud-flow, but it was big enough to destroy a railway bridge across a valley low down on the slopes of Ruapehu. In a situation tragically reminiscent of cinema drama (life imitating art?) a passenger train steamed on to the bridge before it could be stopped. There were many fatalities when it crashed into the valley.

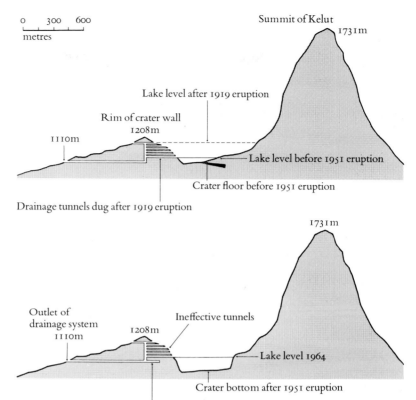

0   300   600

metres

Summit of Kelut

1731m

Lake level after 1919 eruption

Rim of crater wall
1208m

1110m

Lake level before 1951 eruption

Crater floor before 1951 eruption

Drainage tunnels dug after 1919 eruption

1731m

Outlet of
drainage system
1110m

Ineffective tunnels

1208m

Lake level 1964

Crater bottom after 1951 eruption

Single seepage tunnel before 1966 eruption

Fig. 13.19 Profiles through Kelut crater lake and the tunnels excavated to drain it. Above, situation before the 1951 eruption; below, situation before 1966 eruption. (After Zen, M. T. and Hadikusumo, D. (1964). The future danger of Mt. Kelut (eastern Java–Indonesia). *Bull. Volcanol.* **28**, 275–82.

13.2.2 Mud-flows generated by snow melt

Mud-flows caused the most serious volcanic disaster of modern times. A small eruption of Nevado del Ruiz volcano, Colombia, in November 1985, melted snow and ice in the summit region, thereby triggering a series of mud-flows which were channelled down narrow valleys on the flanks of the volcano. Armero, a town of 22 000 people at the mouth of one of the valley, was wiped out. The events at Armero were so dreadful, so recent, and so overlain with subsequent recriminations about what warnings were given and when they were given, that it is difficult to describe them dispassionately. For the most comprehensive account of the catastrophe and the events leading up to it, readers should consult an authoritative set of papers.[22] What follows is a chronological summary of events, as recounted by Darrel Herd and the *Comité de Estudios Vulcanologicos* six months after the

eruption.[23] Herd was a United States Geological Survey geologist who had worked on Ruiz shortly before the eruption; the Comité was an *ad hoc* group of Colombian and international scientists who studied the eruption. Recall that Nevado del Ruiz is a high volcano (5200 m) and that it has a permanent snow and ice cover (Fig. 13.20).

After nearly a year of intermittent precursory activity, Nevado del Ruiz erupted on Wednesday 13 November 1985. The onset of the eruption occurred with little warning. The day before, a group of Colombian geologists had climbed to the summit of Ruiz to collect gas samples from the fumaroles on the floor of Arenas crater. They noted no signs of the eruption that was to follow next day.

3.05 p.m. local time
The eruption of 13 November began abruptly with a strong phreatic explosion in Arenas crater shortly after

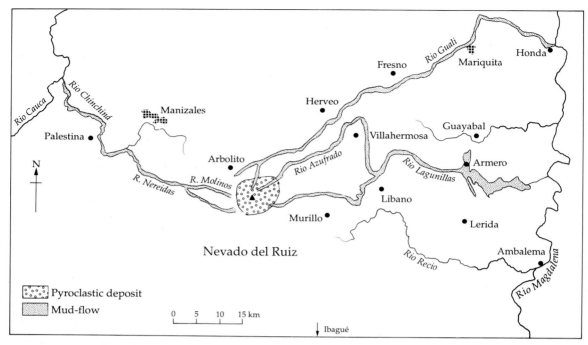

Fig. 13.20 Main drainages around Nevado del Ruiz, Colombia; extent of mud-flows generated by eruption of 13 November 1985; and principal villages and towns mentioned in the text. Area directly affected by pyroclastic flows in summit area is also shown.

3 p.m. Ranchers north of the volcano heard a deep rumbling and observed a black plume rise from the summit of Ruiz. A fall of fine lithic ash occurred northeast of the volcano, falling on Armero at about 5 p.m. Several towns east and north of the volcano reported a strong sulphur odour.

4.00 p.m.
About an hour after the eruption began, the director of the Civil Defense of Tolima, Coronel Rafael Perdomo S. was alerted at his office in Ibagué by radio that Ruiz was erupting ash and sulphurous odours. Perdomo called the INGOMINAS (*National Geological Survey*) regional director Alberto Nunez, telling him of the ongoing eruptive activity. Nunez recommended that Armero and Honda towns should be prepared for immediate evacuation.

5.00 p.m.
Members of the Emergency Committee of Tolima gathered in emergency session at the Red Cross Headquarters in Ibagué, where Nunez informed the committee of the eruptive activity. Referring to the volcanic hazards map of Ruiz, the committee discussed evacuation of Armero, Mariquita, and Ambalema and

the need to post observers along rivers to warn of possible mud-flow activity.

6.00 p.m.
Midway through the meeting, the committee suggested that Police Captain Gomez should alert police stations in Armero and its neighbouring towns.

7.30 p.m.
At the conclusion of the meeting, Nunez and others went to the Red Cross and insisted that Armero, Mariquita, and Honda be prepared for evacuation. It is understood that the Red Cross ordered the evacuation of Armero at 7.30 p.m.

9.08 p.m.
The paroxysmal eruption began with two strong explosions that were heard in Libano, more than 30 km from Arenas crater. A succession of pyroclastic flows and surges erupted from the crater, moving outwards across the surface of the northern summit ice-cap and down the steep flanks of the volcano. The flows and surges flattened grasses and stripped bark from shrubs in the Rio Azufrado valley 5 km northeast of the crater and destroyed the Refugio ski hut on

the west side of the valley. The flows and surges locally scoured and melted the surface of the ice-cap, triggering streams of melt-water, ice, and debris that ran from the summit down the volcano's west, north, and east flanks. The runoff was channelized into rapidly moving debris flows that surged down the Las Nereidas, Molinos, Guali, Azufrado, and Lagunillas rivers. The *lahars* scoured the canyon walls as they passed, sweeping up trees, brush, rocks, and soil.

9.30 p.m.
An eruptive column developed 15–30 minutes after the beginning of paroxysmal eruption, raining andesitic pumice north-east of Arenas crater. Boulder-sized blocks and bombs were hurled several kilometres, impacting with crater forming force. The plinian column rose thousands of metres into the air: the pilot of a passing commercial airliner reported that he saw a thick column of smoke and ash rising over 11 km. A regional ash fall occurred north-east of the volcano, with ash reported as far away as southwestern Venezuela—400 km from the volcano.

9.45 p.m.
Officials of Murillo alerted Civil Defense in Ibagué that Ruiz had exploded. Perdomo directed the Ibagué radio operator to contact Armero and order the town's evacuation. Attempts by Ibagué to contact Armero, however, were unsuccessful. (Survivors of the Armero disaster reported that the electricity went off several times that night.)

10.30 p.m.
Lahars originating in the head-waters of the Molinos and Las Nereidas rivers surged through Chinchiná, destroying 400 homes along the banks of the Rio Chinchiná, and severing the main road linking Manizales and Chinchiná. At about the same time, Ibagué Civil Defense heard Ambalema and Murillo radios warn Armero to evacuate because of 'an approaching avalanche'. Ibagué Civil Defence also overheard a radio conversation between Miguel Angel Perdomo, head of Civil Defense in Líbano, and Margarita Bejarano, secretary of the Civil Defense junta in Armero. Angel told Bejarano to flee Armero because the city was in great danger.

11.30 p.m.
Shortly before midnight, *lahars* invaded Armero and passed Mariquita. (Some authorities reported that the *lahars* reached Armero at 10.30 p.m.). The mud-flows in the Rio Guali carried away houses on the outskirts of Mariquita and destroyed the bridge on the main road to Bogotá. Survivors at Armero reported that successive waves surged into the centre of town, sweeping homes, cars, and debris eastward towards the Rio Magdalena. The multiple surges apparently originated as separate lahars in the head-waters of the Azufrado and Lagunillas rivers, and in the consequential failure of the landslide-dammed lake on the Lagunillas River at Canon de la Vereda El Cirpe, which lay in their path.

No one who saw the media coverage of the ruins of Armero can be unaware of the horror that overtook the town. Some of the photographs taken the following morning must number amongst the most haunting images of the century—a young girl trapped alive in swirling mud and water; her head above water but unable to move, dying slowly of cold and exposure. We shall never know exactly why the disaster occurred, but its essence concerned the response of the various authorities to a known hazard, rather than a purely random 'act of God'. In an insightful study, Barry Voight summarized the main events in the catastrophe, and drew some conclusions about how similar tragedies could be avoided in the future.[24] A curious aspect of his study is that it is punctuated by quotations from Albert Camus' novel 'The Plague'. Voight draws an analogy between the political situation in Armero, and that which prevailed in Oran (Algeria) during an outbreak of the plague, as described by Albert Camus in his book, which was in itself an allegory of the German occupation of France.

In volcanological terms, the Ruiz eruption drove home three lessons that were well-known previously, but which need to be constantly relearned:

First, a minor volcanic eruption can propagate devastating mud-flows. In itself, the Ruiz eruption was small, and of only passing interest to volcanologists. But the multiplying effect caused by melted snow and ice cascading downslope resulted in effects far more serious than those due to the eruption *per se*. At Ruiz, less than five million cubic metres of magma ejected as pyroclastic flows spawned ten to twenty million cubic metres of melt-water and sixty million cubic metres of *lahar* deposits.

Second, topography can focus mud-flows with

lethal effect. Nevado del Ruiz was dissected by numerous deep, steep and long canyons along which the mud-flows were bound to be channelled (Fig. 13.21). Armero was built at a site which in one sense was logical: on flat ground precisely at the mouth of the canyon of the Lagunillas River, a natural confluence of communication routes (Figs. 13.22 and 13.23). This logic was lethal, however, almost as dangerous as building a city in the muzzle of a cannon: mud-flows from the volcano were bound to sweep directly through the city. It is difficult to overemphasize the importance of this topographical focusing. At its maximum, debris swept down the Lagunillas canyon at about 47 000 cubic metres per second, a rate equivalent to about one-fifth that of the Amazon. Had Armero been built on slightly higher ground only a few hundred metres away, disaster would have been averted. A particularly tragic aspect of the

disaster was that *it had all happened before.* Writing in 1846, Jose Acosta described a mud-flow which swept over the site of Armero in 1845:

... then descending along the Lagunillas from its sources in the Nevado del Ruiz came an immense flow of thick mud which rapidly filled the bed of the river, covered or swept away the trees and houses, burying men and animals. The entire population perished in the upper part and narrower parts of the Lagunillas Valley.

In a later letter he wrote:

It is astonishing that none of the inhabitants of these villages, built on the solidified mud of old mass movements, has even suspected the origin of this vast terrain ... although ancient traditions testify to the frequent mud-flows in these regions. (Quoted in Voight 1990.)

Fig. 13.21 A view of the higher reaches of the Lagunillas Valley. Mud-flows confined to the narrow channel scoured the valley sides, exposing bare rock tens of metres above normal channel. People living in small farms above the valley, such as that at top left, were terrified by the noise of the descending mud-flows. US Geological Survey Photo: courtesy of Jack Lockwood.

Fig. 13.22 A view from an aircraft vertically above the site of Armero, two weeks after its destruction, looking westwards up the Lagunillas Valley. The town was located on flat ground where the narrow Lagunillas Valley opens out into the much broader Rio Magdalena Valley. US Geological Survey Photo: courtesy of Jack Lockwood.

Finally, even given the highly vulnerable location of the city, its inhabitants could have been saved if a rudimentary alarm system had existed. At least 90 minutes intervened between the onset of the eruption and the arrival of the mud-flows in Armero. This is not surprising: Armero is more than 50 km distant from the crater, and although mud-flows travel fast, they do not travel at the speed of light. Subsequent analyses showed that the Ruiz flows had travelled at an average speed of only 36 km/hour. A single phone call from an observer high up in the valley could have given time for many, perhaps most, people in Armero to escape to higher ground. Ruapehu volcano in New Zealand is now instrumented such that passage of a mud-flow down the higher reaches of a valley *automatically* triggers alarms for the lower slopes. Of

course, elaborate warning systems are now in place around Nevado del Ruiz . . .

13.2.3 Mud-flow deposits

Unlike those seen in the process of formation, identifying the deposits of prehistoric mud-flows presents problems, since mud-flows range from stiff sludges to highly fluid slurries. Deposits of fluid flows lack an overall shape, but more viscous ones flow like freshly poured concrete, and like concrete, they set rock-hard when they dry out. They have well-defined flow fronts, marginal *leveés*, and patterns of surface ridges (ogives) produced by faster-moving parts of the flow piling up on slower parts (Fig. 13.18).

Like avalanche deposits, mud-flows are unsorted; they contain fragments of all sizes from fine mud to boulders as big as houses (Fig. 13.24).

Fig. 13.23 Dozens of sacks of grain and an automobile form parts of the maelstrom of debris which is all that remained of Armero. Flow was from right to left. Few of the inhabitants caught up in the mud-flows survived. Many victims were never found. US Geological Survey Photo: courtesy of Jack Lockwood.

Fig. 13.24 A prehistoric *lahar* deposit, exposed on Tenerife, Canary Islands. A poorly sorted mixture of different clast lithologies is supported in a fine-grained matrix. This deposit is of modest grain-size. Metre-sized boulders are often found.

There is no trace of bedding, except at the bottom, where a finer-grained, sheared layer may be present. At the extreme top, laminated, finer-grained beds may be present, deposited during the waning stages of the flow, when its concentration has decreased to something approaching that of 'normal' sediment transport. These late, fine-grained layers often give mud-flow deposits noticeably flat tops. In a typical mud-flow, the larger clasts tend to be fairly widely spaced, with much fine matrix in between, so that the deposit has a rather open appearance. As in block and ash and avalanche deposits (and for the same reasons) there may be a marked upwards increase in clast size, so that the largest boulders often stick up through the surface of the deposit.

As Nevado del Ruiz demonstrated, volcanic mud-flows may travel far from their sources, sometimes over 100 km. One of the biggest known deposits is the Osceola mud-flow in the State of Washington, USA, which originated on Mt. Rainier, near Seattle in the USA. Mt. Rainier is a huge volcano, one of the largest in North America. It is 4000 m high and thickly draped in glaciers. The Osceola mud-flow travelled at least 112 km along the west fork of the White River Valley, spread out to cover an area of over 300 square kilometres, and modified the shoreline of Puget Sound.[25–27] That happened about 5700 years ago, but Mt. Rainier is known to have produced more than fifty major mud-flows in the last 10 000 years. A major mud-flow descends the White River Valley about once every 600 years. Since the valleys overrun by the Osceola mud-flow are now densely populated, with many settlements east of the city of Auburn built on the mud-flow deposits, one can only hope that there will be ample warning of future eruptions of Mt. Rainier . . .

13.2.4 *Flow mechanisms*

There is a continuum of debris flows, from thin floods of muddy water to bouldery debris flows. Wet, muddy suspensions are of more interest to sedimentologists than volcanologists, so we will ignore them. Some volcanic *lahars* have significant yield strengths, as the huge boulders on their surfaces confirm. But how do they manage to flow so freely and rapidly, even on gentle slopes? Why do they not flow more like viscous, oozy lavas? This question was addressed in a classic paper by Rodine and Johnson,[28] elegantly titled 'The ability of debris, heavily freighted with coarse clastic materials, to flow on gentle slopes'. They showed that three factors are important:

First, whereas in other varieties of pyroclastic flows, the voids between clasts are occupied only by air, or a fluidizing medium of gas and fine particles, in a mud-flow, the void spaces contain a cohesive slurry of extremely fine-grained clay and water. There is therefore much less of a density contrast between large clasts and the medium that supports them. Thus, the largest boulders can 'float' in a mud-flow more easily, buoyed up by the denser intergranular medium.

Second, the huge range in grain sizes found in mud-flows means that the intergranular medium can act more efficiently in lubricating the mud-flow than would be the case with better-sorted flows. Experimental work by Rodine and Johnson showed that when particles of a single size (sand) were mixed with a water–clay slurry, the flow 'locked' when only 45–55 per cent of solid material was present. When the range of grain sizes was increased, flows continued moving until 90–95% of it was made of solid materials! It follows that a remarkably small proportion of water is needed to make the mud to mobilize a mud-flow.

Third, and most important, the water–clay slurry that surrounds and separates all the clasts provides its lubricating effect by reducing the stress caused by one clast on another, thus reducing the friction between clasts. A fraction of the load of the whole mass is borne by the slurry. If the whole load were borne by the slurry, the entire mass of debris would be virtually frictionless.

— Notes

1. Siebert, L. (1984). Large volcanic débris avalanches: characteristics of source areas, deposits and associated eruptions. *J. Volcanol. Geotherm. Res.* **22**, 163–97.

2. Ui, T. (1983). Volcanic dry avalanche deposits—identification and comparison with nonvolcanic debris stream deposits. *J. Volcanol. Geotherm. Res.* **18**, 135–50.

3. Francis, P. W., Gardeweg, M., O'Callaghan, L. J., Ramirez, C. F., and Rothery, D. A. (1985). Catastrophic debris avalanche deposit of Socompa volcano, north Chile. *Geology* **13**, 600–603.

4. Naranjo, J. and Francis, P. W. (1987). High velocity debris avalanche at Lastarria volcano, north Chile. *Bull. Volcanol.* **49**, 509–14.

5. Francis, P. W. and Self, S. (1987). Collapsing volcanoes. *Sci. Am.* **287**, 90–9.

6. Francis, P. W. and Wells, G. L. (1988). Landsat Thematic Mapper observations of large volcanic debris avalanche deposits in the Central Andes. *Bull. Volcanol.* **50**, 258–78.

7. Siebert, L., Glicken, H. X., and Ui, Tadahide (1987). Volcanic hazards from Bezymianny- and Bandai-type eruptions. *Bull. Volcanol.* **49**, 435–59.

8. Shreve, R. L. (1968). The Blackhawk Landslide. *Geol. Soc. Am. Spec. Pap.* **108**.

9. Voight, B., Glicken, H., Janda, R. J., and Doug-

lass, P. M. (1981). Catastrophic rockslide avalanche of May 18. In *The 1980 eruptions of Mount St. Helens, Washington*, (eds. P. W. Lipman and D. R. Mullineaux), Geol. Surv. Prof. Pap. No. 1250, pp. 347–78.

10. Melosh, H. J. (1987). The mechanics of large rock avalanches. *Geol. Soc. Am. Rev. Eng. Geol.* **VII**, 41–9.

11. Bagnold, R. A. (1954). Experiments on a gravity-free dispersion of large solid spheres in a Newtonian fluid under shear. *Proc. Roy. Soc. Lond.* **A225**, 49–63.

12. Melosh, H. J. (1983). Acoustic fluidization. *Am. Scient.* **71**, 158–65.

13. Campbell, C. S. (1989). Self-lubrication for long run-out landslides. *J. Geol.* **97**, 653–65.

14. Stoopes, G. R. and Sheridan, M. F. (1992). Giant debris avalanches from the Colima Volcanic Complex, Mexico: implications for long-runout landslides (>100 km) and hazard assessment. *Geology*, **20**, 299–302.

15. Crandell, D. R. (1989). Gigantic debris avalanche of Pleistocene age from ancestral Mount Shasta volcano, California, and debris avalanche hazard zonation. *U.S. Geol. Surv. Bulletin*, **1861**, 32 pp.

16. Siebert, L., Glicken, H., and Kienle, J. (1989). Debris avalanches and lateral blasts at Mount St. Augustine volcano, Alaska. *Nat. Geog. Res.* **5**, 232–49.

17. Kienle, J., Kowlaik, Z., and Murty, T. S. (1987). Tsunamis generated by eruptions from Mount Saint Augustine Volcano, Alaska. *Science* **236**, 1442–7.

18. Begét, J. E. and Kienle, J. (1992). Cyclic formation of debris avalanches at Mt. St Augustine volcano.

Nature, **356**, 701–4.

19. Moore, J. G., Clague, D. A., Holcomb, R. T., Lipman, P. W., Norman, W. R., and Torresan, M. E. (1989). Prodigious submarine landslides on the Hawaiian ridge. *J. Geophys. Res.* **B12**, 17 465–84.

20. Moore, J. G. and Moore, G. W. (1984). Deposit from a giant wave on the island of Lanai, Hawaii. *Science* **226**, 1312–15.

21. Zen, M. T. and Hadikusumo, D. (1965). The future danger of Mt Kelut (eastern Java–Indonesia). *Bull. Volcanol.* **28**, 275–82.

22. Williams, S. N. (ed.) (1990). Nevado del Ruiz volcano, Colombia. *J. Volcanol. Geotherm. Res.* **42**.

23. Herd, D. G. and Comité de Estudios Vulcanologicos (1986). The 1985 Ruiz volcano disaster. *EOS* **67**, 457–72.

24. Voight, B. (1990). The 1985 Nevado del Ruiz volcano catastrophe: anatomy and restrospection. *J. Volcanol. Geotherm. Res.* **42**, 151–88.

25. Crandell, D. R. (1971). Post-glacial lahars from Mt. Rainier volcano, Washington, *US Geol. Surv. Prof. Pap.* **677**, 75 pp.

26. Crandell, D. R. and Waldron, H. H. (1956). A recent mudflow of exceptional dimensions from Mt. Rainier, Washington. *Am. J. Sci.* **254**, 349–62.

27. Swanson, D. A., Malone, S. D., and Samora, B. A. (1992), Mount Rainier: A decade volcano. *Eos*, **73**, 171–8.

28. Rodine, J. R. and Johnson, A. M. (1976). The ability of debris, heavily freighted with coarse clastic materials, to flow on gentle slopes. *Sedimentology* **23**, 213–34.

14

Caldera complexes and complex calderas

A horseshoe-shaped amphitheatre two kilometres in diameter was excavated by the May 1980 eruption of Mt. St Helens when 0.6 cubic kilometres of magma were ejected. It was a stunning spectacle. A decade later, photographs of the towering ash column were still widely reproduced. But Mt. St Helens' paroxysm was a damp squib compared to the fireworks that happened

(a)

(b)

Fig. 14.1 (a) 'Old Faithful' geyser, Yellowstone caldera, Wyoming, USA, illustrated by T. G. Bonney in his 1899 volcanology textbook, before the nature of the caldera was understood. (b) A group of admirers watches Old Faithful doing its thing in 1989. Little has changed except for the number of spectators. Photo: courtesy of Peter Mouginis-Mark, University of Hawaii.

600 000 years ago at Yellowstone, 950 kilo-metres to the east. At least 1000 cubic kilometres of pyroclastic material were erupted, forming an elongate caldera 70 kilometres across. The site has since been scoured by glaciers and camouf-laged by vegetation, so there is little to see today. One ebullient vestige is the Old Faithful geyser in Yellowstone National Park (Fig. 14.1). Yellow-stone is a product of volcanic processes operating on the grandest scale: it is a *resurgent caldera*, a vast crater whose floor initially subsided but was heaved up again, many millenia after the erup-tion. Resurgent calderas are the largest volcanic structures on Earth; so large that the eruptions that form them must rank among the greatest natural catastrophes, comparable with the impact of an asteroid.[1]

— 14.1 Cauldron subsidence

Caldera-forming eruptions are of such large scale that they present formidable problems to volca-nologists striving to understand them. When confronted by large craters surrounded by thick blankets of pumice, it was natural for the earliest geologists to think in terms of stupendous explosions, which simply blew the tops off earlier volcanic edifices. Few calderas are formed so straightforwardly. Volcanological thinking about calderas started at Santorini in the Aegean, where the *c.*1620 BC Minoan eruption produced a superb caldera eight kilometres in diameter, whose walls are now defined by the steeply cliffed islands of Thera, Therasia, and Aspronisi, encircling an active vent on Nea Kameni (Fig. 4.2).

In a study which deserves broader recognition among the *incunabula* of volcanology, Ferdi-nand Fouqué, a French geologist, showed that the caldera could not have been formed simply by destruction of an earlier cone, since there was not enough material from the cone present as lithic clasts among the magmatic pumice to recon-struct the pre-existing volcano. In his 1879 book on Santorini, he concluded that the 'missing' part of the volcano had foundered below sea-level.[2]

A further step in caldera studies was made in a sharply different environment, in the gloomy

Fig. 14.2 Cauldron subsidence, as envisaged by Maufe and Bailey. They realized that the phenomenon was a complex one, expressed in a variety of different ways. At left, eruptions take place from ring fractures; second from left, eruptions continue from a central vent after cauldron subsidence along ring fractures. (After E. B. Bailey and H. B. Maufe 1960.)

mists of Rannoch Moor, Scotland. Here, among the fobidding mountains of the Pass of Glencoe, C. T. Clough, H. B. Maufe, and E. B. Bailey showed in 1909 that a caldera eight kilometres in diameter had formed in Glencoe after major explosive eruptions 360 million years ago.[3] They also proposed that the caldera formed because eruption of huge volumes of pyroclastic rocks had left the roof of the magma chamber unsupported, causing it to founder downwards (Fig. 14.2). They termed this process *cauldron subsidence*.

Cauldron subsidence forms calderas varying in diameter from a few kilometres to as much as one hundred. At its simplest, cauldron subsidence can be likened to a piston subsiding within its cylinder: subsurface *ring dykes* are the geological expression of the cylinder, while the subsided block and thick accumulations of welded ignimbrites form the piston. An ancient cauldron subsidence, the Sabaloka Complex in the Sudan, displays these features outstandingly (Fig. 14.3). A caldera 25 by 15 km across has been cut through by the Nile, exposing beautifully preserved ring dykes.[4]

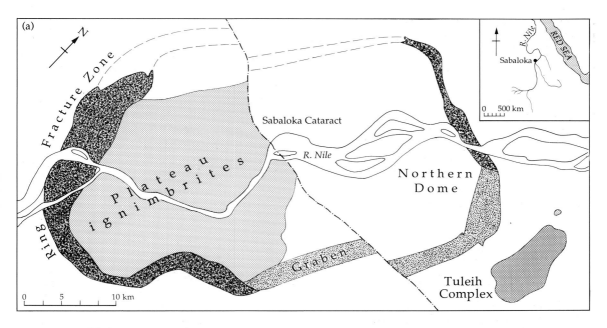

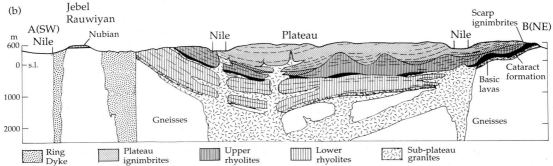

Fig. 14.3 (a) Outline map of Sabaloka igneous complex, Sudan, showing ring dykes and 'plateau' ignimbrites (shaded) filling cauldron subsidence. (b) Cross-section showing ignimbrites, ring dyke, and subsided roof of magma chamber. Sabaloka is thought to be about 700 million years old. After Almond, D. C. (1977). The Sabaloka igneous complex, Sudan. *Phil. Trans. Roy. Soc. Lond.* **A287**, 595–633.

14.1.1 Mount Mazama, 6845 BP

Crater Lake, Oregon, is one of the most exquisite lakes in North America, but its serenity belies a violent origin. Its intensely blue waters fill a caldera nine kilometres in diameter, whose floor is 600 metres below the lake level, while the encompassing walls rise steeply 600 metres above it. Wizard Island, a small volcanic cone, occupies one corner of the lake (Fig. 14.4). Howel Williams, a distinguished American volcanologist of Welsh extraction, concluded in a classic study that the present caldera occupies the site of an older volcanic cone, about 3600 metres high.[5–6] He called this ancestral volcano Mt. Mazama, and suggested that it had been glaciated during the Ice Age, because the remains of valleys filled with glacial debris are terminated by the walls of the present caldera. Some 6800 years ago, a huge plinian eruption occurred, which showered tephra as far as Alberta in Canada.

After the plinian phase, a series of ignimbrites was erupted, filling the valleys radiating from the old cone. While the ignimbrites were being erupted, the main mass of Mt. Mazama foundered downwards by cauldron subsidence to form the present caldera. Williams' original mapping suggested that a total of about forty cubic kilometres of air-fall material and ignimbrite were erupted. He estimated however, that a

volume of *sixty* cubic kilometres would be required to reconstruct the original Mt. Mazama. He concluded that the 'missing' twenty cubic kilometres had to be accounted for by 'withdrawal of magmatic support from below'. His diagrams illustrating the formation of the caldera summarize a standard model for the origin of some calderas, which we can call *Crater-Lake type calderas* (Figs 14.5 and 14.6).

In a more recent study, Charles Bacon of the USGS confirmed that caldera subsidence took place while ignimbrites were being erupted from circular ring fractures, but suggested that the eruption consisted of two distinct phases; an early, single vent phase, and a later ring fissure phase.[7] During the early phase, a plinian eruption column ascended from a small, single vent, located near the centre of Mt. Mazama. Widening of the vent during the eruption caused column instability, and some pyroclastic flows resulted, spreading out mostly to the north and east of the volcano. About 30 cubic kilometres of magma were erupted during this phase, leaving the roof of the magma chamber unsupported. This then foundered downwards along ring-fractures, to form the floor of the present caldera. Formation of the ring-fractures opened up a number of vents around them, and eruption of pyroclastic flows then continued from these multiple vents. About 13 cubic kilometres of

Fig. 14.4 Wizard Island, a young volcanic cone, rises above the waters of Crater Lake, encircled by the walls of the 6800-year-old caldera.

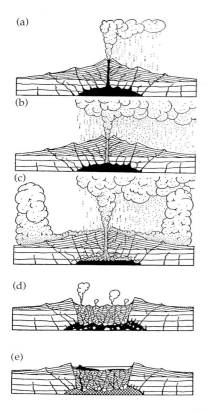

Fig. 14.5 Howel Williams' classic diagrams illustrating the formation of Crater Lake caldera, Oregon. His original captions: (a) Beginning of culminating eruption; magma high in conduit; mild eruption of pumice. (b) Activity increases in violence. Showers of pumice more voluminous and ejecta larger. Magma level lowers to top of feeding chamber. (c) Activity approaches climax. Combination of vertically directed explosions with glowing avalanches (*nueés ardentes*). Chamber being emptied rapidly; roof commencing to fracture and founder. Magma also being drained from the chamber through fissures at depth. (d) Collapse of the cone as a jumble of enormous blocks, some of which are shown sinking through the magma. Fumaroles on the caldera floor. (e) Crater Lake today. Post-collapse eruptions have formed the cone of Wizard Island and have probably covered parts of the lake bottom with lava. Magma in the chamber largely crystallized.

Fig. 14.6 Vertical Space Shuttle photograph of Aniakchak, a magnificent young 11-km-diameter Crater-Lake-type caldera in Alaska, formed by an eruption about 3500 years ago. Ignimbrites spread 50 km from the caldera. An eruption in May 1931 deposited 0.6 cm of ash 200 km distant. NASA STS 17-33 043.

material were erupted during this second stage of the formation of Crater Lake caldera.

Bacon looked again at Williams's conclusion that mysterious 'withdrawal of magmatic support from below' was required to account for 'missing' material. Through some careful geological accountancy, he estimated the volumes of the various pyroclastic components, converted them to dense-rock equivalents, and concluded that the total erupted magma volume of about 50 cubic kilometres matched reasonably closely with the volume of 'missing' rock represented by the caldera and the vanished upper parts of Mt. Mazama, eliminating the need to invoke 'magma withdrawal'. Eruption of huge volumes of pyroclastic material caused the caldera floor to founder into the magma chamber, while the magmatic material that originally occupied the upper part of the magma chamber is now spread thinly over vast areas of the State of Oregon and beyond.

Tephra from Mt. Mazama forms an important time marker in sedimentary deposits over much of the north-western USA. However 'recent' those deposits may be to geologists, in everyday terms the demise of Mt. Mazama is lost in the mists of time. Tambora's eruption in 1815 provides a close parallel in modern times.[8] It had such profound effects world-wide that the following year became notorious as the 'Year without a Summer' (Section 17.4.2). The topmost kilometre of the volcano disappeared during the course of a few hours on the afternoon of 11 April 1815 when 30–50 cubic kilometres of pyroclastic material were erupted (Section. 10.4.4). Cauldron subsidence produced a caldera seven kilometres in diameter and 1.3 kilometres deep, resembling Crater Lake, except that Tambora's caldera is dry.[9] Many other small- to medium-sized calderas around the world formed in much the same way. One magnificent example is the five kilometre-diameter Deriba caldera, deep in the heart of the Sahara Desert, formed around 2000 BC (Fig. 16.25).

— 14.2 Resurgent calderas

Crater Lake-type calderas are genetically related to pre-existing volcanic edifices such as Williams's Mt. Mazama. Resurgent calderas are a different kettle of fish. While some volcanic activity usually precedes their formation, the scale of these calderas is so enormous that any pre-existing ordinary volcano pales into insignificance, and indeed may even be entirely swallowed up by the caldera. Sheer size apart, the definitive feature of a resurgent caldera is slow post-eruption upheaval or *resurgence* of its floor. Vertical uplift of more than one kilometre often takes place. Unlike an ordinary volcano, therefore, a resurgent caldera forms a broad topographical depression with a central elevated massif. Resurgence as a phenomenon was first recognized in 1939 by the distinguished Dutch geologist R. W. van Bemmelen, through his work on the vast Toba caldera in northern Sumatra.[10]

Like their smaller counterparts, many resurgent calderas are filled by lakes soon after their formation. At Toba, van Bemmelen noted that about two kilometres of subsidence had taken place initially to form a deep lake almost 100 kilometres long and 30 km across. Subsequent uplift of the floor formed Samosir, an island covering 640 square kilometres. His evidence for the uplift speaks for itself: sediments deposited below lake level are now found on Samosir hundreds of metres above the lake (Fig. 14.7). Toba is the youngest resurgent caldera in the world. It was formed only 74 000 years ago by a massive eruption which caused ash fall as far distant as northern India, 2000 kilometres away (Fig. 14.8). Youthful though Toba is, 74 000 years is plenty of time for luxuriant tropical vegetation to colonize new volcanic terrain. Toba is therefore not well exposed and difficult to study. Craig Chesner and Bill Rose of Michigan Technical University tackled the problem, and showed that four separate eruptions over the last 1.2 million years were involved in the formation of the Toba Complex, with repose periods of 340 000– 400 000 years between eruptions. Their work showed that the 74 000-year-old Youngest Toba

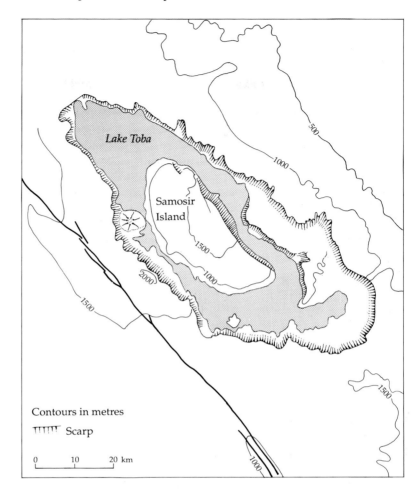

Fig. 14.7 Toba caldera, Lake Toba, and Samosir Island. Surface of Lake Toba is about 900 m above sea-level; the highest point on Samosir Island is 1630 m. Resurgence of about 500 m has taken place. The caldera is elongated parallel to major regional faults, and is not a simple structure—more than one episode of subsidence was probably involved in its formation.

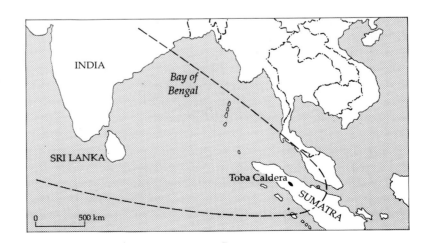

Fig. 14.8 Ash from Toba caldera rained down 75 000 years ago over most of the Bay of Bengal, and parts of Sumatra, Sri Lanka, and India. Oceanographical work by members of Lamont–Doherty Geological Observatory, Columbia University, revealed ash ten centimetres thick in sea-floor core samples up to 2000 km distant.

Tuff has a volume of about 2800 cubic kilometres.[11]

14.2.1 Resurgent calderas of the USA

More favourable conditions than Toba could provide were required to elucidate the mechanisms of resurgent caldera formation. These were found in the south-western USA, where geologists of the United States Geological Survey first unravelled the complexities of some major calderas. Indeed, the term *resurgent caldera*—or cauldron—was itself first coined by Robert L. Smith and Roy S. Bailey of the USGS in 1962, during mapping of the Valles caldera in the Jemez Mountains of New Mexico, a study which influenced a generation of volcanologists.[12]

There are three important young resurgent calderas in the western USA, each in a different tectonic environment. Valles caldera, which last erupted one million years ago, is located on the edge of the Rio Grande Rift, in the heart of the continental USA, where lithospheric thinning and extension have been active for millions of years (Fig. 14.9). In California, Long Valley Caldera, which erupted about 700 000 years ago, is located immediately east of one of the great faults bounding the Sierra Nevada mountains. Wes Hildreth used reconnaissance petrological data on the Bishop Tuff (the ignimbrite erupted from Long Valley Caldera) and an unusually circular argument to suggest that most of the silicic magma chambers underlying calderas are compositionally zoned: silicic tops overlying more mafic bases.[13] He also convincingly advocated a school of thought that argues that *all* volcanism, even the most silicic, is fundamentally basaltic in origin.

Yellowstone, the third of the major young calderas in the USA, last erupted about 600 000 years ago, showering tephra as far distant as Louisiana (Fig. 14.10). Yellowstone is located above a mantle hot-spot trace

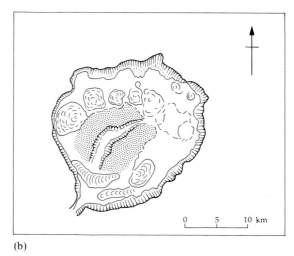

(a)   (b)

Fig. 14.9 (a) Landsat RBV image of Valles Caldera, New Mexico, showing topographical rim, resurgent centre with apical graben, and necklace of post-caldera rhyolite lavas. The youngest eruption within the caldera was the Banco Bonito lava, in south moat of caldera about 200 000 years ago. Prominent radial valleys are incised in outflow ignimbrites from the caldera—the Bandelier Tuffs. A fault scarp is prominent at left. Caldera is 24 km in diameter. (b) Sketch of Valles Caldera delineating main features visible on satellite image. Resurgent centre shaded. Structural margin of caldera probably coincides with lava vents.

Fig. 14.10 Distribution of ash fall from the three major resurgent calderas in the western USA. Mapping was done largely by G. A. Izett and his colleagues in the U. S. Geological Survey.

(Section 2.5.1); it marks the present site of a blow-torch which has been burning its way progressively north-eastwards from the flood basalts of the Snake River Plain (Fig. 14.11). Like Long Valley, Yellowstone is far from extinct. At present, it is the site of much the largest heat-flow anomaly in the USA. Seismic evidence suggests that a magma body is still located beneath the caldera. Changes in lake level since the area was surveyed also indicate continued activity.

Study of these young calderas, and a host of older ones, encouraged Smith and his co-workers to suggest that several well-defined phases could be identified in the evolution of large calderas: pre-caldera volcanism and uplift; eruption of pyroclastic rocks; caldera formation; post-caldera lava extrusion; and resurgence. Typically the resurgent block itself exhibits readily recog-

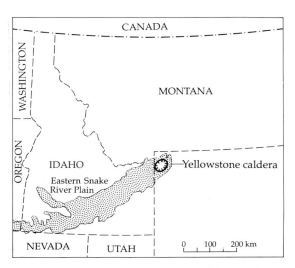

Fig. 14.11 Location of Yellowstone caldera at the end of the hot-spot trace defined by the young volcanics of the Snake River Plain.

nizable tectonic evidence of uplift and extension, such as fracturing and development of graben along the crest of the arch—the keystone often seems to have slipped.

14.2.2 *Cerro Galan, north-west Argentina: case study of a caldera*

Valles, Long Valley, and Yellowstone are the most closely studied calderas in the world simply because they are so accessible. Many Ph.D. theses have been written on them. A major research facility, the Los Alamos National Laboratory is located on the very flanks of the Valles caldera. (Scientists of the Manhattan Project built the A-bomb at Los Alamos. It is arguable that a major caldera may not have been an ideal place to site such a facility.) In recent years, remote sensing satellites have revealed some less-accessible but better-exposed calderas in the central Andes. One of these, Cerro Galan, provides an instructive case study, both of its discovery and its evolution.

Cerro Galan is located on the high-altitude *puna* of north-west Argentina, near the small hamlet of Antofagasta de la Sierra. Although it is the most explicit example of a large caldera in the world, it remained undiscovered until the advent of space craft imagery in the 1970s. Unknowingly, geologists had traversed across the structure, and there was even an excellent set of air photographs covering it, but Cerro Galan remained camouflaged by its own scale. No one mosaiced the air photographs together, so it was not until space craft provided a synoptic view that the caldera jumped into startlingly clear focus. Cerro Galan was first snapped serendipitously by astronauts using hand-held Hasselblad cameras during the Apollo programme. Because these photographs were not systematically studied among all the exhilaration of the Moon missions, the caldera went unnoticed for a few years longer. In 1973, scientists at Los Alamos (a happy coincidence) picked out the caldera as a 35-km-diameter elliptical structure with a central mountain peak, heavily mantled with snow, on photographs from the Skylab missions.[14]

Independently, my colleague M. C. W. Baker and I identified Cerro Galan during a specific search for calderas in the Central Andes using images from the Landsat remote sensing satellite.[15-16] Discovery of the enormous caldera proved to be a mixed blessing, because it consumed more than a decade of research activity. Much more remains to be done. Apart from its great size, the altitude of the caldera presents formidable obstacles to field-work. Its western rim rises over 5200 metres; its floor is at a mean height of 4500 m and the summit of its resurgent centre is at 6000 metres. The logistics of mapping the caldera were so great that an expedition of British and Argentinian geologists joined forces to tackle the caldera in 1981. Support was provided by British and Argentinian army groups, helped by local mule teams and their drivers. War broke out between Britain and Argentina shortly afterwards.

Early history   Like other major calderas, Cerro Galan is the product of a long volcanic evolution. While the existing caldera was formed 2.2 million years ago, the events leading up to its explosive birth commenced more than 10 million years ago. Between about 12 and 7 million years ago, andesitic and dacitic volcanoes began to erupt on what are now the western flanks of the caldera, forming respectably sized 'ordinary' volcanoes such as Cerros Beltran and Colorado, both more than 5000 m high. At the same time, *basaltic* volcanism commenced in the valley of Antofagasta de la Sierra. Lava flows were erupted from numerous small scoria cones and vents localized on prominent north–south faults. They are crisply preserved today in the high desert environment of the valley. Basaltic activity continued intermittently through to the present day: some of the youngest scoria cones may be only a few thousand years old (Figs 14.12–14.15).

About five million years ago, eruption of voluminous ignimbrites commenced. Eruption magnitude increased with time, culminating 4.2 million years ago with the eruption of the Real Grande and Cueva Negra ignimbrites, with volumes of at least 500 cubic kilometres. These vast outbursts most probably produced their own large calderas, subsequently obliterated by the even larger eruption of the Cerro Galan ignimbrite. Precipitous cliffs soaring hundreds of metres high on the western flanks of

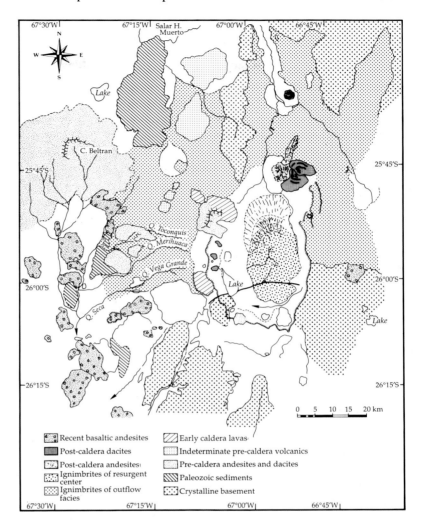

Fig. 14.12 Outline geological map of the Cerro Galan caldera and adjacent region, based on Landsat TM photo-geological studies and field-work. Compare with Fig. 14.13.

Recent basaltic andesites
Post-caldera dacites
Post-caldera andesites
Ignimbrites of resurgent center
Ignimbrites of outflow facies
Early caldera lavas
Indeterminate pre-caldera volcanics
Pre-caldera andesites and dacites
Paleozoic sediments
Crystalline basement

the present caldera now expose the ignimbrites erupted during the preceding episodes. Tinged pink, white, and brown, the cliffs provide stunning vertical sections through ignimbrites (Fig. 14.16).

Caldera formation   Between 4.8 and 4.2 million years ago, some large rhyodacitic lava domes were extruded from fractures on the margins of the early calderas. A 1.8-million-year period of quiet ensued, until 2.2 million years ago when a colossal eruption of a minimum of 1000 cubic kilometres of crystal-rich ignimbrite formed the present caldera. Thanks to superb Landsat Thematic Mapper images, the present outcrop of

the Cerro Galan ignimbrite can be traced easily. Its outflow sheet reaches up to 100 kilometres radially from the caldera, running hundreds of metres up the slopes of older mountains such as Cerro Beltran. Where the ignimbrite ponded tens of metres deep in pre-existing valleys, magnificent palisades of columnar-jointed pink ignimbrite are now exposed (Fig. 14.17). Within the confines of the caldera itself, huge volumes of ignimbrite accumulated—preliminary mapping suggests a *minimum* thickness of 1.2 kilometres.

Interestingly, there is no evidence that eruption of these enormous volumes of pyroclastic flows was preceded by a plinian air-fall episode (cf. Section 10.2.1). At Cerro Galan, caldera

Fig. 14.13 Landsat Thematic Mapper of Cerro Galan caldera. Image is roughly 135 km by 100 km. Caldera is prominent on the right side of the image, surrounded by light-toned, deeply-gullied ignimbrites. Basaltic andesite satellite vents (dark) are prominent in valley of Antofagasta de la Sierra west of caldera. Laguna Diamante, in south-west corner of caldera, is a salty relic of a larger lake.

subsidence probably took place abruptly, on downward widening ring fractures, so that enormous mass eruptions rates were immediately established, preventing the formation of a stable convecting plinian column.

As in most calderas, the *caldera fill* facies and the *outflow* facies show a number of textural and lithological variations as well as differences in thickness. Whereas the relatively thin outflow facies is generally unwelded, the much thicker caldera fill facies is densely welded. (In older volcanic terrains, discovery of huge thicknesses of densely welded ignimbrite is often a good pointer to the site of a caldera.)

In small calderas, such as Crater Lake, there is nothing to be seen in the caldera walls but volcanic rocks; components of the earlier volca-

nic massif. At Cerro Galan, the caldera walls expose Palaeozic crystalline basement rocks, gneisses, and schists, capped with a relatively thin veneer of ignimbrites. This emphasizes the scale of resurgent calderas: they are far larger than ordinary volcanoes, and punch *through* the upper crust, rather than merely being built on *top* of it.

Resurgence and after   For the first few millennia after its birth, the Cerro Galan caldera was probably a simple, though awesomely deep and large, open depression. Avalanching and erosion of its walls soon softened its contours, and sediments accumulated on its floor. At one time, the caldera also enclosed a large lake, of which the present salty, turquoise-coloured Laguna

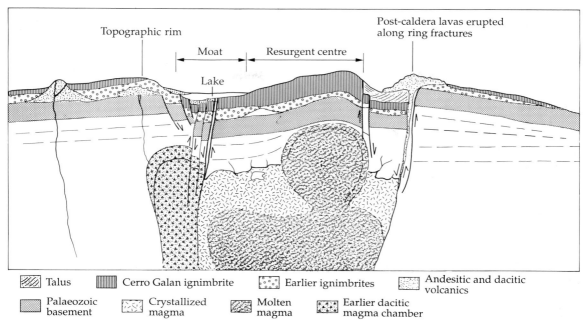

Fig. 14.14 Cross-section of Cerro Galan caldera, showing principal structural and stragraphical units.

Fig. 14.15 Isolated peaks of Palaeozoic metamorphic basement rocks poke up through sediments filling the floor of the valley of Antofagasta de la Sierra, west of Cerro Galan. They are partially drowned by young basaltic andesite lavas from satellite vents. The symmetrical shield in the distance is Carachipampa.

Fig. 14.16 Ignimbrites totalling several hundred metres in thickness are exposed on the western flanks of the Cerro Galan caldera. Most of the thickness seen here is about four million years old; only the topmost crag is the 2.2-million-year-old Cerro Galan ignimbrite.

Diamante is but a shallow relic (Fig. 14.18). Subsequently, resurgence of the centre arched the floor of the caldera and tilted the ignimbrites and their sediment cover. Resurgence was not a simple piston-like rise of the caldera floor along the ring fracture. The caldera floor was elevated asymmetrically, the eastern part being uplifted most, so that the ignimbrites dip away from this highest point. This suggests that resurgence was a local phenomenon, perhaps taking place above a small pluton newly intruded above the one that caused the eruption. In other cases, as at Yellowstone, two separate centres of resurgence are present in one caldera.

Van Bemmelen demonstrated that resurgence of the Toba caldera had taken place by showing that young lake sediments had been raised by hundreds of metres. At Cerro Galan, resurgence of more than one kilometre raised the centre of the caldera to an altitude of six kilometres above sea-level, making it one of the highest mountains in Argentina. Little is known about how long resurgence took, but at Long Valley, a caldera for which a uniquely detailed radiometric chronology exists, the process probably continued for up to 200 000 years.[17]

After resurgence at Cerro Galan, dacitic lavas oozed quietly to the surface about 2.1 million years ago from a vent on the northernmost part of the caldera ring fracture, piling up in thick *coulées*. These lavas may represent the last batches of the same magma that fed the ignim-

Fig. 14.17 Columnar jointing in outflow facies of the Cerro Galan ignimbrite, near Antofagasta de la Sierra. The cliff is about 20 metres high.

Fig. 14.18 The resurgent centre of Cerro Galan seen from the western rim of caldera. The summit of the resurgent centre is about 6000 metres; caldera moat in the middle distance is about 4500 metres altitude. Talus eroded from caldera walls and flat-lying lake sediments form the desert expanse in the middle distance. Centuries before volcanologists reached the summit of Cerro Galan, hardy Incas had left small silver votive offerings there.

brites; magma which was completely 'flat', having lost much of its volatile content. Only one set of lavas was erupted at Cerro Galan, but at Long Valley separate episodes of effusion took place 500 000, 300 000, and 100 000 years ago. Activity at Long Valley may not yet be over. Periodic seismic events and ground deformation point to subterranean magma movements which could cause future eruptions. At Cerro Galan, there are no historical records of seismic activity or ground deformation, but boiling springs at several locations within the caldera (one of them unimaginatively called *Aguas Calientes*) confirm that Cerro Galan should not be written off as a system whose course is run.

14.2.3 Mechanism of resurgence

Although resurgence defines the eponymous class of calderas, the phenomenon itself is poorly understood. Semantically speaking, *resurgence* implies resurrection in the religious sense of being raised from death to life. Many calderas do indeed return to violently explosive life after long periods of deathly quiet, but volcanologists use the term resurgence for the quiet upheaval of the caldera floor after an eruption. Is this rebirth, or merely the twitching of a cadaver as life finally ebbs away?

Two possible causes of resurgence were suggested by Bruce Marsh.[18] He argued that because the process takes between 1000 and 100 000 years to accomplish, it must be regulated by a highly viscous structural member of the crust, not by magma alone. As Smith and Bailey recognized, caldera formation is preceded by a long period of doming, or tumescence, over a broad area, much larger than that of the caldera itself. After eruption *detumescence* may compress magma remaining in the magma chamber, squeezing it up against the caldera floor, which responds by arching upwards.

Arrival of new magma in the magma chamber could also, of course, increase pressure, but Marsh estimates that this is likely to cause uplift of only a few hundred metres. Observed uplifts are far greater than this—more than 1 km—and so Marsh suggested that if magmatic processes are involved in resurgence, they are more likely to involve vesiculation. If magma left in the magma chamber after eruption rises towards the surface, decreasing pressure will permit its water content to exsolve, causing a large increase in volume which will in turn be expressed in an upwards pressure on the caldera floor. Of course, detumescence and magma vesiculation may operate simultaneously.

— 14.3 Caldera complexes —

Large silicic calderas often occur in clusters or *complexes*, a fact first demonstrated in a study of the San Juan Mountains of Colorado by USGS geologists T. A. Stevens and P. W. Lipman.[19] Their mapping of this magnificent mountain terrain revealed at least 18 separate calderas between 22 and 30 million years old. Ignimbrites from the different calderas cover 25 000 square kilometres and are stacked one on top of each other in stratigraphical order. Calderas overlap one another and are nested within one another, often making it difficult to see where one caldera starts and another ends (Fig. 14.19). The largest caldera is the 27.8-million-year-old La Garita caldera, more than 30 km in diameter, and source of the Fish Canyon Tuff, which has a volume estimated at an incredible 3000 cubic kilometres. Best preserved is the Creede caldera, less than 20 km in diameter, which retains a recognizable resurgent centre. Fossiliferous sediments show that a lake was once ponded within the Creede caldera. Silver mineralization associated with the later history of the caldera led to a nineteenth century mining boom, and the town

of Creede sprang up in a narrow canyon on the north flank of the caldera. A flavour of the old West still permeates the town and the weathered woodwork of abandoned mine workings.

Rich in historical associations though they are, the calderas of the San Juans are old, and the forests which mantle the picturesque mountainsides also conceal most of their volcanology. Although the Toba Complex in Sumatra contains by far the largest and youngest resurgent caldera in the world, it also is poorly exposed. To see better-exposed examples of caldera complexes, one has to travel to the Central Andes. In Bolivia, the Frailes Cordillera contains several major calderas known to range in age from 21 million years downwards but so far, the mapping required to relate individual ignimbrite sheets to their source calderas has not been done. One minor plinian explosive eruption has taken place within the last few thousand years. In the western cordillera, there is an even more impressive complex, where modern remote-sensing techniques have been used to outline the main components.

Fig. 14.19 Calderas of the San Juan volcanic field, south-western Colorado, as mapped by T. A. Steven and P. W. Lipman. In order of increasing age, calderas are LC = Lake City, C = Creede, CP = Cochetopa Park, SL = San Luis, B = Bachelor, LG = La Garita, MH = Mount Hope, S = Silverton, SJ = San Juan, UN = Uncomphagre, L = Lost Lake, U = Ute Creek, SM = Summitville, P = Platoro, BZ = Bonanza.

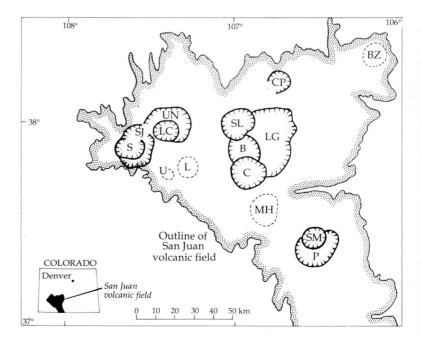

14.3.1 *The Altiplano–Puna volcanic complex*

Ignimbrites dominate Landsat images of much of the central Andes, especially in the remote high plateau region where Chile, Bolivia, and Argentina meet. On the ground, their pinkish, castellated outcrops give a distinctive character to the landscape. In total, ignimbrites are exposed at the surface over at least 50 000 square kilometres, and are undoubtedly present over far larger areas where they are buried by younger deposits (Fig. 14.20). Apart from their sheer extent, the most remarkable feature of these ignimbrites is that so many have been erupted within the last 10 million years: such a prodigious outburst that S. L. de Silva christened it an 'ignimbrite flare up'.[20]

Tracing the extraordinary extent of these ignimbrites proved easier than identifying their source calderas. As early as 1978, Michael Baker and I had identified ignimbrites which covered huge areas of the high, windswept *puna* zone between Chile, Argentina, and Bolivia.[21] On satellite images, the ignimbrites are unmistak-

able, because the soft pumiceous deposits have been carved by the prevailing north-westerly winds into elegant *yardangs*, elongate ridges, shaped like upturned canoes hundreds of metres in length, pointing upwind. From samples collected at a few sites on the ground, we knew that these ignimbrites were about 4.6 million years old. We proposed that they came from a caldera in Bolivia, obvious on satellite images, which we called Cerro Guacha. Later field-work by Chilean geologists Carlos Ramirez and Moyra Gardeweg demonstrated that they actually came from the *La Pacana caldera*, a huge but structurally ill-defined feature 65 km × 35 km in size (Figs 14.21–14.22).

La Pacana illustrates well two aspects of large caldera systems: the mismatch between the *topographic* and structural margins of a caldera, and the structure of a resurgent centre. It is natural to suppose that the margins of a caldera should correspond with the limits of the defining topographic depression. Even in young calderas,

Fig. 14.20 Locations of the centres of the principal calderas in the Altiplano–Puna volcanic complex. Young ignimbrites mantle most structural features, making it difficult to delineate the limits of calderas.

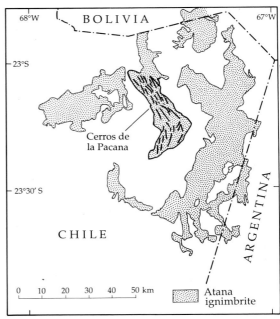

Fig. 14.21 Distribution of the 4.1-million-year-old Atana ignimbrite, and the resurgent centre of the La Pacana caldera, north Chile. There is no distinct topographical rim to the caldera. (After Gardeweg and Ramirez 1987.)

Fig. 14.22 Landsat Thematic Mapper image of the resurgent centre of the La Pacana caldera (lower left) and the outflow facies of the Atana ignimbrite (right). Apical graben faults are prominent along axis of resurgent centre. The gullies in the ignimbrite at top right are largely of aeolian origin. Image is 45 km across.

however, this is not the case. Structural subsidence of the caldera floor along the bounding faults leaves steep, unsupported walls behind. These may collapse in great avalanches, even while the eruption is in progress, so the topographic rim retreats rapidly from the original structural margin. (Evidence for massive landslides from rims of the San Juan calderas has been presented by Peter Lipman.[22]) After a few million years of erosion, the topographic rim may have retreated kilometres from the bounding faults, which are often difficult to locate precisely. In the case of La Pacana, the caldera is so irregularly shaped and so much erosion and later volcanism have taken place, that it is difficult to determine where the orginal structural margins lay (Fig. 14.23).

By contrast, the resurgent centre of La Pacana is unusually well displayed. A large thickness of ignimbrite accumulated in the caldera depression

Fig. 14.23 Pinnacle of Atana ignimbrite north of La Pacana resurgent centre, isolated by rapid aeolian erosion in high, dry conditions. Vehicle gives scale.

during the eruption which was followed by resurgence to form the *Cerros de La Pacana* (4905 m high). This resurgent block is long and narrow, measuring *c.*50 × 12 km, and exhibits antiformally-dipping, densely welded ignimbrites and a clearly defined *apical graben* complex. So plainly exposed is the structure that it is easy to imagine the up-arching of the caldera floor, and the subsequent failure of the keystone block. Two small post-resurgence dacite domes form pimples on its flanks. Many of the North American resurgent calderas are roughly circular, or at least equant, in plan. Like Toba, its larger counterpart in Indonesia, the elongate shape of La Pacana probably reflects regional tectonic stresses.

Calderas of south western Boliva   Huge ignimbrites were erupted from several other vents in and around the La Pacana complex: notably Cerro Guacha, and Cerro Purico, a 1.3-million-year-old ignimbrite shield in the west. A dacitic extrusion on the summit of Cerro Purico which post-dates the last major glacial regression 10 000 years ago was probably the most recent eruptive event associated with La Pacana.

Some 100 km north of Cerro Guacha is the Pastos Grandes Caldera, a resurgent structure about 60 km in diameter. Pastos Grandes was the source of several major ignimbrites, dated at 3.2, 5.6, and 8.1 million years. The latter, known as the *Sifon* ignimbrite, is exceptional because of the huge distances that it travelled—it may be the most worlds' most extensive ignimbrite. It outcrops at many widely spaced localities in north Chile, and performed prodigious feats in over-running mountain ranges to reach them.[23] Although the outflow facies of this ignimbrite has been studied, much remains to be done on its outcrop within the Pastos Grandes caldera itself. Pastos Grandes is a rather subdued feature topographically, but there are several young extrusive lava domes and hot springs to attest to its continued activity. Cerro Chascon, for example,

Fig. 14.24 Landsat Thematic Mapper image of resurgent centre of Pastos Grandes caldera, southwest Bolivia, source of the Sifon ignimbrite, showing radial drainages off elevated block, apical graben, and post-resurgence dacitic extrusions. Cerro Chascon, lower left, adjacent to diamond-shaped lake is youngest extrusion in the caldera, and may be only 10 000 years old. Image is 15 km across.

is a textbook example of an extrusive lava dome or *torta*. Its morphology is so fresh that was probably erupted only a few thousand years ago (Fig. 14.24).

An extensive area of boiling springs and mud pots at Sol de Mañana has been studied as a potential geothermal energy resource by the Bolivian government, in collaboration with the Italian government. Thermal manifestations occur locally over an area of about 120 square kilometres at elevations between 4800 to 5000 m. Some 30 km distant across the border in Chile is the splendid El Tatio geyser field, which has also been explored as a geothermal resource. Owing to political difficulties, the links between these major active hydrothermal systems have never been studied, but both are probably related to the major caldera.

— 14.4 Origins of large silicic caldera complexes

In an influential paper on large silicic systems, Wes Hildreth observed that 'almost all magmatism is fundamentally basaltic'.[13] This dictum reminds us of the ubiquity of basaltic volcanism on the terrestrial planets. In the case of large calderas such as Yellowstone and Cerro Galan, there is often abundant evidence for the role of basalt in the form of young basaltic lavas. At Cerro Galan, these mafic lavas were erupted from satellite vents both before and after the main caldera-forming event. Elsewhere, as at La Pacana, there is no *visible* basaltic volcanism at the surface, but its influence is still manifest in the indelible geochemical fingerprints it has left in the silicic magmas.

There is no need for us to stray too far into the arcana of petrology to understand the origins of silicic magmas. Furthermore, the physics of the process is more interesting than the chemistry.[24]. Large silicic caldera systems are *only* found in

areas of continental crust, usually at continental margins, where basaltic magmas are produced in large volumes from the subducted oceanic plate and the overlying mantle wedge. These magmas rise into the lower continental crust, partially melting it to form large volumes of silicic magmas. Effectively, *basaltic* magmas are swapped for *silicic* ones; the basalts underplating the crust, while the less-dense silicic magmas rise upwards through the crust to form huge cordilleran batholiths, which give birth to calderas. Inevitably, there is some intermixing between basaltic and silicic magmas, hence the distinctive geochemical fingerprint of the former in the latter. Often, small batches of basalt contaminated with crustal components also manage to reach the surface, forming satellite mafic vents and lavas as those seen at Cerro Galan. Where the continental crust is unusually thick, as at La Pacana, the higher-density mafic magma batches

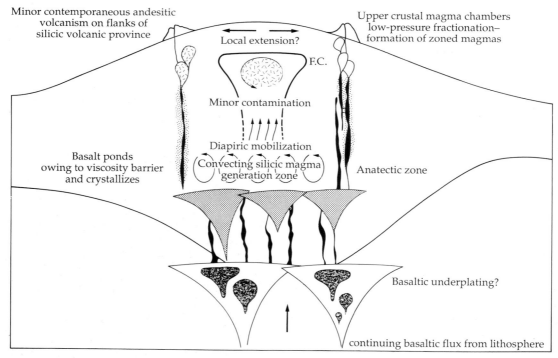

Fig. 14.25 One model for the formation of large-volume silicic magmas through the introduction of basaltic melts into the lower continental crust. After de Silva, 1989.

are unable to struggle all the way through all seventy kilometres of crust; hence the absence there of *surface* basaltic volcanism (Fig. 14.25).

14.4.1 Zoned magma chambers

Influenced by the ideas of Bob Smith and Wes Hildreth of the US Geological Survey, petrologists working on large silicic calderas have emphasized the way that some of them tap zoned magma chambers. In the 'standard' model, the first erupted products are the most evolved, and the later one less so, so that stratigraphically they provide a mirror image of the zonation in the magma chamber (c.f. Fig. 10.11). This model does not apply universally. At Cerro Galan, the dacitic ignimbrites are almost boringly homogeneous, showing only the smallest variations. And at Long Valley caldera itself, where the idea of compositional zonation was first elaborated, analyses of the plinian fall deposit underlying the Bishop Tuff have revealed an intricate story—the first phase of the plinian eruption tapped *less* evolved magma from the margin of the magma chamber. As the intensity of the eruption increased, more evolved magma from the roof of the magma chamber was tapped. Less evolved magma was then erupted for the remainder of the plinian phase, but the first post-plinian ignimbrites again tapped more evolved magma. These complex variations reflect the different levels within the magma chamber tapped at different stages of the eruption, when migration of vents along the ring fault system abruptly opened up new pathways to the surface.[25] Similar complexities are likely to be revealed when other large caldera systems come under close scrutiny.

— 14.5 Calderas on basaltic volcanoes

On Mars, there are no large resurgent calderas, perhaps because silicic magmas are absent. However, Martian basaltic calderas reach extraordinary dimensions—Olympus Mons's summit caldera is more than 60 km in diameter (Fig. 14.26 and Chapter 18). On Earth, the summit of the 4000-m-high Mauna Loa shield volcano in Hawaii is occupied by the 2.6×4.5 km Mokuaweoweo caldera, while the 4 km Kilauea caldera crowns the much smaller shield of the same name (Fig. 14.27a–c). Similar calderas probably topped the other Hawaiian shield volcanoes during their evolution. These Hawaiian calderas differ in several ways from their silicic counterparts: they are smaller; often exhibit 'nesting' of one generation of caldera within another; *never* show signs of resurgence; but do show signs of alignment along summit rifts. In their summit regions, Hawaiian shield volcanoes are exceedingly gently sloping, or flat.

Fig. 14.26 Oblique view from the Viking space craft of Olympus Mons volcano, Mars, showing broad, Hawaiian-shield shape and summit caldera complex. Ridged white areas in background are clouds in Mars' atmosphere.

Their calderas are steep sided, often vertical. Thus, one comes across such a caldera abruptly; the ground in front of one suddenly dropping away to a vast pit.

Instead of resulting from the explosive eruption of huge volumes of pyroclastic rocks, basaltic calderas may be formed when magma is extracted from the magma chamber underlying the volcano, leaving the summit unsupported, and causing it to founder. Rather than failing catastrophically, these calderas often do so in increments, one subsidence following another,

gradually enlarging the caldera, forming a 'nested' structure of pits and benches. Collapse is not initiated by eruption of voluminous lavas directly from the caldera, although lava often leaks through the floor to form long-lived lava lakes. Instead, dykes are propagated laterally from the underlying magma chamber, so that the soon-to-be lavas travel many kilometres horizontally before reaching the surface where they are erupted from subsidiary vents.

This has been a consistent pattern along the East Rift Zone of Kilauea for decades. For much

Fig. 14.27 (a) Summit caldera of Piton de la Fournaise, Réunion Island, Indian Ocean. Eruptions from this volcano have been almost exclusively basaltic effusions; the summit caldera (about 1 km in diameter) has formed by subsidence. Cones and lavas on the floor of the caldera were erupted in 1975. Concentric scarps in background are horseshoe shaped in plan, and probably formed by large-scale flank failure of the volcanic edifice. Photo: courtesy of P. Mouginis-Mark.

Fig. 14.27 (b) A crusted lava lake occupying floor of Kilauea caldera, showing successive bench levels in lake and terrace levels in caldera walls formed by incremental collapse. Hawaii Volcano Observatory on skyline at centre.

(c)

Fig. 14.27 (c) Oblique aerial view looking north-eastwards towards Mokuaweoweo caldera on the summit of Mauna Loa, Hawaii. Mauna Kea is in the background. Circular pit craters in the foreground are located on Mauna Loa's south-west rift zone. US Geological Survey photograph by D. W. Peterson.

of the later part of the twentieth century, there has been little to see in the caldera itself, save for small fumaroles, but the rift zone has been frequently active. Some 17 km along the rift from Kilauea, dyke-fed lava has been welling conti-

nously to the surface at Pu'u O'o since 3 January 1983, flowing down to the sea near the settlement of Kalapana. ('Pu'u' is the Hawaiian word for a peak or a point. It is said that Pu'u O'o was named by an unimaginative helicopter pilot because the vent broke out near the letter 'O' in the word 'south' printed on a topographical map . . .) (Fig. 14.28).

While lava has been erupted at a average rate of five cubic metres per second from Pu'u O'o since 1983, there has been no sign whatever of caldera collapse at Kilauea resulting from withdrawal of all this lava. Other voluminous historic eruptions have also not initiated collapse. This shows that caldera formation in Hawaii is not as simple as one might suppose. Major caldera-forming events may take place in response to broad tumescence or swelling of the shields by intrusion of magma into the rifts. As Williams and McBirney put it 'the shields seem almost to breathe; they swell and subside as underlying magma rises and falls'.[26]

Because Hawaiian calderas show such a poor correspondence between eruption events and subsidence, George Walker sought an alternative mechanism to account for their formation.[27] He noted that Hawaiian calderas are much more dynamic than silicic ones and are 'shaped by frequent small events instead of a few great ones'.

Fig. 14.28 Recent activity on Kilauea volcano has taken place only from points along the East Rift zone, distant from caldera itself. Eruption of *pahoehoe* lavas from Puu Oo vent began in January 1983, and continued at time of writing. Outline of lavas shows situation in 1987; lavas subsequently reached sea, and Kalapana village was overrun in 1990. Mauna Ulu was the scene of 901 days of continuous effusion of lava between 3 February 1972 and 22 July 1974, following an earlier eruptive episode in 1969–71. (Its lava field is not shown.)

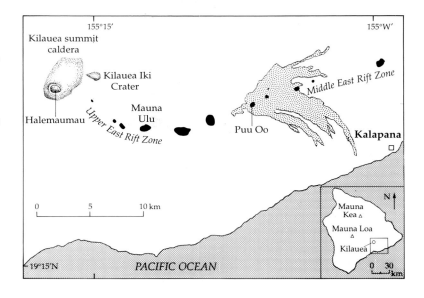

Fig. 14.29 SPOT satellite image of Cerro Azul, Isla Isabela, Galapagos Islands, showing compound summit caldera and radial lava flow fields. The volcano is symmetrical, in contrast to Hawaiian shields which are elongated along rifts. Caldera is 4 km across. Photo: courtesy of Duncan Munroe, Planetary Geosciences, HIG, copyright CNES.

Fig. 14.30 Interior wall of caldera of Volcan Cumbres, Isla Fernandina, Galapagos Islands. Caldera rim is 1200 m above sea-level, almost 1 km above caldera floor. Dark lavas on floor were erupted in 1988. Photo: courtesy of Peter Mouginis-Mark.

He suggested that the cause of caldera subsidence should be sought below the surface, and that it might actually be due to loading of the volcano caused by repeated intrusions into the infrastructure. Through time, the cumulative effect of all the intrusions would be to increase the mass of the volcano, causing it to sag down into the lithosphere, itself weakened and softened by the hot-spot driving the volcanism. Down-sagging of the infrastructure would be expressed in caldera formation at the surface.

14.5.1 Galapagos calderas

Calderas on the Galapagos islands are often identified as representing a separate class of caldera. Whereas the Hawaiian calderas are not more than a couple of hundred metres deep, some of the Galapagos calderas are almost a kilometre deep. Fernandina caldera, for example, has a maximum diameter of 6.7 km and is 845 m deep. The Galapagos calderas are more nearly circular, and do not show such marked elongation along prominent rift zones. They do however show striking circumferential fissures in the summit regions (Fig. 14.29–14.30).

It is not clear exactly how the Galapagos calderas form, but withdrawal of magma from the magma chamber at depth for eruption or intrusion elsewhere, perhaps even below sea level, may be involved. As in Hawaii, collapse is incremental: in 1968, a large part of the floor of caldera of Fernandina abruptly subsided by about 300 metres, although little lava was erupted at the surface.

—— 14.6 Historical unrest at large calderas of the world ——

If the eruption which formed Cerro Galan caldera were to take place in London, cities as far distant as Oxford and Cambridge would be wiped out. Southern England would become uninhabitable. If it were to take place in New York, cities as distant as Philadelphia and Providence (Rhode Island) would have to be evacuated. So extreme are their potential hazards that 'active' caldera systems are anxiously monitored. Although massive, caldera-forming events are themselves rare, even a minor reawakening could have ruinous effects in a populated area. These concerns underlay the compilation of detailed catalogue of events at major calderas, acknowledged in the title of this section.[28] This catalogue summarizes the histories of the calderas and provides information on their 'background' level of activity, making it possible for volcanologists to interpret new events taking place in a caldera in the context of a broader knowledge of events at it and other calderas. Case histories of three major calderas are presented here.

14.6.1 Campi Phlegræi

Familiar to volcanologists since the days of Sir William Hamilton, the Phlegrean Fields have a long and alarming record of historic unrest. Long, because the record stretches back to Roman days. Alarming, because the area at risk includes vast, densely populated areas near Naples. Eruption of about 80 cubic kilometres of the Campanian Tuff about 35000 years ago initiated formation of the Phlegrean Fields, a complex of cones and craters within a 13 km diameter caldera (Fig. 14.31). Some 10,000 years ago, the Neapolitan Yellow Tuff was erupted, forming the warm yellow rock widely exposed around Naples and used in many of its buildings. Several periods of post-caldera volcanism have taken place, most recently on 29 September 1538 when the Monte Nuovo cone was born. Eruption continued for five days, stopped for two days and then culminated in an explosion on 6 October which killed 24 people. Trees five kilometres from the new volcano were blown down. Only 0.03 cubic kilometres of material were erupted.

Monte Nuovo's abrupt birth was important in the history of science because it was the first time in the Renaissance that a 'new' volcano had been born, in full view of scholars and clergy. More important to us, though, is the record of associated ground movements and seismic activity. Vertical ground movements in and around the town of Pozzuoli have been startling. Pillars in the famous Roman marketplace of Serapeo

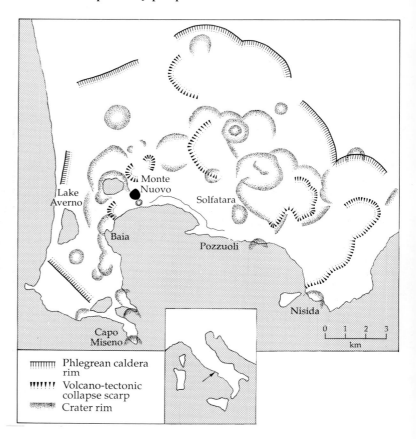

Fig. 14.31 Principal volcanic and tectonic features within the 35 000-year-old Phlegrean Fields caldera complex near Naples, Italy, showing locations of 1538 eruption of Monte Nuovo, and Pozzuoli. From Newhall, C. G. and Dzurisin, D. (1988). Historical unrest at large calderas of the world. *US Geol. Surv. Bull.* **1855**, 1108 pp.

provide illuminating evidence of post-Roman movements. Borings in the marble columns of the temple of Seraphis made by the marine organism *Lithodomus lithophagus* show that the site had sunk no less than 11 metres below sea-level by about AD 1000, after which it rose again (Fig. 14.32, Fig. 14.33). Between 1000 and 1538 (the Monte Nuovo eruption), 12 metres of uplift had taken place, at an average of 2 cm per year. So much new land was exposed along the emerging shoreline that it was deeded to the local University. Extremely rapid uplift accompanied by intense seismic activity took place in the days before the Monte Nuovo outburst: on 26–27 September 1538, the ground at Pozzuoli shot up by four metres.

After sundry other episodes of unrest, a new episode began in 1982 when uplift and seismic activity again disturbed Pozzuoli, possibly trig-

gered by a major (magnitude 6.9) earthquake centred 100 km south-east (Fig. 14.34). Earthquake damage rendered many buildings unsafe, and continuing swarms of tremors caused widespread concern. Between June 1982 and December 1984, 1.8 metres of uplift were measured at Pozzuoli, the peak rate of uplift reaching 5 mm per day during periods of high seismicity in September/October 1983 and March/April 1984.

These events naturally prompted fears that a fresh eruption might be on its way, perhaps a relatively minor, Monte Nuovo type eruption; perhaps something much worse. Fortunately, nothing happened. Seismicity decreased to low levels in 1984, and some subsidence took place, at a rate of 0.4 mm per day. For now, the Campi Phlegræi and the city of Naples are safe. But when will the next large eruption occur, extinguishing the life and commerce of this busy

Fig. 14.33 Observatory in the floor of Solfatara crater, Phlegrean Fields. Sulphurous fumaroles and boiling mud pots have been active since Roman times, and may have been more prolific then.

Fig. 14.32 Darker, lower parts of the famous pillars in the Roman Temple of Seraphis in Pozzuoli (Phlegrean Fields) show effects of depredations by boring marine organisms.

part of Italy? No one can say, but fortunately it is likely that there will be precursors which will be hard to ignore.

14.6.2 Rabaul

Located at the eastern tip of the island of New Britain (Papua New Guinea), the port of Rabaul was prized by the Imperial German Navy as one of the best in the Pacific during Germany's brief experiment in colonialism prior to World War I.

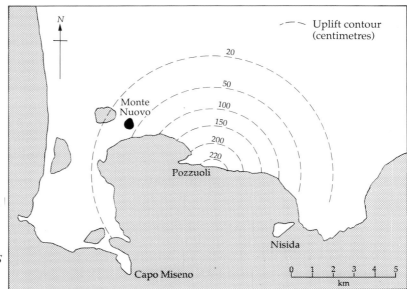

Fig. 14.34 Contours of cumulative uplift around Pozzuloi, in centimetres, for the period 1970–85. Maximum uplift is 240 cm. From Newhall, C. G. and Dzurisin, D. (1988). Historical unrest at large calderas of the world. *US Geol. Surv. Bull.* **1855**, 1108 pp.

During World War II, Rabaul was again strategically vital when the Allies sought to end Japanese imperial ambitions and the extension of their Greater East Asian Co-Prosperity Sphere. Since then, Rabaul has been acclaimed as the site of the most potentially serious volcanic crisis of recent times.

Blanche Bay, the sheltered harbour that made Rabaul so attractive to imperial navies, is actually a caldera, breached on its eastern side. Rabaul town lies on its northern rim. Rabaul lacks the long written record of historic activity that the Phlegrean fields have, but geochronological work shows that there were major eruptions 3500 years and 1400 years ago. During the period of western colonialism, there were many periods of unrest, including several minor eruptions from vents such as Vulcan and Tavurvur. Vertical movements around the shoreline were well known to the locals, who joked about the possibility of the port being cut off from the sea. A major crisis took place in late May 1937, when intense seismic activity and uplift commenced. Between 26 and 29 May, frequent tremors shook Rabaul, and rapid uplift of Vulcan island took place. A comment by a resident conveys an impression of the intensity of the seismic activity: 'For much of the morning it was impossible to write during the maximum periods of vibration . . .'. On 29 May, a violent eruption of Vulcan commenced, causing 500 casualties through pyroclastic flows. Tavurvur and some other minor vents were also active.

Rabaul's most recent episode of unrest began in 1971, when broad uplift of the caldera floor and swarms of tremors commenced. Over a decade, the level of activity gradually increased, but in 1983 it changed gear, and became thoroughly alarming. Although uplift and tilting of the caldera floor were impressive, reaching rates of $1°$ per month during the height of the crisis, it was the level of seismic activity that was most alarming. Tilt is imperceptibly slow, and can lead to amusingly incongruous results, but there was nothing amusing to the inhabitants of Rabaul about the tremors they experienced daily, and the uncertainty of the final outcome. An idea of their ordeal is best conveyed by the record of events shown in Table 14.1.

Table 14.1 Earthquakes and seismic energy release during the crisis at Rabaul, September 1983–July 1985

Date	Number of earthquakes	Seismic energy release (joules)
September 1983	2135	3.8×10^{10}
October 1983	5198	1.2×10^{11}
November 1983	5751	3.2×10^{9}
December 1983	7117	8.6×10^{9}
January 1984	8372	1.7×10^{11}
February 1984	8339	1.7×10^{11}
March 1984	8729	1.9×10^{12}
April 1984	13 749	2.7×10^{11}
May 1984	8938	1.4×10^{10}
June 1984	5304	1.9×10^{8}
July 1984	4404	1.3×10^{10}
August 1984	5285	2.5×10^{9}
September 1984	4048	8×10^{7}
October 1984	6749	5.5×10^{11}
November 1984	3985	6.0×10^{8}
December 1984	2887	1.3×10^{8}
January 1985	1297	4.4×10^{8}
February 1985	1672	3.4×10^{8}
March 1985	2042	6.9×10^{9}
April 1985	1041	$< 10^{7}$
May 1985	723	5.8×10^{9}
June 1985	639	$< 10^{7}$
July 1985	595	$< 10^{7}$

Not all of these were 'felt' earthquakes, but many were. April 1984 must have been quite a hectic month for the townspeople, with two swarms of more than 1000 events on 21 and 22 April. During this peak period, the rate of uplift was also rapid, the southern tip of Matupit Island rising by 7.6 centimetres during a four-week period (Fig. 14.35). Given Rabaul's prehistoric record of massive eruptions, the lethal 1937 eruption and the large number of people now living in and around the caldera, volcanologists at the Rabaul Volcano Observatory were deeply concerned about the possibility of a major disaster. Fortunately, the crisis passed without incident, but their geophysical monitoring and contingency planning were exemplary.

Four levels of unrest are defined for the volcano. Stage 1 is a low, almost background

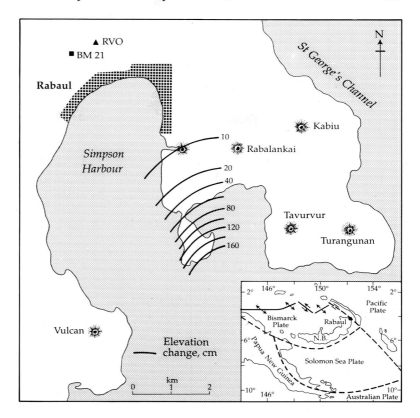

Fig. 14.35 Location of Rabaul city, Papua New Guinea, (shaded) in relation to area affected by uplift during the 1983/84 seismic crisis. Contours show cumulative elevation in centimetres from September 1973 to May 1984, relative to Bench Mark 21, adjacent to Rabaul Volcano Observatory (RVO) on the outskirts of Rabaul. Asterisks are volcanic vents. From Newhall, C. G. and Dzurisin, D. (1988). Historical unrest at large calderas of the world. *US Geol. Surv. Bull.* **1855**, 1108 pp.

level, in which an eruption is possible in years or months. No specific advice is given to the public. Stage 2 is significantly elevated activity, with tens to hundreds of tremors per day and measurable ground deformation. An eruption is judged to be possible within months or weeks. At Stage 2, the public are notified, and low-key precautionary measures are taken. Stage 3 is an alarmingly high level of unrest, with several hundred to thousands of earthquakes per day and marked ground deformation. An eruption is possible within weeks or days. People are advised to prepare for evacuation. At Stage 4, an eruption is imminent within days or hours, with thousands of earthquakes per day and rapid ground deformation. Evacuation is ordered.

A Stage 2 alert was in place at Rabaul from October 1983 to November 1984. Evacuation plans were carefully drafted and rehearsed, but fortunately there was no cause to invoke Stages 3 or 4. It would have been, of course, all too easy to cry 'wolf' too early, and order a premature evacuation, with consequent massive personal and economic disruption to the community. In the face of extremely worrying circumstances, the Rabaul authorities judged the situation with great professional skill, enabling the busy commercial life of Rabaul to continue uninterrupted throughout one of the most remarkable volcanic crises of modern times.

14.6.3 Long Valley

Long Valley caldera was formed by a colossal eruption 700 000 years ago. Since then, many smaller events have taken place in and around the caldera, most recently the formation of a chain of rhyodacitic domes, the Inyo and Mono domes, north of the caldera. Eruptions from this chain took place as recently as 650–550 years ago. Since then, the area has been one of quiet beauty and tranquillity. A prosperous ski resort has grown up at Mammoth, on the slopes of the Sierra Nevada overlooking the caldera. Five large earthquakes disturbed the peace in May

1980, and were followed by four years of seismic swarms, whose foci were located just inside or just outside the south margin of the caldera, at depths of a few kilometres. Since July 1984, most of the seismicity has been outside the caldera. Marked ground deformation was associated with the seismicity. Between October 1980 and February 1983 the resurgent centre of the caldera was elevated by 15 cm, but the rate of deformation subsequently subsided.

From a technical point of view, the unrest at Long Valley is interesting because while it may in part have been due to inflation of a residual silicic magma chamber at a depth of 7–10 km, it may also have been the response of the caldera system to *regional* tectonic conditions—the seismicity and deformation being caused by the system shifting slightly as it readjusted to the external stresses. It is not clear that shallow-level magma movement took place. From a more mundane point of view, the geological unrest was illuminating because of the unrest that it caused in certain quarters of the business community. There were some unpleasant scenes between geologists pointing out hard facts, and real-estate developers who stood to lose their shirts on properties which they suddenly realized might get buried by deposits less profitable than snow.

Notes

1. Lipman, P. W., Self, S., and Hieken, G. (1984). Calderas and associated igneous rocks. *J. Geophys. Res.* **89**, B10, 8219–841.
2. Fouqué, F. (1879). *Santorin et ses eruptions*. Paris, G. Masson, 440 pp.
3. Clough, C. T., Maufe, H. B., and Bailey, E. B. (1909). The cauldron subsidence of Glen Coe, and the associated igneous phenomena. *Q. J. Geol. Soc. Lond.* **65**, 611–78.
4. Almond, D. C. (1977). The Sabaloka igneous complex, Sudan. *Phil Trans. Roy. Lond.* **A 287**, 595–633.
5. Williams, H. (1941). Calderas and their origins. *Univ. Calif. Publn. Bull. Dept. Geol. Sci.* **25**, 239–346.
6. Williams, H. (1942). The geology of Crater Lake National Park, Oregon, with a reconnaissance of the Cascade range southward to Mt. Shasta. *Carnegie Inst. Washington Publ.* **540**. 162 pp.
7. Bacon, C. R. (1983). Eruptive history of Mount Mazama and Crater Lake Caldera, Cascade Range, U.S.A. *J. Volcanol. Geotherm. Res.* **18**, 57–115.
8. Self, S., Rampino, M. R., Newton M. S., and Wolff, J. A. (1984). Volcanological study of the great Tambora eruption of 1815. *Geology* **12**, 659–63.
9. Sigurdsson, H. and Carey, S. (1989). Plinian and co-ignimbrite tephra from the 1815 eruption of Tambora volcano. *Bull. Volcanol.* **51**, 243–70.
10. van Bemmelen, R. W. (1949). *The geology of Indonesia*. General Government Printing Office, The Hague, Netherlands, **1A**, 1–732.
11. Chesner, C. A. and Rose, W. I. (1991). Stratigraphy of the Toba Tuffs and evolution of the Toba caldera complex, Sumatra, Indonesia. *Bull. Volcanol.* **53**, 343–56.
12. Smith, R. L. and Bailey, R. A. (1968). Resurgent cauldrons. In Coats, R. R., Hay, R. L., and Anderson, C. A. (eds.) *Studies in volcanology* Geol. Soc. Amer. Mem. No. 116, 153–210.
13. Hildreth, W. (1981). Gradients in silicic magma chambers: implications for lithospheric magmatism. *J. Geophys. Res.* **86**, 10 153–92.
14. Friedman, J. D. and Heiken, G. (1977). Volcanoes and volcanic landforms. In *Skylab explores the Earth*, NASA Spec. Publ. No. 380. 137–70.
15. Francis, P. W. and Baker, M. C. W. (1978). Sources of two large volume ignimbrites in the Central Andes: some LANDSAT evidence. *J. Volcanol. Geothermal Res.* **4**, 81–7.
16. Francis, P. W., Hammill, M., Kretzschmar, G. A., and Thorpe, R. S. (1978). The Cerro Galan Caldera, Northwest Argentina. *Nature*, **274**, 749–51.
17. Bailey, R. A., Dalrymple, G. B., and Lanphere, M. A. (1981). Volcanism, structure and geochronology of Long Valley Caldera, Mono County, California. *J. Geophys. Res.* **11**, 293–315.
18. Marsh, B. D. (1984). On the mechanics of caldera resurgence. *J. Geophys. Res.* **89**, 8245–51.
19. Steven, T. A. and Lipman, P. W. (1976) Calderas of the San Juan volcanic field, south-western Colorado. *US Geol. Surv. Prof. Pap.*, **958**. 35 pp.
20. de Silva, S. L. (1989). Altiplano–Puna volcanic complex of the central Andes. *Geology*, **17**, 1102–6.
21. Francis, P. W. and Baker, M. C. W. (1978).

Sources of two large volume ignimbrites in the Central Andes: some LANDSAT evidence. *J. Volcanol. Geothermal Res.* **4**, 81–7.

22. Lipman, P. W. (1976). Caldera collapse breccias in the western San Juan mountains, Colorado. *Geol. Soc. Amer. Bull.* **87**, 1397–1410.

23. de Silva, S. L. and Francis, P. W. (1989). Correlation of large ignimbrites: two case studies from the Central Andes of north Chile. *J. Volcanol. Geotherm. Res.* **37**, 133–49.

24. Huppert, H. E. and Sparks, R. S. J. (1988). The generation of granitic melts by intrusion of basalt into continental crust. *J. Petrol.* **29**, 599–624.

25. Gardner, J. E., Sigurdsson, H. and Carey, S. N. (1991). Eruption dynamics and magma withdrawal during the plinian phase of the Bishop Tuff eruption, Long Valley Caldera. *J. Geophys. Res.* **96**, 8097–111.

26. Williams, H. and McBirney, A. R. (1979). *Volcanology.* Freeman Cooper, San Francisco, 379 pp.

27. Walker, G. P. L. (1988). Three Hawaiian calderas: an origin through loading by shallow intrusions. *J. Geophys. Res.* **93**, 14 773–14 784.

28. Newhall, C. G. and Dzurisin, D. (1988). Historical unrest at large calderas of the world. *US Geol. Surv. Bull.* **1855**, 1108 pp.

15

Submarine volcanism

Below the thunders of the upper deep;
Far, far beneath in the abysmal sea,
His ancient, dreamless uninvaded sleep,
The Kraken sleepeth.

Tennyson

If an astronaut bound on some sidereal odyssey were to look back at the receding Earth, she would see a predominantly blue planet, wreathed in white cloud. Dry land would be inconspicuous. Seventy per cent of the Earth's surface is covered by water. The same seventy per cent is covered by basalt lavas. By this measure, submarine basaltic volcanism is much the most important subject for terrestrial volcanologists to address. But the deep ocean floor has only recently become accessible to study. Our exploration of 'inner space' is proceeding concurrently with our exploration of outer space; arguably at a slower pace. Although oceanographers in surface vessels are separated from the ocean floors they investigate by a scant few kilometres of water, their technical problems are often more formidable than those confronted by scientists studying planets tens of millions of kilometres away. In the last decade, new technologies have been developed which permit large-scale imaging surveys of the ocean floors. Just as the radar instrument on the Magellan space craft enabled us to peer through Venus's dense mantle of clouds by illuminating its surface with radio waves, so conceptually similar techniques such as GLORIA (*G*eological *L*ong *R*ange *I*nclined *A*sdic) enable us to image volcanoes on the sea-floor using sound waves. But whereas orbital imaging radars can scan an entire planet in a matter of months, even the most sophisticated side-looking sonars still involve instruments laboriously towed by ships moving at only a few knots.

Trying to imagine the Earth without its oceans, the American naval officer Matthew F. Maury wrote in 1855 'Could the waters of the Atlantic be drawn off to expose to view this great sea-gash . . . it would present a scene the most rugged, grand and imposing . . . the empty cradle of the oceans'. We have in one sense started to 'draw off the waters' but it will still be decades before we know the topography of the Earth's sea-floor as intimately as we know that of Venus.

In volcanological terms, there are two different submarine environments to consider: shallow coastal waters, and the deep waters of the oceans proper. In these two environments the physics of volcanism are quite different. In shallow water, the ordinary range of volcanic phenomena that take place on dry land prevails, with the extra complications introduced by rapid cooling and explosive fragmentation caused when hot magma and cold water come together. In the abyssal ocean depths, rapid chilling is still an important factor, but the range of magma compositions is restricted (overwhelmingly basalt) and the crushing pressure entirely prevents explosive activity.

15.1 Shallow submarine volcanism

15.1.1 Basalts in shallow water

Given the profusion of basaltic volcanoes on Earth, the flow of basalt lavas into the sea is a common enough occurrence. It continued uninterrupted for many years on the Kalapana coast of Hawaii, providing a fine spectacle for tourists. Intuitively, one might suppose that the copious flow of basalt lava at a temperature of over 1000°C into sea-water at 20°C would be powerfully explosive. Remarkably, this is not the case, for *pahoehoe* flows at least. To be sure, a great banner of steam marks the point on the coast of Hawaii where a flow enters the sea, but when the steam is blown momentarily away, one can often see lava tubes a metre or so above wave level, from which glowing basalt streams into the sea almost as smoothly as cream poured into coffee. A highly insulating thin film of steam bubbles forms instantly around the hot lava, forming a barrier against the water, and preventing more violent interaction. (Small explosions take place intermittently, blasting fragments of quenched lava a few metres into the air, so tourists should not get too close.)

Abrupt chilling of the basaltic liquid forms thin sheets of brown transparent glass, like honey-coloured cellophane, called *Pele's limu* (it is supposed to resemble a kind of Hawaiian seaweed). Large flakes of *limu* patter gently down from the steam plume, and some are borne one or two kilometres downwind. At night, the scene is a real devil's kitchen, as glowing lava lights up the steam plume with a ruddy red glare, while hissing, floating globs of incandescent lava bob implausibly in the surf. Brisk evaporation of sea-water forms magnesium chloride (among other things) and then hydrochloric acid. Thus, an acid rain showers down from the steam plume, stinging the eyes and prompting a hasty upwind retreat.

When a lava tube opens below water level, entry of lava into the sea may be so placid that divers can approach safely. During the 1969–74 eruption of Mauna Ulu in Hawaii, tube-fed *pahoehoe* lavas flowed steadily into the sea. In some extraordinary film sequences, divers recorded what happens when hot lava emerges from tubes below sea-level. Their film shows a steep underwater flow front disappearing into the blue haze of the Pacific. All appears motionless. Suddenly, the camera picks out a rounded lobe (30–50 cm acrosss) of hot lava extruding: there is a brief flash of red, almost immediately extinguished. This process is repeated many times, each lobe successively swelling, cracking open abruptly and budding forth a new lobe within two minutes of its predecessor. Dramatic although it was, the process was not violent, allowing the divers to swim to within a metre of the swelling lobes, taking care to avoid pockets of scalding-hot water. Although not explosive, the budding of new lobes was often noisy, *im*plosions frequently taking place, as hot gases cooled and contracted beneath the brittle carapace of each lobe.[1]

Submarine eruption of pahoehoe basalt lavas ultimately produces multitudes of small packets or *pillows* of lava, commonly about one metre across, sometimes accumulating to form deposits hundreds of metres thick. Characteristically, the carapace of each pillow sags into the gap between the ones below it. Pillows also typically have bands of small vesicles parallel to their margins. If exposed in three dimensions, their carapaces reveal linear grooves resulting from their toothpaste-like extrusion (Fig. 15.1). Pillows are therefore easily recognizable, and can often be found in the fossil record (Fig. 15.2). They have been identified in some of the most ancient rocks on Earth, more than 3400 million years old. Wherever one finds them, one can be sure that one is dealing with basaltic lavas erupted under water. It is more difficult to determine the depth of water merely from the shape of the pillows. Although the examples filmed off Hawaii formed at depths of only about 10 metres, photographs taken by the submersible ALVIN at depth of more than 2000 metres in the east Pacific showed almost identical pillows (Fig. 15.1). Other pillows form in only marginally subaqueous environments, for example when a surface lava squeezes into wet sediments in boggy ground (Fig. 15.3).

Fig. 15.1 Bulbous pillow lavas photographed from the submersible ALVIN at a depth of more than 2000 metres on seamount 6-C on the East O'Gorman fracture zone. (East Pacific Rise, off the coast of Mexico, 12°44′N, 102°35′W). Toothpaste-like extrusion and chilling under-water produced the corrugated carapaces of the pillows. Comparable *pahoehoe* lavas erupted on dry land are smooth-skinned and glossy. Compare Figs 2.16 and 7.13. Photo: courtesy of Rodey Batiza, University of Hawaii.

Fig. 15.3 Pillows of the Wilbur Creek flow (part of the Columbia River Basalt) formed when the lava ploughed into wet sediments in the Lewiston basin, Washington State, USA. Photo: courtesy of P. R. Hooper.

Fig. 15.2 Palaeozoic pillow lavas, exposed on dry land in Britain. Concentric bands of vesicles form a roly-poly pattern. Institute of Geological Sciences photograph.

Underwater lava flow of a different kind was seen a few kilometres along the coasts in 1989, when lavas from the Kupaianaha lava pond were streaming into the Pacific along a front less than 250 m wide.[2] Accumulated debris had piled up offshore to form a steep slope at the point where lavas were entering the sea. On numerous occasions, lava flows were observed actually flowing down this slope, occupying narrow channels. The channels were about a metre wide, and the flows within them moved at speed of 1–3 metres per second. According to the divers who saw the moving flows at depths of less than 50 metres, about 5 per cent of their surface area was glowing red through cracks and along the flow margins; the remainder was chilled crust. At any one time, only one channelized flow was observed to be active. From considerations of the size of the flows and the slopes they descended, volumetric flow rates of about 0.7 cubic metres per second were indicated: about 10 per cent of the total flow entering the sea from Kupaian-haha.

Although divers could get close enough to see what was going on, the underwater environment was noisy and far from healthy. Frequent minor detonations took place against a constant background of sizzling and cracking. Resonant concussions could be heard as much as a kilometre distant underwater. Close to the flows, divers observed that some of these concussions took place when bubbles of incandescent lava 50–100 cm in diameter were blown out of the lava stream into the water. Explosions took place sometimes more than once a second, but these periods of high-frequency activity were generally short lived. Clearly as hazardous to divers as small depth-charges, these explosions obliged them to stay at a distance of about 5 metres.

Hyaloclastites As these underwater observations show, flow of *pahoehoe* lavas into the sea is accompanied by continuous, albeit minor, explosions and implosions. Much hot lava is abruptly chilled. Extensive fragmentation inevitably takes place, producing fine-grained, angular detritus. Along the coast of Hawaii, new beaches of black sand are rapidly constructed by lava fragments transported by longshore currents. Even more intensive fragmentation takes place when *aa* flows enter sea-water. Almost the whole volume of the flow may be comminuted into debris consisting of small glassy fragments known as a *hyaloclastite* breccia (hyaloclastite = glassy– broken). Huge amounts of hyaloclastites may form, a large enough heap accumulating for the lava to advance over it, remaining above sea-level, and constructing a lava delta (Section 7.3.6). Hyaloclastite breccias are common in the geological record, but because they are composed of small, unstable glassy fragments, they often weather rapidly into an unrecognizable blur.

Where basaltic magmas are erupted from sufficiently shallow submarine vents, pyroclastic eruptions that break surface may of course ensue, resulting in the formation of Surtseyan tephra deposits (Chapter 9).

15.1.2 *Submarine silicic extrusions*

Although not as common as basaltic eruptions, silicic magmas are occasionally erupted beneath sea-level. They are erupted more slowly than basalts, so their products are not as spectacular. A silicic extrusion immediately becomes covered with a thick mantle of hyaloclastite material, into which more magma continues to be injected. Thus, a variety of lava dome grows on the sea-floor; one in which the hot rock in the core is partially insulated from sea-water by the overlying layer of self-generated fragmental material. As mentioned in Chapter 7, the only known eruption of true rhyolite lavas in recent history was in the Tuluman Islands off Papua New Guinea. This eruption was largely submarine.

Submarine silicic lavas are important in an economic context, because some are associated with economic mineral deposits. Best known of these are the Miocene Kuroko deposits of Japan, where massive sulphide deposits were formed in association with submarine rhyolitic lavas and fragmental rocks, probably at depths in excess of 3000 metres. Most of the lavas appear to have been intruded into hyaloclastite-draped domes, or as cryptodomes intruded into water-saturated sediments.

— 15.2 Deep-water eruptions

15.2.1 *Physical considerations*

Marine geologists like to demonstrate in a chillingly unforgettable way the frightful pressures they work under in their submersibles: they display polystyrene foam coffee cups that have been carried to great depths *outside* their craft. Cups brought up from depths of a thousand metres are tiny, shrivelled relics. Every ten metres of water, of course, is equivalent to one atmosphere pressure (1 bar), so at 1000 metres, the pressure is about 100 times greater than at the surface. Following Boyle's immutable law, the bubbles in the polystyrene are crushed accordingly.

In Chapter 8, it was shown that in order for a pyroclastic eruption to take place, a magma must contain some proportion of dissolved volatiles;

that the vapour pressure of the volatiles in the magma must be sufficiently large for bubble nucleation and vesicle formation to take place; and that vesicles should be able to grow sufficiently to overcome the tensile strength of the magma and fragment it. All of these become more difficult as pressure is increased.

Water is the dominant volatile in basaltic magmas. Owing to its industrial importance, its thermodynamic properties have been investigated in the two centuries since sibilant steam engines first began to pump in Cornish tin-mines. With increasing pressure, the temperature at which water undergoes the phase change from liquid to vapour (steam) increases. But it does not do so in a straightforward way. Beyond a certain pressure, known as the *critical point*, highly compressed steam behaves in the same way as liquid water: it no longer decreases in volume with further increased pressure (Fig. 15.4). For sea-water, the critical point is at 315 bars, equivalent to a depth of about 3000 metres.

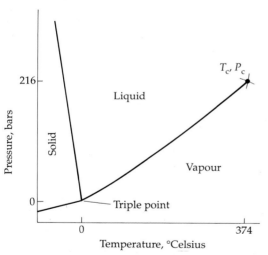

Fig. 15.4 Phase diagram for pure water, showing the position of the critical point (T_c, P_c). At temperatures and pressures above the critical point, vapour, and liquid phases are indistinguishable. This is most obviously manifest in volume changes: below the critical point, water vapour shows a huge volume expansion for a given increment of temperature, whereas liquid water expands far less. Above the critical point, there is no difference.

From a volcanological point of view it follows that below the critical depth, explosive fragmentation of magma is impossible, no matter how volatile rich it is, or how high the temperature. In volatile-rich magmas it is possible for vesicles to nucleate, and indeed deep-water vesicular basalts are common. But the vesicles can *never* expand and grow. Thus, the volcanology of the abyssal ocean floors is essentially that of lavas. There are no pyroclastic rocks. Similar arguments apply to silicic lavas erupted underwater on continental margins. At depths of about 1000 m, dredge samples of volatile-rich dacitic lavas sometimes appear remarkably vesicular; at depths below 3000 m they are dense, dark, and glassy.

15.2.2 Submarine pyroclastic eruptions

Explosive eruptions are theoretically possible at depths intermediate between the surface and the critical depth. Until recently, it was supposed that in practice these rarely occur at depths greater than a few hundred metres, because magma volatile contents were presumed to be inadequate to provide enough fizz to overcome the ambient pressures. In 1991, however, Kathy Cashman and Dick Fiske changed this view when they reported the discovery of abundant pyroclastic rocks at depths of around 1500 m at several sites in the Lau Basin, near Tonga and the Okinawa Trough south of Japan.[3] Deep-sea drilling in the Sumizu Basin revealed four massive deposits of rhyolitic pumice 8 to 40 metres thick which were deposited between 130 000 and 1000 years ago. Beneath these silicic deposits is an even thicker accumulation of highly vesicular basaltic scoria. Because there are no nearby land volcanoes, these thick pyroclastic deposits must have been erupted underwater, at depths of up to 2000 m.

But how, apart from the absence of nearby land volcanoes, can one show that a pumice deposit was indeed of submarine origin, and not just fall-out from an eruption on dry land which accumulated in deep water? This problem is particularly acute, of course, in ancient deposits which have been uplifted and exposed at the surface. In their analysis, Cashman and Fiske showed that the physics of particles settling through water provides a solution. In air, pumice clasts have densities between 0.6 and 1×10^3 kg m^{-3}, whereas lithic clasts have densities of around 2.4×10^3. Thus, in air, a lithic clast which is the 'hydraulic equivalent' of a pumice clast (one which falls at the same terminal velocity), will be about one-half to one-third of the diameter of the corresponding pumice. Under water, pumice clasts will be saturated and have densities of about 1.1 to 1.3×10^3 kg m^{-3}, so the density contrast between water and pumice clasts will be much less than between air and pumice. Consequently, the pumice will fall much more slowly in water than in air. Lithic clasts will also fall more slowly through water than air, but they are still much denser than water. Thus, in water, a lithic clast which is the hydraulic equivalent of a pumice clast will be much smaller, between one-fifth and one-tenth of the diameter of the corresponding pumice.

The upshot of all this is that pumice deposits erupted under water and on dry land will have strikingly different grain-size characteristics *if* the pumice is immediately water saturated. (Pumice from an eruption on land could well fall into the sea, but it would take some time to saturate and sink, separating pumice and lithics). Specifically, a pumice which was erupted underwater should show a *bimodal* grain-size distribution: at any one location, pumice clasts should fall within one size range, while lithic clasts should fall in a much smaller size range, one-fifth to one-tenth the diameter. Such bimodality is much less well-defined in pumice deposits originating above sea-level.

— 15.3 Deep-ocean submarine volcanoes —

Although one might anticipate that the apparently limitless expanses of the ocean floors should be monotonous plains, they are in fact scattered with numerous substantial mountains. Termed 'seamounts' or *guyots* by oceanographers, these features were first recognized as volcanoes by the eminent marine geologist W. H. Menard.[4] There are two distinct types of volcanic seamount. One type forms above hot-spots, constructing giant volcanoes which may rise above sea-level, such as those of the Hawaiian chain. The other type consists of smaller volcanoes, initially formed near ridge axes and subsequently carried away from them by sea-floor spreading. It follows that these volcanoes are only slightly younger than the oceanic crust on which they are built, and get progressively older away from their parent ridge axes (Fig. 15.5–15.6).

15.3.1 Off-axis volcanoes

It is hard to say how many volcanoes there are in the deep ocean. Most of them are inactive, and have been so for a long time. One estimate, based on a statistical survey suggested that there might be about 200 volcanoes higher than about 1000 m high, and a multitude of smaller ones. Most of these volcanoes formed at or near ocean ridges, soon becoming extinct as sea-floor

spreading carried them away from the ridge axis. Since there are so many submarine volcanoes around the world's oceans, it is difficult to generalize about them. A study of 30 representative examples on the Cocos plate by Batiza and Vanko provides a good introduction.[5] Their volcanoes were all located on oceanic crust 1.5 to 7.5 million years in age on the flanks of the East

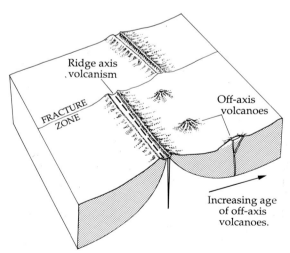

Fig. 15.5 Contrasted sites for submarine volcanism: ridge axis and off-axis. Off-axis volcanoes are progressively older with distance from the ridge axis.

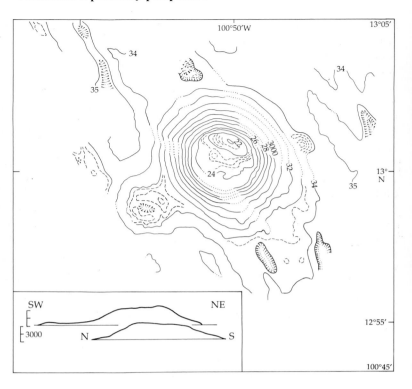

Fig. 15.6 Topographical (bathymetric) map of an unnamed seamount in the east Pacific, with profile cross-sections. Contours are in hundreds of metres below sea-level, with finer contours interpolated for features of interest. A small, shallow crater exists on the crescentic summit bench. From Batiza, R. and Vanko, D. (1983/84). Volcanic development of small oceanic central volcanoes on the flanks of the east Pacific Rise inferred from narrow-beam echo sounder surveys. *Mar. Geol.* **54**, 53–90.

Pacific Rise, and ranged in volume from 1 to 600 cubic km. Most occur in clusters, rather than in isolation, so Batiza and Vanko suggested that their positions are controlled by highly fractured areas of fast-spreading crust.

Of the volcanoes surveyed, the smaller ones tended to be low domes of a wide variety of shapes and floor plans, with both convex and concave side slopes, while the larger ones had less-varied morphologies. They are most commonly steeply sloping, truncated cones with broad, flat summit regions several kilometres in diameter containing one or more well-developed craters, often with smaller pits nested within them. (Fig. 15.7). Many of the larger volcanoes also have smaller cones on their flanks and summits, and sometimes also on the nearby sea-floor. Fresh, young lava fields are commonly associated with them.

Craters on volcanoes may imply explosive activity, or dramatic caldera subsidence. This is not the case underwater. Fully fifty per cent of small deep-sea volcanoes are cratered according to Batiza and Vanko's study. They suggest that

crater formation is a normal process during the whole of a volcano's evolution, rather than a unique event late in its life. Most probably, the craters on submarine volcanoes form in the same way as the calderas on large basaltic shields; that is, by incremental foundering as eruption of lavas from lateral vents, perhaps fed by dykes, removes magmatic support from beneath the volcano.

Lavas found on submarine volcanoes include

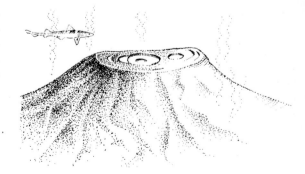

Fig. 15.7 Sketch of a generic volcano seamount such as that in Fig. 15.6. Note the steep side slopes. Shark not to scale.

familiar lobate pillows; extensive, poorly characterized 'sheet' flows, and a rather curious form of
smooth, shelly *pahoehoe*. Broad, amoeboid
swellings of these lavas are often observed from
submersibles. Individual bulbs are often many
metres across. Strangely, they are hollow: where
the shell (only a few centimetres thick) is broken,
it is possible to peer into the cavernous interior,
and see the floor littered with platy fragments of
shell like the broken egg of the Kraken. Pillars
2–3 metres tall support the roof in places. These
exotic structures may have formed when copious
effusions of fluid basalt formed temporary pools
on the ocean floor. Plumes of trapped sea-water
forced out through them from beneath formed
the pillars when they chilled the hot basalt.[6]
Later, the basalt drained away, leaving peculiar
submarine vaulted cathedrals for us to marvel at.
If there are any chambers fit for the Kraken to
sleep his dreamless, uninvaded sleep, they are
these abyssal caverns (Fig. 15.8).

A curious example of hyaloclastite formation
was discovered by Smith and Batiza during dives
in the submersible ALVIN on seamounts near
the East Pacific Rise.[7] They found accumulations
of rubbly hyaloclastite debris around vents,
which they deduced to have been produced by
submarine lava fountaining. Hydraulic pressure
sprayed lava a few metres up into the water above
the vent, where rapid chilling and fracturing of
the lava formed copious quantities of fine glassy

Fig. 15.8 A thin shell of chilled lava forms the
ceiling of a submarine lava pond, which later
drained away. A camera aboard a submersible peers
through a fracture into the fathomless gloom of the
interior abyss. Photo: courtesy of Rodey Batiza,
University of Hawaii.

shards, which were dispersed around the vent
(Fig. 15.9).

15.3.2 *Loihi volcano—an active submarine hot-spot volcano*

Loihi seamount is the youngest addition to the
Hawaiian family of hot-spot volcanoes. As

Fig. 15.9 Direct formation of
hyaloclastites in deep water,
where lava fountains up
through the sea floor.
Hyaloclastite outcrops formed
in this way were found on six
seamount volcanoes near the
East Pacific Rise at depths
between 1240 and 2500 m.
From Smith, T. L. and Batiza,
R. (1989). New field and
laboratory evidence for the
origin of hyaloclastite flows on
seamount summits. *Bull.
Volcanol.* **51**, 96–114.

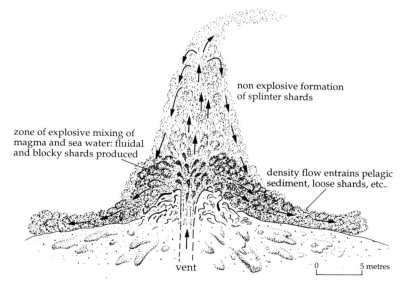

non explosive formation
of splinter shards

zone of explosive mixing of
magma and sea water: fluidal
and blocky shards produced

density flow entrains pelagic
sediment, loose shards, etc..

vent

0   5 metres

outlined in Chapter 2, the Hawaiian Islands form a chain which gets progressively younger in an eastwards direction, the chain marking the imperturbable progress of the Pacific plate over a fixed mantle hot-spot. Loihi is located 50 km south of Kilauea, the southernmost shield volcano on the island of Hawaii, but its location may be more directly influenced by the much larger shield of Mauna Loa, which soars to 4000 m above sea-level. Although it still has to grow by almost 1000 metres before it pokes its head above sea-level, Loihi is already a formidable volcano. Its base is at a depth of more than 4000 metres, but its summit is only 969 metres below sea-level. Like the other Hawaiian shield volcanoes, Loihi is an elongate structure, stretching 30 km along a rift running north-west to south-east (Fig. 15.10).

Alex Malahoff and colleagues at the University of Hawaii have investigated Loihi over several years.[8] They showed that Loihi, like its subaerial counterparts, has a summit caldera, 2.8 km wide by 3.7 km long, containing two pit craters, one 0.6 km in diameter and 73 m deep, the other 1.2 km across and 146 m deep. Loihi's upper parts, in fact, are topographically strik-ingly similar to Kilauea; they just happen to be covered by a lot of water. The lower slopes, however, are steeper than those of Kilauea. Persistent swarms of seismic activity, typical of that associated with the Hawaiian volcanoes, occur close to the summit of Loihi and extend down to depths of about 10 km, suggesting that an active magma body is present beneath the volcano, its top probably at a depth of about 2–3 km. This interpretation was reinforced by the discovery within the summit caldera of active hydrothermal vents, possibly up to 30°C warmer than the surrounding sea-water. Submersible photographs suggest that the hydrothermal vents are associated with roughly conical piles of pillow lavas. Large areas of yellow, red, and green oxides, clay deposits, and bacterial mats are evidence of more extensive hydrothermal activity.

While Loihi is not erupting at present, this is unremarkable. Most volcanoes—for example Mauna Loa—erupt only at long intervals. There has been recent activity on Loihi, however, because photographs reveal young, fresh lava flows, apparently erupted from rifts along the

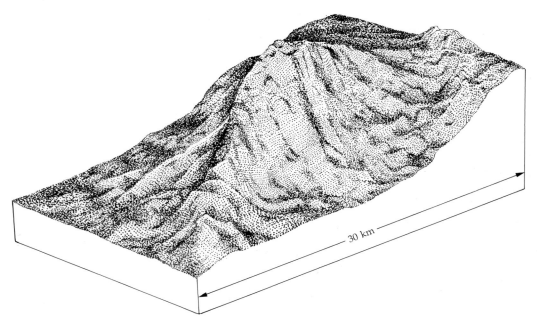

Fig. 15.10 Topographical sketch of Loihi seamount, off the coast of Hawaii, viewed from the south-east. Loihi shows the same sort of elongation along axial rifts as its subaerial counterparts, Kilauea and Mauna Loa. Vertical scale exaggerated. (Hawaii Institute of Geophysics; NECOR.)

southern edges of the summit caldera. Fresh *pahoehoe* lavas overly the upper edge of the caldera rim, while the steeper slopes of the crater are covered by *aa* flows, and the gentler lower slopes by pillow lavas (Fig. 15.11). The longest flows are not more than 2 km long. Malahoff's team suggests that lavas are erupted normally on the upper flanks of the volcano, but break up to form talus on the steep lower slopes. Mass wasting and landsliding on the lower slopes maintain the steep profile. Samples of both tholeiitic and alkali basalt lavas have been obtained from Loihi. Paradoxically, the youngest lavas samples are tholeiitic, and the oldest are alkalic, the reverse of the situation on mature shields like Mauna Kea.

Loihi may become the site of the world's first underwater volcano observatory: plans have been published to site instruments near its summit to detect signs of renewed activity.

Fig. 15.11 Fresh, elongate pillow lavas near the summit of Loihi, the submarine shield volcano off the coast of Hawaii, photographed from the submersible ALVIN. Photo: courtesy of Mike Garcia, University of Hawaii.

— 15.4 Mid-ocean ridge volcanism

It is a curious fact that the most widespread form of terrestrial volcanic activity is also the most unspectacular, at least in its surface expression. Some 70 per cent of the Earth's crust was created via the approximately 65 000 kilometres length of mid-ocean spreading ridges. None of this oceanic crust is more than about 170 million years old, because older crust is subducted at convergent plate margins. Prodigious production of MORB basalts and oceanic lithosphere at mid-ocean ridges is required to maintain the status quo. It is an oft-cited dictum, worth emphasizing, that if the Earth's oceans were to be stripped away, the long linear mountain belts defined by the spreading centres would be the dominant topographical features on Earth. It is also paradoxical that the *fastest* spreading ridges are less impressive topographically than slower spreading ridges: because oceanic lithosphere cools and contracts progressively away from the ridge axis, the lithosphere thickens and subsides closer to the axis of a slow ridge than a fast one (Fig. 15.12). These topographic differences have been formalized to some extent in nomenclature: oceano-graphers speak of fast-spreading axes as *rises*; slow-spreading ones as *ridges*.

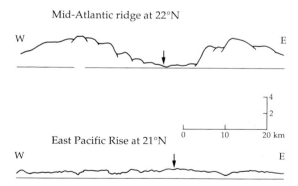

Fig. 15.12 Contrasted topography of slow- and fast-spreading ridge axes. Spreading at the Mid-Atlantic Ridge (top) is slow, of the order of 1–2 centimetres per year, yielding a ridge with pronounced topographical relief, and a 2-km-deep axial rift along its centre, whereas the fast-spreading East Pacific Rise (10 cm per year), below, has much more subdued topography, and lacks a prominent axial rift. The arrow in each case marks the centre.

15.4.1 Ridge axis volcanoes

Soon after Plate Tectonic Theory provided a fuller understanding of the nature of mid-ocean ridges, a concerted effort was made to study them, and the volcanic processes taking place along them. In the 1970s, a historic examination of the Mid-Atlantic Ridge at 36°N (near the Azores) was made by a joint French–American group. Selecting the acronym FAMOUS (French–American Mid-Ocean Undersea Study) to bolster their collective insecurities, the group collected bathymetric, seismic, magnetic, heat-flow, and gravity data over the ridge, as well as diving in submersibles to collect samples and view directly what was going on. Their results have been confirmed and extended by many later studies. We now have a good idea of the nature of volcanism along the 65 000 km of ocean ridges. A special edition of the journal *Oceanus* provides a comprehensive review of recent progress in this rapidly growing, exhilarating field of research and discovery.[9]

Two words summarize the volcanology: *dis-*

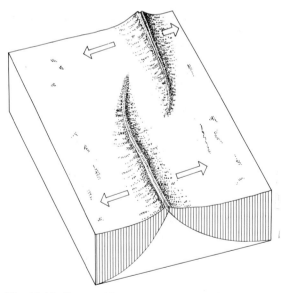

Fig. 15.13 Crests of fast-spreading ridges such as the East Pacific Rise are defined by axial shield volcanoes. Active ridges are not continuous, but segmented. Segments may overlap each other by ten kilometres and be offset horizontally by three kilometres, but these dimensions are variable.

continuous and *linear*. Unsurprisingly, the ocean ridges are not continuously active along their entire length. Individual small segments are active for a while, fade away, and other segments take over. Often, the tips of active segments overlap one another (Fig. 15.13). In total, the entire ridge system probably generates of the order of 10 cubic kilometres of ocean crust per year, not all of it, of course, erupted as lava.

Given the linear nature of the ridge system, it follows that the active volcanic zones should also be linear. But it is remarkable how narrow they are. A typical slow-spreading ridge consists of a broad spine with a central rift valley less than 3 km wide and 400 metres deep. Innumerable faults and fractures parallel the rift axis. Within the axial rift is the active volcanic zone itself, composed of a number of central volcanoes forming a discontinuous central ridge up to 240 m high and 1–2 km wide (Fig. 2.18). These central volcanoes have been likened to elongate Hawaiian shield volcanoes, with gentle slopes and a summit rift. Both pillow lavas and massive sheet lavas have been observed on the flanks of the volcanoes. A study of the median valley floor of the Mid-Atlantic Ridge between 24 and 39°N revealed 481 seamounts; an average of 80 per km, with relief more than 50 m. Here, it is clear that the active ridge axis is constructed entirely of very small, coalesced seamounts.[10]

In detail, the morphology of both volcanoes and rift valleys varies with spreading rate.[11] Medium- and fast-spreading ridges such as the East Pacific Rise are typically characterized by a 2–10-km-wide axial shield volcano, reaching only 100–400 m high, which varies in cross-section from triangular to domed to rectangular. Eruptive activity is focused along a narrow zone a few hundred metres wide at the crest of the rise, while elongate and tubular pillow lavas are present on the slopes. Within the active zone, lava pillows commonly pile up to form haystacks up to 10 m high.[12]

In a detailed study of the Mid-Atlantic Ridge (24°–30°N) Deborah Smith and Jo Cann confirmed earlier observations of the profusion of small seamount volcanoes along the ridge axis, finding about 195 sea mounts 60 metres high per one thousand square kilometres.[13] Such seamounts are virtually absent along the axis of the faster spreading East Pacific Rise. They used the detailed morphology of the volcanic constructs to identify 18 separate spreading segments along their section of the Ridge.

In summary, then, both slow-spreading ridges and fast spreading rises are segmented, but while ridges are characterized by long, narrow axial rifts, rises have elongate axial shield volcanoes. Innumerable small coalesced seamounts make up the active volcanic zone in slow-spreading ridges, whereas it is not possible to distinguish such centres along fast-spreading rises.

To see mid-ocean ridge volcanism at first hand it is necessary to descend in a submersible into the bone-chilling darkness of the deep ocean. But to observe something almost identical, it is only necessary to subject oneself to the searing heat and glare of the Danakil Depression in Ethiopia, where the equivalent of an ocean spreading ridge is exposed on dry land. Erta'Ale is a useful analogue of a rift axis volcano. Like a small Hawaiian shield, it is highly elongated, and bounded by a swarm of parallel faults and fractures. Two convecting lava lakes churn perpetually in pit craters on the summit, and young lavas cover the slopes. Curiously, although the lava lakes have persisted for decades, no lava *flows* have been observed to erupt.

Each year, we learn more about the nature of volcanism beneath the oceans, from side-scan sonar and submersibles. It is hardly likely, though that any future discovery will match in importance that of 'black smokers', plumes of hot, mineral-rich water jetting from vents along ridge axes (Fig. 15.14). These have helped to explain the origin of some major economic mineral deposits, such as the massive sulphide deposits of Cyprus, sources of the copper which gave Cyprus its name. Although a stunning discovery, the black smokers presented few intellectual surprises for volcanologists. Intensely active hydrothermal processes were only to be expected along the mid-ocean ridges, and indeed were required to account for the concentrations of some elements in sea-water. More startling and profound in its implications for the Earth sciences was the discovery of exotic life-

Fig. 15.14 One of the famous 'Black Smokers' of the East Pacific Rise. These powerful jets of hot, mineral-rich solutions are direct evidence of the intense hydrothermal activity associated with ridge axis volcanism. Small sea creatures thrive around the vent. Photo: courtesy of Dudley Foster, Woods Hole Oceanographic Institute.

forms that thrive in pitch darkness around the hot vents, cut off from all normal sources of energy and nourishment.

Science fiction-like images of the giant worms and other bizarre creatures that live around the hot vents have now become familiar, and specimens of them have been brought to the surface. Outlandish though they are, it seems that the life-forms of the Earth's deep ocean ridges are not wholly exotic; they are specialized species of more familiar organisms, which have evolved to fit their unique environmental niche. But their discovery has obliged biologists to reckon with entirely different evolutionary schemes, in which it would be possible for creatures unrelated to those thriving in the realm of sunlight and photosynthesis to evolve. One school of thought argues that life on Earth actually started around hydrothermal vents on the ocean ridges. It has been natural to suppose that the first organic molecules, and life itself, originated in shallow surface waters, bathed by the Sun's beneficent rays. This remains a possibility, but some planetary scientists have used several lines of evidence to argue that life must have evolved extremely early in the history of the Earth—the oldest known fossils are about 3 billion years old, and life must have got under way well before.

During the first 500 million years or so of its early history, the Earth was so severely battered by impacting planetismals that even if nascent life had emerged, it would have been wiped out *if* it were struggling in the vulnerable surface environment. In the deep oceans, the niches provided by mineral-rich hydrothermal vents would have provided a degree of protection and stability, permitting primordial organic molecules to evolve to higher forms. Not all biologists share this view of the deep oceans as the ultimate in air raid shelters, but the concept does open up some wonderful possibilities. Did volcanic warmth incubate the birth of life on other plants? What unimaginable life-forms may be living around mineral-rich warm-water vents beneath Europa's ice cover? Could they resemble mutant turtles?

— 15.5 Deep ocean eruptions

Given the many thousands of kilometres of active spreading ridge beneath the world's oceans, submarine volcanic eruptions should be frequent events. To be sure, there are many accounts of 'submarine' eruptions in the literature, but these are invariably of volcanoes in shallow water; shallow enough for vesiculation and explosive pyroclastic activity to take place. While eruptions in deep water probably are common, to date, few examples have been detected. In January 1991, scientists from the US National Oceanographic and Atmospheric Agency reported the discovery of a line of new volcanoes on the Pacific ocean floor, 450 km off the coast of Oregon, along a segment of the Juan de Fuca ridge. Several of the new volcanoes are more than 30 metres high and over 800 m wide; they are thought to have been constructed by sea-floor eruptions taking place over a ten-year period. They may have been related to huge plumes of hot, mineral-rich water detected by NOAA in 1986 and 1987 on the southern part of the Juan de Fuca Ridge (Fig. 15.15). Apart from their intrinsic volcanological interest, such deep-ocean eruptions are important because they may have climatic effects—variations in sea-water temperature over large areas caused by submarine eruption of large volumes of hot basalt lava might have consequences for global climate that are hard to predict.

Even more compelling direct evidence of a submarine volcanic eruption came from a section of the East Pacific Rise at 9°–10°N. This section of the Rise had been intensively studied in late 1989 through in a series of imaging surveys, aimed at studying active hydrothermal vents and the exotic animal communities living around them. Fifteen months later, dives by the ALVIN submersible to the same sites revealed that volcanic eruptions had wrought significant changes. A caldera on the rise axis which had previously been densely colonized by animal communities was now floored with glassy and ropey lava flows. Thousands of newly-killed vent organisms, plus a few traumatized survivors were found, partly buried by a thin flow of new lava. As the oceanographers succinctly described it: 'carnage at the Tubeworm Barbeque site was so recent that crabs and other scavengers had not yet arrived to consume the tissues of the abundant dead animals . . .'.[14]

15.5.1 Detecting submarine eruptions

There are basically three ways of detecting submarine volcanic activity: acoustically, by 'conventional' seismic signals and by harmonic tremor.[15]

Acoustic waves Readers of modern thrillers may be familiar with the phenomenal range at which submarines can be detected using sonar equipment. Similarly, whales may be able to communicate with each other over distances of

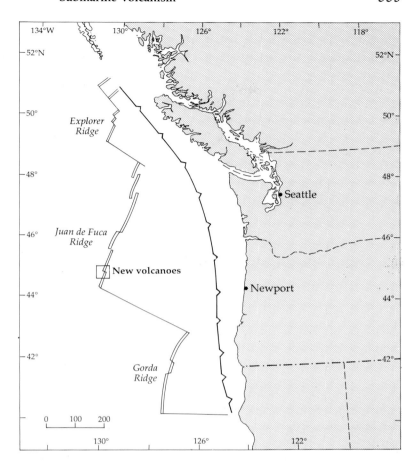

Fig. 15.15 Location of some possible new submarine volcanoes found on the Juan de Fuca Ridge.

tens or even hundreds of kilometres because of the ease with which low-frequency sound waves propagate through water. An explosive volcanic eruption is *much* noisier than any submarine or whale, so it can easily be detected. Eruptions of Macdonald volcano, a submerged mid-Pacific hot-spot volcano in the Austral Islands (29°S, 140°W) have been detected many times since 1977 by instruments in Tahiti and Hawaii. But, as discussed at the beginning of this chapter, explosive eruptions cannot take place at depths greater than the critical pressure, equivalent to depths of about 3000 metres. Macdonald volcano is almost at sea-level, its summit being only 40 metres below the waves. Thus, it lies well within the regime of explosive eruptions. Unfortunately, because eruptions below the critical pressure are quiet, they cannot be detected by these simple techniques.

Conventional seismic signals Active volcanoes are often sites of small fault movements and related fracturing events as the mass of the volcano shifts and adjusts itself to the loads imposed upon it by erupted lavas and by movements of the magma within it. These short, sharp events are caused by brittle failure in the rocks of the volcanic edifice at a shallow level, so they generate seismic shock waves identical to those of ordinary earthquakes. Large earthquakes, triggered by displacements of major fault systems such as the San Andreas in California, propagate waves that are so energetic that they can be easily detected on the far side of the Earth. Volcanic tremors are rarely powerful enough to be detected, or at least usefully interpreted, at distances of more than a few hundred kilometres from the volcano. Thus, in a large ocean such as the Pacific, they could easily go unnoticed.

Harmonic tremor One of the most reliable methods for monitoring a potentially dangerous volcano on dry land is the detection of harmonic seismic tremor. Rather than short, sharp, earthquake-like shocks, this is an almost continuous agitation or vibration that often accompanies the upward movement of magma through the volcanic conduit. Harmonic tremor is distinctive on seismograms, and experienced volcanic seismologists have become confident in using it as a means of predicting eruptions. However, while harmonic tremor is often intense enough on a volcano to be felt (and feared) by the people living there, it is not powerful enough to be detected at great distances.

Thus, unfortunately, we do not have much in the way of techniques to detect submarine volcanic eruptions, particularly those in the deep oceans and along mid-ocean ridges. As any geology text will show, the mid-ocean ridges are sites of major seismic activity, dozens of seismic events being detected along them each year. Some of these may be associated with events that open new fractures in the axial regions of rifts, from which magma may erupt. But we don't really know.

— 15.6 Pyroclastic deposits on the sea-floors —

15.6.1 Submarine tephra fall deposits

As discussed in Section 15.2.2, a few pyroclastic eruptions may take place at depths above the critical depth. Below that, all pyroclastic activity is snuffed out. However, nothing can stop tephra erupted from volcanoes on dry land from raining out over the oceans. When the huge eruption of the Toba caldera took place about 75 000 years ago, tephra fell as far distant as India (Fig. 14.8). Dragoslav Ninkovitch and his colleagues working on piston cores recovered from the sea-floor found as much as 10 centimetres of ash 2000 kilometres from the caldera.[16] Tephra from the great Minoan eruption of Santorini has also been traced over wide areas of the eastern Mediterranean. Since little sediment of terrigenous origin reaches the deep ocean floors, volcanic ash layers form an important part of the stratigraphy of ocean-floor sediments, along with fine clays and oozes. They provide useful marker horizons which are much more continuous laterally than their counterparts on dry land, where rapid erosion operates.

Deep-sea pyroclastic deposits are not much fun to work with. They can only be studied as core samples, and are often mechanically disturbed by the drilling process. Burrowing organisms on the sea-floor may also disrupt the original depositional structure. Worse still, it does not take long, in geological terms, for alteration processes to turn submarine tephra deposits into an unpleasant, clayey mess.

There is more to submarine tephra, though, than shapeless slime. Realizing that tephra particles of different sizes and masses from a pyroclastic eruption would sink through water at different rates, Ledbetter and Sparks derived an ingenious algorithm to determine the *duration* of a large eruption.[17] Two simple concepts underlie their technique: first, the rate of settling of particles with known properties through water can be estimated by Stokes' Law, which relates particle settling rate to the size and density of the particles, and the viscosity and density of the fluid that they are settling through. Second, in an eruption which lasts more than a few hours, the Stokesian settling rates through a deep water-column will be so different for the largest and smallest particles that efficient size fractionation takes place. (These are similar arguments to those used by Cashman and Fiske in relation to settling of lithic and pumice clasts.)

Assuming that all else remains constant during the eruption, large particles will fall through the water column fastest; finer particles following at a slower rate. While the supply of tephra continues, the size of the largest particles accumulating on the sea-floor will remain constant. As soon as the eruption ceases, the size of the largest particles falling through the water will also decrease. This will be observed as a decrease in largest particle size at some point in the ash deposit. The *finest* ash particles at the same level, however, will only just have arrived, after having sifted slowly down the water column *since ash fall commenced*. Thus, knowing the settling rates in air and water of the particles, it is possible to calculate the duration of the eruption.

There are many pitfalls in this technique, of course, but it has been used to suggest that the great Toba eruption lasted about 10 days, consistent with what we would expect from other geological considerations. Although it may seem hard to swallow, a *serious* pitfall is that many of the particles falling through the water column will be swallowed by fish or other marine creatures. Such particles are ultimately excreted almost unchanged in *faecal pellets*, but there is no way of telling how long they were held up ...

15.6.2 *Submarine pyroclastic flow deposits*

Scholarly passions have been fired by controversies over submarine pyroclastic flows to an extent which outsiders might find surprising. Can pyroclastic flows carry on moving from land into water? More worryingly, can volcanologists really get hot under the collar about the origins of anonymous green rocks exposed on bleak, rain-swept Welsh hillsides? Ray Cas and John Wright have provided an illuminating review of why submarine pyroclastic flows are so contentious.[18]

Two volcanological problems, both of them highly combustible, fuel the controversy. As Chapter 10 emphasized, pyroclastic flows include a wide range of phenomena, ranging from dense *nuées ardentes* to insubstantial pumice flows, so it may be difficult to determine what was involved in the first place. Second, much of the evidence for submarine pyroclastic flows necessarily comes from ancient deposits, now uplifted and exposed at the surface. Owing to their antiquity, it is often difficult to determine exactly the geological environment in which the candidate pyroclastic deposits were laid down. This is the problem with the much debated Welsh flows, which were clearly laid down at or near sea-level.[19-20] Deep-sea drill cores have often recovered samples thought to be young pyroclastic flow deposits, but these are difficult to interpret for the same reasons that submarine tephra fall deposits are.

Leaving aside the notion sometimes advanced that pyroclastic flows can actually be erupted under water, the fact that so many major volcanoes are located close to the oceans means that large numbers of pyroclastic flows generated on dry land *must* enter the water. Our problem is to determine what happens to them then. This must be controlled by their densities and volumes.

There is no reason why dense flows, such as nuées ardentes, consisting mostly of unvesiculated lava blocks, should not continue moving under water, provided that the submarine topography remains favourable. To be sure, the density contrast with the surrounding water will be much less than in air, but the flow will probably be transformed into a *turbidity current*, similar to those well known on continental shelves. Non-volcanic turbidity currents give rise to the familiar sequences of graded sedimentary beds known as turbidites or Bouma sequences. According to Lacroix, the dense glowing avalanches (nuées) from Mt. Pelée in 1902 generated underwater equivalents large enough to break submarine telegraph cables nearly 20 km offshore.

Pumiceous pyroclastic flows, which would deposit ignimbrites on dry land, present more difficult problems. Highly expanded pyroclastic flows, with densities less than water, may scoot directly across the surface, though inevitably there is bound to be some interaction at the interface. Tad Ui has suggested that the 6000-year-old Koya flow travelled at least 10 kilometres over water to reach the southern shores of Kyushu Island in Japan. In these circumstances, extensive rafts of floating pumice would doubtless form. After the Krakatau 1883 eruption, floating masses of pumice seriously obstructed shipping. In places they were so solid and extensive that they were mistaken for dry land.

When really large flows enter the sea, the situation is even more complex. If the flows are sufficiently large, they may simply bulldoze the water aside, generating immense tsunamis in the process. This may have happened during the Krakatau eruption. When the eruption was over, the wrecked island was surrounded by a broad apron of pyroclastic debris, breaking the surface in places in shallow shoals, and forming two islands, Steers and Calmeyer (Fig. 4.16). Although the chronological details are obscure, eruption of these ignimbrites and their displacement of sea-water may have caused some of the tsunamis recorded at Batavia (Jakarta).[21]

Whether or not the flows entering the sea are large enough to generate tsunamis, intensive explosive activity is inevitable at the interface. A large convecting cloud will be generated, carrying a lot of fine material high into the air, from whence it will ultimately descend as a fall deposit, as was the case at Tambora (Section 10.4.4). A low-density surge may also spread across the surface of the water. Most of the mass of the flow will probably continue on beneath water, but, like a nuée ardente, will almost immediately lose its unique identity as a gas-fluidized flow, and will move downslope as a water-supported density current.

Many examples of deposits formed in this way are known from marine geology. Some of the best studied are in the eastern Caribbean, in and around the Lesser Antilles. Sigurdsson and his colleagues[22] showed that of the total of 527 cubic kilometres of volcanic material erupted over the last 100 000 years, eighty per cent was deposited as volcanogenic sediments in the adjacent marine basins. Seventy per cent of this material was deposited in the back-arc Grenada Basin in the form of sediment gravity flows. The amount of air-fall tephra is thus relatively small, but the effects of topography and prevailing trade winds are such that while the density currents flowed westwards, the air-fall tephra was blown eastwards, to fall out over the Atlantic.

One particularly extensive submarine pyroclastic flow is the Roseau flow, which has been traced in piston core samples over distances of over 250 kilometres from its presumed source on the island of Dominica[23] (Fig. 15.16). The

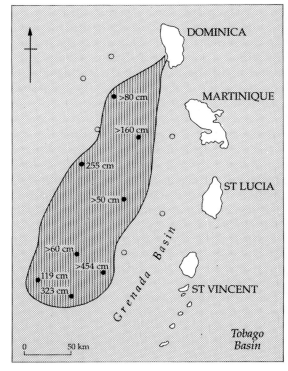

Fig. 15.16 Distribution of the Roseau subaqueous pyroclastic flow deposits. Solid circles are sites of submarine piston core samples with thicknesses of the deposit in centimetres; open circles are core sites where no deposit was found. After Carey, S. N. and Sigurdsson, H. (1980). The Roseau ash: deep sea tephra deposits from a major eruption in Dominica. *J. Volcanol. Geotherm. Res.* **7**, 67–86.

Roseau flow appears to be the submarine equivalent of the Roseau ignimbrite, erupted from a complex of lava domes and pyroclastic rocks on the southern part of the island. Whereas the Roseau ignimbrite was unequivocally hot when emplaced (it is welded in some places) the submarine deposits were not. This led to suggestions that the submarine deposits are not strictly 'pyroclastic' and that they would best be described as 'volcaniclastic debris flow deposits'.

Notes

1. Moore, J. G. (1975). Mechanism of formation of pillow lava. *Am. Sci.* **63**, 269–77.
2. Tribble, G. W. (1991). Underwater observations

of active lava flows from Kilauea volcano, Hawaii. *Geology* **19**, 633–6.
3. Cashman, K. V. and Fiske, R. S. (1991). Fallout of

pyroclastic debris from submarine volcanic eruptions. *Science* **253**, 275–81.

4. Menard, W. H. (1964). *Marine geology of the Pacific*. McGraw-Hill, New York, 271 pp.

5. Batiza, R. and Vanko, D. (1983/84). Volcanic development of small oceanic central volcanoes on the flanks of the east Pacific Rise inferred from narrow-beam echo sounder surveys. *Marine Geology* **54**, 53–90.

6. Francheteau, J., Juteau, T., and Rangna, R. (1979). Basaltic pillars in collapsed lava pools on the deep ocean floor. *Nature* **281**, 209–11.

7. Smith, T. L. and Batiza, R. (1989). New field and laboratory evidence for the origin of hyaloclastite flows on seamount summits. *Bull. Volcanol.* **51**, 96–114.

8. Malahoff, A. (1987). Geology of the summit of Loihi submarine volcano. In Volcanism in Hawaii (eds. R. W. Decker, T. L. Wright, and P. H. Stauffer. US Geol. Surv. Prof. Pap. No. 1350, pp. 133–44.

9. Oceanus (1991–2). Vol. 34, Woods Hole Oceanographic Institution, Woods Hole, Mass. 111 pp.

10. Smith, D. K. and Cann, J. R. (1990). Hundreds of small volcanoes on the median valley floor of the Mid-Atlantic ridge at 24–30°N. *Nature* **348**, 152–5.

11. Macdonald, K. C. (1982). Mid-ocean ridges: fine scale tectonic, volcanic and hydrothermal processes within the plate boundary zone. *Ann. Rev. Earth Planet.* **10**, 155–90.

12. Batiza, R. (1989). Petrology and geochemistry of Pacific Spreading Centres. In *The geology of North America, Vol. N. The Eastern Pacific Ocean and Hawaii*. pp. 145–59. Geol. Soc. Am.

13. Smith, D. K. and Cann, J. R. (1992). The role of Seamount Volcanism in Crustal Construction at the Mid-Atlantic Ridge. *J. Geophys. Res.* **97**, 1645–58.

14. Haymon, R. *et al.* (1991). Active eruption seen on East Pacific Rise. *Eos*, **72**, 505–7.

15. Talandier, J. (1989). Submarine volcanic activity: detection, monitoring and interpretations *EOS, Trans. Amer. Geophys. Union* **70**, 561 and 568.

16. Ninkovich, D., Sparks, R. S., and Ledbetter, M. J. (1978). The exceptional magnitude and intensity of the Toba eruption, Sumatra: an example of the use of deep sea tephra layers as a geological tool. *Bull. Volcanol.* **41**, 286–98.

17. Ledbetter, M. T. and Sparks, R. S. J. (1979). Duration of large magnitude explosive eruptions deduced from graded bedding in deep sea ash layers. *Geology* **7**, 240–4.

18. Cas, R. A. F. and Wright, J. V. (1991). Subaqueous pyroclastic flows and ignimbrites: an assessment. *J. Volcanol. Geotherm. Res.* **53**, 357–80.

19. Cas, R. A. F. and Wright, J. V. (1987). *Volcanic successions modern and ancient*. Allen and Unwin, London, 487 pp.

20. Fritz, W. J., Howells, M. F., Reedman, A. J., and Campbell, S. D. G. (1990). Volcaniclastic sedimentation in and around an Ordovician subaqueous caldera, Lower Rhyolitic Tuff Formation, North Wales. *Geol. Soc. Amer. Bull.* **102**, 1246–56.

21. Francis, P. W. and Self, S. (1983). The eruption of Krakatau. *Sci. Am.* **249**, 172–87.

22. Sigurdsson, H., Sparks, R. S. J., Carey, S. N., and Huang, T. C. (1980). Volcanogenic sedimentation in the Lesser Antilles Arc. *J. Geol.* **88**, 523–40.

23. Carey, S. N. and Sigurdsson, H. (1980). The Roseau ash: deep sea tephra deposits from a major eruption in Dominica. *J. Volcanol. Geotherm. Res.* **7**, 67–86.

16

Volcanoes as landscape forms

Ordinary, non-volcanic landforms are the results of erosion by wind, water, and ice. Erosion is an irreversible process which ultimately reduces even the loftiest mountain range to a flat plain. Volcanic landforms by contrast, are the results of opposing constructive and destructive forces. *Constructive* processes operate only while volcanoes are active. This may be an extremely short period—a matter of days or weeks—or rather long, with activity continuing intermittently over tens of thousands of years. Paricutin, a common-or-garden basaltic scoria cone was born in a Mexican cornfield on 20 February 1943. After a year of activity it was 325 m high; when it finally simmered into silence in 1952 it was 410 m high. About 2 cubic kilometres of lava and tephra were erupted during its nine years of activity. By contrast, Stromboli in the Mediterranean has been erupting throughout history, but is still only 981 metres above sea-level. Many rapidly constructed volcanic landforms are not 'volcanoes' at all—the Valley of Ten Thousand Smokes was buried under 15 cubic kilometres of ignimbrite in less than 60 hours.

When contemplating the impassive grandeur of a mountain range, one intuitively thinks of the destructive processes of erosion as acting infinitely slowly. Erosion is often conceived as the epitome of the slowness of geological processes. But this is a considerable oversimplification. In arid environments such as the Atacama Desert or the Moon's surface, erosion rates are indeed immeasurably slow, but in others, such as the humid tropics, they can be startlingly fast, even by human standards. Catastrophic processes such as avalanching accomplish in a few moments what it might take millennia to achieve otherwise. Whatever their rate, one thing is certain: erosion starts work on a volcano as soon as it starts growing, even before its lavas cool. Erosion never ceases. A large volcano may experience several phases of rapid construction in its lifetime, during which the rate of construction exceeds the rate of erosion, but once eruptive activity wanes, erosion instantly gains the upper hand. On a large volcanic massif, erosion may be proceeding in one part, while new lava is being added to another.

All these variables yield volcanic landscapes that are as richly diverse to analyse as they are pleasing to behold. Strangely, though, volcanic landscapes have been little studied. There *is* a book on the subject, which remains an authoritative account although it was published almost 50 years ago; one which is still capable of supplying volcanologists with insights into why their volcanoes look the way they do. '*Volcanoes as Landscape Forms*' was the title which C. A. Cotton selected for his book; his title is used for this chapter to acknowledge his contribution.[1]

16.1 Monogenetic volcanoes

16.1.1 Scoria cones

A monogenetic volcano is the product of a single eruptive episode. This may last a few hours, or a few years, but the essential point is that once eruption has ceased, the plumbing connecting the vent to its magmatic source freezes over, so the volcano never erupts again. Basaltic scoria cones are good examples of monogenetic volcanoes. They are found in thousands all around the world, in many tectonic environments, either as components of scoria cone fields, like Paricutin in Mexico, or as parasitic vents on the flanks of larger volcanoes—Etna has dozens. All over the world, they have the same distinctive morphology. They are rarely more than two or three hundred metres high, and are often asymmetrical; either elongated along a fissure, or else higher on the side that was downwind at the time of eruption. A breach on one side often marks the site from which lava has flowed. A distinctive feature is their simple geometrical profile, defined by the angle of rest for loose scoria (Fig. 16.1). *All young scoria cones have side slopes close to 33°.*

In a statistical study of scoria cones, Chuck Wood showed that 50 per cent were formed during eruptions that lasted less than 30 days; 95 per cent of them during eruptions that lasted less than one year.[2] Wood looked at the dimensions of 910 scoria cones, and found that their mean basal diameter was 0.9 km. In a sample of 83 *fresh* scoria cones, he found some regular geometrical relationships: the *height* of the cone proved to be 0.18 times the basal width, while the crater diameter was 0.40 times the basal width. This emphasizes a characteristic feature of scoria cones: their craters are large in relation to the size of the edifice as a whole. Naturally, their crisp profiles soften with age, but Wood showed that the ratio of crater width to basal width changes remarkably little. Thus, scoria cones remain easily recognizable, even after millennia of weathering.

16.1.2 Maars

Scoria cones are the results of minor basaltic eruptions taking place in dry conditions. When basaltic magmas interact with water, the nature of the eruption is explosively different, producing *surtseyan* pyroclastic deposits (Section 6.4.1). It is not necessary for the eruption to take place under water to produce explosive consequences —a water-bearing stratum (aquifer) in sedimentary rocks is all that is needed. In the simplest case, shallow phreatic explosions caused by magma–ground-water interactions blast upwards through to the surface, forming large

Fig. 16.1 La Poruña, north Chile, a 300m-high scoria cone. La Poruña appears youthful, but may be many thousand years old, since it is located in a hyper-arid part of the Atacama Desert. In the shadow at the foot of the cone, a train on the Antofagasta–La Paz railroad provides scale. (Compare the air photo in Fig. 7.8.)

holes in the ground. In the Eifel area of Germany, eruptions of this kind formed 30 craters about a kilometre across, now occupied by lakes, which gave their name to the landform: *maars*. *Maar* craters are simple, circular depressions surrounded by low rims of ejected debris. Their walls are steep-sided initially, but are quickly eroded away to gentle slopes. Since they are by definition holes in the ground rather than structures built up above it, *maars* typically fill with water and are thus manifested as lakes.

In his statistical study, Wood showed that *maars* are typically small features, most having diameters of about one kilometre. His preferred examples are from the Pinacate region of northwest Mexico, where eight young *maars* are beautifully exposed in the Sonoran Desert. They are circular to oval, with diameters between 750 and 1750 metres and range in depth between 36 and 245 metres.

In the desiccated heart of the Sahara Desert, an improbable place to look for magma–water interactions, there is an instructive *maar* at Malha, in the Darfur province of the Sudan. About one kilometre in diameter and a hundred metres deep, the *maar* was blasted through a layer of Nubian sandstone, depositing an apron of ejected debris around the crater, now well exposed in the rim (Fig. 16.2). Conspicuous in the debris are rounded boulders of gneiss and granite up to a metre across derived from the metamorphic basement underlying the sandstone. Explosive activity thoroughly comminuted the relatively weak sandstone, while the granites and gneiss were more resistant. At the bottom of the Malha *maar* is a small lake, fed by a series of small springs seeping through the Nubian sandstone. Because they are the only source of fresh water for thousands of square kilometres, these springs are vital to the people of the area. Their meagre flow illustrates how little water needs to be contained in an aquifer for explosive magmatic interactions to take place.

At Malha and other *maars*, there is often only a small proportion of magmatic material in the ejecta. This can lead the unwary into grievous errors of interpretation, since basin-shaped maar craters have been mistaken for meteorite impact craters when there is little obvious volcanic material present, as at Jayu Khota in Bolivia (Fig. 16.3).

16.1.3 Tuff rings

A convenient, but not iron-clad distinction, between *maars* and tuff rings is that *maars* are excavated into the substrate, whereas tuff rings are built up above it (Fig. 16.4). And whereas *maars* are the results of shallow explosions, often involving scanty amounts of juvenile material, tuff rings contain an abundance of highly fragmented basaltic scoria: they are essentially accumulations of surtseyan tephra. Tuff rings are

Fig. 16.2 Camels and goats drinking at the Malha *maar*, Darfur province, western Sudan. White Nubian sandstone is exposed in the wall of the crater, while darker overlying material is ejecta. Animals in the foreground are clustered around small springs which all rise at the same stratigraphic level, probably the contact between the sandstone and the underlying crystalline basement.

Fig. 16.3 Jayu Khota *maar* on the altiplano of Bolivia, initially thought to be an impact crater. Tufa deposits from the high stand of glacial Lake Tauca are present around the crater, which therefore must be more than 10 000 years old. Erosion has subdued the ejecta rim.

Fig. 16.4 A superbly symmetrical tuff ring, near the Erta'Ale volcano, Ethiopia. (Photo: H. Tazieff.)

formed when magma comes near to the surface before being explosively fragmented. Cerro Xico, only 15 km from the heart of Mexico City, is an elegant example which formed in the basin of shallow Lake Texcoco, before it was drained by the Spanish in the sixteenth century. Superbly circular when seen from the air, Cerro Xico is frustrating to photograph from the ground because it is so broad and flat, like all tuff rings.

Diamond Head, a much photographed landmark on the island of Oahu (Hawaii) is another tuff ring formed in prehistoric times (Fig. 16.5). For years it dominated the exotic palm-fringed surfing beach at Waikiki. Now the beach is dominated by high-rise tourist hotels . . . Diamond Head crater is about 1 kilometre in diameter, and for the most part its rim is about 100 metres high. On its south-western extremity, however, Diamond Head itself rises no less than 232 metres. The crater's marked asymmetry was caused by north-east trade winds, which must have been blowing strongly during the eruption,

Fig. 16.5 Diamond Head tuff ring, Honolulu, Hawaii. Trade winds caused maximum ash accumulation downwind, forming the high point on the rim, furthest away from camera.

just as they do through much of the year today. Ejecta inevitably accumulated much more thickly on the downwind side.

16.1.4 Tuff cones

Tuff cones are smaller, steeper versions of tuff rings, composed of similar surtseyan tephra. Morphologically, they resemble scoria cones. Only a few kilometres from Cerro Xico, on the same lake bed near Mexico City, is a prominent tuff cone called El Caldera (Fig. 16.6). It is not immediately obvious why one eruption should form a tuff ring, while another only a short distance away forms a cone, but it probably has to do with the relative amounts of water and magma, and the duration of the eruption. Tephra has to be widely dispersed to form a tuff ring, calling for a violent hydromagmatic eruption, with relatively high column and mass eruption rate. Tuff cones may form from less violent, more prolonged eruptions.

The tephra erupted are hot and water saturated, so wet, muddy deposits are formed. Their stickiness results in tuff cones which develop

Fig. 16.6 El Caldera, a 250-m-high tuff cone constructed above the bed of former Lake Texcoco, near Mexico City. Morphologically, it resembles a scoria cone, but is composed of surtseyan tephra.

startlingly steep slopes on erosion, sometimes steeper than the 33° angle of rest of loose, dry scoria; sometimes so steep that it is barely possible to walk up the dip slopes of the tephra layers. Koko crater, only a few kilometres from Diamond Head on Oahu, is an excellent example of a steep tuff cone (Fig. 16.7). Koko crater is smaller than Diamond Head, only about 800 metres in diameter, but is much higher, 366 metres. As in Diamond Head, the high point on the rim is on the downwind side of the crater.

Unlike ordinary scoria cones, both inward and outward dips are commonly seen in tuff cones, tephra layers mantling the rim and both inner and outer slopes of the crater. Why does the tephra not simply slump off while still wet and muddy? This is not clear. Some slumps do take place, but for the most part the steeply dipping tephra layers appear to have been stable from the moment they were deposited. Probably, the hot, wet pyroclastic material dried quickly, setting as hard as concrete. (Alteration of the glassy shards by hydration to *palagonite* takes place rapidly after deposition, often obscuring original structures.)

Eruptions which construct both tuff rings and cones also yield pyroclastic surges which may travel several kilometres (Section 11.1.1). However, on the slopes of the cone or ring itself, it is often not clear whether the material was deposited by fall or surge. In a thoughtful analysis of the differences between tuff rings and cones, Wohletz and Sheridan suggested that the massive bedding found in tuff cones is due to emplacement in cool, wet conditions, less than 100°C.[3] In contrast, the thinner-bedded deposits of tuff rings are emplaced hot and relatively dry. As increasing amounts of water mix with magma to generate surge blasts, increasing amounts of steam are produced, but the steam is cooler and therefore 'wetter'. Thus, as the water:magma ratio increases, wetter and stickier surges result. Wohletz and Sheridan suggested that there should therefore be a regular progression of minor basaltic landforms, from scoria cones in dry environments, through tuff rings where ground-water is present to tuff cones in open water (Fig. 16.8).

16.1.5 Diatremes

Some of the *maars* that turn up in odd geological contexts around the world are the surface expressions of *diatremes*, vertical pipes blasted through basement rocks, which contain angular rock fragments of all sorts and conditions.[4] Some fragments are lumps torn from the sides of the pipe at depth, others are from shallow levels; some may be of sedimentary origin—even lumps of coal—and some are of obvious volcanic origin. Diatremes are abundant in some regions—in the Schwabian Alps area of southern Germany, more than 300 have been identified over an area

Fig. 16.7 Profile view of Koko crater, Oahu, Hawaii. Trade winds blowing from right to left are responsible for the pronounced asymmetry of this 366-m-high tuff cone, which shows good parasol ribbing.

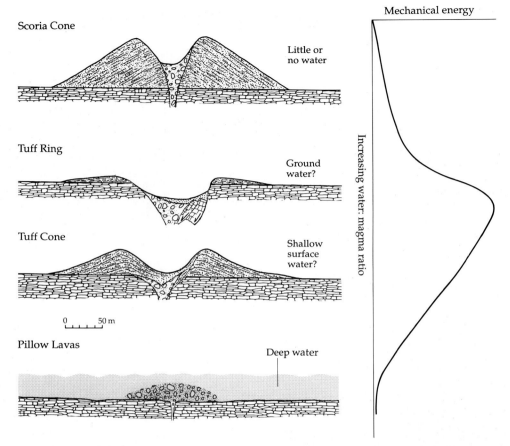

Fig. 16.8 Effect of increasing amounts of water on morphology of small basaltic volcanic constructs. Internal structures are shown only schematically: slumps from steep crater walls are observed in some tuff rings, while inward and outward dips are characteristic of tuff cones. After Wohletz, K. H. and Sheridan, M. F. (1983). Hydrovolcanic explosions II. Evolution of basaltic tuff rings and tuff cones. *Am. J. Sci.* **283**, 385–413.

of about 1600 square kilometres. They are all between 15 and 20 million years old, and so their surface expression is muted. Some are identifiable as vague depressions, but most can only be picked up through geophysical surveys.

At Kimberley in South Africa, a deep mine was excavated to probe the limits of a large diatreme. This expensive undertaking was not carried out in the spirit of scientific enquiry, but because the diatreme contained gem-quality diamonds. These were found in *kimberlite*, a mixed, brecciated rock containing a proportion of peridotite, derived from the mantle, where the diamonds originated. In the deepest part of the mine, the width of the carrot-shaped pipe is only 30 metres, while at the present surface level it is 300 metres. It may have been linked with a maar about 500 metres across at the original surface. South Africa is not now a region of active volcanism: from isotopic evidence, the diamonds themselves appear to be more than one thousand million years old, whereas the kimberlites which brought them to the surface were erupted ninety million years ago. Erosion and subsequent redeposition of the diamantiferous rocks in the upper part of this diatreme and others like it gave rise to the rich alluvial diamond fields of modern South Africa and Namibia.

Fig. 16.9 Hyalo Ridge, Wells Gray Provincial Park, British Columbia, a 1000 m high *tuya* formed about 10 000 years ago by subglacial eruption of alkali olivine basalt, now thickly forested. Photo: courtesy of Catherine Hickson, Canadian Geological Survey.

Fig. 16.10 Detail of a 'pillow' of basalt in hyaloclastite matrix, showing quenched margins and radial jointing, formed by subglacial basalt eruption. Hyalo Ridge *tuya*, British Columbia. Photo: courtesy of Catherine Hickson, Canadian Geological Survey.

16.1.6 Landforms resulting from sub-glacial eruptions

At the present day, permanent glaciers are confined to the polar regions, a few sub-arctic ice caps such as those in Iceland and Patagonia, and shrinking valley glaciers on high mountains around the world. It is easy to forget that huge areas of the Earth were covered by continental ice sheets which receded only 11 000 years ago. When the ice receded, the steep sides and flat, lava-capped tops of subglacial volcanoes were revealed (Section 6.4.3). In British Columbia, easily recognizable 'table mountains' developed by subglacial central vent eruptions are called *tuyas*; in Iceland, where eruptions from fissures commonly produce elongate ridges, they are called *mobergs*[5] (Fig. 16.9–16.10).

— 16.2 Polygenetic volcanoes

Polygenetic volcanoes are those that have experienced more than one eruptive episode in their history. Most of the world's volcanoes fit into this rather loose category, but several different subgroups can be identified, based on the number and location of the vents from which eruptions took place.

16.2.1 Simple cones

Simple cones are overgrown scoria cones; scoria cones which carried on erupting. They have a single summit vent and radial symmetry. These are the volcanoes whose graceful profiles adorn so many calendars and postcards. There are innumerable examples of splendidly symmetrical simple cones around the world. Mt. Mayon (2400 m high) in the Philippines is often cited as the world's most beautiful volcano (Fig. 12.14). Licancabur (5916 m high) has a striking geometrical purity of line, dominating the oasis of

Fig. 16.11 Licancabur, north Chile, a 5916-m-high simple cone. The date of its last eruption is not known, but appears to have been recent, since lavas on its flanks are fresh. There are Inca ruins and a small lake in its summit crater.

San Pedro de Atacama in Chile (Fig. 16.11). In Peru, daily life in the city of Arequipa is carried out beneath the sublime curves of El Misti (Fig. 16.12), while in Guatemala, Agua and Fuego volcanoes form an appropriate backdrop for the strombolian eruptions of Pacaya. It is easy to be

over-impressed by the apparent simplicity of such volcanoes, however. Subtle structural complexities often turn up on closer inspection, so that the volcano should formally be regarded as *composite*—magnificent Mount Fuji is a case in point.

Simple cones are characterized by rather small summit craters; often tiny for the size of the edifice as a whole, and not much bigger than that of a scoria cone—Mt. Mayon's for example, is less than 200 m in diameter. Nestled within Licancabur's summit crater is a miniscule freshwater lake, 90 metres by 70 metres. At an elevation of almost 6000 m, it is probably the world's highest lake, but none the less a planktonic fauna of considerable interest to biologists manages to exist. Weak volcanic thermal emissions probably help to prevent the lake from freezing into a solid mass—stalwart divers have explored its depths during the world's highest altitude dive.

In their summit regions, simple cones are often armoured by lava flows (Fig. 16.13). These provide mechanical strength, so vertiginously steep slopes, over 40°, are possible, as for example on the 1157-m-high volcano St Eustatius in the Dutch Antilles. On the mid-flanks, interstratified scoria, lava, and talus eroded from the lavas predominate, so slopes ease off.

Fig. 16.12 El Misti (5822 m) rises directly above the suburbs of Arequipa, Peru's second city, presenting an obvious hazard to the city (foreground). There have been many eruptions since the Spanish conquest. A lava dome was active in the summit crater as recently as the 1950s. Photo courtesy of F. Bullard, University of Texas.

Fig. 16.13 Acamarachi, north Chile, a 6046-m-high simple cone with steep slopes resulting from extrusion of lava flows from the summit crater. Edifice height is only 1200 m.

Although lava flows may snake down to low levels, the lowest slopes are mostly constructed of talus carried downwards by mass wasting. Thus, the elegant concave profiles of simple cones reflect the interplay between erosion and eruption.

Given a single vent, the shape and height of a simple volcano are immutably controlled by geometry: every additional increment of *height* requires a huge additional increase in *volume*. When a volcano reaches a height of more than 2000 m, each additional metre of height requires the eruption of tens of millions of cubic metres of rock. And as height increases, so erosion becomes more effective, though not in a simple geometrical manner. So, for any given combination of eruption and erosion rates, the height of a simple cone is self-limiting. As volcanoes become larger, approaching 3000 m, the volume increment required for each additional height increment is so huge that the required eruption rates begin to exceed the geologically plausible. This explains why volcanoes on Earth rarely exceed 3000 m in *edifice* height. (Note that the *summit* heights of many volcanoes are much higher, but these edifices are constructed on elevated basements.)

These geometrical relationships can easily be appreciated by experimenting with the formula for the volume of a simple right cone:

$$v = 1/3\pi r^2 h$$

where v = volume, r = radius, and h = height.

In reality, the shape of a 'conical' volcano such as Mt. Mayon is actually expressed by a more complex exponential relationship of the form:

$$r = Be^{Mh}$$

where B and M are constants. Its volume can be found by integrating this expression. Geologists with a mathematical bent established these relationships more than a century ago, but since the pioneering work of Milne[6] and Becker,[7] the subject has been largely ignored, so we still do not understand many of the subtleties underlying the geometrical form of volcanoes.

A factor which further complicates consideration of the height of a volcano is that once a volcano grows large, it begins to deform under its own weight. When volcanoes are built on thin oceanic lithosphere, the mass of the volcano also causes the lithosphere to sag downwards into the asthenosphere. In the case of the large Hawaiian volcanoes, this subsidence has been very considerable. According to J. G. Moore, as much as a half or two-thirds of the upbuilding of the volcanoes may be offset by lithospheric subsidence. Thus, if the lithosphere were not so flexible, the Hawaiian volcanoes would be kilometres higher.[8]

16.2.2 Composite cones

Composite cones have had more than one evolutionary stage in their existence, but still retain an overall radial symmetry. Throughout their complex eruption history, the locus of activity has been essentially confined to a single site. Vesuvius is an example of a volcano where an earlier edifice was wrecked by an eruption (AD 79) and a younger one built up in its place, such that from some vantage points the ruins of the older edifice (Monte Somma) are not obvious, and the volcano appears to be simple and symmetrical. It is surprising how extensively a volcano may be modified, yet still retain an overall symmetrical shape. At Parinacota in Chile, a huge debris avalanche eviscerated the western flank of the volcano 13 000 years ago (Fig. 16.14). So much reconstruction has taken place since then that no trace of the amphitheatre remains. Parinacota *appears* to be a simple cone. But the hummocks and hollows of the vast avalanche deposit to the west confirm that its history has been anything but simple.

An average composite cone is an edifice only about 2 km high, but its structure may exhibit a complex history (Fig. 16.15). Mt. Etna is an example of a huge composite volcano, 3308 m

high. It has broad radial symmetry, although there are several summit vents and innumerable parasitic monogenetic vents. On its northern flanks, vents are aligned along a rift-like extension, while on the eastern flanks a great amphitheatre (the Valle del Bove) takes a great bite out of the edifice. From the ground, these features make the volcano look distinctly complex, but from the perspective of space, the volcano takes on a symmetrical shape, and its identity as a single conical edifice is obvious. All its many vents have tapped a single mantle source.

Etna also exhibits conveniently another property of many apparently simple major volcanoes: the edifice we see now is only the youngest of a series constructed on more or less the same site. Present-day Etna (also called Mongibello), appears to have been constructed over the last 34 000 years, but several edifices existed previously, stretching back more than 100 000 years. For example, between 80 000 years and 60 000 years ago, an older volcano (Trifoglietto) existed slightly east of the present Mongibello.

16.2.3 Compound volcanoes

Nevado Ojos del Salado is a compound or *multiple* volcano, in the terminology used by Cotton. It is not an individual cone, but a massif covering an area of about 70 square kilometres, formed of at least a dozen young cones intermin-

Fig. 16.14 Large mounds in the middle distance of this photography are *toreva* blocks, emplaced by a catastrophic debris avalanche taking place about 13 000 years ago on the 6348 m high Parinacota volcano, north Chile (background). No trace of an avalanche scar remains on the cone. Pomerape volcano is at the left.

Fig. 16.15 Vertical air photograph of Ceboruco volcano, Mexico, an excellent example of a composite volcano with a complex history but an individual identity. Many different episodes of activity are identifiable, including nested calderas and lava-flow fields of different ages. Ceboruco is 2164 m high, and its outermost caldera about 4 km in diameter. It last erupted in 1872.

gled with domes and craters. Remarkably, the detailed anatomy of this, the world's highest volcano, has never been studied, so it is not certain exactly how many components make up the massif, or how their magmatic plumbing systems are interconnected. Although the summit of the massif is very high (6887 m) the individual cones are not especially large: base level in the area is around 4500 metres, so the volcanic edifice is not much more than 2000 m high (Fig. 2.4).

Nevado Ojos del Salado's many vents do not seem to reflect any obvious tectonic control. But many compound volcanoes do—they commonly form elongate massifs, with many cones, vents, and craters aligned to form a ridge, probably the surface expression of a dyke. Aucanquilcha in Chile is a fine example. Topographically, it forms a ridge extending about 10 km in an east–west direction, and composed of several individual cones reaching over 6000 m (Fig. 16.16). In this

case, as in many others, there is evidence for activity migrating through time along the ridge.

16.2.4 Volcano complexes

Compound, or multiple, volcanoes have an individual identity. On a map, one could point to the volcano and say *there*. Volcano complexes are so jumbled that the best one can do is to draw a line around them. Formally, they can be defined as extensive assemblages of spatially, temporally, and genetically related major and minor centres with their associated lava flows and pyroclastic rocks. More practically, one could get lost in the confusing tangle of similar-looking flows and cones that make up a volcano complex. Cordon Punta Negra in Chile is a beautiful example, where at least 25 small cones with well-developed summit craters are present over an area of some 500 square kilometres. None of the cones is more than a few hundred metres high, and some of the older ones are

Fig. 16.16 Aucanquilcha, north Chile. View of north part of the 10-km-long east–west-trending compound volcano, which does not possess obvious volcano morphology. Active fumaroles have deposited sulphur which is extracted from the world's highest mine at the summit at over 6000 m. Plant in the foreground is at an elevation of about 5000 m and is part of world's highest permanently inhabited settlement.

almost buried under a confusion of lavas. It is difficult to tell which lavas came from which cone.

Volcano complexes like Cordon Punta Negra represent a form of 'distributed' volcanism. If erupted from a single vent, all the magma found in a complex would make a decent large cone, but rather than erupting through a single conduit, several closely spaced conduits were active more or less contemporaneously. It remains to determine the length of time required to create a complex like Punta Negra, the range of compositions present, and how these varied through time.

Scoria cone fields, like those of Central Mex-

ico, represent another type of distributed volcanism. In an area near the centre of the Mexican Volcanic Belt, directly south of Mexico City, one hundred and forty-six Quaternary scoria cones were counted within an area of about 1000 square kilometres, a cone density of 0.15 per square kilometre. Basal diameters varied from 0.1 to 2 km.[9] For years, many of the buildings on the campus of the University of Mexico have been renowned for their exuberantly colourful murals. It is less well known that the campus is constructed on the lava field from one of the most recently active cones in the field. Cerro Xitle erupted only 2400 years ago.

— 16.3 Shield volcanoes

Whereas cones are either straight sided (scoria cones) or concave (most large cones), shield volcanoes are *convex* upwards. And while cones may be rather steep, sometimes reaching above 40° in their summit regions, shields are gently sloping, often less than 10°. Simple and composite cones include lavas, pyroclastics, and talus, but shields are constructed almost entirely of lavas. Finally, rocks with compositions ranging from basaltic to rhyolitic turn up in cones, but shields are almost exclusively basaltic. Whereas erosion and mass wasting play important parts in

shaping growing cones, the geometry of shield volcanoes is dictated only by the rheology of the lavas of which they are made. Thus, a young shield volcano is a rather subtly shaped construct, with a gently swelling profile, made entirely of basalt lavas. Whereas artists respond readily to the sweeping, uplifting curves of a Fuji-like cone, the bland profile of a shield volcano offers less inspiration.

16.3.1 Hawaiian shields

Mauna Loa and Mauna Kea are shield volcanoes in a class of their own, rising nearly nine kilometres from the floor of the Pacific. Mauna Loa reaches 4169 metres above sea-level. It has a total volume of about 40 000 cubic kilometres, a hundred times greater than a typical composite conical volcano like Mt. Fuji. Despite its huge proportions, Mauna Loa is an unpretentious volcano, its smoothly arching whale-back profile more reminiscent of the gentle contours of the Dorset downs than a mountain of alpine altitude (Fig. 2.26). Only when the snowline defines the summit region with sharp whiteness unexpected in a tropical island is the magnitude of its edifice apparent.

While Maunas Loa and Kea are the largest and youngest shields in the Hawaiian Islands, there are many others of different ages and degrees of erosion. Each of the eight major islands represents one or more dissected shields, getting progressively older westwards. A young shield has gentle slopes of only 2–3° at its base, steepens slightly to about 10° in its middle slopes, and then flattens off again in the summit region. Each shield is composed of myriads of individual flows, many of them compound *pahoehoe* flows, averaging only a few metres thick. While each major shield probably had a summit caldera in which some activity was focused, a characteristic feature of the Hawaiian shield volcanoes is that they are elongated along rift zones, from which most lavas were erupted. Dykes propagating laterally from the central magma chamber carry basalt magma laterally until it emerges from a parasitic vent on the flanks of the volcano. These dyke-fed rift zones are prominent topographic features, extending for tens of kilometres, and are marked by many small spatter cones, pit craters, and fissures. Kilauea, the smallest but most active shield at present, has experienced almost continuous lava effusion since 1983 from the Pu'u O'o vent, 17 kilometres from the summit caldera. During the period 1969–74, activity was centred on the Mauna Ulu ('growing mountain') vent, 10 km along the rift. In 1955, lavas spewed out from a vent 25 km along the rift, while in 1960 a major outburst engulfed the village of Kapoho, almost at sea-level, and nearly 30 km distant along the rift.

When a typical rift eruption begins, lava flows spread rapidly over the surface, at a rate of about 50 cubic metres per second, burning their way through *ohia* rain forest and marijuana plantations indifferently. Once the eruption is well established, a large fraction of the lava (perhaps as much as 80 per cent) flows through lava tubes. This dramatically extends the distances the flows can reach, enabling lavas initiated high on Kilauea's east rift to debouch into the sea, extending the coastline. Apart from creating opportunities to photograph sizzling red lava and white steam clouds against an azure ocean, these tube-fed flows account for the gently sloping profiles of shields: most of the erupted volume ends up a long way from its point of origin, adding to the flanks of the volcano, rather than the summit region. Eruptions of more viscous magmas, which do not flow in tubes, lead to steeper, conical volcanoes.

Maunas Loa and Kea are similar in altitude. But whereas Mauna Loa is active (it erupted in 1984), Mauna Kea has not erupted in historic times. One reason why astronomers have been willing to risk building huge and hugely expensive telescopes on Mauna Kea is that Mauna Kea appears to be more mature. This conclusion is based on the presence on its upper flanks of abundant small scoria cones and lavas of alkalic composition; markedly different from the tholeiites making up most of the volume of the volcano (Section 3.2.1). Morphologically, these alkalic lavas somewhat resemble andesites, in that the flows are thicker and chunkier than ordinary basalts. In the summit region of Mauna Kea where a small ice-cap existed during the last Ice Age, glaciers scoured the surface of the flows, creating topography more reminiscent of the high Andes than a tropical ocean island volcano.

Mauna Kea has reached what is termed the 'alkali cap' stage in its evolution; a stage which the older shields on the Hawaiian Islands have also reached (Fig. 16.17). Unfortunately for the astronomers, reaching the alkali cap stage does not necessarily mean the volcano is extinct—on the neighbouring island of Maui, Haleakala volcano, also a mature shield, erupted as recently as 1790. Should an eruption take place on Mauna Kea, the damage done to astronomical research programmes would be incalculable.

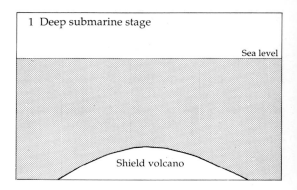

Fig. 16.17 Scoria cones cluster (background) at 4000 m at the summit of Mauna Kea, Hawaii. During the Ice Age, a small ice-cap occupied the summit. Boulders in the foreground show evidence of glacial transportation. Photo: courtesy of P. Mouginis-Mark.

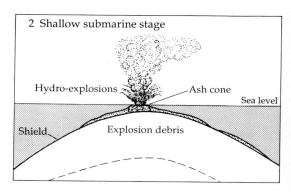

However, the volcanological record suggests that eruptions are exceedingly infrequent—the last one appears to have been about 4500 years ago. Thus, although there is a slight risk, the incomparable viewing conditions provided by a high mountain in the mid-Pacific easily justify it. None the less, if an eruption *were* to take place, astronomers would inevitably take a somewhat jaundiced view of their volcanological colleagues . . .

During their active lifetimes, construction of Hawaiian shield volcanoes keeps ahead of erosion (Fig. 16.18). At any one location on a volcano, the intervals between successive sets of lava flows may be long enough for soil horizons to develop, but there is relatively little lateral transport of material. Once activity slows, deep canyons are rapidly incised, leading to some startling topography on the older islands. On Haleakala (Maui) late-stage scoria cone eruptions have taken place within a vast summit amphitheatre excavated by erosion. Even during their active lives, the morphologies of shields can be drastically modified by landsliding. Kilauea's entire south-east flank is slumping slowly into the ocean, slipping down along a series of great fault scarps or *pali*. Some catastrophic collapses may also take place, generating huge tsunamis. Fortunately, none has occurred in historic times (Section 13.1.4).

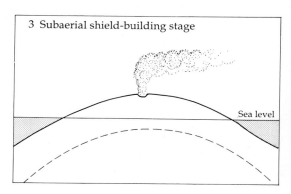

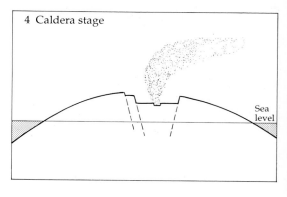

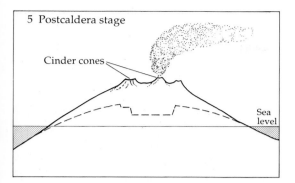

5 Postcaldera stage

Cinder cones

Sea level

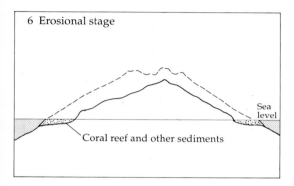

6 Erosional stage

Sea level

Coral reef and other sediments

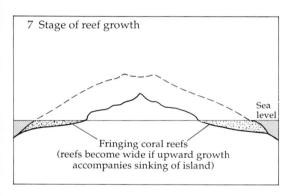

7 Stage of reef growth

Sea level

Fringing coral reefs
(reefs become wide if upward growth
accompanies sinking of island)

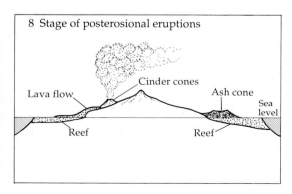

8 Stage of posterosional eruptions

Lava flow

Cinder cones

Ash cone

Sea level

Reef

Reef

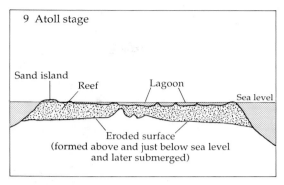

9 Atoll stage

Sand island

Reef

Lagoon

Sea level

Eroded surface
(formed above and just below sea level
and later submerged)

Fig. 16.18 Stages in the morphological evolution of Hawaiian shield volcanoes. (After Macdonald, G. A., Abbott, A. T., and Peterson, F. L. (1970). *Volcanoes in the Sea.* Univ. Hawaii Press, Honolulu. 517 pp.)

Each major Hawaiian volcano is an enormous shield. But individual rift eruptions may themselves construct small, gently sloping lava shields. Mauna Ulu, in the Volcanoes National Park was born in 1969. Innumerable *pahoehoe* lavas welled up and over the rim of the vent, spreading out to form an apron of anastomosing flows. In less than a year, a shield almost 100 metres high and a kilometre in diameter had grown. Subsequently, its summit crater was occupied by a small lava pond which persisted for several years. *Pahoehoe* lava erupted from Mauna Ulu between 1969 and 1974 ultimately reached the coast, covering almost 45 square kilometres.

16.3.2 *Galapagos shields*

Clustered on the Equator 1100 km west of Ecuador, the Galapagos Islands are better known for their contribution to Darwin's ideas on the origins of species than for their volcanoes. None the less, the volcanoes occupy a hot-spot setting similar to, but more complex than, the Hawaiian volcanoes. Each island is either a shield volcano or a coalescence of several shields, each 45–80 km across, which rise about 1500 m above sea-level. In detail, the Galapagos shields differ from those of Hawaii in three ways: First, they lack the gentle, whale-back profiles of Mauna Loa, but instead have profiles traditionally likened to upturned soup-plates, with a

Fig. 16.19 Volcano Cumbres, Isla Fernandina, Galapagos, showing the inverted soup-plate profile characteristic of Galapagos shields. Cumbres is 1250 metres high. Photo courtesy Peter Mouginis-Mark.

marked change of slope from gentle to steep ($>10°$) on the mid-flanks, and flattish tops (Fig. 16.19).

Second, whereas the active Hawaiian shields Mauna Loa and Kilauea have summit calderas several kilometres across, these are shallow, less than two hundred metres deep. On the Galapagos Islands, by contrast, summit calderas are spectacularly deep. That on Fernandina, for example, is more than 880 m deep.

Third, the Galapagos volcanoes are more nearly radially symmetrical than the Hawaiian shields. While there is some evidence for dyke intrusion, linear rift zones like those of Kilauea and Mauna Loa are subdued. Surrounding the summit calderas are prominent sets of circumferential fissures. These circumferential fissures are almost unique to the Galapagos. On Earth, a few terrestrial volcanoes exhibit comparable fissures, for example Deception Island (Antarctica) and Niuafoou Atoll ('Tin Can Island'; Tonga), but the best analogues are on Mars.

Is is not clear what factors are responsible for the marked differences in topography between the Hawaiian and Galapagos shields. One widely held hypothesis is that their internal architecture is different, with ring dykes in the Galapagos and rectilinear dykes in the Hawaiian shields. In some ways, the Galapagos shields resemble overgrown seamount volcanoes.

16.3.3 Icelandic shields

Iceland's shield volcanoes are modest in size, but elegantly symmetrical. Twenty have been constructed in post-glacial times. They are topographically subdued and are usually only a few hundred metres high. Some have slopes of as little as $1°$, reflecting the low viscosities of the basaltic lavas involved. They resemble the small Hawaiian lava shields such as Mauna Ulu. Skjalbreidur, the classic Icelandic shield, has uniform slopes of $7–8°$, is 600 m high, and has a diameter of about 10 km. Its total volume is only about 15 cubic kilometres, a mere pimple compared with Mauna Loa. George Walker has suggested that the Icelandic shields were built up quickly, by almost continuous eruption of thin *pahoehoe* basaltic lavas from the central vent. Some of these fluid flows were able to travel long distances over gentle slopes. A flow from Trolladyngja may have travelled more than 100 km over a $1°$ slope.

Small, flat lava shields of Icelandic type are characteristic of many fluid basalt provinces, such as those of the Snake River Plain.[10] They provide important clues to the morphologies to be expected in the source regions of extraterrestrial basaltic volcanism, for example on the smooth plains of Mars (Section 18.5.3).

— 16.4 Volcanic landforms resulting from erosion —

16.4.1 *Stages in the erosion of cones*

The topography that a volcanic cone displays as it succumbs to the depredations of erosion depends on the climate, and what it is made of. Scoria cones are essentially heaps of loosely packed, porous pyroclasts, and therefore they absorb water like sponges. Even under conditions of heavy tropical rainfall, water soaks immediately into a scoria cone rather than running off, giving erosion little chance of taking hold. Scoria cones thus remain recognizable for millennia. In cones which contain a high proportion of welded spatter, and in tuff cones where the tephra are fine grained or muddy, porosity is much less, and runoff is enhanced. Once runoff gets under way, it radically reshapes the volcano. On a symmetrical cone, the first stage in this process is the development of *parasol ribbing*: evenly spaced V-shaped radial gullies separated by ridges (Fig. 16.20).

It is unusual for parasol ribbing to remain intact for long. For one reason or another, usually reflecting subtle variations in original topography, 'master' gullies begin to prevail, capturing the headwaters of lesser gullies, and cutting rapidly downwards into the heart of the volcano. Youthful volcanic materials are often poorly consolidated, and so this process can take place amazingly rapidly, especially in tropical conditions. After the great Krakatau eruption of 1883, the Dutch geologist Verbeek reported that only two months after the eruption, gullies *40 metres* deep had been cut into the pyroclastic deposits. Even in the more temperate conditions of the USA, deep gullies had been cut in the Mt. St Helens debris avalanche deposit within a few weeks of the eruption of May 1980. Within a few years, gullies tens of metres deep in places presented formidable obstacles to movement.

When two or three master gullies are active on a conical volcano, widening and deepening

Fig. 16.20 Parasol ribbing on surtseyan tuff-ring, west of Lake 'Abhe, Ethiopia. Photo: H. Tazieff.

themselves, there will inevitably come a point when the heads of two gullies intersect. This isolates a triangular, flat-surfaced facet of the original cone, known as a *planeze*, a term originally adopted by the French geographer E. de Martonne (Fig. 16.21). Their distinctive triangular shapes also earned *planezes* the more colloquial term 'flat-irons'. As erosion continues, *planezes* get whittled away, lavas on the flanks of the volcano become progressively degraded, and its summit level is reduced. Ultimately, all that is left of a volcano is a gently rounded hill. Table 16.1 shows one possible sequence of stages in the erosional history of a cone.

In regions with different climatic regimes, it will take different periods of time for a volcano to pass through these successive stages. In the hyper-arid conditions of the Central Andes it may take several million years to reach stage 3; in tropical areas such as Indonesia, it may take only twenty thousand years.[11]

Necks and dykes Even after a volcano has been deeply dissected, anatomically distinct landforms may remain. Feeder vents through which lava reached small composite volcanoes are often preserved long after the rest of the volcano has disappeared, surviving as massive pillars of rock more resistant to erosion than the lavas and pyroclastic rocks of the cone. Travel posters often feature the oddly situated churches in the Auvergne area of France, which are perched on volcanic necks or *puys*, protruding above the countryside as steep, craggy eminences. Recogni-

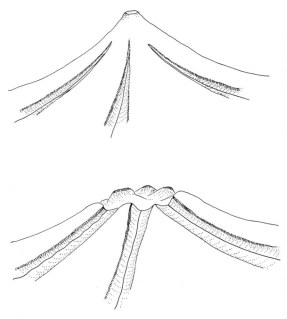

Fig. 16.21 Formation of *planezes* or 'flat-irons' on a volcanic cone subjected to gullying erosion.

tion that *puys* such as the Rocher Saint Michel and the rocks surrounding them were of volcanic origin played an important role in the history of geology and science as a whole. As Archibald Geikie put it in 1897:

To France, which has led the way in so many departments of human enquiry, belongs the merit of having laid the foundations of the systematic studies of ancient volcanoes. Her group of Puys furnished the

Table 16.1 Stages in the erosional history of a volcanic cone

Stage	Morphological forms
1	Fresh, young cones, pristine lava flows and summit craters. No glacial moraines
2	Small gullies on flanks; lavas and summit crater discernible but degraded; cone still sharp, moraines present in glaciated areas.
3	Individual lava flows barely visible; no crater; well-established gullies; constructional surfaces dwindling; *planezes* initiated.
4	No lavas visible; deeply incised gullies; *planezes*, but original surfaces left. Considerable relief. Major U-shaped valleys in glaciated areas
5	Barely recognizable; low relief; radial symmetry main clue to volcanic origin.

earliest inspiration in this subject, and have ever since been classic ground to which the geological pilgrim has made his way from all parts of the world.

Nicolas Desmarest (1725–1815) and George Poulett Scrope (1797–1876) were able to demonstrate that several episodes of eruption and erosion had taken place amongst the volcanoes of the Auvergne. Scrope used these observations, seemingly trivial to us today, to emphasize the continuity of geological processes, thus helping to overturn the prevailing 'catastrophist' view of the history of the Earth. Influenced by religious belief in the Creation, and particularly in the great Deluge, catastrophists had explained mountain belts and the gorges that incise them as the products of single-short lived events.[12]

In Britain, Arthur's Seat, a famous Scottish landmark which looms over the city of Edinburgh, was the site of a Carboniferous volcano, now exhumed. Eroded necks and vents form the highest points, while on the flanks outlines of some of the original lava flows can still be picked out. In the United States, Ship Rock in New Mexico jags up in an astonishing pinnacle 430 metres above the desert surface. Dykes radiating out from the centre are beautifully displayed, standing up starkly like walls. Ship Rock owes its soaring, perpendicular architecture to a series of vertical joints in the breccia-filled volcanic neck, a feature common to other necks around the world. Interestingly, the finest necks and puys all seem to have been formed from *small*, probably monogenetic volcanoes. When large composite volcanoes are eroded, comparable necks are not revealed, probably because their cores are deeply altered and decayed by hydrothermal activity.

In New Mexico, the dykes around Ship Rock stand up as walls in the desert because they are more resistant than the rocks around them. (In English, the word *dyke* itself originally meant wall; dykes in Holland are walls to keep the water out.) When dykes are less resistant than the rocks they cut, the opposite happens. In the north-west Highlands of Scotland, ancient gneisses more than 1700 million years old are cut by a swarm of dykes about 56 million years old, associated with the Tertiary volcanic centres on Skye and Mull. On land, many of these dykes have been eroded

into parallel-sided depressions; where they cut the coastline, long, deep grooves result. These are such common features of the coastline that they have acquired their own name in the local Gaelic language: *slocs*.

Dykes are often thought of as rather minor igneous phenomena. Most are indeed rather small, less than a metre wide. There are many huge exceptions, though. In the Tertiary dyke swarm of north-west Britain, the Cleveland dyke (about 100 metres wide) can be traced discontinuously from its source in the Hebrides to the North Sea coast of Yorkshire, a distance of almost 400 km. Calculations of rates of injection and cooling suggest that the dyke must have been intruded very quickly, zipping across Northern England in less than 5 days. The Cleveland dyke is only one of a swarm, all of broadly the same age, which formed in response to rifting movements related to the opening of the North Atlantic Ocean.

16.4.2 Erosion of lava flows

Lava flows have superlatively rough surfaces, so they are natural traps for wind-blown dust. Their glassy outer skins also break down rapidly when exposed to water and air. Thus, soils quickly form in crevices, allowing plants to colonize the flow. Obviously, the rate of this process depends critically on local climate, so it may vary dramatically even on a single volcano. New lava flows on the wet, trade-wind side of the Hawaiian Islands are overgrown by rain forests in only a few decades, whereas those on the parched leeward side remain prominent for centuries. In regions where higher forms of plant life fail to thrive, lichens colonize the surface of flows. Some of the lichens on the lavas of the Craters of the Moon lava field, Idaho, are so exuberantly colourful that they almost distract attention from the lavas—almost. Even in the most arid areas, accumulation of wind-blown dust subtly changes surface features of lavas, producing a general lightening in albedo, significant in remote-sensed images which 'see' homogenized pixels covering tens of square metres.

Basalt plateaux the world over are composed of lavas of similar physical properties. Thus, the stepped topography of basaltic *traps* is similar the world over (Section 2.5.2), whether in the dismal, rain-sodden moors of the Hebrides or the grandeur of the Drakensberg Escarpment. Because their lavas and the scoriaceous horizons between them are so porous, surface drainage does not easily become established on lava plateaux. Water seeps instead downwards through the flows, moving laterally when it reaches impermeable, clay-rich horizons.

In the Snake River Plain (Idaho), the Snake River itself receives no actual tributaries from the expansive lava plain it cuts through, but its flow is substantially increased by springs soaking out from the lava and scoria horizons in the canyon. A consequence of this is that erosion of the basalt plain takes place by *sapping*, rather than stream flow, producing deep, steep, dead-end canyons, terminated by *alcoves* or *amphitheatres*. Headward erosion of these alcoves takes place as springs seeping out from the lavas undermine or sap the cliff line, so that it eventually collapses. Removal of this talus is caused by chemical disintegration and solution of the basalt by water, and partly by mechanical transport when the flow rate is great enough. A typical example is the Blue Lakes alcove, which heads a short canyon three kilometres long and 100 metres deep. Some Hawaiian shield volcanoes display similar alcove-headed valleys, probably also formed by sapping.

Spring sapping of lava plateaux may seem to be of less than riveting interest to volcanologists. It deserves our attention because similar processes may have played a key role in shaping huge areas of the surface of Mars. Areas of basalt lavas such as the Snake River Plain provide excellent opportunities to explore these.[13]

Columnar jointing   Basalt lavas like those in the Snake River Plain often exhibit spectacular columnar jointing in cliffs and canyons. Two of Britain's best-known landmarks, the Giant's Causeway in Antrim and Fingal's Cave on the island of Staffa, originated when the Brito-Arctic or Thulean basalt province was formed (Fig. 16.22). Both are famous for the elegance of the thousands of polygonal columns that stand out in cliffs and on shore. In California, the Devil's Post Pile is another popular natural landmark,

Fig. 16.22 Part of the Giant's Causeway, Antrim, Northern Ireland, where a thick columnar jointed basalt flow is exposed on the sea coast.

high in the forests of the Sierra Nevada. From a distance, the columns in these eroded lavas resemble the pipes of a great cathedral organ.

Most lava columns are hexagonal, though four-, five-, seven-, and eight-sided examples also occur. Each column is terminated by either a smooth concave or a convex surface. These surfaces are joint planes, sometimes called *ball-and-socket* joints, which divide the columns up into innumerable stacked tablets, like piles of poker chips (Fig. 16.23). Both the columns themselves and the curved joint surfaces are caused by contraction and fracturing in the cooling flow. In detail, a flow may have one or more sets of regular columns, separated by an irregularly fractured layer. The regular columns constitute a *colonnade*, and the irregular layer an *entablature*.[14] Columnar jointing is a consistent feature of some large lava flows, and can be

Fig. 16.23 Smoothly curved joints form the surfaces of the columns of the Giant's Causeway. Concave and convex surfaces appear equally abundant.

Fig. 16.24 On the small Hebridean island of Eigg, Scotland, a rhyolite lava flow which originally filled a river valley cut in older basalt lavas is now preserved as the Scuir of Eigg, an imposing crag. It is a classic example of volcanic inversion of relief. From an 1897 sketch by Archibald Geikie.

traced for many kilometres in the flows of the Columbia River Plateau.

Inverted relief   Obedient to the laws of physics, lavas flow down valleys, sometimes filling them completely. Distinctive landforms result if the lavas are more resistant to erosion than the underlying rocks in which the valleys were cut, as often happens when lavas flow over sedimentary strata. The valley-filling lava forms a thick, resistant mass, while the sediments on either side are more rapidly removed. In the fullness of time, the original valley will be expressed as a resistant ridge of lava standing above the surrounding sediments, in which new valleys have been cut. In Britain, a beloved example of *inversion of relief* is found on the obscure Hebridean island of Eigg, where a Tertiary rhyolite lava which filled an older valley now forms a 400 metre high, five kilometer long ridge. This ridge, the famous 'Scuir of Eigg' dominates the tiny island, otherwise renowned only because its nearest neighbour is Muck. (Fig. 16.24). Inversion of relief can be seen on many volcanic cones, where flows which originally flowed in gullies cooled to form thick, confined wedges. After erosion, the lavas end up as resistant caps to spurs and *planezes* on the dissected volcano.

16.4.3 Erosion of pyroclastic deposits

In the short term, the topographical effects of a heavy ash fall are straightforward: tephra blankets the pre-existing surface, smoothing and subduing the previous humps and hollows. Where large-volume pyroclastic flows are concerned, involving thousands of cubic kilometres of magma, the pre-existing topography may be *completely* buried, leaving an unbroken plateau many thousands of square kilometres in extent. This situation does not last long. Unwelded pyroclastic deposits are soft and friable, and so they are rapidly eroded. Deep gullies are cut at an astonishing rate as soon as the first rains fall. Within a few years, a distinctive topography emerges. Flat-topped relics of the original plateau remain, separated from one another by a dendritic maze of narrow, vertical-sided gullies. (Wadis, canyons, quebradas, gulches, or gorges, depending on local idiom.) These gullies are often so deep and steep that they are mere slits, less than a metre wide and ten or more deep, so constricted that they are difficult to walk along (Fig. 16.25).

Intensive gullying of this kind is common in soft rocks of all kinds, not only tephra. It happens distressingly quickly as a result of overgrazing or clearing of timber in areas subject to occasional heavy rainstorms. Development of extremely steep (vertical) sides of gullies cut in pyroclastic rocks is an expression of the fact that they are incoherent, structureless deposits. Unlike sedimentary deposits, unwelded pyroclastic rocks often lack layering, and because they are unlithified, cannot form boulders. All they can do is

Fig. 16.25 An eruption about 3000 years ago formed the Deriba Caldera, Darfur Province, western Sudan and its apron of pyroclastic fall and flow deposits. Erosion of the soft pyroclastic deposits has yielded a dendritic maze of steep, deep gullies.

form vertical cliffs, with scree slopes of loose pumice beneath them. If the deposit is welded or shows vapour phase alteration (Section 10.5.1), then boulders form, allowing a talus slope to develop at the foot of the cliff.

Where welding or vapour-phase alteration are present, cooling joints and fractures also develop; usually, but not always, perpendicular to the original surface. On erosion, elegant sets of columnar joints are sometimes exposed; not as commonly as in basalt lava plateaux, but more pleasing because the rocks are a warm, reddish ochre colour, rather than drab basalt black (Fig. 14.17). Vertical jointing accentuates the vertical cliffs developed in pyroclastic rocks. Where

ignimbrite plateaux are eroded in arid environments, box canyons are common. As the edge of the plateau recedes, flat-topped islands of ignimbrite are left, their walls rising as steep, high, and forbidding as castle ramparts above talus slopes of fallen boulders. These *mesas* and *buttes* make appropriate backgrounds to Western movies.

Jointing in ignimbrites is often so perfectly developed that the fracture surfaces look and feel artificially clean. They are so smooth that they provide irresistible temptations for people to express themselves in *grafitti*. In both north and south America, prehistoric Native Americans used ignimbrite joint surfaces to draw petroglyphs showing animals that they hunted, and abstract mystical symbols (Fig. 16.26). Twen-tieth-century Americans use the same surfaces for more rudimentary graffiti.

Yardangs In the Central Andes, ignimbrites are exposed over huge areas at high altitudes. Precipitation is slight in the region, taking place mainly in the form of snow which ablates rather than melting. Surface runoff is therefore minimal, so erosion by flowing water is inconsequential. By contrast, fierce winds blow from the north-west for much of the year. As a consequence, much erosion takes place through aeolian pro-cesses, producing a wind-sculpted topography of *yardangs* and deflationary hollows. *Yardangs* are long, wind-eroded ridges, somewhat resembling upturned canoes, their prows facing the prevail-ing wind (Figs. 16.27–16.28). Andean ignimbrites

Fig. 16.27 SPOT satellite image of *yardangs* developed in the 4.1-million-year-old Atana ignimbrite on the frontier between Chile and Argentina. *Yardangs* are pointing into the prevailing north-westerly wind. Image is about 4 km across. Courtesy of CNES.

Fig. 16.26 Pre-Columbian petroglyphs on smooth joint surfaces on the 9-million-year-old Sifon ignimbrite, Rio Loa valley, north Chile. Llamas and geometrical symbols are abundant.

are silicic, and so they contain generous quanti-ties of quartz phenocrysts. On erosion, they therefore liberate the agents of their own destruc-tion: quartz grains freed from their matrix and picked up by the wind rapidly abrade the remaining rock. Absolute rates of erosion have never been determined, but the 'half-life' of an ignimbrite subject to aeolian erosion in the central Andes may be about a million years: after a million years, only half the initial volume remains.

Why worry about aeolian erosion rates of ignimbrites in some obscure part of South America? One reason is that some planetary scientists believe that huge areas of Mars are covered by ignimbrites. There is no doubt that there are enormous areas of superb *yardangs* on Mars. But how did these *yardangs* form? And what are they eroded in? Are they ignimbrites, or an older aeolian deposit? To help answer these questions, terrestrial *yardangs* must be better understood. The high, dry, cold Andean plateau provides an excellent terrestrial counterpart to the even drier and colder Martian deserts, so further studies of Andean *yardangs* may throw some light on an important aspect of Martian

Fig. 16.28 Prow of a *yardang* 50 km south of those in Fig. 16.27. Aeolian erosion is most intense within about one metre of the ground; thus the ignimbrite is rapidly undercut, and boulders fall to the valley floor, where they are rapidly abraded away.

geology. One intriguing problem is this: if ignimbrites exist on Mars, they are more likely to be of mafic composition than silicic. If they are mafic, they will lack quartz. What, then, provides the abrasive material to help the Martian wind sculpt such outstanding *yardangs*?

Wigwams and tent rocks From Cappadocia in Turkey to Los Alamos in New Mexico, erosion of pyroclastic rocks consistently yields wigwams or tent rocks: conical pinnacles or spires of dazzling white rock that may be tens of metres in height (Fig. 16.29). These can form in two ways. When a pyroclastic plateau is dissected by dendritic drainage, two branches of the complex system commonly intersect each other, isolating a block of the deposit which is then trimmed into a pinnacle. More often, wigwams are formed like earth pillars: where the edge of a plateau is being eroded away, resistant blocks of lava or other rocks within or on top of the deposit (perhaps lag breccias) protect the underlying deposit from erosion, while the surrounding material is rapidly carved away. For a while, the protecting

Fig. 16.29 Tent rocks formed by erosion of the Bandelier Tuff (ignimbrite) on the flanks of the Valles caldera, New Mexico.

boulder remains perched precariously on top of the pinnacle, but eventually it topples, leaving only the pinnacle itself. As in other pyroclastic erosional phenomena, formation of wigwams depends on the deposit being homogeneous, lacking marked horizontal or vertical variations in strength.

Fumarole mounds and ridges Ignimbrites are hot when first deposited, maybe even close to magmatic temperatures. A consequence of this is that after they are emplaced, they sit and stew in their own magmatic volatiles, and in any steam that may be liberated from underlying groundwater. Apart from causing generalized vapourphase alteration, this process may cause localized effects which have striking morphological consequences on erosion. Where the ignimbrite is broken up by vertical joints, hot fluids moving up the joints cause alteration and deposition of silica along the joints. This may affect the ignimbrite around the joint for a few centimetres or as much as a metre, resulting in formation of *armoured*

joints. On erosion, armoured joints, originally fractures in the rock, stand up as resistant walls, often several metres high. In places in Chile, these walls are so steep and the jointing so regular that from a distance the weathered ignimbrite looks like a ruined city. Where hot fluids escape to the surface via pipes rather than joints, armoured *chimneys* are produced.

Where more extensive ignimbrite–groundwater reactions took place, steam blasts may break through to the surface of the ignimbrite, producing fumarole vents similar to those in the Valley of Ten Thousand Smokes. On erosion, the most common expression of these fossil fumaroles is in myriads of shallow depressions or pits, better seen on aerial photographs than on the ground, but in some circumstances, swelling mounds of blisters are exposed. These positive relief features are probably the result of erosion revealing differences in hardness around the fumarole at depth within the ignimbrite, and are not original surface features[15] (Fig. 16.30).

Fig. 16.30 'Fumarole' mounds and ridges developed on the surface of the Carcote Ignimbrite, Rio Loa valley, north Chile. Regular alignment of the ridges suggests that they were developed above joints in the main body of the ignimbrite.

— 16.5 Topographical by-products

So far, we have considered volcanoes in isolation. But volcanoes can also cause drastic topographical changes to their surrounding landscape,

usually by interfering with local drainage. On a small scale, lava flows commonly block valleys, impounding lakes behind a lava dam. This can have serious consequences if the dam should suddenly fail, releasing large volumes of water. An instructive example is Sabancaya volcano in Peru. Lavas from the north flank of this volcano once dammed the mighty Majes Canyon, more than two thousand metres deep, and one of the grandest in the world. A substantial lake must have existed for a time, because flat-lying sediments deposited in the lake are preserved behind the dam, which also forms a prominent nickpoint in the canyon profile. At some unknown time, the dam was breached: there is no lake at the present day.

Also in the Andes, a large debris avalanche from Parinacota volcano 13 500 years ago blocked the existing drainage to the Pacific, resulting in the formation of Lake Chungara, 10 km across, which at 4550 m is the highest lake of respectable dimensions outside Tibet (Fig. 16.14).

On a larger scale, Lake Van (70 km across) in Armenia is said to have been ponded by lavas from Nemrut volcano. And on an even larger scale in central Africa, Lake Kivu flooded part of the Western Rift valley when it was empounded by construction of the volcanoes of the Virunga Mountains about five million years ago. Formerly, drainage was northwards to join the Nile via Lake Albert. When the volcanoes interposed themselves, thwarted rivers initially drained into and filled Lake Kivu, which eventually overflowed, draining southwards at the southern end of the rift via the Ruzizi River into Lake Tanganyika. Lake Tanganyika itself is connected via the Lukuga River to the great Congo (Zaïre) River. Thus, the Kivu drainage was switched from the Nile and the Mediterranean to the Congo and Atlantic[16] (Fig. 16.31).

Any map of rivers around the Rift is, of course, of only ephemeral value. When Haroun Tazieff

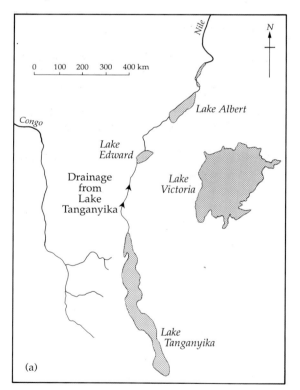
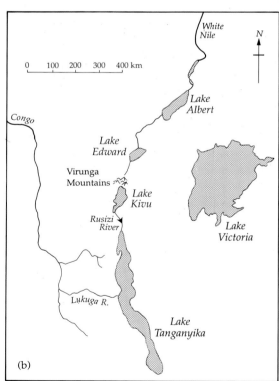

Fig. 16.31 (a) Sketch of the Nile drainage through the Western Rift before eruptions which built the Virunga Mountains, ponding Lake Kivu. At that time, drainage from Lake Tanganyika flowed northwards via the Nile into the Mediterranean. (b) After formation of the Virunga Mountains, Lake Kivu formed, the Ruzizi River drained southwards, and Lake Tanganyika drained via the Lukuga into the Congo and the Atlantic.

visited Lake Kivu in 1948, lava from the Kituro volcano was streaming steadily into the lake, reshaping its northern shores, and providing a bonanza to local boatmen in the form of shoals of parboiled fish. Future eruptions along the rift will eventually reshape drainage patterns once again.

On the very largest scale, the topographical swell above a mantle hot-spot may influence topography over a region thousands of kilometres in extent. Radial drainage patterns incised as a result of hot-spot uplifts have been mapped on several continents. If we were to include the topography of the sea-floors, we could conclude that most all the world's landscape is of volcanic origin.

— Notes ——————————————

1. Cotton, C. A. (1944). *Volcanoes as landscape forms*. Whitcombe and Tombs, Christchurch, New Zealand, 415 pp.
2. Wood, C. A. (1980). Morphometric evolution of scoria cones. *J. Volcanol. Geotherm. Res.* **7**, 387–413.
3. Wohletz, K. H. and Sheridan, M. F. (1983). Hydrovolcanic explosions II. Evolution of basaltic tuff rings and tuff cones. *Am. J. Sci.* **283**, 385–413.
4. Lorenz, V. (1986). On the growth of maars and diatremes and its relevance to the formation of tuff rings. *Bull. Volcanol.* **48**, 265–74.
5. Jones, J. G. (1969). Intraglacial volcanoes of the Laurgavatn region, southwest Iceland, *Q. J. Geol. Soc. Lond.* **124**, 197–211.
6. Milne, J. (1878). On the form of volcanoes. *Geol. Mag.* **5**, 337–45.
7. Becker, G. F. (1885). The geometrical form of volcanic cones. *Am. J. Sci.* **30**, 283–93.
8. Moore, J. G. (1987). Subsidence of the Hawaiian Ridge. In *Volcanism in Hawaii*, US Geol. Surv. Prof. Pap. 1350, pp. 85–100.
9. Martin del Pozzo, A. L. (1982). Monogenetic vulcanism in Sierra Chichinautzin, Mexico. *Bull. Volcanol.* **45**, 9–24.
10. Greeley, R. (1977). *Volcanism of the Eastern Snake River Plain, Idaho*. NASA CR 154621, 308 pp.
11. Ollier, C. D. *Volcanoes*. Australian National University Press, Canberra, 177 pp.
12. Scrope, G. P. (1858). The geology of the extinct volcanoes of Central France. John Murray, London.
13. Pieri, D. (1980). Martian valley: morphology, distribution, age and origin. *Science* **210**, 895–7.
14. Tomkieff, S. T. (1940). Basalt lavas of the Giant's Causeway. *Bull. Volc.* **6**, 89–143.
15. Sheridan, M. F. (1970). Fumarolic mounds and ridges of the Bishop Tuff, California. *Geol. Soc. Amer. Bull.* **81**, 851–68.
16. King, L. C. (1942). *South African Scenery*, pp. 153–4.

17

The golden glow of volcanic winter

A volcanic eruption does not take place in an environmental vacuum. It may have a variety of side-effects; some with serious implications for society. A few of these are predictable; others are less so. Some are confined to the immediate vicinity of the volcano; others are perceptible world-wide. We worry today about the fragile environment of our planet, which two decades of exploration of the solar system have helped us to perceive as 'Spaceship Earth', carrying a finite supply of pure air and water. Chlorinated fluorocarbon gases released from refrigeration systems are destroying stratospheric ozone, while burning fossils fuels is increasing the amount of carbon dioxide in the atmosphere, causing a 'greenhouse' effect. Consequently, the atmosphere and its role in global change have become pre-eminent subjects for research.

Atmospheric studies are extraordinarily complex. It is difficult to progress from the simple observation that the amount of carbon dioxide in the atmosphere is increasing to the consequences that this increase may have. For example, will the greenhouse effect lead to an increase or decrease in globally averaged cloud cover? And does cloud cover have a positive or negative feedback on globally averaged temperatures? These problems make it hard to predict the effect of atmospheric perturbations on *climate*, the regionally averaged combination of meteorological conditions, and still harder to predict local *weather*—whatever happens to be going on outside at the moment.

Such complexities require that we understand smaller components of the total atmospheric system before we can understand the whole. Large volcanic eruptions have an important role here, because they provide distinct, quantifiable signals whose atmospheric effects commmonly affect whole hemispheres. It would be comforting to be able to say that volcanic eruptions provide a key to understanding global change. But unfortunately we are still a long way from being able to understand the effects of even a single large eruption on the atmosphere, let alone the longer-term climatic impacts of volcanism. However, the record of major eruptions provides illuminating insights into the way the atmosphere works, and it was instrumental in the discovery of important atmospheric phenomena such as stratospheric circulation.

— 17.1 Making the connection

To those whose noon sunshine has been replaced by cimmerian darkness, the volcanic origin of their misery may be suffocatingly obvious if falling ash is sifting down around them. But it may be difficult for someone observing a small anomaly in his daily circumstances to relate that anomaly to events taking place on a volcano thousands of miles away, of whose very existence he may be unaware. Benjamin Franklin, the great American polymath, first made such a connection.[1] In 1784, while serving in Paris as the first diplomatic representative of the newly formed United States of America he wrote:

During several of the summer months of the year 1783, when the effects of the Sun's rays to heat the Earth in these northern regions should have been the greatest, there existed a constant fog over all Europe, and great part of North America. This fog was of a permanent nature; it was dry, and the rays of the sun seemed to have little effect towards dissipating it, as they easily do a moist fog, arising from water. They were indeed rendered so faint in passing through it that when collected in the focus of a burning glass, they would scarcely kindle brown paper. Of course, their summer effect in heating the earth was exceedingly diminished.

Hence the surface was early frozen.

Hence the first snow remained on it unmelted, and received continual additions.

Hence the air was more chilled, and the winds more severely chilled.

Hence, perhaps the winter of 1783–4 was more severe than any that had happened for many years.

The cause of this universal fog is not yet ascertained. Whether it was adventitious to this earth, and merely a smoke, proceeding from the consumption by fire of some of those great burning balls or globes which we happen to meet with in our rapid course round the Sun and which are sometimes seen to kindle and be destroyed in passing our atmosphere, and whose smoke might be attracted and retained by our Earth; or whether it was the vast quantity of smoke, long continuing to issue during the summer from Hecla in Iceland, and that other volcano which rose out of the sea near that island, which smoke might be spread by various winds, over the northern part of the world, is yet uncertain.

The year 1783, of course, was the date of the great Laki basaltic fissure eruption in Iceland (not Hecla, which was much better known in Europe); its effects are discussed shortly.

In Britain, the renowned diarist Gilbert White, author of 'The Natural History of Selbourne' also noted that conditions during the summer of 1783 were distinctly unusual. He termed the summer 'amazing and portentous', but as a provincial English cleric, lacked Franklin's synoptic knowledge, and was unaware of the eruption in Iceland. White noted the presence of a 'peculiar haze or smoky fog' from 23 June to 20 July, but

also recorded that the summer in England was exceptionally *hot*, emphasizing the difficulty of generalizing about local climatic effects.

White's brother in law, Thomas Barker, an English squire from the ambrosial county of Rutland, was a pioneer of scientific weather observing who compiled a unique set of records for sixty years from 1733 onwards; invaluable in modern studies of climate change. He noted the thick haze, which he recorded as persisting throughout the summer until Michaelmas. He described the haze as being dense enough to make the Sun appear the colour of 'rusty iron'. A gentleman well read in the classics, Barker commented that the haze resembled one described by Plutarch in 44 BC (Section 17.3.1), but pointed out that 1783 was unusually warm, whereas 44 BC was cold.

By all accounts, the summer of 1783, and the month of July in particular, were exceptional. Tremendous electrical storms raged across the country, prompting the 'Gentlemen's Magazine' to write that 'there is no year on record when the lightning was so fatal in this island', and to collate reports from all the afflicted shires. This resulted in many splendid vignettes of English rural life— two men who were mowing in a field near Shaston in Worcestershire 'took shelter under an elm tree where one was presently struck dead, and the other very much hurt'. On 21 July, 'a body of electric fire entered the porch of the White Hart Inn at Whitchurch in Hampshire . . . it struck down the landlord, his wife and a maid-servant . . . (and) shivered the kitchen chimney-piece to atoms, penetrated a wall near it two feet deep, and otherwise damaged the house considerably'. It is likely that this wealth of electrical phenomena was due to high concentration of volcanic aerosol particles in the atmosphere, forming the 'peculiar haze or smoky fog'.

Other records confirm that the summer was exceptionally hot, leading to an unprecedently early harvest. This warm weather may have been linked to the Laki eruption, but may also, of course, have been merely a statistical accident. Much of Scandinavia suffered much more seriously than Britain, crop harvests being ruined by the acid haze. For example, in the Eidsberg parish of Sweden, a letter of 4 September 1783 states that in many fields it was not even possible to determine what crops had been planted there and continues . . . 'many farmers who normally harvest two barrels expect not even a single bushel this time, and the same is the case with the oats'.

— 17.2 On the unusual optical phenomena of the atmosphere, 1883–6, including twilight effects, coronal appearances, sky hazes, coloured suns and moons, etc.

This was the subtitle selected for the 312-page section of the Royal Society's Krakatau Committee Report, which dealt with atmospheric effects of the 1883 eruption. These were so dramatic that this section of the Report is much longer than the account of the eruption itself. Its authors were no less impressive: the Honourable F. A. Rolo Russell and Mr E. Douglas Archibald. Their long-winded, Victorian title is adopted here to celebrate their seminal contribution, and to underscore the importance of Krakatau in all volcanic studies. Observations of Krakatau's effects prompted two Swiss scientists, the cousins P. and F. Sarasin, and later the distinguished American meteorological physicist W. J. Humphreys to enquire whether volcanic eruptions could influence global climate in a systematic way, rather than merely suggesting a connection, as Franklin had done.

In his prescient 1913 paper, Humphreys[2] considered the effects of two factors, one obvious, the other less so. Volcanic 'dust' in the upper atmosphere scatters, reflects, and absorbs the Sun's radiation. An ash pall shuts out some of the incoming short wavelength solar radiation, but does not significantly reduce the longer-wavelength energy re-rediated from the ground surface. Humphreys calculated that a 'dust' pall might reduce ground temperatures by several degrees. He also realized that less obvious *indirect* effects might be more important for world climate. By upsetting temperature distributions around the world, established atmospheric circulation patterns might be disturbed, leading to cold, stormy winters and cool, cloudy summers. Given enough large eruptions, it was a logical step to suggest that prolonged volcanic activity could trigger Ice Ages. Unfortunately, Humphrey's hypothesis could not be supported by observational data—there were not enough to show any *significant* climatic effect after the Krakatau eruption. Two years later, Henryk Arctowski published a compilation of weather records from stations all around the world, concluding that the 'dust veil produced by the Krakatau eruption affected atmospheric temperature very greatly'.[3]

After the two great Wars, Harry Wexler, chief scientist at the US Weather Bureau, returned to the problem in 1952.[4] He realized that Humphreys and Arctowski had been working on the right lines, and that a large eruption might displace southwards the normal pattern of westerly winds, such that the July weather chart would be more like that for mid-May. Later, he argued that the effects of *single* eruptions would be too subtle to detect; instead he suggested that longer-term trends should be looked for. According to him, the latter part of the nineteenth century was marked by an overall cooling trend, which he related to several major eruptions (Krakatau 1883; Tarawera, New Zealand 1886; Bandai-San, Japan, 1888; Bogoslof, Alaska 1890; and Awue, Indonesia 1892; Fig. 17.1). By contrast, he suggested that overall the world's climate had been warming up during the first half of the twentieth century because of the *absence* of any large eruptions. Like Humphreys, he also thought that there was a connection between long periods of volcanic activity and the onset of Ice Ages.

It is easy to speculate aimlessly on the vagaries of wind and rain. Residents of the British Isles do it all the time, their intellectual vigour doubtless sapped by the monotony of too many grey, wet winters. Meaningful debate about climate and weather suffers from a lack of good records covering wide regions and long time-scales. It

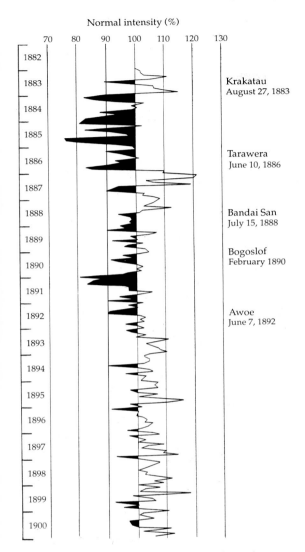

Normal intensity (%)

Krakatau
August 27, 1883

Tarawera
June 10, 1886

Bandai San
July 15, 1888

Bogoslof
February 1890

Awoe
June 7, 1892

Fig. 17.1 Variations in solar radiation intensity measured at noon at Montpellier, France from 1882 to 1900. Wexler argued that the cumulative effects of several large eruptions caused global cooling, but his conclusion was based on some shaky data—for example, 1888 eruption of Bandai-San was of minimal importance in terms of aerosol production. From Wexler, H. (1952). Volcanoes and world climate. *Sci. Am.* **220**, 3–5.

was, fortunately, a Briton of great intellectual vigour who sought to redress this problem in the

matter of volcanic impact on climate. Inspired by the gorgeous sunsets seen around the world after the eruption of Agung (Indonesia) in 1963, H. H. Lamb[5] of the British Meteorological Office examined records of all known volcanic eruptions since AD 1500 to estimate what he termed the 'Dust Veil Index' (DVI), a measure of the amount of fine ash lofted into the atmosphere by each eruption. Lamb concluded that there was *some* correlation between years with high DVI values (one or more large eruptions) and climatic cooling, but that it was not a particularly robust one. His work was of immense importance though, because it drew attention to the essential problem: documenting the records of both volcanic eruptions *and* climatic variations.

Lamb also drew attention to a second critical point: ash or 'dust' particles are not the only things that matter. Even more important are *acid aerosols* injected into the atmosphere by volcanoes, particularly those with sulphur-rich magmas. Soon after entering the stratosphere, sulphur dioxide reacts with hydroxyl radicals (OH^-) formed by photo-dissociation of water vapour in the upper atmosphere. Via chemical routes not fully understood at present, tiny droplets of sulphuric acid, H_2SO_4, are formed. The process is not instantaneous. Rather, photo chemical reactions may continue for months, replenishing the aerosol cloud for some time; new particles forming while older, larger ones settle out of the atmosphere. Because the stratosphere does not vary much in temperature, little vertical mixing takes place, which means that acid aerosols tend to remain at the same level at which they were injected.

Lamb's work marked the beginning of modern studies of volcano–climate interactions. His 'dust veil index' was replaced in 1982 by the 'Volcanic Explosivity Index', a related measure estimated for 6000 historic eruptions by Newhall and Self.[6] Subsequent research has followed three different directions: refinements of the record of volcanic eruptions; studies of individual eruptions, and investigations of how volcanic aerosols actually affect atmospheric circulation.

— 17.3 Paper, ice, and wood: the global record of volcanic activity

17.3.1 Written records

Among events of divine ordering there was . . . after Ceasar's murder . . . the obscuration of the Sun's rays. For during all the year its orb rose pale and without radiance, while the heat that came down from it was slight and ineffectual, so that the air in its circulation was dark and heavy owing to the feebleness of the warmth that penetrated it, and the fruits, imperfect and half ripe, withered away and shrivelled up on account of the coldness of the atmosphere.

Volcanologists are not much at home amongst the mildewed tomes of library archives. The quotation from Plutarch above, written about AD 100, however, shows how valuable such archives can be. Richard Stothers and Michael Rampino of the NASA Goddard Space Center made an exhaustive search of the classical European literature,[7] scanning the equivalent of a quarter of a million modern pages, excavating a gold mine of information on early eruptions. Inevitably, many of the references they turned up were tantalizingly oblique, but some are highly informative. Plutarch was writing about an eruption that took place in 44 BC, 1700 years before the one that Franklin described, but the similarities are obvious. The Caesar referred to was, of course, the late-lamented Julius, erstwhile conqueror of Britain. When the young Octavian entered Rome that year, a spectacularly coloured halo was observed around the Sun. This was such an unusual phenomenon that it was noted by several classical authors such as Seneca, Suetonius, and Pliny.[8] Stothers and Rampino concluded that the 44 BC eruption may have been of Mt. Etna, because literary sources indicate it was active then. Virgil's *Aeneid* described Etna's activity at the time:

Etna thunders with terrifying crashes, and now hurls forth to the sky a black cloud, smoking with pitch-black eddy and glowing ashes, and uplifts balls of flame and licks the stars—now violently vomits forth rocks, the mountain's uptorn entrails, and whirls molten stones skyward with a roar . . .

Stirring though this account is, it does not describe an *exceptional* eruption of Etna. No pyroclastic deposits of the right age have been found around Etna, so the identification of Etna as the source may not be correct. Another more distant volcano was probably involved. Wherever it was located, the eruption was probably a massive one, because the contemporary Chinese literature also has some important clues—in chronicles of the Han dynasty, there are references to the April sun in the same year being bluish white, and casting no shadows.

Of all the references turned up by Stothers and Rampino, none is more intriguing than that of the 'mystery cloud' of AD 536.[9] Procopius, a Byzantine historian who lived in Rome at the time wrote that:

the Sun gave forth its light without brightness, like the Moon, during this whole year, and it seemed exceedingly like the Sun in eclipse, for the beams it shed were not clear nor such as it is accustomed to shed . . . and it was the time when Justinian was in the tenth year of his reign.

This was clearly no local phenomenon, because John Lydus, living in Constantinople confirmed in an independent account that:

The Sun became dim . . . for nearly the whole year . . . so that the fruits were killed at an unseasonable time

A much later chronicler, Michael the Syrian, wrote about conditions at Constantinople:

The Sun became dark and its darkness lasted for eighteen months. Each day it shone for about four hours, and still this light was only a feeble shadow. Everyone declared that the Sun would never recover its full light. The fruits did not ripen and the wine tasted like sour grapes.

From these and other literary sources, Stothers deduced that the eruption which caused these widespread effects was *far* bigger than any other in the last three millennia. But where was the volcano? It cannot have been in the Mediterranean, for such a large eruption taking place there could scarcely have gone undocumented. Of the many possible candidate volcanoes around the world, Stothers suggested that Rabaul in New Guinea may have been the culprit—it had an eruption radio-carbon-dated at AD 540 ± 90. An alternative candidate, located at a high northerly latitude and therefore more capable of affecting the northern hemisphere, was the great White River eruption in the Yukon Territory of Canada, thought to have had such widespread ecological effects that it caused the diplacement of the Athapaskan Indians from their ancestral homelands. Although not precisely dated, this eruption took place in the right time frame.[10]

17.3.2 Ice cores

Humans are notoriously unreliable witnesses, whose written archives extend back only a couple of thousand years. Where, therefore, are we to find an *objective* record of volcanic eruptions reaching back continuously through time? Although it may not be intuitively obvious, the best records of the Earth's fiery volcanic history are written in acid script in the polar ice caps. Recall that volcanic acid aerosols, mainly H_2SO_4 are more important in their climatic effects than ash particles. When these acid aerosols fall on to permanent ice fields, they may be sealed beneath the next snowfall. Successive winter and summer ice layers are easily identifiable in favoured locations where accumulation and ablation rates are right, providing unbroken records of the Earth's climate extending back thousands of years.

Much the best record of eruptions during the last 10 000 years is contained in acidity profiles in cores obtained by drilling through the Greenland ice-cap, in a study by C. U. Hammer and his colleagues at the University of Copenhagen.[11-13] Such acidity profiles are useful as guides to variation in global volcanism, and provide a means for refining dates for large eruptions not securely described in the historical literature (Fig. 17.2). The difficulty with ice core acidity profiles lies in understanding which eruptions are recorded, and why, and which are not. Two cores have been studied, one from Crete in central Greenland and the other from Camp Century in north-west Greenland. An acidity signal has been obtained for every year from AD 1970 back to AD 550, but only the largest signals have been identified for the interval 50 BC to 8000 BC. Dating uncertainty is ± 1 year back to AD 1050 and increases to ± 3 years at AD 550.

Antarctica has a far larger ice-cap than Greenland, but conditions there are not as favourable (accumulation rates are much slower), so good acidity profiles have yet to be measured. Some data are available for microparticles deposited by volcanic eruption clouds which provide a means of cross-checking the records and dates of some, but not all, eruptions.[14] There are a few ice-caps on high mountains in tropical regions which also contain potentially excellent records of equatorial eruptions. In the Andes of Peru, the Quelccaya ice-cap has proved particularly valuable. Ice cores, therefore, have the potential to provide long term records of world-wide eruptions. But how should we read these curious records, inscribed in acid on ice?

Eruption volume Other factors being equal, large eruptions are obviously likely to produce bigger acid 'spikes' in the ice core record than small ones (Table 17.1). For an Icelandic eruption to register a clear signal in the nearby Greenland ice, cap an eruption volume greater then 2–3 cubic kilometres is required. A complicating factor here is that there may be a limit to the amount of aerosol that the stratosphere can sustain. Thus, correlation between strength of acidity signal and eruption volume need not necessarily be linear: a massive eruption may yield a signal not much bigger than a modest one.[15]

Eruption style This is a critical issue. Rain quickly washes small particles from the troposphere, but they can remain in the stratosphere for months, allowing time for them to diffuse to high latitudes, to be deposited on the Greenland or Antarctic ice-caps. Plinian eruption columns easily penetrate the tropopause, the level in the atmosphere above which temperature ceases to decline with altitude. They can inject sulphuric acid droplets directly into the stratosphere, and

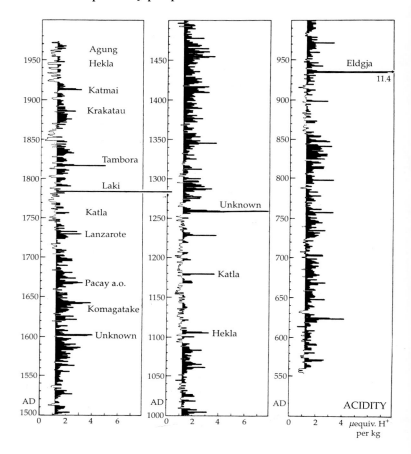

Fig. 17.2 Acidity profile through the ice core from Crete, central Greenland, showing H_2SO_4 'spikes' due to acid aerosols from historic eruptions. Several spikes have been related to known eruptions; others are more debatable. The 'unknown' spike at 1601 may have been due to a major eruption of Huaynaputina, Peru. Acid concentration is measured in terms of hydrogen ions per kilogram. From Hammer *et al.* (1980). Greenland ice sheet evidence of post-glacial volcanism and its climatic impact. *Nature* **288**, 230–5.

Table 17.1 Erupted volumes, atmospheric particle, and aerosol loadings and climatic effects of three large eruptions

Eruption year and volume	Total dust less than 2 microns (kg)	Total aerosol (kg)	Global SO_2 fall-out (kg)	Mid-latitude temperature change (°C)
Tambora 1815 150 km³	1.5×10^{10} to 1.2×10^{12} (150)	? (?)	1.5×10^{11} (7.5)	-1 (c.3)
Krakatau 1883 20 km³	2×10^{9} to 1.6×10^{11} (20)	3×10^{10} (c.3)	5.5×10^{10} (c.3)	-0.6 (2)
Agung 1963 c.1 km³	1×10^{8} to 8×10^{9} (1)	9×10^{9} (1)	2×10^{10} (1)	-0.3 (1)

Data from Rampino and Self (1982). Figures in parentheses represent quantities of dust, aerosol and sulphate fall-out, with Agung set to 1. Aerosol data are estimates from atmospheric opacity, SO_2 fall-out is calculated from ice core records so it might seem logical that explosive eruptions should be more likely to leave a record in the ice-caps than lava effusions, which superficially at least, are surface phenomena. But eruption of 14 cubic kilometres of basalt lava flows from Laki in 1783 left a massive signal in the Greenland ice-cap, demonstrating that voluminous effusive activity can also generate huge amounts of atmospheric aerosols. *If* these aerosols can reach the stratosphere, then their long-term effects could be profound. We return to this issue shortly.

Volcano latitude   Only six major non-Icelandic eruptions can be identified in the Greenland ice cores (Agung 1963, Katmai 1912, Krakatau 1883, Tambora 1815, an eruption about 1500 BC—possibly Thera—and Mazama about 7000 BC) It is natural that an eruption taking place close to an ice-cap should leave a larger acid signal than a similar one at greater distance. But the issue is more one of latitude, rather than distance alone. Atmospheric circulation takes place in quite tightly confined latitudinal zones, which is why navigators in the days of sail knew exactly in which latitudes they would find trade winds and their opposites, the Horse Latitudes and Doldrums, where calms and fickle winds predominate. Upper atmosphere circulation patterns, which affect stratospheric aerosols, are different from those at the surface, but similar considerations apply. Aerosols from an eruption taking place in one hemisphere will ultimately affect the whole of that hemisphere, and aerosols erupted in the tropics may also spread to cover both Northern and Southern hemispheres. But aerosols from an eruption taking place *outside* the tropics (at latitudes greater than about 20° north or south) find it extremely difficult to diffuse into the opposite hemisphere. Thus, aerosols from an eruption taking place at high southern latitudes will not easily leave an acid signal in the Greenland ice core.

Unfortunately, we hava a rather sinister means of quantifying the effects of latitude on the magnitude of the acid signal. During the 1950s and 60s many nuclear weapon tests were carried out in the atmosphere. All of us unwillingly contain traces of the isotope ^{90}Sr in our bones from these tests. Dispersal of radioactive fall-out from the tests was carefully monitored, and ^{90}Sr was used to monitor the diffusion of fall-out around the globe. (Nevil Shute's popular novel 'On the Beach' describes the implacable spread of radiation from nuclear war in the northern hemisphere to Australia). Hammer and his colleagues used the bomb fall-out data to estimate 'acidity multiplicative factors' for eruptions at different latitudes—a mid-latitude eruption yields a signal half that of a high-latitude one, and a low-latitude eruption one-third. Much smaller factors are probably needed for mid- to high-latitude volcanoes in the opposite hemisphere.

Magma composition   Sulphuric acid signals in arctic ice cores originate from sulphur dioxide gas erupted from the volcano. Thus, the magnitude of the acid signal is controlled by the amounts of SO_2 erupted, and ultimately by the sulphur content of the original magma. Solubility of sulphur in a magma depends upon its ferrous iron content, FeO, which generally decreases with increasing SiO_2. Thus, silicic magmas are less likely to have high SO_2 contents than mafic ones. Dacitic lavas erupted from Mt. St Helens during 1980 contained only 100 parts per million of sulphur, whereas the Laki basalts of 1783 contained 800–1000 ppm, and the 1982 El Chichón magma a spectacular 12 500 ppm.

Completeness of the record   The interplay of four variables (eruption volume, style, magma sulphur content, and volcano location) combine to make it difficult to identify the source volcanoes causing individual acidity peaks. Another less well-understood variable is the rate of conversion of volcanic sulphur dioxide gas to sulphuric acid aerosols—several reactions go on at different rates. Because the sulphur content of the magma and location of the volcano are such important factors, some large historic eruptions have not been recorded in the Greenland ice cores at all. Of the 20 eruptions during the last 1000 years with putative volumes greater than one cubic kilometre, nine have *not* left an identifiable signature in the ice cores. These include eruptions in high northerly latitudes (Bezymianny 1956); low latitudes (Santa Maria 1902), and far south latitudes (Quizapu 1932; and Tarawera 1886, the latter despite a high measured magmatic sulphur content). For whatever reasons, *half* of known large eruptions have not been recorded in the Greenland cores. Thus, the ice core record, although much longer and more objective than the historical record, is also frustratingly fragmentary.

17.3.3 Tree-rings

Trees are the oldest living things on the planet. Through their longevity and sensitivity to environmental change, they provide proxy witnesses of volcanic eruptions. Trees in temperate latitudes record faithfully the passage of the seasons in their annual growth rings, and so the technique of *dendrochronology* can provide remarkable accurate dates for archaeologists, against which the high-tech radio-carbon technique has been calibrated. A long-lived tree contains within itself a record of all the seasonal changes it has lived through. Even when the tree has been long dead, its history can be read if points of overlap between its record and those of younger trees can be identified, thus extending the record further back into the past.

Bristlecone pines are the oldest known trees, some of them veterans of nearly 5000 winters. *Pinus longaeva* and *Pinus aristata*, the two species concerned, are found in numerous localities in the western USA from California to Colorado, flourishing at high altitudes where the growing season is short. Dendrochronological studies showed that these trees consistently recorded frost damage at certain clearly defined dates. Frost damage to the wood of mature trees is a rare phenomenon, not caused by ordinarily severe winters, but by temperatures well below freezing sustained *after* the end of winter and *during* the growing season. Two successive nights with temperatures reaching $-5°C$ and an intervening days at about freezing are sufficient to leave their mark. In the North American tree-ring record, LaMarche and Hirschboeck found that several large eruptions, such as Krakatau 1883 appeared to have been associated with frost-damaged annual rings[16] (Fig. 17.3). In a later study, Lough and Fritts used tree-ring data to try and estimate variations in temperature

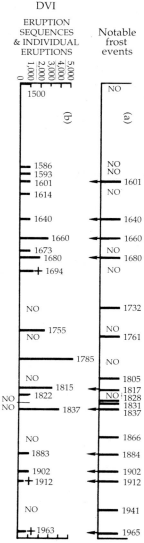

Fig. 17.3 (right) Dates of notable frost-rings events detected in western USA bristlecone pine trees and (left) major historical eruptions with their Dust Veil Index. Arrows indicate good correlations, but 'NOs' indicate frost rings for which there are no correlative eruptions, and vice versa. From LaMarche, V. C. and Hirschboeck, K. K. (1984). Frost rings in trees as records of major volcanic eruptions. *Nature* **307**, 121–6.

through time for North America for period 1602–1900.[17] They concluded that volcanic eruptions had significant climatic effects, but that large eruptions were followed by lowering of the annual average temperatures in the central and eastern USA, while the western states experienced *warming*.

This study highlights the complexity of volcano–climate interactions. Self-evidently, it is only eruptions severe enough to cause frost damage that leave a record. Thus, as with the acidity record, many eruptions may leave no tree-ring record. There is also an increasing body of evidence which suggests that large eruptions may be followed by winter *warming* in some regions. None the less, tree rings are an invaluable source of strictly *chronological* value in volcanological work—for example, they have been used to support a date of 1620 BC for the great eruption of Santorini, although this remains controversial.

— 17.4 Effects of some great historical eruptions

17.4.1 *Krakatau 1883*

Krakatau's culminating eruption took place on 26 and 27 August, 1883. It ejected about 20 cubic kilometres of dacitic pyroclastic material in a plinian column which reached an altitude of about 40 kilometres. Tenebrous darkness ensued when the ash cloud blotted out the sun. Although the larger particles quickly fell out, finer ones remained airborne longer and were wafted away westwards by the upper atmosphere winds. During the afternoon of 27 August, Krakatau haze was observed in Ceylon; by 28 August it had reached Natal in South Africa; by 30 August it was over the Atlantic, continuing its westward drift, spreading and being diluted the whole time. By 2 September the haze had reached the west coast of South America; by 4 September it was south of Hawaii in the Pacific, and on 9 September it was approaching the East Indies again, having been carried *right round the world* by the upper atmosphere winds (Fig. 17.4). It did not stop there though, but continued to be wafted westwards, making several trips round the world. During its first trip, it was confined to a narrow tropical belt, but later it gradually seeped up to higher latitudes in both hemispheres, until the atmospheric effects of the eruption were visible over most of the world.

Many unusual optical phenomena were observed. Most memorable were glorious sunsets, which even found their way into a poem by Tennyson:

Had the fierce ashes of some fiery peak
  Been hurl'd so high they ranged about the globe?
For day by day, thro' many a blood-red eve,

In that four-hundredth summer after Christ
The wrathful sunset glared against a cross . . .
  (St Telemachus)

Sunsets are normally red, of course: the daytime sky looks blue because scattering of the Sun's rays by the atmosphere permits only shorter wavelengths to reach our eyes. At dawn or dusk, forward scattering means that we see mostly the red wavelengths. Aerosols or solid particles of submicron size (as in smoke) amplify the scattering effect, so we see blood-red sunsets after events such as forest fires and volcanic eruptions which inject copious quantities of small particles into the upper troposphere and stratosphere.

Krakatau also caused some more bizarre effects. On 17 September 1883, the *Ceylon Observer* carried the following account:

The Sun for the last three days rises in a splendid green when visible about 10° above the horizon. As he advances he assumes a beautiful blue, and as he comes further on looks a brilliant blue, resembling burning sulphur. When at about 45° (above the horizon) it is not possible to look at him with the naked eye, but even at the zenith, the light is blue, varying from a pale blue to a light blue later on, somewhat similar to moonlight, even at midday. Then, as he declines, the Sun assumes the same changes, but vice versa. The Moon, now visible in the afternoons, looks also tinged with blue after sunset, and as she descends assumes a very fiery colour 30° from the zenith.

Similar phenomena were visible all over the world. The Honourable F. A. Rollo Russell, no

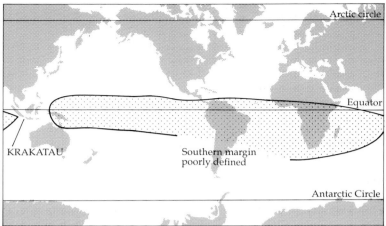

7 September 1883

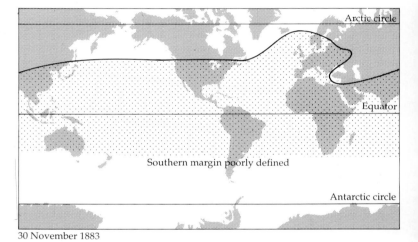

Fig. 17.4 Gradual spread of the Krakatau eruption cloud around the world, from the Royal Society's compilation of eyewitness reports of atmospheric phenomena. Distribution is shown on 7 September 1883, two weeks after the eruption, and 30 November 1883.

30 November 1883

less, made a note in his diary of what he saw from his home in Surrey, worlds away from the volcanic holocaust of Krakatau:

9 November. Sun set in very slight haze bank of cirrus; remarkable whitey-greenish opalescence above the Sun at sunset. About fifteen minutes after sunset the sky in the WSW, from near the horizon up to about 45° was of a brilliant but delicate pink. Below this a very curious opalescence going off into bronze-yellow, and that to the green tint . . . The sight was altogether an extraordinary one . . .

and in Gluckstadt Germany, Herr Dr A. Greber wrote more fulsomely of what he saw on 29 November, and initially thought was a display of the *aurora borealis;*

. . . its overwhelming magnificence presents itself to me as if it had been yesterday. When the Sun had set about a quarter of an hour there was remarkably little red (or ordinary) afterglow, yet I had observed a remarkably yellow bow in the south, about 10° above the horizon . . . this arc rose pretty quickly, extended itself all over the east and up to and beyond the zenith. The sailors declared, 'Sir, that is the northern lights!' and I thought that I had never seen the northern lights in greater splendour. After about five minutes the light had faded, though not vanished, in the east and south east, and the finest purple-red rose up in the SW; one could imagine oneself in Fairyland. The SW sky was bathed in an immense sea of light red and orange, and till more than one and a half hours after sunset the colouring of the sky was much more intense than it is half an hour after a very fine sunset in ordinary conditions.

Apart from these memorable crepuscular light shows, a different phenomenon was also seen—a halo encircling the Sun. This closely resembled the halo seen when the Sun shines through filmy cirrus cloud. It was first reported by the Reverend Sereno E. Bishop in Hawaii on 5 September. In his own words:

Permit me to call special attention to the very peculiar corona or halo extending to 20° to 30° from the Sun, which has been visible every day with us, and all day, of whitish haze with pinkish tint, shading off into lilac or purple against the blue. I have seen no notice of this corona observed elsewhere. It is hardly a conspicious object.

This effect was subsequently observed all round the world in the next two or three years, becoming known as 'Bishop's Rings', in recognition of the first observer. Bishop's rings have a subtly different origin from the ordinary solar haloes that they resemble. 'Ordinary' haloes are produced by *refraction* of light through the myriads of tiny, transparent ice crystals that constitute cirrus clouds, whereas Bishop's rings are the result of *diffraction* by opaque volcanic particles. In ordinary haloes the reddish or pinkish tinge appears on the *inside* of the ring, whereas in a diffraction halo, it is on the *outside*.

These optical phenomena have a deeper significance. While the world was revelling in gloriously hued sunsets, astronomers at the Montpellier Observatory in France recorded a more sinister effect: the intensity of the radiant energy from the Sun reaching ground level dropped by almost 20 per cent when the eruption cloud first arrived over Europe. Their measurements remained about 10 per cent below normal for many months (Fig. 17.1).

Climatic effects of the Krakatau eruption Many scholars have scrutinized the climatic effects of the Krakatau eruption. Identifying the consequences of even so obvious a cause is difficult, because climate is such a complex entity. What does 'average temperature' actually *mean*? Can nineteenth-century weather data of variable quality from many different countries round the world really be meaningfully combined to produce a synoptic picture? How does one recognize a volcanic 'signal' against the background noise of inevitable meteorological variability? Throughout history, there have been winters of brutally unexpected severity, for which there is no question of volcanic links, and for which meteorologists cannot account.

Notwithstanding these reservations, there is a consensus that Krakatau did have a climatic effect: cooling in the Northern Hemisphere of about 0.25°C for one or two years after the eruption.[18] A crucial advance in interpreting the climatic impact of eruptions was made by Self *et al.*,[15] who showed that temperature excursions within latitudinal bands or *zones* may be more significant than those averaged over entire hemispheres. Their analyses revealed that while there was a maximum temperature decrease of only about 0.3°C from 1882 to 1884 in tropical latitudes of the northern hemisphere, the decrease was 0.6°C in middle latitudes (30–60°N) but a full 1°C at high northern latitudes (60 to 90°N). This zonal effect has been noted in other large eruptions (Fig. 17.5).

17.4.2 Tambora 1815

Tambora 1815 was the largest eruption in modern history. But although it took place only 68 years before Krakatau, we know far less about what happened. Science progressed at an extraordinary pace during the nineteenth century, so while we are fortunate to have the meticulous Royal Society and Dutch government studies of Krakatau, there is nothing remotely equivalent for Tambora. As discussed in Section 10.4.4, the great eruption released over 50 cubic kilometres of trachyandesite magma, some of it in a plinian eruption column reaching 43 km altitude, but most of it in the form of pyroclastic flows and the fine co-ignimbrite ash elutriated from them. So great an eruption has naturally attracted a great deal of recent scholarly study. There are two key papers, the volcanological study by Sigurdsson and Carey already reviewed,[19] and a study of the world-wide aftermath of the eruption by Stothers.[20]

Optical effects Krakatau's glorious sunsets

Fig. 17.5 Temperature changes for latitudinal zones for the period 1880–1890. Zone 1 corresponds with high northern latitudes; Zone 2 to temperate northern latitudes; Zone 3 to tropical northern latitudes and Zone 4 tropical southern latitudes. K denotes Krakatau eruption. Largest temperature change following Krakatau was in Zone 1 (about 1°C) but there was little detectable effect at the same latitude as Krakatau (Zone 4). From Self, S., Rampino, M. R., and Barbera, J. J. (1981). The possible effects of large 19th and 20th century volcanic eruptions on zonal and hemispheric surface temperatures. *J. Volcanol. Geotherm. Res.* **11**, 41–60.

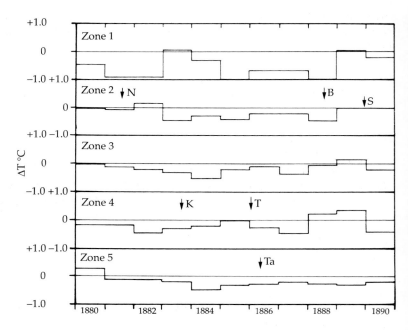

inspired religious fervour in Tennyson's poetry. Tambora may have also elicited a poetic response, but this time a rather more downbeat one by Byron, experiencing a wet and miserable summer on the shores of Lake Geneva in 1816:

The bright Sun was extinguish'd, and the stars
   Did wander darkling in the eternal space
Rayless and pathless, and the icy earth
   Swung blind and blackening in the moonless air;
Morn came and went—and came,
     and brought no day . . .

Darkness

(Byron and his contemporaries knew little about Tambora, unlike Tennyson, who was able to read about Krakatau in the press—the *Gentleman's Magazine* even carried an article discussing the causes of the sunsets).

Brilliantly coloured sunsets and prolonged twilights were noted near London in the summer and early autumn of 1815. These displays were recognized as distinct from the familiar red sunsets seen through London's smoke-laden atmosphere. In the following year, 1816, there were reports from the north-eastern part of the United States of a persistent 'dry fog', which was dispersed by neither wind nor rain. According to a New York report, the fog reddened and dimmed the Sun to such an extent that sunspots became visible to the naked eye. It has been suggested that the stunning sunsets and luminescent twilights enjoyed in the northern hemisphere that summer influenced the work of J. M. W. Turner, often called 'the painter of light'. While this may be true, it is difficult to confirm. Whereas artists such as William Ascroft drew explicit renderings of Krakatau sunsets seen from London, Turner left only his glorious paintings for us to wonder over.

Stothers estimated the opacity (*optical depth*) of the Tambora stratospheric veil by consulting the records of astronomers and other scientific observers. For example, he concluded from the naked-eye observation of sunspots in New York that the visual extinction due to the aerosol was about 1.0 magnitude. (Magnitude in this sense is used in relation to the brightness of stars. In good seeing conditions an observer can ordinarily see stars as faint as the sixth magnitude with the naked eye. This an extinction of 1 mag. is the equivalent of saying that the volcanic obscuration would make a star of mag. 1 brightness appear as a 2 and so on.) From lunar eclipse and ordinary stellar observations, he concluded that the stratospheric veil produced by Tambora in April 1815 reached England in about three months. During the summer of 1815, the northward spread of haze was halted temporarily by the meridional wind barrier, caused by Hadley circulation, near 30°N. Northward spread continued during the winter, and by the middle of 1816 the visual extinction in northern hemisphere mid-latitudes was probably about 1 magnitude at the zenith. *Two and a half years after the eruption, some haze still remained, causing visual extinction equivalent to about 0.4 magnitude* (Fig. 17.6).

Weather effects—the Year without a Summer Many ill effects have been attributed to Tambora's massive eruption. The year 1816 was so thoroughly miserable in the Northern Hemisphere that it has become known as the 'Year without a Summer'. Daily temperatures (especially the daily minima) were often abnormally low from late spring to early autumn; frequent north-west winds brought snow and frost to northern New England and Canada, and heavy rains fell in Western Europe. One report speaks of there being only three or four days without rain between May and October in

Merionethshire, Wales. There were widespread crop failures and delayed harvests. In Europe, the latest grape harvest for four *centuries* between 1484 and 1879 was in 1816. Not surprisingly, famine was widespread.[21]

Byron may have been gloomy and depressed when he wrote 'Darkness'. (Feeling a little under the weather, perhaps?) While his poem sank into well-merited obscurity, Mary Shelley made use of the dismal weather which kept her indoors to write a short story that led ultimately to her novel *Frankenstein*, which has survived in bastardized form to become part of the popular culture of horror movies. But was Frankenstein really born in the bowels of Tambora?

As with Krakatau, demonstrating a causal link between the eruption and distant weather effects involves complex statistical manipulation of global meteorological data. And it is much more difficult with Tambora than Krakatau because systematic meteorology did not exist then. Frosting-ring records in trees and the written record demonstrate beyond doubt that the 1816 summer weather was awful in the Northern Hemisphere, but was this due to the volcano, or merely coincidental? Not all regions experienced low temperatures, and in some the winter of 1815/16 was mild. Few adverse effects have been des-

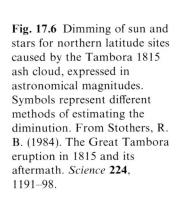

Fig. 17.6 Dimming of sun and stars for northern latitude sites caused by the Tambora 1815 ash cloud, expressed in astronomical magnitudes. Symbols represent different methods of estimating the diminution. From Stothers, R. B. (1984). The Great Tambora eruption in 1815 and its aftermath. *Science* **224**, 1191–98.

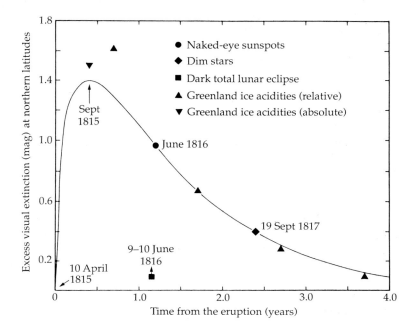

cribed from the Southern Hemisphere where one might have expected to find an imprint of the eruption, but that may reflect the sparseness of available data there. In southern South America, Argentina, and Chile were in the throes of revolutionary wars, so archives for the critical years are in disarray. Australia and New Zealand had barely been colonized. A further complication is that because there is less land in the southern hemisphere than the northern, hemispheric temperatures are stabilized by the vast southern oceans.

Stothers averaged data over latitudinal bands to determine if Tambora really had caused a climatic effect. He concluded that the average deviation from 'normal' over the whole of the northern hemisphere for 1816 was $-0.7°C$. This sounds significant, until one learns that in 1814 the deviation was *also* $-0.7°C$, and that in 1812, it was $-1°C$! By smoothing the data over a 6 year period, Stothers derived a 'corrected' temperature depression in 1816 of $-0.4°C$. Unfortunately, there is a further confusing factor: a decline in global temperature appears to have started *before* the eruption, possibly related to the 'Little Maunder Minimum', (a decline in solar radiance and sunspot activity), making it even more difficult to identify the Tambora signal.

Given the statistical uncertainties of weather records, what is one to make of all this? Perhaps only this: even if the cooling effect of a volcanic event is numerically quite small (half a degree), its climatic implications could be widespread. If *sustained*, they might be profound. We return to this issue in a later section.

17.4.3 El Chichón, 1982

Until its eruption in 1982, El Chichón was an obscure volcano in southern Mexico. Reaching only 1260 m above sea-level, El Chichón was covered by vegetation except for a couple of locations where the warm, acid breath of fumaroles inhibited growth. After a reconnaissance by a Mexican scholar, Frederico Mullerried in 1930, it was ignored. Thus, its violent eruption in April 1982 took volcanologists by surprise, obliging them to turn to atlases to discover where it was (Section 11.4). El Chichón's eruption is impor-

tant in a climatic context because it was the first eruption whose atmospheric effects could be studied with the panoply of modern instrumentation, both on the ground and from space. To be sure, Mt St Helens' 1980 eruption hit the headlines, but its atmospheric effects were negligible.

El Chichón's effects were far-reaching not because it was an exceptionally large eruption, but because it yielded exceptionally large volumes of acid aerosols.[22] Less than one cubic kilometre of trachyandesite tephra was erupted, trivial compared with Tambora's 50 cubic kilometres, also of trachyandesitic composition. But the tephra contained up to 2 per cent by weight of sulphates, some of it in the form of crystals of anhydrite ($CaSO_4$), a mineral not usually found in volcanic rocks. It is uncertain where all this sulphur originated: one possibility is that it is genuinely magmatic, derived from source region of the magma; the other that it was accidental, caused by the magma assimilating large quantities of sulphates from evaporite deposits known to underlie the volcano. Whatever its origin, much of the sulphur caught up in the magma was released during the eruption as sulphur-rich gas, mostly sulphur dioxide (SO_2), and to a lesser extent hydrogen sulphide (H_2S), which is rapidly oxidized to SO_2, yielding an eruption cloud unusually rich in acid aerosols.

Observations on the eruption cloud   All three of El Chichón's plinian eruptions sent gas and dust high into the stratosphere. The third event, on 4 April, appears to have been the largest, pumping a massive cloud up to 26 km. Satellites such as NOAA 6 and GOES *West* provided a stream of images of the steady westward drift of the eruption cloud. Like Krakatau's eruption cloud, El Chichón's cloud drifted rapidly round the world, returning to Mexico on 26 April. Its average westerly drift was measured at 20 metres per second (Fig. 17.7). More importantly, data serendipitously obtained from the *Solar Mesosphere Explorer* satellite also enabled the dispersion of the cloud into higher latitudes to be tracked. Unexpectedly, most of the cloud remained south of 30°N latitude for more than

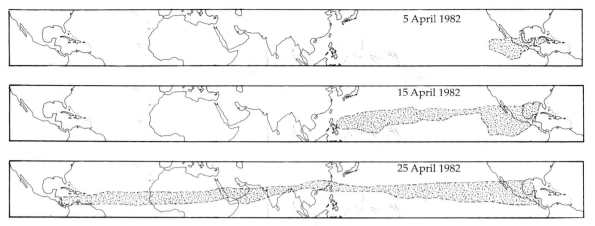

Fig. 17.7 Remote-sensing satellites tracked the westerly drift of the El Chichón eruption cloud continuously and precisely. Distribution of the cloud on three selected dates is shown. Westerly drift averaged 20 metres per second. Note the limited dispersion to north and south—the cloud was confined by atmospheric circulation patterns. Cf. Fig. 17.4. From Rampino, M. R. and Self, S. (1984). The atmospheric effects of El Chichon. *Scientific Americ.* **250**, 48–57.

six months after the eruption, blocked by atmospheric circulation cells.

One of the weather satellites, *Nimbus 7*, carried an instrument called TOMS (Total Ozone Mapping Spectrometer), capable of measuring the atmospheric SO_2 burden from the eruption. Arlin Kreuger from NASA'a Goddard Spaceflight Center estimated that the eruption had injected 3.3 million tonnes of gaseous SO_2 into the atmosphere, and that it had all been converted into sulphuric acid within three months of the eruption.

Ground-based studies also made major contributions in studying the eruption cloud. While Stothers was obliged to infer the opacity of the Tambora cloud by tortuous indirect means, modern laser technology provided a means of doing this *directly*. LIDAR, as its acronym suggests is somewhat akin to radar: Light Detection and Ranging. Pulses of light from a powerful laser of a particular wavelength are directed up through the atmosphere. If the atmosphere contains aerosols, a fraction of the light is back-scattered to a detector. By comparing the amount back-scattered with that from 'normal' atmosphere, the amount of aerosol can be estimated. Better yet, by timing the returned signal, the amount of light back-scattered from

different altitudes can be determined, providing a means of mapping vertical density variations within the eruption cloud. Many sets of LIDAR observations were made. Some startling measurements were made at the observatory on Mauna Loa on Hawaii, which was directly downwind from El Chichón and thus witnessed passage of the cloud within a few days of its eruption. Readings obtained were the highest ever recorded there; 140 times denser than the Mt. St Helens eruption plume.

Direct samples of the aerosols obtained by aircraft and balloon studies in the USA provided unique data on the changing composition and dimensions of the aerosol particles through time. In May and early June of 1982, 85 per cent of the cloud consisted of angular glass shards, coated with droplets of sulphuric acid, with average diameters between 3 and 6 microns. By July, the larger ash particles had settled out of the atmosphere, and the average size had decreased to between one and two microns. Furthermore, rather than single particles, the recovered samples showed that *aggregates* of ash and sulphuric acid aerosol were forming; tiny particles clumping together to form large, low-density clusters before finally falling out. Aggregates as large as 80 microns, with densities of $0.1 \times$

10^3 kg m^{-3} were measured. Balloon data were also used to estimate the mass of sulphuric acid aerosol in the atmosphere: a month after the eruption there may have been as much as 20 million tonnes; after a year, it was less than 8 million tonnes.

Climatic effects   Like its larger predecessors, El Chichón gave rise to the usual gamut of crepuscular light shows, Bishop's Rings, and so on. But its climatic effects are difficult to pin down, because although sulphur rich, the eruption was only of modest proportions. One quantifiable result was soon recorded: absorption of a fraction of the incoming solar radiation caused a warming of the equatorial stratosphere by 4°C; the highest reading recorded since continuous stratospheric measurements were begun in 1958.

Effects on the ground are less easy to quantify. As usual, the signal due to the eruption has to be distinguished from the 'noise' of the non-volcanic vagaries in conditions. According to one study, El Chichón may have caused a decrease of about 0.2°C in the Northern Hemisphere in June 1982. If this correct, there are interesting implications: the satellite data demonstrated that the eruption cloud did not stray north of 30° for several months, well after June. Thus, if the cooling effects is genuine, it must result from an *indirect* rather than a direct, mechanism of climatic perturbation. If nothing else, this reinforces the difficulty inherent in probing the workings of the Earth's climate.

17.4.4 *The 1991 eruptions of Mt. Hudson, Chile and Mt. Pinatubo, Philippines*

1991 was an unusual year for volcanology. When this chapter was being written, aerosol plumes from two major eruptions were diffusing around the globe. Ten days after the 15–16 June 1991 plinian eruption of Mt. Pinatubo in the Philippines aerosols had begun their drift around the world; the aerosol cloud stretching 11 000 km from Indonesia to Central Africa. By 11 July, its effects were so widespread that they degraded observations of a long-awaited total solar elipse at Hawaii. Two weeks after the 15 August eruption of Mt. Hudson in south Chile, satellite

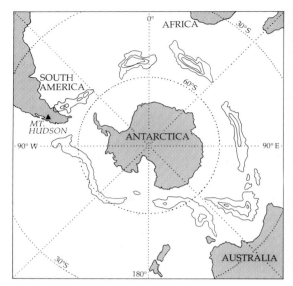

Fig. 17.8 Successive daily images from the Total Ozone Mapping Spectrometer (TOMS) aboard the Nimbus-7 weather satellite were used to chart the progress of the Mt. Hudson aerosol cloud between 15 and 21 August 1991. Preliminary estimates suggested that the cloud covered an area of 270 000 sq km, and totalled 250 million kilograms of sulphur dioxide. In the images, the cloud lacked an extended trailing edge, showing that it was ejected in a single burst, rather than in a continuous plume, as the composite image may imply.

observations showed that the aerosol plume had already girdled the Earth (Fig. 17.8).

About 3 cubic kilometres of material were erupted from Mt. Pinatubo, about 6 times as much as were erupted El Chichón, making it the third largest eruption of the century, after Santa Maria, Guatemala (1902) and Novarupta, Alaska (1912). Its sheer magnitude ensured that it would have world-wide effects, but the Mt. Pinatubo magma startled volcanologists because, like El Chichón, it contained abundant anhydrite, thus ensuring that huge amounts of sulphuric acid aerosols would be injected into the atmosphere. Satellite observations suggest that about 20 million tonnes of SO_2 were erupted. It will take several years for the climatic effects of the Mt. Pinatubo eruption to be clearly established, but scientists at the US National Oceanic and Atmospheric Admistration (NOAA) estimated that average global temperatures could drop by 0.5°C for between 2 and 4 years. Mt. Pinatubo erupted at a time when global warming due to carbon dioxide accumulation in the atmosphere was beginning to become of increasing concern. Consequently, some scientists welcomed Pinatubo's cooling effect, seeing it as a counter-balance to global warming.

El Chichón revealed a further unexpected global effect of volcanic eruptions: it apparently caused a marked decrease in stratospheric ozone. At that time, the 'ozone hole' was not of widespread concern, but a decade later, when Mt. Pinatubo erupted, it had reached the headlines, and become one of the preeminent environmental issues of the decade. Because Mt. Pinatubo was so much bigger than El Chichón, its effects on stratospheric ozone are bound to be even more worrying.

Much research has gone into ozone chemistry because the loss of the Earth's protective ozone layer could have such far-reaching consequences. Unfortunately, the chemistry is complex, and it is not clear how volcanic aerosols actually cause ozone depletion. Apparently, sulphuric aerosols help to reduce the amount of reactive nitrogen in the stratosphere, which in turn allows more of the active chlorine which catalyses decomposition of ozone to be generated. Immediately after the eruption of Mt. Pinatubo, a 30–40 per cent decrease in the amount of stratospheric nitrogen dioxide was observed in New Zealand. However, no direct impact on ozone was detected. It may be that serious loss of ozone only takes place in the cold air at polar latitudes, where production of free chlorine may be accelerated.

17.4.5 The Skaftár Fires—Laki 1783

Some volcanological aspects of this eruption, one of the most consequential in history, were summarized in Chapter 7. About 14 cubic kilometres of basaltic lava, and some tephra were erupted over an eight-month period: this was emphatically *not* a short, explosive event, abruptly injecting an ash cloud into the stratosphere.[23] Benjamin Franklin drew the connection between the 'dry fog' that crept southwards to blanket much of Europe, and the events unfolding in Iceland. Some accounts suggest that the haze reached as far as Syria and western Siberia. In Europe, where its effects were most obvious, the haze was so dense that it dimmed the Sun. In Pavia, northern Italy, people could look at the Sun with the naked eye, while in southern France the setting Sun was sometimes invisible once it had sunk below 17° from the horizon. Acid rain fell in Norway and Scotland. In Iceland itself, the 'dry fog' was a dense bluish acid haze which destroyed most of the summer crops. Famine followed with dreadful inevitability, leading to the deaths of 75 per cent of Iceland's livestock, and 25 per cent of the human population.

Studies of the acid deposited in the Greenland ice-cap suggest that between 13 and 63 million tonnes of sulphur dioxide were erupted into the atmosphere. From petrological considerations, Sigurdsson concluded that the Laki magma was rich in sulphur, containing between 800–1000 parts per million. Of this, about 85 per cent was released to the atmosphere during the eruption, while the remainder was contained in the lavas themselves. He calculated that about 80 million tonnes of sulphuric acid aerosols were formed from the eruption, consistent with the figure estimated from the Greenland ice data. It would be useful to put these figures in a broader context by comparing them with the amounts of acid gases belched into the atmosphere from power stations, automobiles, and other human activities every year. Unfortunately, this figure is poorly known, but most estimates suggest that it is about the same order, or a bit larger, than the amount emitted by Laki. Thus, to have lived in Iceland in 1783 would have been tantamount to suffering *all* the modern world's atmospheric pollution in concentrated form.

Apart from sulphur dioxide, large volumes of other gases were emitted, including hydrogen chloride and hydrogen fluoride. Fluorine caused the death of cattle that grazed on grass contaminated by the acid haze. Early signs of the fluorosis that killed the animals were blotchy, discoloured teeth; later, their joints and bones became diseased.[24] Subsequent experiments have shown that if the fluorine content of dry grass exceeds 250 parts per million, animals grazing on it die in 2–3 days. Curiously, fluorine is only a trace constituent of basaltic lavas, present in proportions of a few hundred parts per million, and its abundance in erupted volatiles as hydrogen fluoride is difficult to estimate. But the highly reactive gas binds firmly to the surface of erupted particles and aersols, so when hungry cattle grazed on contaminated grass—as starvation compelled them—they ingested lethal doses of the element, even at considerable distances from the volcano.

To achieve long residence times in the atmosphere, acid aerosols must reach the stratosphere—they are washed rapidly out of the troposphere. Although the tropopause is at lower altitudes at high latitudes, it is still above 10 km at the latitude of Iceland. In a basaltic eruption, fire fountains may spray liquid lava more 1.5 km in height. as they did at Oshima in 1986. Convection will carry hot gases much higher. While some recent research suggests that Laki may have had fire fountains 1.4 km high, it is not likely that much material was carried high enough to enter the stratosphere. Apart from the possibility of stratospheric injection, the haze that spread over Europe may owe its longevity to its being constantly renewed during the eight months of the eruption. Thorvaldur Thordarsson, an Icelander who has studied contemporary records of the Skaftar fires, has pointed out that Iceland may have suffered exceptionally badly because low-level cyclonic wind patterns kept volcanic fumes circulating anticlockwise around the island, rather than simply dissipating downwind.

Climatic effects of the Laki eruption Benjamin Franklin's remarks about the bitter winter of 1783–4 in Paris have already been noted. On the other side of the Atlantic, other well-known scholars were keeping temperature records— Thomas Jefferson, third president of the USA, and Ezra Stiles, president of Yale College, an educational institute in the state of Connecticut. These formed part of a unique 243-year-long time series of data for the eastern USA. Analysis of these records shows that an abnormal temperature decline began in autumn 1783, hitting rock-bottom between December and February 1784, when the lowest ever winter average temperature for this region was recorded; 4.8°C

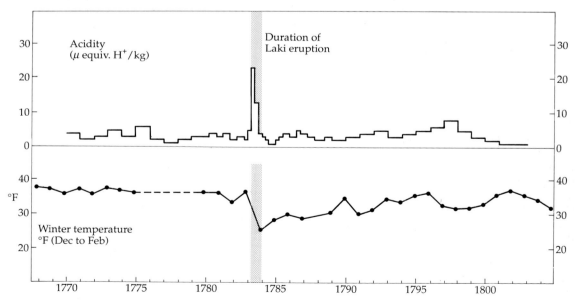

Fig. 17.9 Winter temperature records for the eastern United States and the acidity profile through the Greenland ice-cap, showing the pronounced anomaly associated with the Laki eruption. From Sigurdsson, H. (1982). Volcanic pollution and climate: the 1783 Laki eruption. *Eos* **63**, 601–602.

below the 225 year average (Fig. 17.9). On 28 September, the Vermont correspondent of the *Newport Mercury* was prompted by the bitterly unseasonable conditions to write: 'the weather has been the most extraordinary in these parts for some time past, that the oldest man among us can remember'. Taking the Northern Hemisphere as a whole, a temperature decrease of about 1°C seems to have taken place. Table 17.2 demonstrates the severity of conditions following both

the Laki and Tambora eruptions.

Laki's long-drawn-out eruption therefore teaches two important lessons: first, the insidious environmental effects of an eruption can be far more serious than the more dramatic and immediately obvious manifestations of volcanism, such as fire fountains. Second, an effusive eruption of modest dimensions can have climatic effects comparable with those of violently explosive eruptions (Table 17.3).

Table 17.2 Exceptionally cold years in Brunswick, New Jersey, in the period of more than 200 years, 1738–1976 (°F). From Reiss *et al.* 1980

Year	Year overall	Spring	Summer	Autumn	Winter
1783	average	warm	ave./warm	ave./cool	*coldest by 3°*
1784	5th coldest	coldish	ave./warm	ave./cool	2nd coldest
1785	9th coldest	3rd coldest	ave./warm	ave./cool	cool
1816	ave.	normal	cool	ave.	ave.
1817	4th coldest	*coldest*	*coldest by 2°*	ave.	ave.

Table 17.3 Sizes, eruption column heights, stratospheric aerosol loadings, and Northern Hemisphere temperature decrease for some major historic eruptions

Eruption (lat.), date	VEI	Magma volume (km³)	Eruption column height (km)	H_2SO_4 aerosols (from optical properties) (kg)	H_2SO_4 aerosols (from ice core data) (kg)	Northern Hemisphere temperature decrease (°C)
Tambora (8°S), 1815	7	>50	>40	2×10^{11}	1.5×10^{11}	0.4–0.7
Krakatau (6°S), 1883	6	>10	>40	5×10^{10}	5.5×10^{10}	0.3
Santa Maria (15°N), 1902	6	*c.*9	>30	$<2 \times 10^{10}$	2×10^{10}	0.4
Katmai (58°N), 1912	6	15	>27	$<2 \times 10^{10}$	$<3 \times 10^{10}$	0.2
St Helens (46°N), 1980	5	0.35	22	$c.3 \times 10^8$		0–0.1
Agung (8°S), 1963	4	0.3–0.6	18	$1–2 \times 10^{10}$		0.3
El Chichón (17°N), 1982	4	0.3–0.35	26	$1–2 \times 10^{10}$		0.4–0.6
Laki (64°N), 1783	4	0.3*			$<1 \times 10^{11}$	*c.*1.0

Data from Rampino and Self, 1984. VEI = Volcano Explosivity Index. *0.3 is estimated volume of tephra for Laki, lava volume was *c.*14 km³

17.5 Volcanic winter?

In the early 1980s a group of influential scientists, including Carl 'Cosmos' Sagan, became concerned about what they considered to be an horrific but overlooked consequence of nuclear war: fires ignited by nuclear explosions might generate huge quantities of dark smoke when devastated cities and forests burned out of control. As they saw it, a pall of dense smoke from such a nuclear holocaust would spread over at least one hemisphere, blocking out the Sun, and leading to prolonged surface cooling. Those who had survived the nuclear holocaust would succumb during the 'nuclear winter' that would follow.[25]

Early estimates of the severity of nuclear winter were based on over-simple one- and two-dimensional models of the atmosphere. As more sophisticated models of atmospheric circulation came into use, it emerged that the effect would not be as drastic as first feared. This work did, however, draw attention to the difficulties of atmospheric modelling, and the need to understand the effects of large volcanic eruptions on climate. Drawing an analogy with the nuclear scenario, Rampino, Self, and Stothers, suggested that 'volcanic winter' might be a possible consequence of the largest eruptions.[26]

There are many major differences between a volcanic aerosol pall and a nuclear smoke pall—in particular, smoke may be dark and sooty, with profound thermal implications. At the time of writing in the wake of the Gulf War, no less than 600 oil wells were burning in Kuwait, sending oily black smoke clouds as far distant as the Himalayas. Fortunately, this appalling smoke cloud remained with the troposphere. Immediately around Kuwait the smoke pall cast an evil shadow, sharply reducing temperatures. But its effects were not widespread.

Future great volcanic eruptions are certain to inject aerosols into the stratosphere. What then will be their climatic effects? Stratospheric aerosols affect the global radiation budget by absorbing and back-scattering incoming solar radiation, which should cause a cooling of the lower atmosphere and surface, and an increase in stratospheric temperatures. The volcanic signal

from eruptions in historic times, however, is hard to pick out against the background of 'normal' variations in temperature. When statistical smoothing and compositing techniques are used, a statistically significant average temperature decrease of about 0.2 to 0.5°C for one to three years after the eruption emerges (Fig. 17.10).

Sulphur is the key to the magnitude of the climatic effects caused by an eruption: a small eruption of a sulphur-rich magma (such as El Chichón) can produce relatively large volumes of stratospheric aerosols, whereas the the eruption of a larger volume of sulphur-poor magma (such as Mt. St Helens) will generate less. Acid aerosols are rapidly flushed from the troposphere by rain, and so must be injected into the stratosphere to have long-term global effects. However, if the eruption continues over a long period, wide distribution is possible even within the troposphere, as Laki demonstrated.

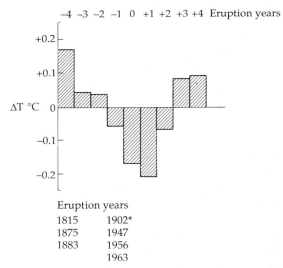

Eruption years

1815	1902*
1875	1947
1883	1956
	1963

*1902 includes close-spaced sequence of 5 eruptions in 9 mths.

Fig. 17.10 Temperature effects (ΔT°C) in the Northern Hemisphere averaged over several major eruptions. The composited data suggest a 0.2° drop in the year following an eruption. From Self, S., Rampino, M. R., and Barbera, J. J. (1981). The possible effects of large 19th and 20th century volcanic eruptions on zonal and hemispheric surface temperatures. *J. Volcanol. Geotherm. Res.* **11**, 41–60.

17.5.1 Large explosive eruptions

Outstanding though it was, the Tambora event was an order of magnitude smaller than the largest eruptions known from the geological record, which formed giant calderas such as Cerro Galan in Argentina and Toba in Sumatra. Such eruptions involve thousands of cubic kilometres of pyroclastic rocks of silicic composition. Toba is particularly relevant, since it erupted 2000 cubic kilometres only 74 000 years ago. Could such a huge eruption have triggered 'volcanic winter'?

It is hard to say. Toba's huge magnitude is not in doubt—the Youngest Toba Tuff has a volume of about 2800 cubic kilometres. Studies of the tephra on the Indian Ocean floor suggest that about 20 000 metric tonnes of fine ash (less than two microns in size) were erupted (Section 15.6.1). If this all entered the stratosphere, conditions of total darkness could have existed over a large area for weeks. But how much actually did so? Work on Tambora suggested that about 30 per cent of the fine-grained material produced in the eruption was formed when pyroclastic flows entered the sea all around the volcano, causing secondary explosions. Much of this material may have risen convectively to the stratosphere.

Owing to its silicic composition, the Toba magma did not contain a high proportion of sulphur. By simply scaling up the effects of historical eruptions, Rampino et al. estimated that between 1000 and 5000 million tonnes of sulphuric acid aerosols were liberated. This stratospheric loading would cause severe effects, comparable to some of the nuclear winter scenarios, but longer lasting. They estimated that the volcanic haze would block at least a tenth of the incoming sunlight, perhaps more. As they admit, however, their figures are subject to all sorts of uncertainties—for example, it is probable that the stratosphere cannot sustain high aerosol loadings—the particles would simply accrete together and fall out. Thus, scaling up from smaller eruptions is almost certainly inaccurate.

There are traces of the Toba ash in the Vostok ice core, obtained by Russian scientists from the Antarctic ice-cap. This core is one of our best

sources of temperature (and atmospheric CO_2) data over the last 100 000 years, and it is known that the Toba eruption coincided approximately with the ending of a period of relative warmth during the Ice Age (the Brørup interstadial) and a fresh glacial advance.

But this coincidence does not prove a connection. The time resolution in the Vostok core is rather poor, and the effects of the Toba eruption would have been exceedingly brief compared with the length of the glacial advance. It is none the less tantalizing to speculate that the Toba eruption might have triggered the advance.

17.5.2 Basaltic flood eruptions

Laki was modest compared to the largest flood basalt eruptions known from the geological record. In the Columbia River Plateau, lavas as huge as the Roza flow (an order of magnitude larger than Laki), may have been erupted in only a few days. By scaling up the Laki data, Rampino et al. estimated that about 6000 million tonnes of sulphuric acid aerosols were injected into the atmosphere. Distributed world-wide, these aerosols would form a dense haze, cutting out all but a small fraction (less than one ten-thousandth) of the sunlight: noon would resemble a moonlit night. But the same problems arise. Can the stratosphere sustain such dense aerosol loadings? And can basaltic fissure eruptions inject them into the stratosphere anyway? It is difficult to resolve the first issue, but Stothers et al. have tackled the physics of the second.

Although high fire fountains were a feature of the Skaftar Fires, maximum convective plume heights were about 6–8 km, not enough to drive aerosols above the tropopause, which is about 11 kilometres above Iceland in summer. This concurs with the Greenland ice core record: because there is a massive signal in 1783 but none in 1784, aerosols are unlikely to have reached the stratosphere. If they had, the effects would have shown up in the year following the eruption. For massive eruptions such as those in the Columbia River Plateau, Rampino et al. calculated that fire fountains more than a kilometre high might be formed, and that under the right combinations of mass eruption rate, fine ash content, and particle cooling rate, it would be possible to generate

convective plumes capable of reaching the stratosphere, even in the tropics. Although there are
still considerable uncertainties in this approach,
it is obvious that basaltic flood eruptions could
have quite disagreeable climatic and environmental effects (Fig. 17.11).

17.5.3 Mass extinctions?

At the end of the Cretaceous Period 65 million
years ago, there was a world-wide extinction of
many animal species, both in the sea and on dry
land. Of these, the demise of the dinosaurs has
captured the public imagination, but huge
numbers of less notorious species also became
extinct, notably microscopic foraminifera living
in the oceans. This mass extinction was one of the
most important events in the history of the Earth,
and is used by palaeontologists to define the
boundary between the Mesozoic (Middle Life)
Era, and the Cainozoic (New Life, or Tertiary)
Era.

An increasingly solid body of evidence demonstrates that a major impact event spread a film
of iridium-rich ejecta over the Earth precisely
at the Cretaceous–Tertiary (K–T) boundary.
Researchers have even zeroed in on the site of the
impact, at Chicxulub near the city of Mérida on
the Yucatan peninsula of Mexico. Blobs of

impact melt glass (tektites) found in nearby Haiti
have yielded dates of 64.5 million years; *precisely*
the age of the K–T boundary. Analyses of the
tektites also suggest that the impact melted huge
amounts of evaporite deposits, rich in sulphates,
thus injecting sulphur dioxide into the atmosphere. According to one recent calculation, the
resulting aerosols may have caused a temperature decrease of 4°C.

But within exactly the same time frame as the
Chicxulub impact, at least a million cubic
kilometres of flood basalts gushed from the
mantle to form the Deccan Traps of India. How
were these two remarkable events connected, if at
all? Did the impact extinguish the dinosaurs? Or
was it the basaltic flood eruption: Laki on a
global scale? Did the impact event perhaps
trigger the basaltic flood? Or did the two
extraordinary events reinforce each other in
some lethal way? These are highly contentious
issues, which have sparked vigorous but produc-
tive debates amongst scientists concerned with
global issues.

As long ago as 1972, P. R. Vogt argued that
widespread basaltic volcanism could have serious consequences, and might have been impli-
ated in the extinction of the dinosaurs. More
recently, the volcanic hypothesis in the K–T

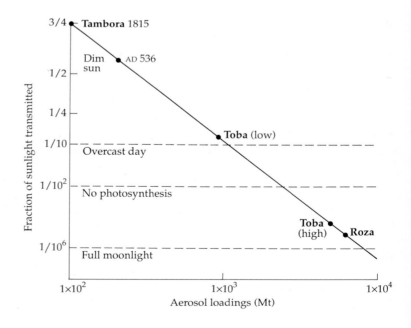

Fig. 17.11 Fraction of sunlight
transmitted through
stratospheric aerosols from
various great historic
eruptions. AD 536 refers to the
'mystery' cloud of that date,
and the Roza data point to
one of the Columbia River
flood basalt lava units. Two
estimates for likely effects of
the Toba eruption are shown.
From Rampino, M. R., Self,
S., and Stothers, R. B. (1988).
Volcanic winters. *Ann. Rev.
Earth Planet. Sci.* **16**, 73–99.

boundary debate has been espoused by a group led by Officer and Drake.[27] Palaeontologists disagree about exactly how abrupt the massive extinctions were, and so much of the debate is too controversial to summarize here. Two of the principal protagonists have usefully summarized their positions in *Scientific American*.[28].

A round number estimate of the volume of the Deccan basalts is one million cubic kilometres. Simple linear scaling from Laki (always risky) suggests that a total of about 8×10^{12} tonnes of sulphuric acid was liberated. An alternative method based on observed eruption data suggests a figure of 17×10^{12} tonnes. These huge amounts of acid aerosol would undoubtedly have severe effects if liberated at one time. But there is another factor to consider. How long did it take to erupt the Deccan Traps? It was clearly not a single event, since there are soil horizons between some flows. Some estimates suggest that it took at least half a million years, but activity probably continued at a reduced level over a much longer period. Assuming half a million years, the annual H_2SO_4 flux rate would be about 1.6×10^7 tonnes, less than that of Laki's single eruption. Given the rapid rate at which aerosols are washed out of the tropical troposphere, it therefore seems unlikely that the Deccan volcanism could have contributed to world-wide mass extinctions. It is too early to discount its indirect effects however. Some fraction of the acid may have entered the stratosphere, causing long-term cooling. It is difficult to determine what global climatic effects might have resulted from repeated and sustained injections of acid aerosols into the atmosphere. Perhaps, when the catastrophic impact event took place, it did so in an environment already perturbed by massive climatic dislocations. We shall probably never know for sure, but there can be little doubt that the massive Deccan eruptions had far-reaching consequences, and probably made more than one tyrannosaurus retch.

A good many other beasts had reason to feel a little off-colour 245 million years ago. This was the occasion of the greatest mass extinction in the Earth's history. It defines the Permian–Triassic boundary; the milepost between the Palaeozoic (Early life) Era and the Mesozoic (Middle life) Era. Huge numbers of species of familiar fossils such as trilobites became extinct by the end of the Permian, while the Mesozoic saw the great blossoming of the Age of Reptiles, including, of course, the dinosaurs. (Millions of years after their own extinction, these grotesquely overgrown lizards still keep many a museum curator in gainful employment. If they had never existed, it might have been necessary to invent them.) Is it purely coincidental that the Palaeozoic–Mesozoic boundary coincides approximately with the vast outpouring of the Siberian flood basalts? A preliminary estimate suggests that more than 1.5 million cubic kilometres of basalt may have been erupted within a period of less than a million years, a rate of more than a cubic kilometre per year.[29]

— 17.6 Other global interactions

17.6.1 *The El Niño–Southern Oscillation (EN–SO)*

For thousands of kilometres along the west coast of South America, the rollers of the Pacific ocean break against long beaches, empty of swimmers. In 1810, Alexander von Humboldt identified the northward current which now bears his name and whose frigid temperatures keep bathers out of the water. For decades, the current's cold but nutrient-laden waters supported Peru's fishing industry, one of the world's largest, so that its behaviour was thus closely monitored. Every year, at about Christmas time, warm but barren water spread down from the north, putting an end to the fishing season. Because its arrival coincided with the nativity, the warm current was termed 'El Niño' by fishermen. Every few years the El Niño current is exceptionally warm, causing catastrophic disruption of the marine ecosystem, and bringing torrential rainfall to the desert coasts of Peru.

El Niño is not merely a local phenomenon, but rather is one part of a world-wide pattern of pressure and temperature changes in the equatorial regions, the *Southern Oscillation* (SO). As first described, the SO was characterized in these terms: 'when pressure is high in the Pacific Ocean it tends to be low in the Indian Ocean from Africa to Australia; these conditions are associated with low temperatures in both these areas, and rainfall varies in the opposite direction'. When an El Niño event takes place, pressure over the central Pacific is lower, and the south-easterly trade winds fail, permitting an easterly flow of anomalously warm surface sea-water along the Equator across most of the Pacific; one-quarter of the Earth's circumference, in fact. It is now realized that the EN–SO is one of the most complex components of the interconnected oceanic and atmospheric circulation system, so it is the subject of intensive research.

Recognizing the importance of unusually warm sea-water temperatures in the EN–SO cycle, H. R. Shaw and J. G. Moore of the US GS raised the intriguing possibility that the anomalous water temperatures could be the result of volcanic heating, consequent on the eruption of large volumes of lava on the ocean floor.[30] As emphasized repeatedly throughout this book, the ocean floors are the sites of the most voluminous volcanism on the Earth, so their role in warming sea-water is intuitively reasonable. As Shaw and Moore showed, however, elementary physics dictates that the effect is unlikely to be a *direct* one: cooling and crystallization of one cubic kilometre of basalt lava liberates enough heat to raise the temperature of one thousand cubic kilometres of sea-water by 1°C. By contrast, the volume of anomalously warm water involved in an El Niño is about 100 000 cubic kilometres. Thus, it follows that even a large basaltic eruption—Laki sized—would only cause about 10 per cent of the warming required for an El Niño. Such large eruptions may occur, but it is unlikely that they are focused on individual submarine centres. It is more plausible to envisage eruptions taking place from a number of sites distributed along a ridge over a period of a few years.

None the less, their effects would still be indirect. It is becoming obvious that the EN–SO represents a subtle balancing of atmospheric and ocean factors over huge areas. Small inputs of volcanic heat could perhaps tip it one way or another. Although appealing, this idea still falters against the problem that the EN–SO is periodic; it *is* an oscillation, and exceptional El Niños take place every 3–7 years. It is difficult to see how such repetitive behaviour could be related to the notoriously irregular character of volcanic eruptions.

In a theoretical study of the effects of stratospheric volcanic aerosols, Paul Handler[31] suggested that eruptions taking place at the right time and place could reinforce the effects of EN–SOs. He pointed out that temperature changes due to stratospheric aerosols would be manifested more quickly over landmasses than the oceans. The land cools more quickly, causing pressures to become slightly higher; to keep things in balance, pressures over the ocean would decrease. Most of the Earth's land masses are in the Northern Hemisphere. As a result of this asymmetry, cooling would cause air to be transferred from high-pressure regions (anticyclones) over the oceans in both Northern and Southern Hemispheres to the continents, especially the vast mass of southern Eurasia. In the right circumstances, pressure gradients in the tropics ranging from relatively high over southern Eurasia to relatively low over the Pacific would resemble those of EN–SO events.

If an eruption happens to take place in the tropics during an EN–SO event, the effects of the EN–SO would be enhanced. The eruption of El Chichón took place in the same year (1982) as an especially large EN–SO, and other large eruptions this century appear to have been followed by similar events. According to Handler's theory, this is not coincidence, if volcanic eruptions *cause* amplified El Niños. His theory will be tested by the great 1991 eruption of Pinatubo.

All of this may sound rather abstract. However, one should never underestimate the subtle interactions that take place in our environment. Who would have thought that worn-out refrigerators could harm the ozone layer? A good example of how a relatively minor local change propagated far-reaching effects through much of the Earth system is provided by the Younger Dryas episode about 11 thousand years ago. In

1988 Broecker and his colleagues [32] suggested that this one-thousand-year-long reversion to cold conditions during late-glacial warming was related to an increase in cold water flowing into the North Atlantic, which in turn caused an albedo increase due to sea-ice formation and a reduction of the trans-equatorial heat transport by shutting off the North Atlantic Deep Water flow. They suggested that this change could have been the result of the abrupt diversion of the cold melt-waters of the great continental ice sheet which covered much of northern north America. Prior to the Younger Dryas, the icy waters drained to the Gulf of Mexico, but during the event they were temporarily diverted down the St Lawrence Valley and into the North Atlantic.

Thus, a local event produced a change in climate felt over the whole of the North Atlantic region, and possibly much further afield. Who knows what curious consequences a massive episode of volcanic activity might have?

17.6.2 Do climatic changes cause volcanic activity?

This chapter has addressed itself to the issue of how large volcanic eruptions might affect climate. To turn the issue on its head might seem merely frivolous, but the possibility of a reverse connection has been seriously considered at least since 1969, most recently by Rampino *et al.*[33] They looked carefully at the timing of large volcanic eruptions and glacial advances, and particularly at a data set initially compiled by J. R. Bray to demonstrate that the advances were *caused* by volcanic episodes. Rather than *preceding* global cooling and glacial advances, they found that the available data suggested that volcanic events in many cases appeared to *follow* them. There are all sorts of problems with the chronologies of glacial and volcanic events, making connections between the two conjectural. Even if the timing of an eruptive episode after a glacial episode could be securely demonstrated, it would still remain to demonstrate a *causal* link between the two—sceptics would want to see a plausible mechanism at work.

Owing to the sheer volume of ice involved in an Ice Age, there are some important geophysical consequences that might be involved in triggering volcanic activity. The mass of ice loading on the Greenland ice-cap during an Ice Age is so huge that the global spin axis is perturbed, causing subtle changes to the geoid (the three-dimensional shape of the Earth's surface). Stresses set up during this world-wide realignment *might* trigger volcanic eruptions. Furthermore, during glaciation, such large amounts of sea-water are removed from the oceans to form ice that the oceans basins are unloaded slightly. As they readjust to this new configuration, it is again possible that eruptive activity could be triggered.

While these connections may seem tenuous, it is worth concluding with a restatement of how sensitive many of the Earth's systems are to small changes. Environmental scientists are increasingly concerned with studying chaotic, or nonlinear systems. In the atmosphere, a favourite example they quote is that of the butterfly and the hurricane. Could the gentle flutter of a butterfly's wing in Japan trigger self-propagating changes through the Earth's atmosphere that eventually cause a hurricane in the Carribbean? Such possibilities are hard to deal with, but we should not discount them on that account alone.

— Notes

1. Franklin, B. (1784). The Meteorological imaginations and conceptions. *Mem. Lit. and Phil. Soc. Manchester* **3**, 373–7.
2. Humphreys, W. J. (1913). Volcanic dust and other factors in the production of climatic changes, and their possible relation to ice ages. *Bull. Mt. Weather Obs.* **6**, 1–34.
3. Arctowski, H. (1915). Volcanic dust veils and climatic variations. *Annals N. Y. Acad. Sci.* **26**, 149–74.
4. Wexler, H. (1952). Volcanoes and world climate. *Sci. Am.* **April 1952**, 3–5.
5. Lamb, H. H. (1970). Volcanic dust in the atmosphere; with a chronology and assessment of its

meteorological significance. *Phil. Trans. Roy. Soc. Lond.* **A266**, 425–533.

6. Newhall, C. G. and Self, (1982). The Volcanic Explosivity Index (VEI): an estimate of explosive magnitude for historical volcanism. *J. Geophys. Res.* **87**, 1231–38.

7. Stothers, R. B. and Rampino, M. R. (1983). Volcanic eruptions in the Mediterranean before AD 630 from written and archaeological sources. *J. Geophys. Res.* **88**, 6357–71.

8. Forsyth, P. Y. (1988). In the wake of Etna, 44 BC. *Classical Antiquity* **7**, 49–57.

9. Stothers, R. B. (1984). Mystery cloud of AD 536. *Nature* **307**, 344–5.

10. Moodie, D. W., Catchpole, A. J. W., Abel, K., and Wortley, J. (1991). Northern Athapaskan oral traditions and the White River volcanic eruption. *Ethnohistory* in press.

11. Hammer. C. U., Clausen, H. B., Dansgaard, W., Gundestrup, N., Johnsaen, S. J., and Reeh, N. (1978). Dating of Greenland ice cores by flow models, isotopes, volcanic débris, and continental dust. *J. Glaciol.* **20**, 3–26.

12. Hammer, C. U., Clausen, H. B., and Dansgaard, W. (1980). Greenland ice sheet evidence of post-glacial volcanism and its climatic impact: *Nature* **288**, 230–35.

13. Hammer, C. U. (1981). Past volcanism and climate revealed by Greenland ice cores. *J. Volcanol. Geotherm. Res* **11**, 3–10.

14. Thompson, L. G. and Mosley-Thompson, E., (1981). Temporal variability of microparticle properties in polar ice sheets. *J. Volcanol. Geotherm. Res.* **11**, 11–27.

15. Self, S., Rampino, M. R., and Barbera, J. J. (1981). The possible effects of large 19th and 20th century volcanic eruptions on zonal and hemispheric surface temperatures. *J. Volcanol. Geotherm. Res*, **11**, 41–60.

16. LaMarche, V. C. and Hirschboeck, K. K. (1984). Frost rings in trees as records of major volcanic eruptions. *Nature* **307**, 121–6.

17. Lough, J. M. and Fritts, H. C. (1987). An assessment of the possible effects of volcanic eruptions on north American climate using tree-ring data, AD 1602–1900. *Climatic Change* **10**, 219–39.

18. Rampino, M. R. and Self, S. (1982). Historic eruptions of Tambora (1815) Krakatua (1883) and Agung (1963), their stratospheric aerosols and climatic impact. *Quaternary Res.* **18**, 127–43.

19. Sigurdsson, H. and Carey, S. (1989). Plinian and co-ignimbrite tephra from the 1815 eruption of Tambora volcano. *Bull. Volcanol.* **51**, 243–70.

20. Stothers, R. B. (1984). The Great Tambora eruption in 1815 and its aftermath. *Science* **224**, 1191–98.

21. Stommel, H. and Stommel, H. (1983). *Volcano Weather*. Seven Seas. Newport, Rhode Island.

22. Rampino, M. R. and Self, S. (1984). The atmospheric effects of El Chichon. *Sci. Am.* **250**, 48–57.

23. Sigurdsson, H. (1982). Volcanic pollution and climate: the 1783 Laki eruption. *Eos* **63**, 601–2.

24. Thorarinsson, S. (1979). On the damage caused by volcanic eruptions, with special reference to tephra and gases. In *Volcanic activity and human ecology*, (eds P. D. Sheets and D. K. Grayson), pp. 125–60. Academic Press, London.

25. Turco, R. P., Toon, O. B., Ackerman, T. P., Pollack, J. B., and Sagan, C. (1983). Nuclear winter: global consequences of multiple nuclear explosions. *Science* **222**, 1283–92.

26. Rampino, M. R., Self, S., and Stothers, R. B. (1988). Volcanic winters. *Ann. Rev. Earth Planet. Sci.* **16**, 73–99.

27. Officer, C. B., Hallam, A., Drake, C. L., and Devine, J. D. (1987). Late Cretaceous and paroxysmal Cretaceous/Tertiary extinctions. *Nature* **326**, 143–8.

28. Alvarez, W. and Asaro, F. (1990). An extraterrestrial impact *and* Courtillot, V. E. (1990). A volcanic eruption. *Sci. Am.* **256**, 44–60.

29. Renne, P. R. and Basu, A. R. (1991). Rapid eruption of the Siberian Traps flood basalts at the Permo-Traissic boundary. *Science*, **253**, 176–9.

30. Shaw, H. R. and Moore, J. G. (1988). Magmatic heat and the El Nino cycle. *Eos* **69**, 1553 and 1564.

31. Handler, P. (1989). The effect of volcanic aerosols on global climate. *J. Volcanol. Geotherm. Res.* **37**, 233–249.

32. Broecker, W. S. *et al.* (1988). The chronology of the last glaciation: implications to the cause of the Younger Dryas event. *Palaeoceanography*, **3**, 1–19.

33. Rampino, M. R., Self, S., and Fairbridge, R. W. (1979). Can rapid climatic change cause volcanic eruptions? *Science* **206**, 826–9.

18

Extraterrestrial volcanism

Over the last two decades, spacecraft have revealed evidence of volcanic phenomena throughout the solar system. Finding basalt lavas on our Moon was a tonic for volcanologists, but the stunning discovery of perennial eruptions on Jupiter's satellite Io marked a profound change in perspective for them. Volcanologists are obliged to raise their sights. They must now consider volcanism as a solar system-wide process, involving sources of heat other than radioactive decay, and melting of exotic materials other than familiar silicates. This tremendously increases the scope of their work, presenting intellectual challenges that are as intimidating as they are exciting.[1]

— 18.1 General considerations

There are two basic requirements for volcanism: a source of heat, and something to melt (Section 1.1). Taking the second point first, things are neither as complex nor as weird as they might appear in the realms of science fiction. Only two kinds of material need concern us: silicates in the dense planets and satellites; and ices, in the low-density satellites of the outer planets. Ice is a lot easier to melt than silicates, so extraterrestrial volcanism falls into two broad categories: the familiar, hot silicate variety, and a cold variety, *cryovolcanism*.

Taking the solar system as a whole, there are three significant sources of heat, and one oddball. *Primordial heat*, left over from accretion, still leaks away from the larger planets, but *radiogenic heat*, released by continuing decay of radioactive isotopes, is far more important (Section 1.1). For icy satellites, there is a trade-off between the smaller quantities of heat required to cause melting, and the tiny quantities of radioactive isotopes they contain. Some of the larger icy satellites contain enough radioactive isotopes in small silicate cores for radiogenic heating to have played a role in their early histories, but for most *tidal heating* has been the key factor. *Solar* heating is the oddball: surface temperatures are so low on Triton, Neptune's frigid satellite (about 38 K during daylight), that a bizarre solid-state greenhouse effect may be enough to volatilize frozen nitrogen.

18.1.1 Gravity and the effects of atmospheres

Ascent of a magma to the surface of a planet and its eruption there are controlled by both intrinsic and extrinsic physical factors. Intrinsic factors are those governed by the stuff that melts and the magma that results: its temperature, composition, rheology, and content of dissolved gases. Extrinsic factors are dictated by the planetary environment—gravity and atmospheric pressure especially.

Before they drifted off into fantasy, science fiction writers of the 1950s and 1960s padded many pages by portraying the effects of surface gravity on different planets—astronauts bounding over tall buildings and driving golf balls improbable distances. Gravity provides the buoyant forces that cause a magma to rise through the crust of a planet. Consider a magma which is slightly less dense than the crust surrounding it. The magnitude of the buoyant force acting on the magma depends on how big the body is, how much less dense it is than its surrounding crust, and the surface gravity. On a planet with high gravity, the buoyant force will be large, whereas on a low-gravity planet the force will be small.

While gravity is important, its role should not be overstated. To reach the surface, a buoyant blob of magma has to overcome resisting forces. These resisting forces are not gravity controlled, but are mostly dictated by rheology. In squeezing an ascending magma through a conduit towards the surface, *rheological variables are more important than gravity.*[2] Squeezing toothpaste out of a tube requires the same effort everywhere, whether in the bathroom at home or aboard the Space Shuttle.

Given that most magmas on the silicate planets are basaltic, they have similar rheological properties. Thus, some general conclusions can be drawn on the effects of low surface gravity. Reduced gravity means reduced buoyant forces, which in turn requires that the magma bodies must be proportionally larger to rise to the surface, forming larger magma chambers in the lithosphere. A further consequence is that dykes connecting magma chambers to the surface would be wider, permitting more rapid eruption of lava at the surface.

Because the laws of physics really are universal, it is also possible to make preliminary estimates of the properties of extraterrestrial pyroclastic eruptions by slotting appropriate planetary parameters into the right equations.[3] Under low gravity, vesicle nucleation and gas exsolution will begin at greater depths than on the Earth, encouraging efficient volatile release. Magmas containing even small quantities of dissolved volatiles will be sprayed from volcanoes in fire fountains. Discounting atmospheric effects, under low gravity, ejected clasts will be propelled higher and follow longer ballistic paths than on Earth, like the golf balls beloved of science fiction hacks.

But pyroclastic activity can only take place at all if there are sufficient dissolved volatiles to expand and disrupt an ascending magma. Lionel Wilson and Jim Head estimated that on Mars, where atmospheric pressure is only 7 millibars, a *minimum* volatile content of only c. 0.01 weight per cent is required; for Earth the minimum is c. 0.07 per cent; while for Venus, no less than 3 per cent is required to overcome the crushing atmospheric pressure (90 times the Earth's). On Mars, there is abundant evidence of volatiles, as we shall see, so pyroclastic eruptions should be favoured. On hot, dry Venus, things are more problematic. Bearing in mind that terrestrial basalts typically contain less than 1 per cent volatiles (mostly water), it is difficult to conceive that volatile concentrations of as much as 3 per cent could be achieved in Venus's magmas. Possibly, carbon dioxide also plays a role. However, we know little at present of the extent to which Venus's atmosphere is recycled through its volcanoes.

On both Mars and Venus, the presence of an atmosphere critically affects the character of pyroclastic eruption columns. As shown in Chapters 8 and 10, the maximum height to which a column will rise is proportional to the fourth root of the eruption rate. Furthermore, there is a critical transition between convecting eruption columns, which give rise to plinian fall deposits, and unstable columns from which pyroclastic flows collapse. As atmospheric pressure increases, it becomes more difficult for pyroclastic eruptions to take place at all. Venus's hot, dense atmosphere not only inhibits vesiculation, it also makes it difficult for an eruption column to acquire the buoyancy necessary to rise convectively—the eruption column cannot entrain and heat up enough of the surrounding atmosphere. For the same mass eruption rate, eruption clouds would rise only one-third as high on Venus as on Earth, and column collapse to form pyroclastic flows would be more common. Lower column heights imply that pyroclastic deposits would be less widespread on Venus than the Earth.

On Mars, by contrast, the atmosphere is so

tenuous that eruption columns would climb five times higher than comparable terrestrial cases, so pyroclastic fall deposits would be far more widespread. But again, sustained convection is unlikely because reasonable mass eruption rates would cause overloading and collapse of the eruption column to form pyroclastic flows.

On the numerous small bodies such as the Moon which lack atmospheres, convection is impossible because there is nothing to provide buoyant support to the column, and no gas to be entrained and heated. Erupted gases therefore expand into the vacuum of space, driving ejected clasts along ballistic trajectories.

— 18.2 The Moon

Because astronauts, including a geologist (Harrison 'Jack' Schmitt), have walked on the Moon's surface and sampled its basalts, the Moon provides a frame of reference against which to compare volcanism on more distant, yet unvisited bodies. Two distinct terrain types exist on the lunar surface: light-toned, cratered highlands and the dark, blotchy mare basins which outline the 'Man in the Moon' (Figs 3.5 and 18.1). Geologically speaking, the cratered highlands are made up of brecciated *anorthosites*, plutonic rocks rich in plagioclase feldspar. Overwhelmingly the most important aspect of the lunar highlands is their extreme antiquity. They are among the oldest rocks ever dated, yielding ages between 3.9 and 4.46 thousand million years.[4]

Only meteorites have yielded older ages: 4.6 thousand million years, the age of the solar system itself.

18.2.1 Bulk composition and thermal evolution

According to current consensus, the Moon formed when a Mars-sized body smashed into the early Earth, spraying off and vaporizing part of its mantle. Condensing debris which settled back into orbit around the Earth accreted to form the Moon. The debris was derived largely from the Earth's mantle, and so the Moon's bulk composition is therefore similar, particularly in things like the heat-producing radioactive isotopes of potassium, uranium, and thorium. A critical difference caused by the Moon's high-

Fig. 18.1 A Lunar Orbiter IV view of the Orientale Basin, a multi-ringed impact structure 1400 kilometres in diameter. Orientale is one of the youngest large impact basins on the Moon, and has not been flooded with basalt lavas, so its bulls'-eye structure is explicit. Part of the Oceanus Procellarum is visible in the top right corner. Orientale lies on the extreme western limb of the Moon so its structure is masked when seen from Earth.

temperature birth was that volatile components were driven off, leaving it bone dry. This resulted in rocks which seem startlingly fresh to geologists used to dealing with dismally weathered rocks on Earth. More to the point, it also meant that there were scarcely any volatiles to drive pyroclastic eruptions.

One of the key objectives for astronauts of the later Apollo missions was to measure the Moon's surface heat flow. Successful measurements were obtained at the Apollo 15 and 17 landing sites, but an astronaut unfortunately tripped over a crucial cable at the Apollo 16 site. An average value of about 18 milliwatts per square metre was obtained. (Recall that heat flow through the Earth's oceanic crust is about 80 milliwatts per square metre.) This value is consistent with a bulk lunar uranium concentration of about 46 parts per billion, close to estimates of the uranium concentration in the Earth's crust and mantle.[5]

Although endowed initially with concentrations of heat-producing radioactive isotopes roughly similar to those of the Earth, the Moon is plainly inactive at present—if a volcano were to erupt there, we could scarcely fail to see it. It might even make evening TV news shows. But the Apollo data show unequivocally that even the sparsely cratered lava plains filling the maria are exceedingly ancient by terrestrial standards. All the dated samples are more than 3.2 thousand million years old; older than all but a few patches of rock on Earth. On Earth, the oldest respectable fossils are only 600 million years old, while dinosaurs became extinct a mere 65 million years ago. By that time, lunar volcanism had long since been extinguished.

It is easy to explain this situation in terms of size: the Moon is about one-third the diameter and one-hundredth the mass of the Earth (Table 18.1). This means that far less thermal energy was locked into the Moon when it formed. Athough it contained broadly similar *concentrations* of radioactive isotopes, total *amounts* were a lot less, and because of its much larger surface area relative to its mass, both its primordial and radiogenic heat leaked away to space much more rapidly.

As the Moon cooled, lunar thermal gradients gradually waned, so that the depth of partial melting got deeper and deeper, until about 3 thousand million years ago it was so deep that melts could no longer reach the surface. As the zone of melting deepened, there were progressive changes in the compositions of erupted basalts (Chapter 3). At present, the solidus is at a depth of about 500 km.

Of course, this oversimplifies the situation. There are dated lavas from only a handful of landing sites to confirm the theoretical models. On the basis of crater size–frequency distribution statistics, some lunar scholars have asserted that some small areas of mare basalts may be less than one thousand million years old. More importantly, all the thermal models suggest that there should have been widespread basaltic volcanism *before* eruption of the mare basalts; that is, there should be lots of basalts more than four thousand million years old. But evidence for the existence of such old lavas is fragmentary, literally as well as figuratively. One mission, Apollo 16, was specifically targeted to investigate a supposed area of ancient highland volcanism, but the returned samples proved to be complex breccias, not lavas. Some breccia clasts were derived from earlier basalt lavas, but no extensive suites of early lavas have been found.

18.2.2 Lunar volcanoes

Samuel Johnson once quipped to Boswell that the shortest chapter in any book he knew was one in a tome on the animals of Iceland. According to Johnson, the chapter on snakes read, in its entirety: 'There are no snakes to be met with throughout the whole island'. It is tempting to write here that 'There are no volcanoes to be met with throughout the whole Moon', because although huge areas of the Moon are covered by basalt lavas, there are no volcanoes worthy of the name (Fig. 18.2). The voluminous mare basalts all welled up into the mare basins from nondescript vents and obscure circumferential fissures, filling the basins from below like a bath tub filled through the plug hole.

There are a few features on the maria which *may* be true volcanoes: they are low, flat circular shields or domes with convex shapes and slopes less than about 5°, often with summit craters.

Table 18.1 Basic solar system statistics

	Mercury	Venus	Earth	Moon	Mars	Jupiter	Saturn	Uranus	Neptune	Pluto
Distance from Sun (million kilometres)	57.9	108.2	149.6	( )	227.9	778.3	1427	2870	4497	5900
Period of revolution	88 d	224.7 d	365 d	27.3 d	687 d	11.86 yr	29.46 yr	84 yr	165 yr	248 yr
Axial rotation period	59 d	243 d retro	23 hr 56 min	27.3 d	24 hr 37 min	9 hr 55 min	10 hr 40 min	17.3 hr 14 min retro	18 hr 30 min	6 d 9 hr retro
Axial inclination	0°	3°	23°27′		25°12′	3°5′	26°44′	97°55′	28°48′	?
Inclination of orbit to ecliptic plane	7°	3.4°	0°	5°	1.9°	1.3°	2.5°	0.8°	1.8°	17.2°
Orbital eccentricity	0.206	0.007	0.017	0.06	0.093	0.048	0.056	0.047	0.009	0.254
Equatorial diameter (km)	4880	12 100	12 756	3476	6794	143 200	120 000	51 800	49 500	3000?
Mass (kg)	0.33×10^{24}	4.9×10^{24}	6.0×10^{24}	7.4×10^{22}	6.5×10^{23}	1.9×10^{27}	5.7×10^{26}	8.7×10^{25}	1.0×10^{22}	1.6×10^{22}
Density ($\times 10^3$ kg m^{-3})	5.4	5.2	5.5	3.3	3.9	1.3	0.7	1.2	1.7	1.5
Atmosphere (main components)	Virtually none	CO_2	N_2 O_2	None	CO_2	H_2 He CH_4	H_2 He	He H_2	H_2 He	None detected
Satellites	0	0	1	not applicable	2	16$^+$	17$^+$	5	8$^+$?	1
Rings	0	0	0		0	1	1000?	10	3	0?

d = days

They range in size from 3 to 17 km in diameter, are up to several hundred metres high, and have been compared in size and morphology with small terrestrial lava shields such as Mauna Ulu, but there is nothing comparable, for example, with Mauna Loa or Mauna Kea. There are also a few rather undistinguished lumps and bumps, some with summit craters, usually interpreted as lava domes of slightly more silicic compositions than basalt. The Gruithuisen domes, on the rim of the Mare Imbrium are examples (Fig. 18.3). Most terrestrial volcanologists, however, would not lose any sleep over these rather dubious landforms.

18.2.3 *Lunar lava flows*

Lunar basalts were hot and sloppy. Geochemi-

cally they are primitive: more like terrestrial komatiites than MORBs (Table 3.1). They therefore formed through high degrees of partial melting of the lunar mantle, probably at temperatures well over 1400°C. Such high melt temperatures ensured that the lavas were erupted with very low viscosities, perhaps as low as 1 Pa s, approaching those of familiar fluids such as olive oil. (Terrestrial basalts usually have viscosities of 100 to 1000 Pa s.) This fact explains many of the features of lunar lavas: they flowed huge distances, often measured in hundreds of kilometres, and are thin relative to their lengths. For the most part, lunar lavas are so thin that they are hard to find, but a few exquisite photographs taken at low sun angles by the Apollo astronauts reveal the lobate edges of some

Fig. 18.2 Although basalt lavas of the Mare Procellarum cover most of this photograph, there are few obvious volcanic features. A small area of highland terrain is present at bottom left; fresh young crater at top is Wichmann. Basalt lavas have largely drowned the incomplete crater (70 km in diameter) at right which protrudes 300–400 m above mare surface. Wrinkle ridges cross diagonally on left side; a sinuous rille crosses ridges at lower right centre. Apollo Metric 16-2991.

Fig. 18.3 Gruithuisen domes (20 km in diameter) rise 1.7 km above the surface of the Mare Imbrium. A crater (3 km diameter) is prominent on the summit of the left dome. Morphologically, it resembles an impact crater and seems unrelated to the volcanic construct. Although possibly of extrusive volcanic origin, the domes may also be blocks of ancient lunar crust, uplifted during the impact that formed the Imbrium Basin. Photo: Lunar Orbiter.

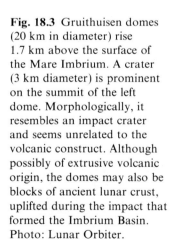

flows spreading over the expansive mare surfaces (Figs. 18.4–18.5). Their flow fronts appear to be about 20–30 metres high. If this figure seems large for low-viscosity flows, recall that only the highest flow fronts are likely to be still detectable after billions of years of meteorite bombardment. Furthermore, because the Moon's surface gravity is only one-sixth that of the Earth's, a flow on the Moon would be much thicker than a comparable flow on Earth (Section 7.2.1).

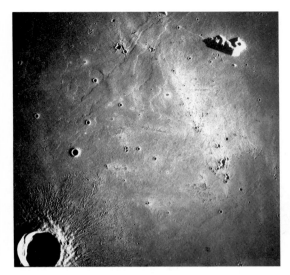

Fig. 18.4 Low sun illumination reveals thin lava flows snaking 150 km over the surface of the Mare Imbrium. Although clearly young, source of these flows is obscure. Crater at bottom left is Euler, 25 km in diameter. Mountain peak is La Hire. AS 15 1701.

Apart from the occasional flow front, two other kinds of volcanic phenomena are seen on the Moon: *wrinkle ridges* and *sinuous rilles*. Wrinkle ridges are low ridges many kilometres long but only a few tens of metres high, that occur all over the maria, often in irregular concentric patterns (Fig. 18.6). Although common, their origin is not clear. They probably formed as a result of compressive forces acting during the cooling, contraction, and settling of a mare lava crust. More than one generation of ridge can sometimes be discerned, the older ones deflecting the passage of younger lavas. This helps to explain the nature of the mare basins: the lava piles they contain may be 6–8 km thick at the centre, but only 1–2 km thick around their margins. Given that each flow was only a few metres thick, flooding of the mare basins probably continued over a long period, with plenty of time for crustal downsagging, cooling, and contraction.

Sinuous rilles are enigmatic structures on the Moon, Mars, and Venus. They are meandering

Fig. 18.5 Margin of a lava flow on surface of Mare Imbrium. Each strip in this Lunar Orbiter V frame is 5 km wide. Visible part of flow margin extends at least 100 km. Most prominent part of flow front (centre) rises only about 20 m above the mare surface.

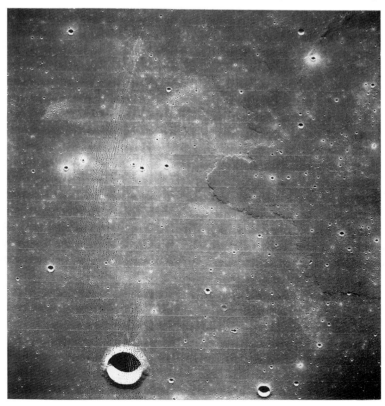

Fig. 18.6 Enlargement of lower left part of Fig. 18.2, emphasizing wrinkle ridges and sinuous rilles. Two rilles start in ill-defined crater-like depressions and cut wrinkle ridges. Longest rille shows little decrease in width along its length. Less-prominent (perhaps older) rille heads off at right angles from same source region. Photo is about 80 km across.

channels a few metres to several kilometres in width which wind over the surface for distances of as much as several hundred kilometres. Some rilles originate in crater-like hollows, others in elongate depressions; some are continuous, while others break up into small segments or crater chains. Many taper in width downstream (Fig. 18.7). Some large rilles have narrower rilles nested within them. Lunar sinuous rilles are so easily seen through Earth-based telescopes that their origin provoked debate from the earliest days. It was natural for early telescopic observers to think that the meandering rilles, which resemble gently winding river valleys, had indeed been shaped by flowing water.

When it was realized that the Moon is and always has been bone dry, this hypothesis evaporated. Various volcanic suggestions were proposed instead, many drawing parallels between rilles and the collapsed lava tubes found on terrestrial volcanoes. Rilles presented so many intriguing problems that Hadley Rille was

selected as the landing site for the Apollo 15 mission.

Hadley Rille is located at the foot of the Appenine Mountains near the margin of the Mare Imbrium. It is about 135 km long, one kilometre wide and 370 metres deep. Traces of what appeared to be rock outcrops in the upper walls of the rille were identified on orbital photographs, prompting hopes that these would provide new perspectives on lunar geology—earlier missions had sampled only shattered, out-of-place samples from the surface of the maria. When they reached it, the Apollo 15 astronauts were able to confirm that parallel, stratiform layers were exposed in the walls of the rille, which were probably counterparts of the lavas underlying Mare Imbrium, but they were not able to sample the outcrops directly.

Although the mission returned some remarkable photographs from the banks of Hadley Rille, together with a wealth of petrological data, it did not decisively resolve the origin of sinuous rilles.

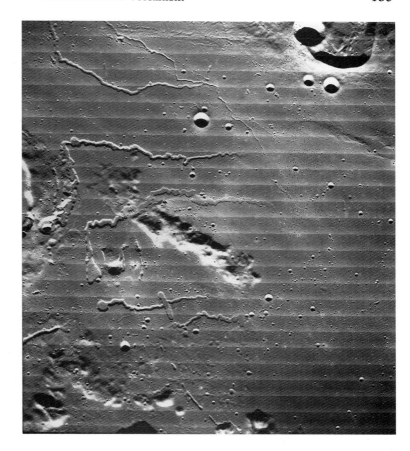

Fig. 18.7 Many puzzling features of lunar sinuous rilles are displayed by this group, forming part of the Rima Prinz system. Rilles taper in width and are less deeply incised down channel. Meandering rille at top centre originates in 2-km-diameter crater, and a narrower rille is incised in its floor. Same rille shows a remarkable incised meander, with prominent spur. Photo is about 114 km across. Lunar Orbiter V.

While some rilles may indeed be collapsed lava tubes, most may be the results of thermal erosion.[6]

Thermal erosion Freshly erupted basalt or komatiitic lavas are hot (*c.* 1200–1400°C) and fast flowing; they therefore transport enormous amounts of thermal energy. On a lunar mare, the most probable surface that a new lava will flow over is an older lava. Initially, the new lava chills against the old, but if the eruption continues long enough, and the flow rate is sufficiently large, the temperature in the underlying rock will be raised to melting point, allowing the overlying lava to cut its way downwards by *thermal erosion.*

Owing to the low thermal conductivity of lavas, and the high flow rates required, terrestrial basalt lavas are rarely capable of thermal erosion. It is best developed if the melting point of the substrate is much less than that of the flow, a condition which did prevail when some ancient komatiite flows were erupted on Earth. Mapping of komatiite flows in Australia and Canada showed that they eroded channels deep into the underlying rocks. Cross-sections of the lava channels are similar to those of lava tunnels, having a pronounced overhang on both sides.[7]

Thermal erosion phenomena in ancient komatiites can only be laboriously deciphered in underground workings—they are minor curiosities. By contrast, lunar sinuous rilles are huge. Truly prodigious effusion rates would be required to cut such large channels. Hulme and Fielder[8] estimated rates as high as 10^8–10^9 kilograms per second, or about one-third of a million cubic metres per second. (Terrestrial basaltic eruption rates rarely exceed 3×10^6 kilogrammes per second, or 1000 cubic metres per second.) Thermal erosion is most efficient if the lava is flowing turbulently, which enormously increases the rate at which it can transfer heat into the substrate. This is logical enough, but observations from an unexpected quarter suggest that turbulent flow may not be *essential* for thermal erosion to take place. Small *carbonatitic* lava flows on the Oldoinyo Lengai volcano on Tanzania were observed to be thermally eroding their substrates. In this well-documented case, flow was laminar.[9]

18.2.4 *Lunar pyroclastic deposits*

Because the Moon lacks so comprehensively the volatiles needed to drive explosive eruptions, pyroclastic deposits are bound to be scarce. However, if even a trace of volatiles were present, given a large enough body of magma and enough time, it might have been possible for sufficient volatiles to accumulate in the upper parts of magma chambers to power explosive eruptions. According to theoretical studies by Lionel Wilson and Jim Head, carbon monoxide, CO, may have been the most likely volatile species.[2] Owing to its different molecular weight, this behaves somewhat differently from the more familiar H_2O. Accumulations of *c.* 0.07 weight per cent carbon monoxide would have been necessary for lunar pyroclastic eruptions to occur. Because such eruptions took place in a vacuum, the magma was thoroughly fragmented, spraying out into space as a shower of tiny pyroclasts with high ejection velocities.

For lunar eruptions approximating hawaiian fire-fountaining, Wilson and Head suggested that velocities were of the order of 40 metres per second, propelling material up to one kilometre radially around the vent. For strombolian-style explosive bursts, velocities were as much as 500 metres per second. Extraordinarily widespread but thin deposits up to 200 km in diameter of submillimetric pyroclasts would have resulted from the largest such eruptions. Lunar scoria cones could *not* be formed, because the ejecta would be far too widely dispersed for that. At high mass eruption rates, optically dense eruption columns would have generated clastogenic lava flows around the vent. As the torrents of incandescent lava drained away, they may have thermally eroded their bases.

So much for theory. In practice, there are features on the Moon which correspond well with Wilson and Head's modelling. *Dark halo craters* occur in many localities, notably on the floor of Alphonsus crater, 118 km in diameter. Dark halo craters are small, about 2 km in diameter; are often found located along well-defined cracks or graben; and are surrounded by blankets of dark ejecta extending out 3–4 km from the vent. They are excellent candidates for small-scale lunar pyroclastic deposits (Fig. 18.8*a*).

Fig. 18.8*a* The small dark patches near the rim of Alphonsus crater (centre) are dark halo craters, thought to be small lunar pyroclastic vents. There is not much to distinguish the craters themselves from ordinary impact craters, except that they are slightly elongated along prominent fissures, and are surrounded by mantles of conspicuous ejecta, which may resemble that in Fig. 18.8*b*. Alphonsus is 118 km in diameter. NASA Lunar Orbiter photograph.

Dark mantle deposits are a second possible variety of lunar pyroclastic deposits. They cover huge areas—over 40 000 square kilometres in the Sinus Aestuum area, for example. Although ill defined, these may be tephra deposits related to the eruption of the lavas that filled the mare basins. Samples collected on the lunar surface support the notion that lunar fire-fountains sprayed lava high into space: minute, perfectly spherical glass beads of basaltic composition have been found at several sites. Apollo 15 'green glass' beads have become particularly valuable in lunar petrological studies (Fig. 18.8*b*).

A footnote for space buffs: because lunar pyroclastic deposits should be fine grained, lacking the bouldery meteorite impact debris which litters most of the Moon's surface, dark halo deposits have been identified as optimal sites for lunar colonization. Initially, astronauts would scrape up the deposits with bulldozers and then bury their inhabitable domes with the material, providing a cheap and effective means of thermal insulation and protection from radiation. Later, titanium-rich pyroclastic deposits might form a source of useful raw materials to build equipment and vehicles on the Moon.

Fig. 18.8*b* Tiny spherules of green glass sprayed into space by hawaiian-style fire fountains on the Moon, collected at the Apollo 15 landing site. Field of view is about 3 mm across; most spheres are less than 200 microns in diameter. Resembling tiny Pele's tears, these glass droplets have rather primitive basaltic composition. Photo: courtesy Graham Ryder, LPI, Houston.

— 18.3 Mercury

Mercury is a small planet, not much bigger than the Moon. It orbits so close to the Sun (only 58 million kilometres) that it is usually lost against the solar glare. Telescopic observations were frustrating and unproductive, because apart from it being hard to observe the planet visually, there was always a danger that excessive solar radiation would fry delicate instruments. Remarkably, even its rotation period remained unknown until the 1960s, when Earth-based radar studies showed it to be 58.6 days, thereby revealing an elegant example of celestial mechanics. Mercury's rotation period is exactly two-thirds of its orbital period (88 days)—the planet rotates three times around its axis while completing two orbits around the Sun. This sophisticated celestial dance is a more elaborate example of the gravitational phenomenon that keeps the same face of the Moon always turned towards the Earth. Like the Moon, Mercury is not perfectly spherical, but has a slight bulge on one side. Mercury's orbit is also elliptical, and tidal forces acting between Mercury and the huge mass of the Sun keep the planet facing the Sun while it is closest to it, but when further away, it makes a complete rotation. This may seem an abstruse piece of celestial mechanics, unlikely to titillate volcanologists. It is relevant, however, because dissipation of energy from the tidal interactions involved may have been an important source of heat in Mercury's interior.

18.3.1 Composition and thermal evolution

At a casual glance, a photograph of Mercury could easily be confused with one of the Moon (Fig. 18.9). Scorched by its proximity to the Sun, which bathes it in an intense flux of short-wavelength radiation, Mercury is an uninviting world, whose crater-scarred surface is even more hostile to life than the Moon. Like the Moon, Mercury began life with a giant impact, but in Mercury's case the planet formed from what was left behind after the impact, rather than the material blasted off. This accounts for Mercury's anomalously high density: originally, the planet probably had an ordinary structure of metallic core and silicate mantle, but the impact stripped off much of the mantle, leaving behind a planet with an unusually high ratio of metal to silicate. Some 60–70 per cent of Mercury's mass is probably metal—the planet's core is about the same size as the Moon.

Geochemists think that Mercury is enriched in refractory elements, notably calcium, aluminium, uranium, and thorium, and that its mantle has a higher ratio of magnesium to iron than the Earth. In practical terms, this means that the interior of Mercury would have had to be hotter in order for melting to take place, but the planet was also well endowed with heat-producing isotopes. Given all the other possible sources of heat (accretion, core formation, short-lived isotopes and of course tidal dissipation), it seems likely that Mercury's interior was extremely hot early in its evolution, and that there was extensive melting of its mantle. The planet is so small, however, that it cooled quickly, and the zone of mantle melting sank deep into the interior.

18.3.2 Volcanic surface features on Mercury

Mercury orbits so close to the Sun, that it is difficult for a space craft to escape from the Sun's deep gravitational well to enter orbit around Mercury. Most of our knowledge of Mercury came in 1974 and 1975 when the Mariner 10 space craft followed a complex, looping trajectory, finely calculated to enable it to make three brief encounters with the planet. So elegant was the trajectory that the mission was termed an 'exquisite celestial slingshot'. Notwithstanding the subtlety of its trajectory, the spacecraft was still able to obtain coverage of less than half the planet's surface. Our knowledge of Mercury therefore remains in its infancy. (But almost everything that is known is summarized in an excellent 47-author book[10].)

Although Mercury superficially resembles the Moon, there are no Moon-like, lava-filled mare basins. Most of its surface resembles the entirely non-volcanic lunar highlands. It is hard, in fact, to find any *obvious* evidence of volcanism on Mercury. There is indirect evidence, however. Crater size–frequency statistics reveal a dearth of

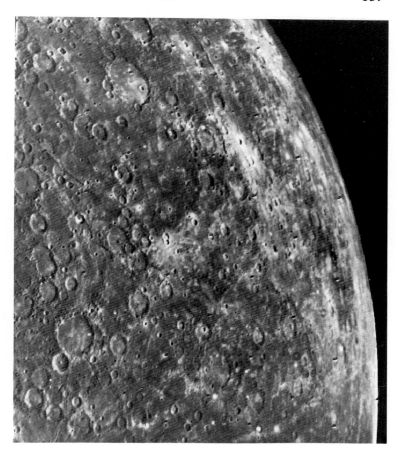

Fig. 18.9 Mercury's cratered surface resembles the lunar highlands at first glance, but mare basins are absent, and Mercury's large craters are widely spaced, not crowded together as on the Moon. Smooth intercrater plains may have been formed during an early episode of volcanic resurfacing (Mariner X, 14465).

craters less than 50 km in diameter on Mercury. Large craters are separated by expanses of featureless *inter-crater plains*.[11] Using the lunar cratering chronology as a bench-mark, the crater statistics suggest that these plains are probably 4–4.2 thousand million years old. Volcanic resurfacing probably smoothed out many small craters to produce the plains. There are no lavas to be seen, or even significant variations in brightness. But a volcanic explanation seems to be most plausible, and fits well with models for Mercury's thermal evolution.

About 3.85 thousand million years ago, a massive impact formed the *Caloris Basin*, a multi-ring structure so immense that it dominates much of the geology of Mercury (Fig. 18.10). Its outermost ring may be as much as 3700 km in diameter. Caloris resembles the huge impact basins on the Moon, but unlike them it was not subsequently flooded by basalt lavas.

Sufficiently large quantities of lava were erupted to form some extensive *smooth plains*, which resemble the lunar maria surfaces, but no flow features can be resolved. These lavas were erupted about 3.8 thousand million years ago.

After that, Mercury appears to have been volcanically extinct. On the Moon, volcanic activity lingered on until about 3.2 thousand million years ago. It is not clear why Mercury became inactive so early. There is evidence in the form of an extensive systems of scarps that the planet actually contracted significantly as it cooled, its crust crinkling like the skin of a dried apple. This contraction was undoubtedly accompanied by lithospheric thickening, which may have permanently sealed Mercurian magmas below the surface. In the case of our Moon, tides aroused in its interior by the Earth may also have helped to keep things warm. Mercury lacks such an intimate companion.

Fig. 18.10 Part of the largest impact structure on Mercury, the 3000-km-diameter Caloris Basin, comparable with the Moon's Orientale Basin (Fig. 18.1). Radial ejecta patterns are prominent at top. Smooth, sparsely cratered areas may have been resurfaced by basalt lavas.

— 18.4 Venus

Mercury is hard to see. Venus, by contrast, often blazes so brightly in the evening (or morning) sky that it is hard to avoid. Reports of Unidentified Flying Objects turn out with tedious frequency to be Venus, despite the fact that planet glitters in the same part of the sky every evening for weeks. Observed through a telescope, Venus is disappointingly blank, its cloud-laden atmosphere concealing the topography beneath. Venus owes its brilliance to its unbroken cloud cover, and to its proximity—its orbit brings it within 40 million kilometres of the Earth. In terms of size, mass, and density, Venus is so similar to the Earth that early astronomers thought of the two as twins, sharing similar properties.

For a frustratingly long time, Venus's cloud cover veiled its surface from prurient scientists' prying eyes. In the three centuries since telescopes were first turned towards Venus, no progress at all was made in determining so simple a parameter as its axial rotation period. All sorts of guesses were made, from 24 hours to 225 days (the orbital period). Venus's modesty remained unassailed until radar astronomers first caressed its surface with radio beams in 1962. They found that Venus lumbers slowly round on its axis once every 243 days, in the *opposite* sense to every other planet in the solar system. The 243-day period revealed another elegant but enigmatic solar system statistic: the Earth's orbital period and Venus's axial rotation period are in the ratio exactly 3:2.

In the 1970s and 1980s, a positive shower of Russian and American space craft descended on Venus. In December 1978, no less than seven probes reached the planet within a few days of each other. These space craft had many different objectives: profiling through the atmosphere;

mapping the surface from orbit using radar altimeters; and direct analysis of the surface by landing craft. In August 1990, the Magellan space craft went into orbit around Venus to commence high resolution radar mapping of its surface. As a result, our knowledge of Venus has increased enormously, and continues to do so.

Naming names was one of the first problems to crop up. How on earth (!) does one find names for a complete planet, starting from scratch? Inevitably, a committee was set up to tackle the task. Sowing the seeds for some future grief, a committee of the International Astronomical Union decided to follow romantic precedent by naming *all* features on Venus after women, real or mythical, from many cultures. So, Venus's topography is replete with polysyllabic feminine names; some familiar (Aphrodite); others hopelessly obscure (Sacajawea—Sacajawea was a native American who assisted Lewis and Clark in their exploration of the Oregon territory). They made one deliberate exception to this: they named Venus's highest mountain after a man, James Clerk Maxwell, whose work on the physics of electricity and magnetism made radio science possible.

Orbital radar data show that Venus has far less topographical relief than the Earth. Eighty per cent of surface is covered by endless rolling plains, with elevations within 1 kilometre of the modal value. There are a few higher regions, of which two (Ishtar Terra and Aphrodite Terra) have been compared with terrestrial continents, although they are much smaller. Significantly, many large, crisp impact craters scar the surface. Their size–frequency distribution suggests that the average age of Venus's surface may be as much as one thousand million years, significantly older than the Earth's but much younger than the Moon's.

Images from Venera 9, the first space craft to survive landing on the planet, revealed a grim prospect: a flat, rock-strewn terrain stretching unbroken from horizon to horizon. 10 000 km away, the view from Venera 10 was similar, although the rocks there were smaller and slabbier. Gamma-ray and X-ray fluorescence analyses of surface rocks carried out on board five surface landers confirmed what had been

previously suspected: the rocks were basalts, broadly similar to terrestrial tholeiitic and alkali basalts (Chapter 3). No surprises there. Unquestionably the most important measurements from the landings concerned surface conditions. Temperatures are searingly hot (around 500°C), and the atmospheric pressure crushing: 90 times greater than the Earth's.

Venus's atmosphere itself is no less unappealing, consisting of 95 per cent carbon dioxide and a few per cent of nitrogen with small traces of sulphur dioxide and water. Three distinct cloud decks are present, separated by clear layers. The cloud base of the lowest and thickest clouds is at about 50 km altitude. Although these clouds are about as dense as Earth's beneficent clouds, they are probably composed of droplets of sulphuric acid, not water. Thus, as one distinguished American scientist wrote: 'Venus is astonishingly hot, with an oppressively dense atmosphere containing corrosive gases, with a surface glowing dimly by its own red heat . . . Venus . . . seems very much like the classical view of Hell'.

18.4.1 *Composition and thermal evolution*

Venus's diameter is only 605 km less than the Earth's and its density is only 2 per cent less. So similar are the two planets in basic properties that the huge environmental differences between them are all the more thought provoking. Venus probably started with the same bulk chemical composition as the Earth, so its thermal evolution should have been similar. Specifically, the surface landers showed that the potassium:uranium ratio in surface rocks is much like the Earth's. It is a reasonable presumption, therefore, that Venus has a hot, vigorously convecting silicate mantle, much like the Earth's. The extraordinary divergence in the geological history of the two planets must be bound up with their different atmospheric evolutions. Determining how the atmospheres of Earth and Venus evolved is a major goal for atmospheric scientists. For example, both planets contain similar amounts of potassium, but Venus's atmosphere contains far less radiogenic argon (formed by radioactive decay of a potassium isotope) than the Earth's. This implies that their evolutions diverged at an early date; early enough to be manifested in different abundances of slowly accumulating radiogenic argon.

On a broader scale, determination of how their atmospheric evolutions were interlinked with their lithospheres presents some of the most challenging questions in science today, because it bears crucially on issues of global climatic change. Ultimately, the differences between the two planets may be traced back to fact that Venus is closer to the Sun than Earth, and therefore receives more solar radiation. Venus's equilibrium black-body temperature (a measure of solar heating which ignores complexities of atmospheres *et cetera*) is 49°C higher than Earth's.[12] Water could not form stable, wet oceans on Venus's primordial surface, but was lost to the atmosphere, where it was ultimately broken down by photodissociation. Hydrogen was lost to space, while oxygen was fixed in surface rocks. Similarly, whereas most of the Earth's carbon dioxide resides in carbonate rocks such as limestones, on Venus it remains in the atmosphere.

18.4.2 Heat-loss mechanisms on Venus: plate tectonics?

Carbon dioxide is an effective greenhouse gas. Consequently, surface temperatures on Venus are maintained at oppressive levels. This has some subtle but important implications for the internal processes on the planet. There are three ways in which internal heat can reach the surface of any planet (Section 2.2):

1. By plate recycling, or plate tectonics, as we know it on Earth.

2. Via 'heat pipes' or hot-spots; sites where mantle plumes pump magma to the surface through volcanoes.

3. By conduction through the lithosphere. This is the most general case, found on bodies such as the Moon, Mercury, and the planetary satellites.[13]

Plate recycling   Our understanding of Cytherean geophysics is poor at present. It will undoubtedly improve as a result of the Magellan mission, but what we really need are seismic and heat flow data, not obtainable from orbit. One important insight into Venus's geophysics comes from its gravitational field. In the simplest case, a high mountain rising above the surface of a planet represents an excess of mass; thus an overflying satellite should sense a slight acceleration, and register the existence of a positive gravity anomaly. On Earth, this does *not* happen, by and large. High mountain ranges such as the Himalayas are made of relatively less dense rocks than their surroundings, and the parts that stick up into the air are *compensated* by 'roots' which protrude into the mantle. Thus, mass differences are smoothed out, so there is no strong correlation between gravity and topography.

Existing orbital data show that gravity and topography *are* strongly correlated on Venus, but that the observed gravity anomalies are four to five times smaller than would be obtained from wholly uncompensated topography. This indicates that the depth to the level of compensation is rather large (more than 100 km) and that because the topography is not fully compensated, there must be a dynamic component to support it—probably mantle convection. These simple but important observations show that Venus's geophysical constitution is different from the Earth's, and thus that we should anticipate different styles of volcanism.

In a pre-Magellan study, Jim Head and his Brown University colleagues argued that plate tectonics of a sort works on Venus: they identified spreading ridges and convergence zones on Venera radar images.[14–15] In particular they argued that Aphrodite Terra in Venus's equatorial region is a symmetrical spreading ridge system 7500 km long, spreading at a rate of 1.5–3 cm per year. This identification was not borne out by the Magellan data. By contrast, Kaula and Phillips used comparisons with the topography of terrestrial spreading ridges to argue that only 15 per cent of the probable heat flow from Venus could be expressed in sea-floor spreading, compared with 70 per cent for the Earth.[16]

So plate recycling does not appear to be important on Venus. This has an important corollary. Apart from transferring heat to the surface, plate recycling on Earth also recycles volatiles. Along with the oceanic lithosphere, large amounts of water are subducted into the mantle, with huge petrological implications, such as permitting granitic melts to form. On 'dry' Venus, this cannot happen. Kaula suggests that magmas formed at depths greater than about 250 km on Venus would be *more dense* than their surroundings, and thus would sink, carrying volatile components and heat-producing elements with them. This would 'dry out' the upper mantle, depleting it of volatiles.[12]

Heat pipes   Heat pipes are an obvious method for Venus to transfer its internal heat to the surface. What could be simpler than to have a few large volcanoes pumping huge quantities of basalt lavas on to the surface? 'Hot-spot' volcanoes exist on Earth, and radar images reveal similar volcanoes on Venus (next section). But there are several difficulties with this hypothesis. If Venus's heat flow is accommodated by pumping magma, then the volcanic flux ought to be of the order of 200 cubic kilometres of basalt per year. This is a lot of basalt, and would cause rapid resurfacing of the planet. Against this, the presence of many impact craters on the surface suggests that large parts of it are geologically old, and that the rate of crust formation may be only one-tenth as fast as on Earth. Furthermore, eruption of such large amounts of lava ought to be accompanied by prodigious volumes of sulphur dioxide, but there is no evidence for such large quantities in Venus's atmosphere. In fact, the limited available evidence suggests that the rate of volcanism on Venus is quite small, about two cubic kilometres per year.

Thermal conduction   If heat pipes were the dominant form of heat transport of Venus, the lithosphere ought to be extremely thick (more than 100 km), and rather rigid. In fact, this does not seem to be the case. Studies of Venus tectonics suggest that the lithosphere is only 60–80 kilometres thick on average, and may be a lot thinner locally. (Recall that the Earth's lithosphere is 120 km thick beneath the oceans.) This thin lithosphere permits loss of substantial amounts of heat by conduction. Average heat flows may be about 50–70 milliwatts per square metre, compared with *c.* 80 for the Earth, while the temperature at the base of the lithosphere may be about 100 degrees hotter than on Earth. Although Venus and Earth are similar in size, Venus's slightly smaller radius means that its mass is barely four-fifths of the Earth's, and its supplies of heat correspondingly less. Thus, although everything about Venus suggests it is a hot, active planet, it probably loses most of its internal heat in the most boring manner possible, by conduction through the lithosphere.

18.4.3 *Volcanic surface features on Venus*

At the time of writing, the Magellan space craft was orbiting Venus, and had obtained high-resolution radar images of about fifteen per cent of the surface. Wonderful though these images are, it would be unwise to say too much about Venus's volcanoes before seeing the remaining eighty-five per cent. However, work by Head and Wilson,[17] Venera images and preliminary observations by the Magellan mission team provide some appetizing hors-d'œuvres.[18]

Perhaps the most stunning observation is simply the sheer number of volcanic edifices that have been identified—twenty-two *thousand* were mapped on the 25 per cent of Venus surface imaged by Venera 15 and 16. Most of these are rather small circular domes or shields, 100–200 metres in height and 2–8 kilometres in diameter. They may be analogues of terrestrial seamounts, liberally scattered around the ocean floors.

Four types of small volcanic edifice were identified on the preliminary Magellan data. Of these, one type was by far the most abundant: shields a few kilometres in diameter with small central craters. These may resemble lava shields such as Mauna Ulu in Hawaii, formed by numerous eruptions of rather small fluid lava flows welling up through a central vent. In the area named Guinevere Planitia, a cluster of 55 such shields was observed, all apparently of broadly similar age.

Larger volcanoes are also common—800 edifices between 20 and 100 kilometres in diameter were mapped in the area covered by Venera 15 and 16 images, and 50 edifices between 100 and 350 kilometres diameter on the rather larger area for which both Venera and Earth-based radar images are available. Two of the best-known

Venusian volcanic constructs, Theia and Rhea Mons, form part of a larger elevated construct prosaically known as Beta Regio, which may be a major rift-like region of extensional tectonics. This pair of volcanoes rise about 5 km above their surroundings and are surrounded by radial lava flows; Theia Mons also has a summit caldera.

Lava domes   Of the hundreds of superb images obtained by Magellan, one has a magnetic attraction for volcanologists—it shows what appear to be a crisp, fresh set of extrusive lava domes, similar to terrestrial *tortas* (Section 7.5.1). Seven steep-sided lava domes are clustered together east of the highland named Alpha Regio (Fig. 18.11). They are all about 25 km in diameter, strikingly circular in outline, 100 to 600 m in height and have volumes of 50–250 cubic kilometres. They have flat tops, with clear patterns of radial and concentric fractures. Two have summit pits. In every way but one, these domes closely resemble terrestrial lava domes formed when viscous dacitic or rhyolitic magmas

are extruded on a level substrate, like the Mono domes, associated with the Long Valley caldera in California. The exception is scale: the Alpha Regio domes are *far* bigger than terrestrial domes. In Chile, typical large dacite domes have volumes of about 5 cubic kilometres. Magellan scientists suggest that the Venus domes may have been able to grow so large because the high surface temperature on the planet permitted the lava to cool more slowly and thus accumulate more easily.

It would be invigorating if one could demonstrate the existence on Venus of lavas other than the interminable basalts that dominate the solar system. Morphologically, the Alpha Regio domes provide evidence for this, although one might also make an argument that they are very shallow-level basaltic intrusions. Some circumstantial evidence supporting a silicic composition comes from Venera 8 lander analyses. Venera 8 landed a long way from the Alpha Regio domes, but it did land near a similar pancake-like dome. Its instruments indicated a potassium content of four per cent in the surface rocks, uranium 2.2

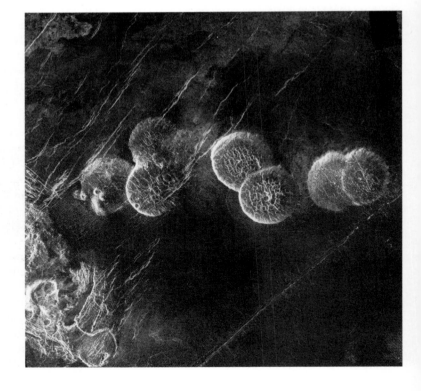

Fig. 18.11 Seven remarkably circular dome-like massifs averaging 25 km in diameter and 750 metres high dominate this radar image of a part of Venus called Alpha Regio. They resemble flat-topped extrusive lava domes found in areas of silicic volcanism on Earth (Chapter 7), but are several times bigger. Alternatively, they may be sites of shallow magma intrusions which have bulged up overlying surface materials. Small craters on summits of central dome pair are enigmatic. Magellan P 37125.

parts per million and thorium 6.5 parts per million. These values are higher than would be found in basalt, and are similar to latites and trachyte lava on Earth (Section 1.4).

Sif Mons—a typical Venusian volcano   Sif Mons is a large Venusian volcano. It is 300 km in diameter, and rises 1.7 km above an elevated area known as Eistla Regio (Fig. 18.12). Magellan images of this vast volcano are dominated by radar-bright lava flows. One which streamed down the northern flank is 250–300 km in length and 15–30 km wide. Structurally, Sif Mons is analogous to a terrestrial shield volcano constructed above a rift zone. A NW–SE trend is well developed on the flanks of Sif Mons, and chains of 30–10 km diameter craters and elongate depressions follow this trend. At the summit of the edifice is a large caldera 40–50 km in diameter, apparently filled almost to the brim with smooth-textured materials, perhaps lava flows. According to preliminary assessments by the Magellan scientists, the large caldera was formed by lateral draining of magma along the rift, or by eruption of large volumes of lava on the flanks. They attribute the smaller (but still Kilauea-sized!) craters along the NW–SE rift to collapse consequent on the eruption of smaller, more local flows.

In a planetary context, Sif Mons, sitting atop the already elevated area of Eistla Regio, may be the surface expression of a mantle thermal plume which uplifted and fractured western Eistla radially, and locally along a NW–SE trend. Voluminous effusions of lava from the plume head constructed Sif Mons, following the sort of hot-spot pattern of volcanism which on Earth led to construction of the Galapagos and Cape Verde Island massifs. Lamentably, it is difficult to say much about the age of Sif Mons and its long lavas—there is a single impact crater on the apron of volcanic material. One impact crater does not make a chronology.

Paterae   Sacajawea Patera is not an elevated massif, but a vast caldera depression, 200 by 300 kilometres in size and 1 to 2 km deep. Magellan images show that a ring of volcanic plains 120 km by 215 km occurs within the depression, surrounding some mottled, radar-bright deposits. Few radial flow features are present. Unlike an ordinary caldera, the depression is not bounded by a simple set of scarps or cliffs, but by a wide belt of concentric faults, many of which define graben 4 to 100 km in length, 100 km wide and several kilometres apart.

Current thinking suggests that Sacajawea, and similar *paterae*, are huge down-sag calderas, formed above large crustal magma chambers in the upper crust. Magma drained laterally away from the magma chamber—it is not clear where it went—leaving its roof unsupported, so it sagged downwards. Spacing of the graben within the bounding annulus suggests that the sagging crust may have been a few kilometres thick. In the later stages of sagging, lava was erupted from vents in

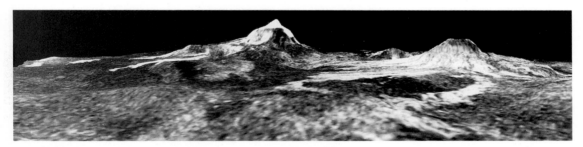

Fig. 18.12 Sif Mons, a 2200 m high Venusian shield volcano (right) and Gula Mons, 4200 m high (centre) seen in a three-dimensional radar visualization prepared by Bruce Campbell, University of Hawaii. Image was created by combining a 1988 radar image from Arecibo Observatory, Puerto Rico (1–2 km spatial resolution) with a digital elevation model obtained from Pioneer Venus data. Radar-bright lava flows extend radially down the flanks of both volcanoes. Vertical exaggeration of 70 times masks gentle slopes of these broad shields—the two volcanoes are 600 km apart. Photo: courtesy of Bruce Campbell.

the annulus of graben. Some flooded the floor of the depression, perhaps filling it hundreds of metres deep, before partly draining away again. This interpretation of Sacajawea's origin may well change, but one thing is clear: nothing remotely resembling a down-sagging *patera* caldera is present on Earth.

Coronae   Coronae are even larger than calderas such as Sacajawea, reaching up to 1000 km in diameter. They are complex structures, with no simple earthly analogues. Quetzalpetlatl corona, for example, rises 2 km above the surrounding plains, is surrounded by an annulus of concentric ridges, and is filled with lava flows, most of which seem to have been erupted from a caldera-like depression in the south-east part of the corona. This depression is itself about 150 by 200 km and 400 metres deep, and contains many smaller volcanic edifices, similar to the shields seen on the plains. A few of them have steeper slopes, suggesting more viscous, silicic magmas.

Coronae are so large, complex, and contain such a diverse range of volcanic phenomena that it is hard to characterize them straightforwardly. Later, perhaps, when the Magellan data have been more fully digested, a more coherent picture will emerge, but, at present, it seems probable that they are the surface manifestations of large mantle plumes. Their diversity of volcanic structures probably reflects variations in temperature and composition of the erupted products: hotter, more primitive basalts and komatiites may be present at the centre, while tholeiites may be erupted from the cooler margins of the plume.

Lava flows   Lava flows are common on the Magellan images. They show almost all the morphological features that one might expect such as central channels, levées, and evidence of lava tubes (Fig. 18.13). While it is not possible to make fine-scale interpretations of the lavas from

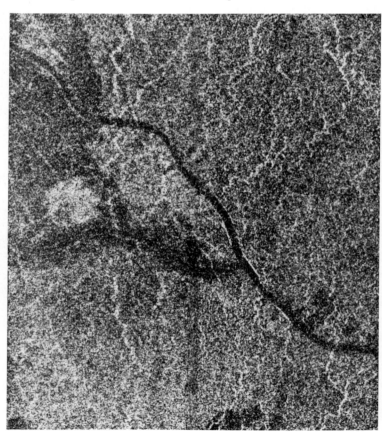

Fig. 18.13 The thin, radar-dark ribbon in this radar image may be a Venusian lava channel 30 km long and 1–2 km wide. Magellan P 36698.

radar images, one fact is brutally obvious: some of the Venus flows are *huge*. In an area known as Mylitta Fluctus, flows between 300 and 800 km in length are present. Massive outpourings of fluid lavas are obviously indicated. From a study of lengths and widths of flows Wilson and his colleagues concluded that they were most probably mafic magmas, more similar to lunar lavas or terrestrial komatiites than MORBs, erupted at enormous effusion rates, perhaps of the order of 10^5 cubic metres per second.

Such torrents of hot lava obviously have implications for thermal erosion, and indeed sinuous rilles similar to those seen on the Moon have been mapped on several parts of Venus's surface. Of course, the high surface temperature on Venus keeps lavas hot and mobile for longer than anywhere else, so flows have longer to sculpt their surroundings. In Lada Terra, there is an area of landscape that looks almost as though it has been shaped by floods coursing down a large river: braided channels, streamlined structures, teardrop shaped islands, and so on. But because there is no water, and plenty of evidence for lava flows, it seems pretty certain that this remarkable terrain was shaped by truly massive outpourings of fluid lavas. Here, perhaps better than anywhere in the solar system, 'flood basalts' really lived up to their name.

Pyroclastic eruptions By their very nature, pyroclastic deposits are hard to identify on radar images. Furthermore, because the immense atmospheric pressure on Venus suppresses vesiculation, widespread pyroclastic activity is unlikely. Magma volatile contents of several weight per cent would be needed to cause bubble growth and magma disruption. Given Venus's dry environment, this seems unlikely. However, as in the case of the Moon, a long-lived magma chamber could fractionate enough volatiles to power an explosive eruption. Since there would be less opportunity for pyroclasts to cool on Venus, welded deposits and clastogenic lavas should be more common than on Earth.

In Guinevere Planitia, there is direct evidence of pyroclastic deposits. There, small shield volcanoes with 1-km-diameter summit craters appear to have erupted radar-dark tephra which mantles

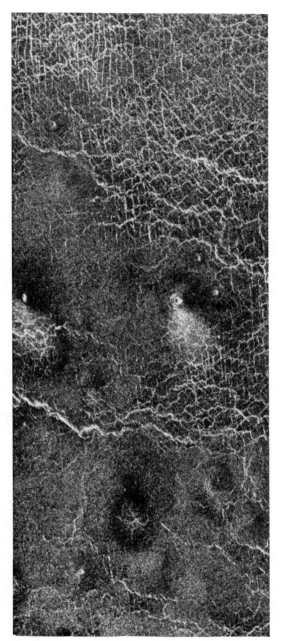

Fig. 18.14 A radar-bright surface deposit extends downwind away from a 1-km-diameter crater, reaching about 10 km before fading away. This deposit is consistent with a small Venusian pyroclastic deposit, with tephra dispersed downwind. An intriguing lattice-work of fractures decorates the surface of the surrounding terrain. Magellan radar image covers 40×112 km. Magellan P 36698.

the topography downwind of the volcano, in an expanding plume of exactly the sort one would expect from a pyroclastic deposit (Fig. 18.14). The visible deposits are so extensive, reaching up to 20 km from the volcano, that they may have been formed from plinian style eruptions; coarse material falling out near the vent, while finer material is wafted further. There are probably many fascinating pyroclastic deposits like these on Venus. Sadly, it will always be difficult to study them, even on the best orbital radar images. We may *never* know much more about them than we do now.

18.4.4 Active volcanism on Venus

Given its similarity to Earth, and the wealth of evidence for volcanism on its surface, there ought to be active volcanism on Venus. But how to detect it? If mapping missions like Magellan could continue indefinitely, it would of course be possible to see changes taking place—to watch lava flows oozing over the surface. Until this happens, it will be impossible to get direct evidence. But there are some indirect suggestions. Between 1978 and 1984, the amounts of atmospheric sulphur dioxide measured by orbiting space craft dropped by 90 per cent, hinting at variations in the rate of injection of the gas from volcanoes. In 1991, the Galileo space craft, which flew past Venus *en route* to Jupiter, detected abrupt electromagnetic pulses from the atmosphere that were probably caused by lightning strokes. Venus's stable atmosphere does not support the powerful convective storms that generate lightning on Earth, so it is possible that the lightning detected on Venus could arise from small, dust-laden eruption columns, though other explanations are also plausible.

— 18.5 Mars

Mars moves rapidly against the background of fixed stars. It also changes rapidly in brightness. Occasionally, its unmistakable fiery red glare dominates the night sky. A few months later, it may have faded to a feeble point of light. These variations, which result from the relative motions of Earth and Mars in their orbits, were for millennia a rich source of myth for astrologers. In the heyday of telescopic observation, Mars was a natural target for study, and it still fires the imagination of scientists and public alike. Its magnetism may stem from the fact that the other planets are so hostile to life. Of all the planets, it is only Mars for which one can construct even faintly plausible arguments for the existence of life. And on Mars alone could lightly protected astronauts step out of their space craft to find an environment faintly reminiscent of their home planet.

As early as 1666, Cassini used Mars's easily visible surface features to show that the planet has an axial rotation period of 24 hours 40 minutes, extraordinarily close to the Earth's. Early observers also discovered Mars's polar ice-caps, and diffuse dusky markings which appeared to show seasonal variations, shrinking and expanding in rhythm with the polar caps. This prompted speculation that the variable markings were huge tracts of vegetation and that Mars would be an agreeable abode for life. At the beginning of this century, many observers, notably the American Percival Lowell, believed that they could see spidery networks of 'canals' on the surface, fuelling beliefs that Mars was inhabited by an intelligent race, struggling to stay alive on their desiccating planet by building vast canal systems from the poles across the ochreous desert wastes, providing irrigation for scattered oases.

All this, of course, was merely the product of fevered imaginations. Romantics such as Lowell were unwilling to accept the physical limitations of terrestrial telescopes. Owing to Mars's vast distance from Earth (never less than 80 million kilometres) and atmospheric turbulence, it is impossible to resolve features on Mars which are less than hundreds of kilometres across, even with the best telescopes. Thus, Mars remained essentially unknown until the era of spacecraft exploration, which revealed that the networks of 'canals' painstakingly drawn by Lowell and others simply did not exist.

Lacking canals, the first space craft views of

Mars were unimpressive. In 1965, Mariner 4 transmitted 22 television pictures, which revealed a cratered surface, depressingly like the Moon's. Subsequently, Mariner 9 and the Viking 1 and 2 missions obtained far more data, revealing Mars as a richly diverse planet, geologically far too multi-faceted to encapsulate satisfactorily in a brief review.

In summary, Mars exhibits a dichotomy between its northern and southern hemispheres; the southern being elevated and rugged, the northern lower and smoother. Huge impact basins, such as Hellas and Argyre (hundreds of kilometres in diameter) dominate its heavily cratered southern hemisphere, which resembles the Moon. This is a key point of Mars's geology: much of the crust is ancient, perhaps four thousand million years or more old, comparable with the lunar highlands. The northern hemisphere is less heavily cratered and therefore less old, but craters are none the less numerous enough to imply that these surfaces also are ancient by terrestrial standards. Despite this dichotomy between northern and southern hemispheres, Mars is effectively a *one-plate planet*. Its heat-loss mechanisms therefore must be either through lithospheric conduction or hot-spot volcanism.

A second key point of Mars's geology is that its climatic and atmospheric histories (the two are inseparable) have been complex. While surface conditions are at present exceedingly cold and dry, with an atmospheric pressure of only 7 millibars, the presence of numerous fluvial channels on the highlands shows that at various times in its history Mars was a warmer, wetter, and altogether milder planet than it is now. There is even an accumulating body of evidence that its low-lying northern hemisphere was intermittently drowned by ephemeral oceans.[19]

18.5.1 *Composition and thermal evolution*

Mars is a small but ordinary silicate planet, with core, mantle, and crust. Unfortunately, the relative sizes of the core and mantle are not well known. The distribution of mass within a rotating planet dictates its *moment of inertia*, which can in part be inferred from its shape—the greater the fraction of mass concentrated in the core, the more nearly spherical the planet will be, and the less polar flattening it will exhibit. Unfortunately, Mars's moment of inertia remains controversial. This controversy may seem adscititious in a volcanological study, but is actually crucial because the moment of inertia is influenced by the amount of iron present in the core. If most of the planet's iron is in the core, then there will be less iron in the mantle, and vice versa. This influences the compositions, viscosities, and thus volcanic behaviours of magmas derived from the mantle. As a first approximation, we can assume that Mars has an iron–iron sulphide core with a radius of about 1900 km, and a fairly iron-rich mantle. Most estimates suggest that the Martian mantle has 16–27 per cent FeO, compared with 8–11 per cent for the Earth. If Mars's mantle is iron rich, the lavas derived from it should have low viscosities.[20]

In 1976, two Viking landers scooped up soil samples of Mars for analysis during their abortive search for life. Their compositions were similar to those of iron-rich clays, consistent with the weathering of the basaltic-looking rocks visible in the Lander surface images. As discussed in Chapter 3, the dog-killing Nakhla meteorite and the other SNC meteorites probably came from Mars. If indeed they did, then they provide useful pointers to Mars's mantle composition. Their compositions are unremarkable, but they do indicate that Mars's lavas may be iron rich— Shergotty contains 18 per cent FeO, much more than ordinary terrestrial basalts (Table 3.1 and Figs. 3.9 and 3.10).

Mars probably contains chondritic abundances of heat-producing radioactive isotopes. Its thermal evolution is difficult to constrain, however—because it is such a small planet, small changes in assumed conditions have large effects on thermal models. A particularly important variable is its starting temperature. While Earth and Venus probably started life hot, or even molten, Mars may not have done so, because it is so much smaller.

According to one popular model Mars started cool, at below solidus temperatures, and homogeneous.[21] After accretion, radioactive heating warmed the planet above melting point of iron–iron sulphide alloys, so that about 500 million years after accretion a dense liquid core had separated out. Core formation caused a further 200°C rise in temperature, permitting solid-state convection in the mantle. Further radiogenic heating warmed the mantle until about 2.5 thousand million years ago, when widespread partial melting of the mantle prevailed. Volcanism and outgassing reached a peak about two thousand million years ago. Concomitant thermal expansion within the planet may also have fractured the crust in thousands of extensional surface fissures. After two thousand million years ago, efficient mantle convection caused lithospheric cooling and thickening, which in turn caused the magma sources feeding surface volcanoes to recede deep into the interior. In the last thousand million years, lithospheric thickening has continued, reaching perhaps 200 km, but depths to melt zones in the mantle are now so great that surface volcanism is extinct.

This model matches well the chronology of Martian volcanism deduced from crater size–frequency distribution studies, but if Mars formed at a higher temperature than the model assumes, then the various thermal stages would have to be pushed further back in time. Regardless of that

debate, Mars is a volcanic morgue. Youthful-looking volcanoes loom above its frigid, sterile surface. But they are quite dead.

18.5.2 Volcanoes on Mars

A planet-wide dust-storm was raging when the Mariner 9 space craft first went into orbit around Mars in 1971. All that could be seen on early images were the tops of four great volcanoes rearing above swirling dust clouds. One of these correlated with a whitish spot noted by telescopic observers and named by them *Nix Olympica*, the Snows of Olympus (Fig. 18.15). The others were unknown, so were initially labelled North, Middle, and South Spots. Later, when time was available for such refinements, they were less prosaically christened Ascreus, Pavonis, and Arsia Mons, respectively. During this spate of name-giving, Nix Olympica lost its snows and became merely Olympus Mons. In Greek mythology, Olympus was the home of Zeus and all the Gods. With or without its snows, Olympus Mons is a fitting home for the Gods, since it is the highest volcano (and mountain) in the entire solar system.

Such a wealth of high-resolution images of

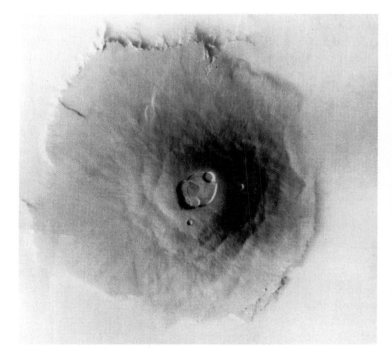

Fig. 18.15 A historic image in both planetology and volcanology: this Mariner 10 image revealed the largest volcano in the solar system. Olympus Mons is more than 600 km across, rises 25 km above its surroundings, and is crowned by a summit caldera complex 65 km across. Although later Viking missions obtained many much higher-resolution images, Olympus Mons still presents many unsolved problems, such as the origin of the massive 4–8-km-high scarp that surrounds it.

Mars were obtained by Mariner 9 and the later Viking missions that much has been written about Mars's volcanoes.[22-24] Three kinds of volcanic construct were recognized on the early Mariner 9 images: *shields, tholi, and paterae.* These names have stuck, although they are difficult to apply to the higher-resolution Viking images.

Shields   Olympus Mons and the three ex-Spots form a group known as the Tharsis volcanic province. Their morphology is reminiscent of the Hawaiian lava shields, but they are much bigger. Olympus Mons has a basal diameter in excess of 600 kilometres, a nested summit caldera complex more than 60 km in diameter and 2 km deep, and rises *25 kilometres* above the surrounding plains. It is natural to enquire why Olympus Mons grew so large. One important factor is Mars's thick rigid lithosphere. A volcano borne above a Martian mantle hot-spot remains fixed above it, rather than being carried away, as happens to volcanoes like those of Hawaii, riding on their 125-km-thick lithospheric plate on Earth. If the Pacific plate could have been riveted to the mantle for the last thousand million years, huge volcanoes would rise in place of the modest Hawaiian shields. Furthermore, the Martian lithosphere is so thick that it bears the load of the great volcanic massifs easily, whereas the lithosphere under the Hawaiian shields actually sags under their weight, reducing the volcanoes' apparent height.

Tholi   Tholi are smaller volcanoes than the great shields, and are generally dome shaped with steep flanks. Some tholi rise abruptly out of the surrounding plains, leading to suggestions that they were partially 'drowned' by hugely extensive flood-basalt lavas (Figs. 18.16–18.17). Such lavas may be up to 4 km thick in the Tharsis area. If the tholi are indeed partially submerged, then it is difficult to determine their original morphology. Whereas the great shields appear to be entirely lava constructs, one tholus, Hecates, may have experienced an episode of relatively recent explosive activity.

Paterae   Loosely speaking, paterae are low, flat

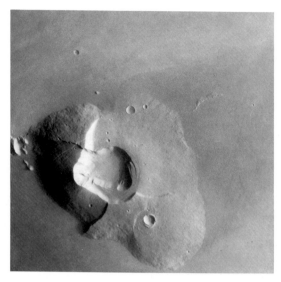

Fig. 18.16 Tharsis Tholus is a 170 km by 110 km 'dome' which protrudes abruptly above the surrounding smooth, sparsely cratered plains, suggesting that its lower flanks have been drowned by lava. Its deep, steep summit caldera complex is 45 km across. Viking image: courtesy of Mark Robinson, University of Hawaii.

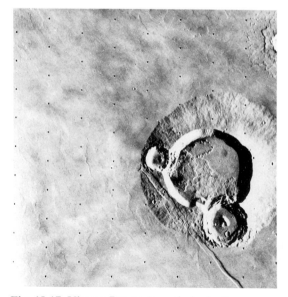

Fig. 18.17 Ulysses Patera (morphologically more like a tholus than a patera) was almost drowned by younger lavas. Its 50-km-diameter caldera appears huge compared with the diameter of its shield. Two impact craters and a younger graben decorate its summit. The Viking image is 190 km across.

shields found on the ancient Martian highlands. They may have been related to the formation of the major impact basins such as Hellas and Argyre. While some paterae are actually shield like (Fig. 18.18), 'true' paterae are very different from the great lava shields. Earth-based radar studies of Tyrrhena Patera, for example, show that it has low slopes (<0.25 degrees) over a horizontal distance in excess of 100 km, and that its summit is probably less than 1 km above the surrounding plains (Fig. 18.19). While the paterae all have summit calderas, these are irregular and subdued compared with those of the lava shields. Most interpretations of the paterae suggest that they are probably not lava constructs, but more like great ignimbrite shields. Two lines of evidence support this suggestion: their remarkably flat topographical profiles, and the laminated, un-lava-like erosional morphologies on their flanks. Whatever their origin, the paterae are clearly old, and may be the oldest constructs on Mars. Alba Patera is an interesting anomaly. It is a huge structure, many

hundreds of kilometres in diameter, but rises only 3–4 km above the surrounding plains. Its lower part may be composed of pyroclastic deposits, but its upper part is dominated by lavas, suggesting that a transition from volatile-rich to volatile-poor eruptions happened at Alba during its evolution.

18.5.3 Lava flows

Superb high-resolution Viking images show myriads of fresh flows draping the flanks of the giant volcanoes such as Olympus Mons (Fig. 18.20) and winding for hundreds of kilometres across the lowland plains (Fig. 18.21). Flows on the plains are about 10 kilometres wide and 20–30 m thick. Their near-pristine morphologies demonstrate that *some* flows are relatively youthful. With only crater size–frequency statistics to go on, however, assigning absolute ages is difficult. Some crater-counters suggest that it is almost a thousand million years since the last eruption, others that the youngest flows may be

Fig. 18.18 Apollinaris Patera rises 5000 m above its surroundings. Its summit caldera complex (65 km across) is wide in relation to the diameter of the shield (about 180 km). Lavas or pyroclastic flows form a large fan which radiates out from a breach in the southern wall of the caldera and mantles the basal scarp. Some channels on the south-west flank may have been cut by erosion. Apollinaris is more like a middle-aged shield volcano than a true patera. Viking image: courtesy of Mark Robinson, University of Hawaii.

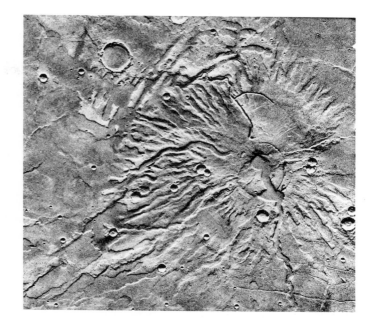

Fig. 18.19 Tyrrhena Patera is an unusual, gently sloping shield, dissected by erosion to a spidery network of channels radiating from the summit caldera, which is about 50 km across. Its low relief (*c.* 1 km) and style of erosion have led to suggestions that it is constructed largely of pyroclastic flows.

Fig. 18.20 High-resolution Viking image, showing innumerable lava flows which cascaded down the south-western flanks of Olympus Mons, mantling its great four-kilometre-high scarp (cf. Fig. 18.15). Visible features include levéed channels and collapsed lava tubes reminiscent of modern Hawaiian shields. Image is about 80 km across. (Photo: courtesy of Mark Robinson, University of Hawaii.)

Fig. 18.21 Viking
photomosaic showing a lava
flow cut by many graben
south-east of the summit
caldera of Alba Patera.
Emplacement of the flow
clearly pre-dated extensive
faulting. Venting of subsurface
volatiles may have formed the
crater chain on the floor of
graben at centre. Image is
112 km across. Part of Viking
Mosaic MTM 35102. Photo:
courtesy of Peter Mouginis-
Mark, University of Hawaii.

only a few hundred million years old, younger
than the youngest flows on the Moon.

Although there are innumerable flows tangled
together in anastomosing flow fields on the
major volcanoes, there are few obvious vents.
Magma was probably transported laterally from
the sub-volcanic magma chamber in dykes which
breached the surface low on the flanks of the
volcano. Lava flowed quietly away from obscure
boccas, flowing within tubes for long distances,
just as happens on terrestrial basaltic shields.
Some dykes may have propagated for distances
of several hundred kilometres from the central

volcanoes. Thus, many extensive lava flows
appear to originate far from any other obvious
volcanic constructs. Some of the flows associated
with the Tharsis volcanoes have extremely large
volumes, of the order of 100 cubic kilometres,
indicating that the magma chambers beneath the
volcanoes must be large.

Vast areas of the smooth northern plains are
covered with lavas. In the Elysium region flood
basalts cover more than 100 000 square kilometres (forming the Cerberus Plains) and are so
sparsely cratered that they may be only a few
hundred million years old. Apart from those

related to major volcanic constructs, many of these lavas were probably erupted from small, gently sloping basaltic shields similar to those found in Iceland and the Snake River Plain (Section 16.3.3). These have such subdued contours that they are easily missed on Mars images, and their significance has probably been underestimated. In places, the surfaces of the smooth plains are peppered with dozens of small dome- and cone-like features, clearly not of impact origin. These may be analogous with the Icelandic pseudocraters, formed by explosive activity when lavas advanced over wet lake beds.[25]

Given the abundance of splendid images of lava flows on Mars, it has been possible to measure parameters such as channel widths, levée heights, flow front heights et cetera (Fig. 18.22). From the equations introduced in Chapter 7, it follows that lava flows should be thicker and have wider levées on low gravity planets such as Mars. In particular:

$$W_b = \tau/2g\rho \ (\tan \alpha)^2$$

Where W_b is the width of the levée, α the slope, and τ the yield strength (Section 7.2.2). Several planetary volcanologists measured levée widths and slopes of Martian lavas to estimate their yield strengths, and thus their composition. Sadly, the exercise has been a frustrating one, since we are still far from understanding fully all the relationships between lava composition, temperature, shear rate, viscosity, and crystallinity, and how these control the final shape of a lava flow. A range of compositions has been postulated for individual Martian lavas, ranging from komatiitic to andesitic. From cosmochemical considerations, it is likely that most

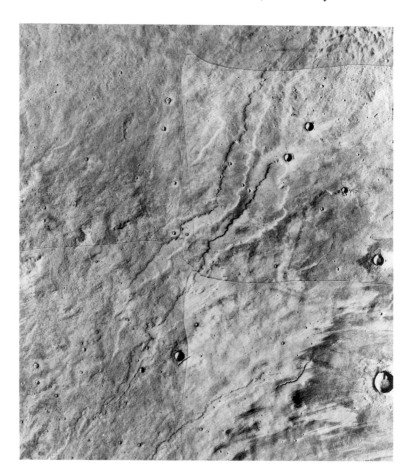

Fig. 18.22 Lava flows almost 100 km in length, located about 120 km north of the summit caldera of Alba Patera. Flows are peppered by small impact craters. Mosaic of 4 Viking images is about 80 km across. Photo: courtesy of Peter Mouginis-Mark, University of Hawaii.

Martian lavas are primitive iron-rich basalts, but given the large, long-lived magma chambers that presumably existed, small volumes of more evolved magmas could have been erupted. It follows from the huge lengths attained by many lavas that their effusion rates were enormous by terrestrial standards, probably greater than 10 million cubic metres per second.[26]

18.5.4 Martian pyroclastic deposits

Given the presence of even small quantities of volatiles, Mars's low atmospheric pressure would favour pyroclastic disruption of magmas. And there is every reason to suppose that both H_2O and CO_2 are *abundant* on Mars. Thus, Martian pyroclasts should be common. The low atmospheric pressure should cause eruption velocities for gas and small pyroclasts to be about 1.5 times greater than on Earth, and eruption clouds should rise about five times higher, at the same mass eruption rate. Pyroclastic deposits, therefore, should be finer grained and much more widely dispersed than their terrestrial counterparts. Unfortunately, pyroclastic deposits are always difficult to identify on spacecraft images. When they are widely dispersed, the problem is doubly difficult. But what should one look for?

Probably the most common Martian pyroclastic deposits would be formed from relatively low effusion rate eruptions of basaltic lavas. These would cause strombolian eruptions constructing gently sloping low scoria cones, several times wider than their equivalents on Earth. More sustained eruptions would produce low cones up to 10 km in diameter. Possible examples of features of this sort have been noted at one or two localities on Mars. Higher mass eruption rate events, which would yield plinian air-fall deposits on Earth, would yield thin, widely dispersed deposits on Mars. Millimetre-sized particles could be carried 100 km downwind under typical Martian meteorological conditions. A thin deposit mantling the summit region of Hecates Tholus is a promising candidate for this kind of eruption[27] (Fig. 18.23).

Convecting eruption columns are less stable in Mars's thin atmosphere than on Earth, and so column collapse and formation of pyroclastic flows are more likely at high mass eruption rates

Fig. 18.23 Hecates Tholus, a shield volcano about 170 km in diameter. This high-resolution Viking frame shows the compound summit caldera, 10 km across, and some radial rilles and channels. Area west of the summit is devoid of craters, suggesting that it has been resurfaced, probably by a mantle of pyroclastic material. Viking image: courtesy of Mark Robinson, University of Hawaii.

on Mars. Thinly laminated materials on the flanks of the ancient patera volcanoes have been tentatively identified as pyroclastic flow deposits, a reasonable enough interpretation. However, whereas the combinations of mass eruption rate and volatile content required to yield pyroclastic flows is almost exclusively found in *silicic* eruptions on Earth, producing voluminous ignimbrites, on Mars, *basaltic* ignimbrites are likely to predominate.

During 1991, evidence for the existence on Mars of large areas of pyroclastic deposits came from a somewhat unexpected source: Earth-based radar was used to study the surface properties of Mars. Satisfactory return signals were obtained from the Olympus Mons area, consistent with rough lava flows, and from the polar ice-caps. No echoes at all, however, were obtained from a huge region of the surface beyond the Tharsis area. Inevitably, this area was christened 'Stealth' by scientists used to thinking about how to conceal fighter and bomber aircraft from probing radar beams. They suggested that this unexpected result could best be explained by an extensive deposit of radar-attenuating material, with a density of less than 500 kilograms per cubic metre, particle size of less than one centimetre, and at least one metre thick. These properties, of course, are consistent with those of

a pyroclastic fall deposit, blown downwind from the Tharsis volcanoes.

While the flanks of the paterae may indeed be built of basaltic ignimbrites, and there may be huge tracts of pyroclastic deposits near Tharsis, we won't know for sure until someone goes there. Conceivably, that someone might be an intelligent mobile robot named HAL . . .

— 18.6 Io

While working on a Voyager 1 image of Io obtained in March 1979, Linda Morabito, one of the space craft navigation team at the Jet Propulsion Laboratory, California, was puzzled when an automated computer program consistently had problems in matching a curve to the profile of Io on an image. On examining the image, she noticed a curious distortion of the edge of the satellite against the blackness of space. This observation was a milestone in volcanology: discovery of active *extraterrestrial* volcanism. Follow-up work showed the 'distortion' to be a huge volcanic plume spraying material into space.[28] Eight others were found later (Fig. 1.4). Io is by far the most volcanically active object in the solar system. Serious scientists have described it as 'a wonderland of physics and chemistry'.[29–30]

Io is a large satellite, 3640 km in diameter, similar in size to our Moon, and almost large enough to be a planet in its own right. (Two of Jupiter's other Galilean satellites, Ganymede and Callisto, are in fact bigger than Mercury.) Well known for its reddish colour, Io was a natural target for telescopic observation. Spectroscopic work showed that its surface is devoid of water ice, making it unique among satellites in the outer solar system, but the nature of the surface materials and the reason for its reddish colour could not be determined.

18.6.1 Composition and thermal evolution

Io's density of 3500 kilograms per cubic metre shows that it belongs to the family of dense objects in the solar system. Thus, it should have an iron or iron sulphide core and silicate mantle. It should also have a 'normal' chondritic comple-

ment of heat-producing isotopes and heat left over from its accretion. But these familiar sources of heat are irrelevant! Io orbits only 421 000 kilometres away from the colossal mass of Jupiter. Like the Earth's Moon, tidal forces keep one hemisphere of Io facing towards the planet. Furthermore, Io is locked into a complex series of orbital resonances: Europa, the next satellite out, orbits Jupiter in exactly *twice* the time (3.55 days) that Io takes, and Ganymede, the next satellite out beyond Europa, orbits in exactly *four* times (7.15 days). These mathematically elegant Laplacian resonances maintain Io in a nearly circular orbit, and cause it to oscillate (librate) about its rotation axis, like the head of a Wimbledon spectator. Huge tidal forces deform its shape into a prolate spheroid, which is constantly being squeezed and stretched by all the pushes and pulls acting on the satellite in its orbit around Jupiter. The inevitable consequence of this ceaseless gravitational kneading is that its silicate interior is perpetually hot, and may well be molten all through.

This state of affairs had been predicted from theoretical considerations months before Voyager images stunningly confirmed the existence of intense volcanism on Io. In a brilliant paper, Peale, Cassen, and Reynolds estimated that energy liberated by tidal dissipation is about 4×10^{13} watts, two orders of magnitude larger than that released by radioactive decay (about 5×10^{11} watts).[31] Voyager data suggest that the surface heat flow on Io is of the order of 1–2 watts per square metre.[32] To put this in perspective, recall that the heat production in the vastly more massive Earth is also about 4×10^{13} watts, giving an average terrestrial heat flow of only 0.08 watts per square metre, while our Moon has a heat production of a mere 0.02 watts per square metre. So much energy is being pumped into Io that the satellite resembles an apple over-baked in a microwave oven: the interior fruit pulp gets so hot that it is extruded from all sort of interesting orifices (worth watching if you have a strong stomach).

18.6.2 Volcanoes on Io

Once volcanism had been spotted on Io, the satellite loomed large on the list of priorities for scrutiny by both Voyager 1 and 2. Voyager images showed that the entire surface is covered by young volcanic materials—there are *no* impact craters. Three basic types of landforms have been distinguished on Io: volcanoes, mountains, and plains.

Volcanoes   Volcanic vents cover about five per cent of Io's surface area. Over 300 volcanoes (not all of them active) have been identified. Most are low shields or paterae, less than 100 km in diameter, though some in the equatorial region

attain more than 250 km. Radial laval flows snake away from the summit regions, some reaching hundreds of kilometres from the volcano (Fig. 18.24). Although the topography of Io is not well constrained, the slopes of the shields appear to be gentle, perhaps less than one degree. Many shields have shallow summit calderas ranging from 2 to 200 km in diameter. A few calderas such as Loki (source of the first plume detected by Linda Morabito) contain features thought to be lava lakes (Fig. 18.25). Others, such as Creidne, have dark lava flows covering the caldera floor, reminiscent of the lava flows filling terrestrial calderas (Fig. 18.26). Creidne was identified as a hot-spot (a mere $-40\,°C$!) during the Voyager 1 encounter. These observations raise a burning question: what is the nature of the lava flows? Are they silicates, or something more exotic?

Mountains   Haemus Mons is one of the few mountains on Io. It rises abruptly above the smooth plains to a height of about 10 km, and is about 200 km across. Haemus Mons looks a bit

Fig. 18.24 Ra Patera, one of the low shield volcanoes on Io. Dark lavas extend radially for over 200 km. These may be of silicate composition. Photo: courtesy Al McEwen, USGS *Flagstaff*.

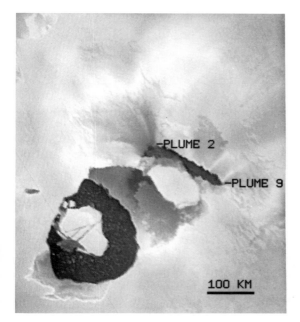

Fig. 18.25 Loki Patera is the site of the most powerful thermal anomaly on Io. Dark fractured material at centre resembles a terrestrial lava lake, with a shifting, overturning crust. If this is true, then Loki's lava lake is an order of magnitude larger than any terrestrial example. Photo: courtesy Al McEwen, USGS *Flagstaff.*

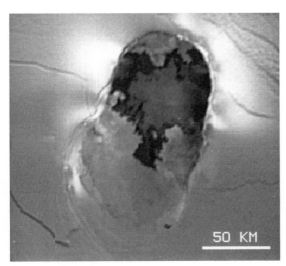

Fig. 18.26 Caldera of Creidne Patera, 100 km × 180 km, contains dark features strongly suggestive of lava flows. Creidne was detected as a hot-spot (230 K) during the Voyager encounter. Photo: courtesy of Al McEwen, USGS *Flagstaff.*

like one of the *nunataks* found in the polar regions of the Earth: isolated mountains which protrude above a sea of ice. Like *nunataks*, Io's visible mountains are probably only the peaks of larger structures, buried beneath thick deposits of smooth plains-forming material. Existence of major peaks such as Haemus Mons demonstrates that Io must have a substantial rigid lithosphere, probably about 30 km thick.

Plains Most of Io's surface is covered by featureless plains, most likely pyroclastic deposits from the eruption plumes, but there may be some extensive lava flow surfaces as well. In several areas, the plains are obviously layered, the layers forming tabular, smooth-topped plateaux, commonly outlined by escarpments 150 to 1700 m high. At present, it is not clear what processes operate to sculpt these layered terrains (Fig. 18.27).

18.6.3 Plumes

During the Voyager 1 mission, a total of nine distinct umbrella-shaped plumes were observed above the surface of Io. They reached heights of between 70 and 280 km, implying eruption velocities between 500 and 1000 m/sec. Some changed visibly over the course of a few hours; others remained apparently unchanged for days. Most were located in the equatorial region of the satellite, between 30°N and S. The largest umbrella plume was 1000 km in diameter, and was fed from a fountain-like vent which appeared to be 35 km across, but was probably much narrower. Eight of the plumes were re-observed during the Voyager 2 encounter. In the four months between the two missions, there were considerable changes on Io. It became clear that eruption plumes are common but short-lived phenomena which leave an unmistakable imprint: erupted material falls back to form symmetrical light-toned haloes around the vent areas (Figs. 18.28–18.29).

McEwan and his colleagues distinguished two different kinds of plume: *Prometheus*-type plumes are 50–120 km high; form bright halo deposits 200–600 km in diameter; contain optically thick, dark jets; erupt at velocities of about 500 metres per second; are active for long periods

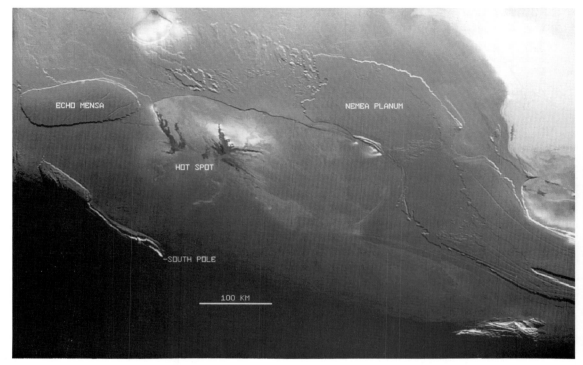

Fig. 18.27 Layered and eroded terrain forms smooth plains in the southern polar regions of Io. Layers are probably pyroclastic deposits. Photo: courtesy of Al McEwen, USGS *Flagstaff*.

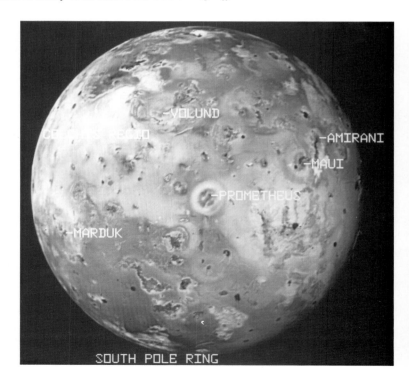

Fig. 18.28 Full disc Voyager image of Io, showing sites of several named Prometheus-type plumes. A symmetrical halo of plume deposits surrounds Prometheus. Photo: courtesy of Al McEwen, USGS *Flagstaff*.

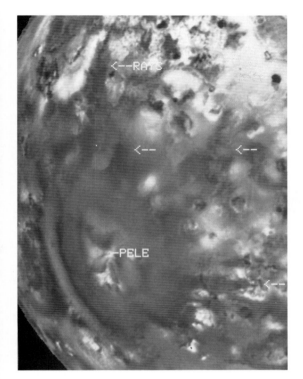

Fig. 18.29 Detail of plume deposit around Pele. Deposit is not circular, but heart-shaped, implying asymmetry in vent geometry. Ill-defined rays of ejected material extend up to 800 km from the volcano. Photo: courtesy of Al McEwen, USGS *Flagstaff*.

(months to years); and are found mostly in the equatorial regions of the satellite. *Pele*-type plumes are much bigger; reach heights of up to 300 km; are optically thin (transparent); produce relatively dark, reddish-coloured annular deposits 1000–1500 km in diameter; erupt at up to one kilometre per second, occur in a restricted longitudinal belt; and are short lived (days to months). Prometheus-type plumes also appear to be associated with lower temperature 'hot-spots' (*c.* 450 K) than the Pele type (*c.* 600 K).

18.6.4 Sulphur or silicates on Io?

All of the above discussions beg the important question: *what is it that is being erupted so prodigiously from Io?* Io's distinctive colour and spectral characteristics prompted initial specula-tion that the surface was covered by sulphur, and that the volcanism was entirely sulphurous—the lava flows and lava lakes were presumed to be made of sulphur, and the active plumes to be driven by sulphur dioxide. This obsession with sulphur was helped by dissemination of so-called 'pizza pictures' of Io, in which image processing techniques were used to colour the satellite in appetizing shades of tomato red and mozzarella yellow.

There are sound reasons for suspecting that sulphur plays an important role in Io's volcanism. First, there is direct spectroscopic evidence. Large amounts of sulphur dioxide and lesser amounts of hydrogen sulphide and carbon dioxide have been detected in gaseous form above the surface, with a 'frost' of solid sulphur dioxide on the surface. Elemental sulphur is much more difficult to identify because it is spectrally featureless, but ionized sulphur has been detected in a torus above Io, and the surface materials are consistent with sulphur.

Second, there is circumstantial evidence. Io appears to have been so hot for so long that *all* of its water has been lost, driven off into space by the incessant volcanism. The 'escape' velocity for Io is about 2560 metres per second. At Io's likely surface temperature of minus 173°C, the root-mean square velocity—a sort of average velo-city—of water molecules is about 371 metres per second, while for sulphur dioxide it is only 197 metres per second. Because some fast-moving molecules are always being lost to space, water will be lost at a much faster rate than sulphur dioxide, because its lighter molecules whizz around faster. Thus sulphur dioxide is probably the volatile component driving Ionian volca-nism. After being sprayed as vapour hundreds of kilometres into the chill of space, it condenses to the solid form, falling as a snow of tiny particles to the surface, to form a thin, even mantle. Ultimately, the deposit will become so thick that the lowest layers become warm and mobile enough to be recycled.

It is worth recalling here that elemental sulphur, which melts at only 113°C, does play a minor role in terrestrial volcanism (Section 7.8). 'Lava' flows of molten sulphur have been observed on some volcanoes, and in Poas, Costa

Rica, small lakes of molten sulphur appeared in the crater floor *after* the water of the crater lake had been boiled away.

In a thermodynamic study of the plumes of Io, it was suggested that *both* sulphur dioxide and elemental sulphur might be involved, each driving a different kind of plume. In the case of the small Prometheus-type plumes, sulphur dioxide may be the working fluid, heated by liquid sulphur at temperatures near its melting point, whereas in the more powerful Pele-type eruptions, the working fluid might be elemental sulphur heated to high temperatures ($c.\,700\,°C$) by silicate magmas. While this is conjectural, it is consistent with the spectral properties of the plume deposits.

So much for the volatile phase. But what of the lavas? Exotic early ideas of lakes of molten sulphur within the calderas of volcanoes such as Loki have been reconsidered in the light of two constraints:

First, the existence of topographical features on Io argues that silicates must be present. Sulphur is simply not strong enough to form mountains as high as Haemus Mons, or the 2-kilometre-high cliffs forming the walls of calderas such as Maasaw (Fig. 18.30). If made of sulphur, these would simply deform glacier-like under their own weight and fade away. The warmer the temperature, the more quickly they would deform.

Second, some thermal phenomena on Io are *too hot* to be sulphur. This discovery has been a triumph for Earth-based infra-red telescopic astronomy. One study showed that a 'hot-spot' on Io, probably one of the lava lakes, was about 30 km in diameter, with a 'model' temperature of $600\,°C$.[33] (In fact, the actual temperature may be much higher, since much of the hot spot is likely to be covered by a chilled crust, so that most of the observed radiance comes from a smaller, hotter region.) Since the boiling point of sulphur in a vacuum is only $442\,°C$, it cannot form the *dominant* magma on Io, although it could form interesting sideshows, as it does on terrestrial volcanoes. Some of the lower temperature 'hot-spots' detected by the Voyager space craft might be expressions of sulphur volcanism. But for now, the high temperatures indicated by the

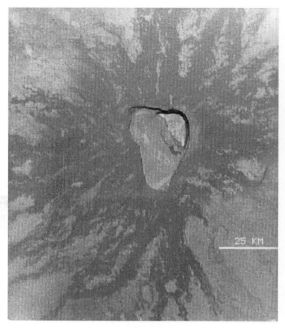

Fig. 18.30 Maasaw *Patera* has a summit caldera measuring about 25 km by 35 km in diameter. Steep, 2-km-high cliffs at northern end of caldera suggest the existence of strong silicate rocks, rather than weak sulphur. Photo: courtesy of Al McEwen, USGS *Flagstaff*.

telescopic data suggest that it is silicate volcanism which is so impressively active on Io.

In December 1989, another set of remarkable images of Io was obtained by John Spencer and his colleagues at the University of Hawaii, using the ground-based NASA Infra-red Telescope Facility on Mauna Kea (Fig. 18.31). Spencer was able to obtain images of Io's active volcanoes by observing the satellite while it was in the shadow of Jupiter. Then, shielded from the sun's glare, Io's hot-spots show up, glowing fiercely in the infra-red. Apart from identifying a major hot-spot corresponding with the known site of Loki volcano, the infra-red images showed at least one other hot-spot in a position that did not correspond with any volcano indentified by Voyager.

Telescopic observations will enable us to monitor activity on Io to some extent. Fresh insights into exactly what kind of volcanism is

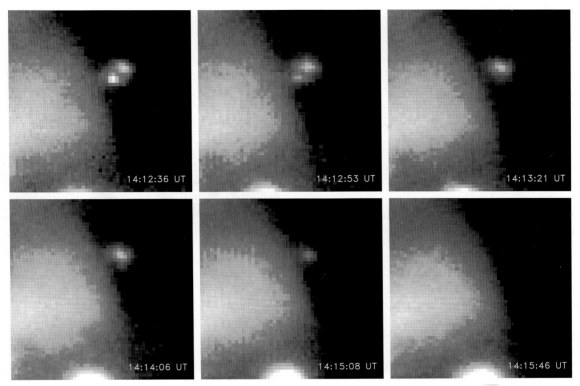

Fig. 18.31 Volcanic hot-spots glow in the infra-red in these images obtained by John Spencer at the NASA Infra-red Telescope Facility, Mauna Kea, Hawaii. Images were acquired at 3.8 micron wavelength while Io was in Jupiter's shadow, so that no reflected sunlight was detected. Two distinct hot-spots are present, each with apparent temperatures of about 500 K and 30 km in diameter. The right-hand spot has been correlated with Loki, a major volcano known to be active during the Voyager encounter. The left-hand spot has been christened Kanehekili, after an Hawaiian thunder god. The bright object at the bottom of the frame is Ganymede, in sunlight. By 14.15.08, Io had been eclipsed by Jupiter. UT = Universal Time. Photo: courtesy of John Spencer, University of Hawaii.

going on will undoubtedly arise from the Galileo mission, if that ill-starred space craft is still in working order when it reaches Jupiter in December 1995. Current plans call for it then to spend at least 22 months observing the planet and its retinue of satellites.

18.7 Europa

Europa is the smoothest, blandest satellite in the Solar System. Devoid of topographical features higher than about 50 m, it looks like an ivory billiard ball, mottled with age. Is Europa the prime candidate for Most Boring Object in the solar system? This would be unfair. Its very featurelessness is a point to the operation of some intriguing 'volcanic' processes.

18.7.1 Composition and thermal evolution

Europa is a little smaller than Io, with a diameter of about 3000 km. Its density is much the same as Io's (about 3.4×10^3 kg m^{-3}), so Europa must have a similar silicate composition, with a similar content of radioactive heat-producing isotopes. Although further away from Jupiter than Io, it must also experience some tidal heating. But whereas Io is so hot that it is devoid of water, Europa is entirely covered with a smooth layer of ice. This was first indicated by telescopic spectral studies, and subsequently confirmed by Voyager images which showed a blank icy surface, marked only by a few impact craters and a large number of subdued, criss-cross fractures resembling large crevasses.

Thermal modelling suggests that the ice layer is quite thin.[34] Enough heat leaks out from the rocky inner part of the satellite to maintain a 'mantle' of liquid water. Thus, Europa may consist of a rocky core, 1500 km in diameter (by far the largest component) surrounded by a mantle of liquid water perhaps 50 km thick, and capped by a 'crust' of ice about 25 km thick. Fascinating plate tectonic possibilities arise from such a structure—the whole icy crust may be mobile with respect to the liquid mantle. Fracturing of the icy crust would allow 'dykes' of water ice to be intruded, and may account for the linear grooves and fractures seen on the surface.

Watery volcanism may be active on Europa! Voyager studies revealed a faint brightening above Europa's limb, consistent with the eruption of an Io-like plume of water vapour or ice particles from the fissure systems. A single telescopic observation, made on 23 April 1981 gives it some credence. Observers using the giant Infra-Red Telescope Facility atop Mauna Kea in Hawaii measured anomalously powerful radiation from Europa in the infra-red, consistent with a warm source on the surface. Confirmation of this possibility will have to wait until the Galileo spacecraft reaches Jupiter, but *if* there is active volcanism on Europa, it is clearly much more low key than on Io.

Ganymede, and the other icy satellites of Jupiter and Saturn The motions of planets around the Sun and of satellites round planets are as smooth and orderly as clockwork. It was not ever thus. As our knowledge of celestial mechanics has improved, and the power of computers increased, it has emerged that the perturbations of the orbits of smaller bodies caused by larger ones can be mathematically chaotic; that is, a

small change in a single variable could cause a gross change in a satellite's orbit. This probably led to innumerable massive collisions in the early days of the solar system. In the case of Ganymede (next satellite out from Europa), mathematical modelling suggests that during an early period of chaotic motion, Ganymede may have gone through a phase in which it was in 3:1 resonance with Europa. Tidal heating sufficient to cause extensive internal melting may have resulted, leading to the curious jigsaw of surface features seen on the satellite today, which suggest continental fragmentation and drift. Similar orbital effects probably led to extensive icy volcanism early in the histories of several satellites of Saturn and Uranus. If these bodies contained ammonia as well as water ice, the low melting temperature of the mixture would have made volcanic manifestations even easier to generate.

In a fascinating theoretical study of cryovolcanism, Jeffrey Kargel of the University of Arizona arged that two varieties of cryovolcanism might operate in the solar system.[35] In truly low-temperature cryovolcanism, ammonia would indeed be present. But ammonia is only likely to be present in the satellites of the outer solar system; nearer the Sun, temperatures would always have been too high for it to be retained. In satellites such as Europa and Ganymede, it is unlikey that pure water or water-ice is present. Up to one fifth of the mass of C1 chondrite meteorites is composed of materials that are potentially water soluble. Since there is every reason to suppose that Jupiter's large satellites are of broadly chondritic composition, it follows that the soluble salts would have been leached from the interior by any circulating water warmed by radiogenic heating.

Kargel therefore argued that the active ingredient in the cryovolcanism on Europa and Ganymede is not water, but *brine*, rich in dissolved magnesium and sodium sulphates. (Sodium chloride, the familiar component of terrestrial brines, is unlikely to be important, because chlorine is much less abundant than sulphur in cosmochemical terms.) If Kargel is right, and his thinking appears sound, then there are many implications for cryovolcanic phenomena, given the differences in melting temperatures and densities of brines compared to pure water.

Voyager images of the icy satellites of Jupiter, Saturn, and Uranus provided tantalizing glimpses of what may be icy volcanic phenomena (Fig. 18.32). Although the concept of cryovolcanism must appeal to any volcanologist, the sad facts are that our understanding of it is still rudimentary, and the images just too tantalizing to warrant detailed elaboration. With luck, Galileo's views of Europa and Ganymede may resolve many cryovolcanic issues, so a little patience might be appropriate.

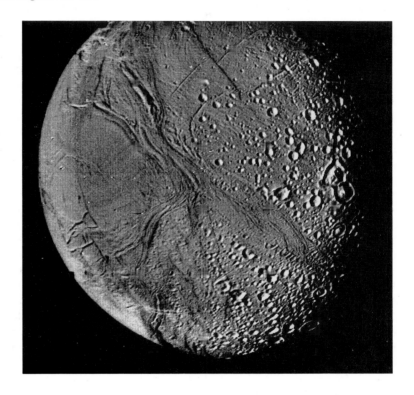

Fig. 18.32 Clear evidence of cryovolcanism is evident on this Voyager spacecraft image of Enceladus, an icy satellite of Saturn only 500 km in diameter. One half of the satellite is heavily cratered; the other is almost free of craters, and must have been re-surfaces by icy volcanism. Belts of linear grooves and ridges at the centre of the image are strongly reminiscent of rifting and extension in terrestrial plate tectonics.

— 18.8 Triton

Neptune orbits the Sun on the desolate fringes of the solar system, a region so remote that the planet was not dicovered until 1846. Even then, its discovery came only as the result of the celebrated mathematical analyses of its gravitational perturbations of Uranus's orbit by John Couch Adams in Cambridge and Urbain Le Verrier in France. These calculations showed astronomers where to point their telescopes to find the planet. Neptune's largest satellite, Triton, was discovered only three weeks later.

Triton is not an obvious candidate for active volcanic processes. It has a diameter of 2700 km, a fairly dark surface (albedo 0.6–0.9) and orbits quite close to Neptune (355 000 kilometres). It is difficult to observe even with a powerful telescope, so little was learned about it in the 150 years since its discovery. Even its size, mass, and density remained obscure. Consequently, Triton was something of a non-entity until the Voyager 2 space craft flew past Neptune in August 1989 at the end of its epic voyage. The extraordinary images it obtained of Triton's polar cap of frozen nitrogen and of exotic volcanic eruptions provided a fitting climax to one of the greatest voyages of exploration in scientific history.

18.8.1 *Composition and thermal evolution*

Voyager 2's observations enabled Triton's mass and density to be resolved. With a density of 2075 kg m^{-3}, Triton turned out to be one of the densest satellites in the outer solar system, exceeded only by Io and Europa. It must consist of a mixture of ice and silicate materials. Different structural models have been proposed, but all have a silicate core overlain by a mantle of ice, forming only about 30 per cent of the total mass of the satellite.[35] This proportion is consistent with the hypothesis that Triton formed directly from the solar nebula, and was later captured by Neptune, an argument which also helps to explain why Triton is in *retrograde* orbit around the planet. Curiously, spectroscopic data do not show direct evidence of water ice on Triton, but instead suggest that its surface is covered by solid nitrogen and methane. Topographic features seen on the images, however, have a relief of about 1 km. Such relief can be supported only by fairly strong materials, and, while the strengths of solid nitrogen and methane are poorly known, it seems unlikely that they would be strong enough. It is more probable that Triton's crust is made of water ice (strong at extremely low temperatures), covered by a thin layer of nitrogen ice. Voyager 2 showed that the satellite is surrounded by an highly tenuous nitrogen atmosphere, with a surface pressure of about 0.014 millibars.

More to the point, Triton's size, bulk composition, and unusual orbit all suggest that the satellite heated up early in its history as a result of accretional and tidal energy, and that its content of heat-producing radioactive isotopes is sufficient to have kept its interior slightly tepid up to the present day. Scientists on the Voyager team estimated that a core of chondritic composition 2000 km in diameter would generate enough heat to keep a mantle made of pure water-ice unfrozen close to the core–mantle boundary. If Triton's mantle is made of a water–ammonia mixture, it might be 'molten' from the core to within 200 kilometres of the surface. Thus all other considerations apart, we could expect to see evidence of extensive volcanic resurfacing of Triton, its record of early cratering having been wiped out by more recent cryovolcanism.

18.8.2 *Active volcanism on Triton*

There is abundant evidence of resurfacing on Triton's icy surface. On some parts of the surface, particularly the region known as the 'Canteloupe Terrain', true impact craters are few and far between. Elsewhere, the crater size–frequency distribution approaches that found on the lunar maria. But volcanism on Triton seems to be quite different from 'conventional' cryovolcanism. Voyager 2 scientists studying stereo images of the satellite discovered two dark, geyser-like plumes, rising about eight kilometres above the surface, and extending almost 150 km downwind.[36] Active venting of some gaseous material appeared to be in progress, with the escaping gas entraining dark, particulate matter. Other images showed numerous dark streaks on the surface, resembling wind-streaks seen on the Earth and Mars. These wind-streaks were probably formed by older, now inactive, geysers.

Many different explanations could be advanced to explain the geysers, and many different forms of heating invoked. While most planetary scientists would have groped for an explanation involving internal heat sources, Voyager scientists put forward a brilliantly original hypothesis: a solar-driven 'greenhouse'. Summarizing their thinking, they argued that a highly efficient way to construct a greenhouse is to cover a dark, radiation-absorbing layer with a transparent layer. In the case of Triton they proposed that a dark substrate is overlain by a thick layer of clear nitrogen ice, which is both volatile and has a low thermal conductivity. Radiation from the distant Sun passes through the transparent nitrogen ice, to be absorbed in the dark substrate. The temperature rises until the thermal gradient reaches a point where the excess heat is conducted and reradiated back to the surface.

Nitrogen's vapour pressure increases rapidly with increasing temperature. A 10 K rise above Triton's surface temperature of 38 K would cause a roughly 100-fold increase in pressure. If the transparent layer of nitrogen ice is thick enough (more than a couple of metres), the subsurface vapour pressure will steadily increase, ultimately reaching a point where it blasts through to the surface, to erupt as a geyser, spraying a plume of nitrogen gas, ice, and entrained dark particles from the vent high into space.

This is certainly a novel concept of volcanism. Unfortunately, no other space craft will reach Neptune for decades, and it will probably be more than a century before one lands on Triton. It will be a long time before volcanologists really understand what is happening on that dim, distant, and frigid world.

— Notes

1. Cattermole, P. (1981). *Planetary volcanism*, Ellis Horwood, Chichester, England, 443 pp.
2. Wilson, L. and Head, J. W. (1981). Ascent and eruption of basaltic magma on the Earth and Moon. *J. Geophys. Res.* **78**, 2971–30001.
3. Wilson, L. and Head, J. W. (1983). A comparison of volcanic eruption processes on Earth, Moon, Mars, Io and Venus. *Nature* **302**, 663–9.
4. Taylor, R. S. (1975). *Lunar science; a post-Apollo view*. Pergamon Press, New York, 372 pp.
5. Langseth, M. G., Keihm, S. J., and Peters, K. (1976). Revised lunar heat flow values. *Proc. 7th Lunar Sci. Conf.*, 3143–71.
6. Hulme, G. (1973). Turbulent lava flow and the formation of lunar sinous rilles. *Mod. Geol.* **4**, 107–17.
7. Huppert, H. E., Sparks, R. S. J., Turner, J. S., and Arndt, N. T. (1984). Emplacement and cooling of komatiite lavas. *Nature* **309**, 19–22.
8. Hulme, G. and Fielder, G. (1977). Effusion rates and rheology of lunar lavas. *Phil. Trans. Roy. Soc. Lond.* **A285**, 227–34.
9. Pinkerton, H., Wilson, L., and Norton, G. (1990). Thermal erosion: observations of terrestrial lavas and implications for planetary volcanism. *Abstracts, Lunar Planet. Sci. Conf. XXI*, 964–5.
10. Vilas, F., Chapman, C. R., and Matthews, M. R. (1988). (eds.) *Mercury*. Univ. Arizona Press, 794 pp.
11. Trask, N. J. and Guest, J. E. (1975). Preliminary geological terrain map of Mercury. *J. Geophys. Res.* **80**, 2461–77.
12. Kaula, W. M. (1990). Venus: a contrast in evolution to Earth. *Science* **247**, 1191–96.
13. Solomon, S. C. and Head, J. W. (1991). Fundamental issues in the geology and geophysics of Venus. *Science* **252**, 252–9.
14. Head, J. (1990). Venus crustal formation and evolution: an analysis of topography and crustal thickness variations. *Abstracts, Lunar Planet. Sci. Conf.* **21**, 477–8.
15. Head, J. W. and Crumpler, L. S. (1990). Venus geology and tectonics: hot spot and crustal spreading models and questions for the Magellan mission. *Nature* **346**, 525–33.
16. Kaula, W. M. and Phillips, R. (1981). *Geophys. Res. Lett.* **8**, 1187.
17. Head, J. W. and Wilson, L. (1986). Volcanic processes and landforms on Venus: theory, predictions and observations. *J. Geophys. Res.* **91**, 9407–46.
18. Head, J. W., Campbell, D. B., Elachi, C., Guest, J. E., McKenzie, D. P., Saunders, R. S., Schaber, G. G., and Schubert, G. (1991). Venus volcanism: initial analysis from Magellan data. *Science* **252**, 276–96.
19. Baker, V. R., Strom, R. G., Gulick, V. C., Kargel, J. S., Komatsu, G., and Kale, V. S. (1991). Ancient oceans, ice sheets and the hydrological cycle on Mars. *Nature* **352**, 589–94.
20. McGetchin, T. and Smyth, J. R. (1978). The mantle of Mars: some possible geological implications of its high density. *Icarus* **34**, 512–36.
21. Toksoz, M. N. and Husi, A. T. (1978). Thermal history and evolution of Mars. *Icarus* **34**, 537–47.
22. Carr, M. H., Greeley, R., Blasius, K. R., Guest, J. E., and Murray, J. B. (1977). Some Martian volcanic features as viewed from the Viking Orbiters. *J. Geophys. Res.* **82**, 3985–4015.
23. Greeley, R. and Spudis, P. D. (1981). Volcanism on Mars. *Rev. Geophys. Space Phys.* **19**, 13–41.
24. Mouginis-Mark, P. J., Wilson, L., and Zuber, M. T. (1991). The physical volcanology of Mars. University of Arizona Press, in press.
25. Frey, H., Lowry, B. L., and Chase, S. A. (1979). Pseudocraters on Mars. *J. Geophys. Res.* **84**, 8075–86.
26. Pieri, D. C. and Baloga, S. M. (1986) Eruption rate, area and length relationships for some Hawaiian flows. *J. Volcanol. Geotherm. Res.* **30**, 29–45.
27. Mouginis-Mark, P. J., Wilson, L., and Head, J. W. (1982). Explosive volcanism on Hecates Tholus, Mars: investigation of eruption conditions. *J. Geophys. Res.* **87**, 9890–9904, 1982.
28. Morabito, L. A., Synott, S. P., Kupferman, P. N.,

and Collins, S. A. (1979). Discovery of currently active extraterrestrial volcanism. *Science* **204**, 972.

29. McEwen, A. S., Lunine, J. I., and Car, M. H. (1989). Dynamic geophysics of Io. In *NASA SP 494, Time variable phenomena in the Jovian system*, NASA, Washington DC, pp. 11–46.

30. Nash, D. B., Carr, M. H., Gradie, J., Hunten, D. M., and Yoder, C. F. (1986). *Io*. In *Satellites*, J. A. Burns, and M. S. Matthews (eds.), University of Arizona Press, 629–88.

31. Peale, S. J., Cassen, P., and Reynolds, R. T. (1979). Melting of Io by tidal dissipation. *Science* **203**, 892–4.

32. Matson, D. L., Ransford, G. A., and Johnson, T. V. (1981). Heat flow from Io (J1). *J. Geophys. Res.* **86**, 1664–72.

33. Johnson, T. V., Veeder, G. J., Matson, D. L., Brown, R. H., Nelson, R. M., and Morrison, D. (1988). Io: evidence of silicate volcanism in 1986. *Science* **242**, 1280–83.

34. Ojakangas, G. W. and Stevenson, D. J. (1989). Thermal state of an ice shell on Europa. *Icarus* **81**, 220–41.

35. Kargel, J. S. (1991). Brine volcanism and the interior structures of asteroids and icy satellites. *Icarus*, **94**, 368–90.

36. Smith, B. A. *et al.* (1989). Voyager 2 at Neptune: imaging science results. *Science* **246**, 1422–49.

Index

Italics denote references to figures